Pressure-Induced Phase Transformations (Volume II)

Pressure-Induced Phase Transformations (Volume II)

Editors

**Daniel Errandonea
Enrico Bandiello**

Basel • Beijing • Wuhan • Barcelona • Belgrade • Novi Sad • Cluj • Manchester

Editors
Daniel Errandonea
Departamento de Física
Aplicada-ICMUV,
MALTA Consolider Team,
Universidad de Valencia,
València, Spain

Enrico Bandiello
Instituto de Diseño para La
Fabricación y
Producción Automatizada
MALTA Consolider Team
Universitat Politècnica
de València
Valencia, Spain

Editorial Office
MDPI
St. Alban-Anlage 66
4052 Basel, Switzerland

This is a reprint of articles from the Special Issue published online in the open access journal *Crystals* (ISSN 2073-4352) (available at: https://www.mdpi.com/journal/crystals/special_issues/Pressure_Induced_Phase_Transformations_VolumeII).

For citation purposes, cite each article independently as indicated on the article page online and as indicated below:

Lastname, A.A.; Lastname, B.B. Article Title. *Journal Name* **Year**, *Volume Number*, Page Range.

ISBN 978-3-0365-8564-2 (Hbk)
ISBN 978-3-0365-8565-9 (PDF)
doi.org/10.3390/books978-3-0365-8565-9

© 2023 by the authors. Articles in this book are Open Access and distributed under the Creative Commons Attribution (CC BY) license. The book as a whole is distributed by MDPI under the terms and conditions of the Creative Commons Attribution-NonCommercial-NoDerivs (CC BY-NC-ND) license.

Contents

About the Editors vii

Preface ix

Yixian Wang, Hao Wu, Yingying Liu, Hao Wang, Xiangrong Chen and Huayun Geng
Recent Progress in Phase Stability and Elastic Anomalies of Group VB Transition Metals
Reprinted from: *Crystals* **2022**, *12*, 1762, doi:10.3390/cryst12121762 1

Vinod Panchal, Catalin Popescu and Daniel Errandonea
An Investigation of the Pressure-Induced Structural Phase Transition of Nanocrystalline α-CuMoO$_4$
Reprinted from: *Crystals* **2022**, *12*, 365, doi:10.3390/cryst12030365 21

Valeriya Yu. Smirnova, Anna A. Iurchenkova and Denis A. Rychkov
Computational Investigation of the Stability of Di-*p*-Tolyl Disulfide "Hidden" and "Conventional" Polymorphs at High Pressures
Reprinted from: *Crystals* **2022**, *12*, 1157, doi:10.3390/cryst12081157 33

Kamil Filip Dziubek
On the Definition of Phase Diagram
Reprinted from: *Crystals* **2022**, *12*, 1186, doi:10.3390/cryst12091186 43

Ying Ma, Zhili Qiu, Xiaoqin Deng, Ting Ding, Huihuang Li, Taijin Lu, et al.
Chinese Colorless HPHT Synthetic Diamond Inclusion Features and Identification
Reprinted from: *Crystals* **2022**, *12*, 1266, doi:10.3390/cryst12091266 59

Tarik Ouahrani and Reda M. Boufatah
Understanding the Semiconducting-to-Metallic Transition in the CF$_2$Si Monolayer under Shear Tensile Strain
Reprinted from: *Crystals* **2022**, *12*, 1476, doi:10.3390/cryst12101476 71

Li Xiang, Dominic H. Ryan, Paul C. Canfield and Sergey L. Bud'ko
Effects of Physical and Chemical Pressure on Charge Density Wave Transitions in LaAg$_{1-x}$Au$_x$Sb$_2$ Single Crystals
Reprinted from: *Crystals* **2022**, *12*, 1693, doi:10.3390/cryst12121693 83

Oliver Tschauner
Corresponding States for Volumes of Elemental Solids at Their Pressures of Polymorphic Transformations
Reprinted from: *Crystals* **2022**, *12*, 1698, doi:10.3390/cryst12121698 97

Pricila Betbirai Romero-Vázquez, Sinhué López-Moreno and Daniel Errandonea
Stability of FeVO$_4$-II under Pressure: A First-Principles Study
Reprinted from: *Crystals* **2022**, *12*, 1835, doi:10.3390/cryst12121835 111

Khaldoun Tarawneh and Yahya Al-Khatatbeh
Phase Relations of Ni$_2$In-Type and CaC$_2$-Type Structures Relative to Fe$_2$P-Type Structure of Titania at High Pressure: A Comparative Study
Reprinted from: *Crystals* **2023**, *13*, 9, doi:10.3390/cryst13010009 125

Alfonso Muñoz and Plácida Rodríguez-Hernández
First-Principle Study of Ca$_3$Y$_2$Ge$_3$O$_{12}$ Garnet: Dynamical, Elastic Properties and Stability under Pressure
Reprinted from: *Crystals* **2023**, *13*, 29, doi:10.3390/cryst13010029 135

Javier Gonzalez-Platas, Ulises R. Rodriguez-Mendoza, Amagoia Aguirrechu-Comeron, Rita R. Hernandez-Molina, Robin Turnbull, Placida Rodriguez-Hernandez and Alfonso Muñoz
Structural and Luminescence Properties of Cu(I)X-Quinoxaline under High Pressure (X = Br, I)
Reprinted from: *Crystals* **2023**, *13*, 100, doi:10.3390/cryst13010100 153

Enrique Zanardi, Silvana Radescu, Andrés Mujica, Plácida Rodríguez-Hernández and Alfonso Muñoz
Ab Initio Theoretical Study of $DyScO_3$ at High Pressure
Reprinted from: *Crystals* **2023**, *13*, 165, doi:10.3390/cryst13020165 173

Meera Varma, Markus Krottenmüller, H. K. Poswal and C. A. Kuntscher
Pressure-Induced Structural Phase Transitions in the Chromium Spinel $LiInCr_4O_8$ with Breathing Pyrochlore Lattice
Reprinted from: *Crystals* **2023**, *13*, 170, doi:10.3390/cryst13020170 191

Jai Sharma, Henry Q. Afful and Corinne E. Packard
Phase Transformation Pathway of $DyPO_4$ to 21.5 GPa
Reprinted from: *Crystals* **2023**, *13*, 249, doi:10.3390/cryst13020249 211

Nourou Amadou, Abdoul Razak Ayouba Abdoulaye, Thibaut De Rességuier and André Dragon
Strain-Rate Dependence of Plasticity and Phase Transition in [001]-Oriented Single-Crystal Iron
Reprinted from: *Crystals* **2023**, *13*, 250, doi:10.3390/cryst13020250 223

Saheli Banerjee, Amit Tyagi and Alka B. Garg
Pressure-Induced Monoclinic to Tetragonal Phase Transition in $RTaO_4$ (R = Nd, Sm): DFT-Based First Principles Studies
Reprinted from: *Crystals* **2023**, *13*, 254, doi:10.3390/cryst13020254 235

Estelina Lora da Silva, Mario C. Santos, P. Rodríguez-Hernández, A. Muñoz and F. J. Manjón
Theoretical Study of Pressure-Induced Phase Transitions in Sb_2S_3, Bi_2S_3, and Sb_2Se_3
Reprinted from: *Crystals* **2023**, *13*, 498, doi:10.3390/cryst13030498 251

Vinod Panchal, Laura Pampillo, Sergio Ferrari, Vitaliy Bilovol, Catalin Popescu and Daniel Errandonea
Pressure-Induced Structural Phase Transition of Co-Doped SnO_2 Nanocrystals
Reprinted from: *Crystals* **2023**, *13*, 900, doi:10.3390/cryst13060900 273

Sanjay Kumar Mishra, Nandini Garg, Smita Gohil, Ranjan Mittal and Samrath Lal Chaplot
High-Pressure Vibrational and Structural Studies of the Chemically Engineered Ferroelectric Phase of Sodium Niobate
Reprinted from: *Crystals* **2023**, *13*, 1181, doi:10.3390/cryst13081181 285

About the Editors

Daniel Errandonea

Daniel Errandonea, Professor Dr., is a full professor with the Department of Applied Physics of University of Valencia (Spain). He is an Argentinean-born physicist (married with two sons) who received an M.S. from the University of Buenos Aires (1992) and a Ph.D. from the University of Valencia (1998). He has authored/co-authored over 350 research articles in refereed scientific journals including *Prog. Mat. Sci.*, *Nat. Commun.*, *Phys. Rev. Lett.*, and *Advanced Science*, which have attracted 11000+ citations. His work on materials under extreme conditions of pressure and temperature has implications for fundamental and applied research. Among other subjects, during the last decade, Prof. Errandonea has comprehensively explored phase transitions in ternary oxides, semiconductors, metals, and related materials. Some of his accomplishments include the determination of high-pressure phase transitions, high-pressure and high-temperature phase diagrams (including melting curves), and the study of their implications in the physical properties of materials. Prof. Errandonea is a fellow of the Alexander von Humboldt Foundation, winning the Van Valkenburg Award and IDEA Prize, among others. He is presently a member of the MALTA Consolider and GREENMAT teams, the Editorial Board of *Crystals*, and the executive committee of the International Association for the Advancement of High Pressure Science and Technology (AIRAPT).

Enrico Bandiello

Enrico Bandiello, Ph.D., is a researcher at the Technical University of Valencia—UPV (Spain). He is an Italian-born physicist (married with one son) who received an M.S. (2012) and a Ph.D. (2017) from the University of Valencia. He has authored/co-authored over 38 research articles in refereed scientific journals including *Materials Today Advances*, *JACS*, and *Advanced Materials*, which have attracted 2000+ citations. His work on materials under extreme conditions of pressure and temperature has implications for fundamental and applied research. Among other subjects, during the last decade, Dr. Bandiello has comprehensively explored phase transitions in ternary oxides, semiconductors, and related materials. Some of his recent accomplishments include the determination of the high-pressure phase transitions of topological semiconductors like GaGeTe. Dr. Bandiello performed postdoctoral stays at the University of Valencia and the University of Verona. He is presently a member of the MALTA Consolider Team.

Preface

The study of phase transitions in solids and liquids under high-pressure and high-temperature conditions is a very active and vigorous research field. In recent decades, thanks to the development of experimental techniques and computer simulation methods, a plethora of important discoveries have been made. Many of the achievements accomplished in recent years affect various research fields, from solid-state physics to chemistry, materials science, and geophysics. They not only contribute to a deeper understanding of solid–solid phase transitions but also to a better understanding of the properties of matter. This Special Issue presents and discusses these recent achievements, and is addressed to physicists, chemists, materials scientists, and researchers working in geophysics and Earth and planetary sciences.

Daniel Errandonea and Enrico Bandiello
Editors

Review

Recent Progress in Phase Stability and Elastic Anomalies of Group VB Transition Metals

Yixian Wang [1,2,*], Hao Wu [1], Yingying Liu [1], Hao Wang [2,3], Xiangrong Chen [3] and Huayun Geng [2,*]

[1] College of Science, Xi'an University of Science and Technology, Xi'an 710054, China
[2] National Key Laboratory of Shock Wave and Detonation Physics, Institute of Fluid Physics, CAEP, Mianyang 621900, China
[3] Institute of Atomic and Molecular Physics, College of Physical Science and Technology, Sichuan University, Chengdu 610064, China
* Correspondence: lsdwyx@xust.edu.cn (Y.W.); s102genghy@caep.cn (H.G.)

Abstract: Recently discovered phase transition and elastic anomaly of compression-induced softening and heating-induced hardening (CISHIH) in group VB transition metals at high-pressure and high-temperature (HPHT) conditions are unique and interesting among typical metals. This article reviews recent progress in the understanding of the structural and elastic properties of these important metals under HPHT conditions. Previous investigations unveiled the close connection of the remarkable structural stability and elastic anomalies to the Fermi surface nesting (FSN), Jahn–Teller effect, and electronic topological transition (ETT) in vanadium, niobium, and tantalum. We elaborate that two competing scenarios are emerging from these advancements. The first one focuses on phase transition and phase diagram, in which a soft-mode driven structural transformation of BCC→RH1→RH2→BCC under compression and an RH→BCC reverse transition under heating in vanadium were established by experiments and theories. Similar phase transitions in niobium and tantalum were also proposed. The concomitant elastic anomalies were considered to be due to the phase transition. However, we also showed that there exist some experimental and theoretical facts that are incompatible with this scenario. A second scenario is required to accomplish a physically consistent interpretation. In this alternative scenario, the electronic structure and associated elastic anomaly are fundamental, whereas phase transition is just an outcome of the mechanical instability. We note that this second scenario is promising to reconcile all known discrepancies but caution that the phase transition in group VB metals is elusive and is still an open question. A general consensus on the relationship between the possible phase transitions and the mechanical elasticity (especially the resultant CISHIH dual anomaly, which has a much wider impact), is still unreached.

Keywords: phase stability; elastic anomalies; structural transition; group VB transition metals; high pressure and high temperature

1. Introduction

The group VB transition metals (vanadium, niobium, and tantalum) have extensive applications because of their excellent mechanical and physical properties. Vanadium is well known as a kind of metallic "vitamin", which is commonly added to iron or steel to increase their toughness, strength, and abrasion resistance. Moreover, vanadium plays an excellent role in titanium alloys, which greatly promotes the development of the aerospace industry. With the progression in science and technology, there are higher requirements for advanced materials, with the application of vanadium becoming more and more extensive, which covers batteries, pharmaceuticals, optics, and many other fields. Niobium and tantalum belong to refractory metals because of their melting point is higher than 2700 K [1]. One can increase a material's strength at high temperature and improve the processing performance by adding niobium or tantalum as alloying elements. On the other hand, tantalum is also a widely used pressure standard and high technology material due to its strong stability in chemistry and mechanics, as well as the very high melting point (3269 K) [2].

Due to their important applications, group VB transition metals have always attracted much attention. However, most studies in early days mainly focused on their superconducting properties [3–10] because their superconducting transition temperature is quite high at ambient conditions. Theoretical studies show that the superconducting transition temperature of vanadium and niobium depends on the pressure; and the density of electron states at Fermi level has a significant impact on the relationship [11]. In 1997, Struzhkin et al. [10] measured the superconducting T_c of niobium and tantalum up to 100 GPa through a highly sensitive magnetic susceptibility technique. The results showed that T_c in tantalum remains nearly constant at 4.38 K in the range of 0–45 GPa. However, they observed T_c anomalies in niobium around 5–6 GPa and 60–70 GPa, where T_c increases by 0.7 K and decreases by about 1.0 K, respectively. Struzhkin et al. suggested that the anomalies in niobium arise from stress-sensitive ETT. Later, Tse et al. [12] calculated the Fermi surface of niobium by the density functional theory (DFT) method and found that the Fermi surface does undergo a topological transformation under pressure, confirming the results of Struzhkin et al. [10].

Unlike niobium and tantalum, vanadium has a large positive pressure coefficient of T_c. Smith [13] measured the T_c of vanadium under pressure up to 2.4 GPa and found a linear increase in T_c with $dT_c/dP = 0.062$ K/GPa. Subsequently, Brandt et al. [14] also observed a monotonic increase in T_c up to 18 GPa using ice bombs and mechanical presses. After that, Akahama et al. [15] carried out electrical resistance measurements of vanadium up to 49 GPa to investigate the upper bound of the monotonically increased T_c with pressure. They found that T_c increases linearly with a coefficient of $dT_c/dP = 0.096$ K/GPa and reaches a value of 9.6 K at about 18 GPa, the maximum pressure in their experiment. This value of T_c is comparable to that of niobium at ambient pressure. With the development of experimental technology, Ishizuka et al. [16] studied the superconducting properties of vanadium under higher pressure by using a vibrating coil magnetometer. The results showed that T_c increases from 5.3 K to 17.2 K from zero pressure to 120 GPa, and the superconducting transition temperature increases almost linearly.

In order to understand these experimental results, Suzuki et al. [17] further studied the superconducting properties of vanadium under pressure by first-principles calculations. It was found that the T_c of vanadium shows an obvious upward trend with the change of pressure, and the increasing rate of T_c decreases gradually at about 80 GPa, which is qualitatively consistent with the experimental results. Moreover, they interpreted such characteristic behavior of T_c under pressure by attributing it to a significant frequency softening of the transverse mode near the Γ-H line as pressure increased. According to their results, when the pressure is above 130 GPa, the transverse acoustic mode (TA) even has imaginary frequency, indicating that the BCC phase of vanadium becomes dynamically unstable and there is an opportunity of structural phase transformation. However, at that time, the research on vanadium was limited to its superconducting properties and did not pay much attention to its structural instability under high pressure. This softening was subsequently confirmed by calculated elastic constants [18,19], in which C_{44} continuously decreased to negative values, indicating mechanical instability of the BCC structure, but the crystalline structure of the high-pressure phase was not proposed at that time.

Until 2007, Ding et al. [20] performed X-ray diffraction (XRD) experiments on vanadium using diamond anvil cell (DAC) up to 150 GPa. They discovered a novel rhombohedral (RH) phase appearing around 63–69 GPa. This novel high-pressure structural phase transition had not been detected in any of the earlier experiments. After that, the focus of investigation on group VB transition metals has been gradually shifted to study their structural stability and phase transition.

In addition to the structural transformation, the group VB transition elements also exhibit striking anomalous elastic softening under pressure, which is quite different from other transition metals. According to existing literature reports [18,21,22], the pressure-induced shear elastic softening of vanadium, niobium, and tantalum originates from the electronic structure, which is closely related to the FSN and ETT. Furthermore, the elastic softening in vanadium and niobium were discovered to gradually diminish with increased temperature, effectively giving rise to a heating-induced *hardening* phenomenon, which is very rare (if any) to our knowledge. These elastic anomalies might be taken as due to the phase transition. However, some experimental and theoretical facts that are incompatible with this scenario.

During the past two decades, the structural stability and elastic anomalies of the group VB transition elements have been the subject of numerous theoretical and experimental studies [23–76]. The progress is tremendous; regardless, there are still many controversies on some key issues. In this paper, we review the current understanding of the structural and mechanical anomalies of these important metals under high temperature and high pressure. It is hoped that this paper will significantly promote the understanding of the physical properties for more broad types of metals under extreme conditions.

2. Phase Stability and Elastic Anomalies in V, Nb, Ta

2.1. Pressure Effect on Structure Stability

The first direct evidence of a phase transition in vanadium came from the static compression experiment by Ding et al. [20]. They observed a new type of high-pressure structural transition from BCC to an RH phase at 63–69 GPa, which once was thought of as a second-order transition and was not found in any of the earlier experiments with elements or compounds. In general, the phase transition sequence of transition metals under pressure is hexagonal close packed (HCP)→body centered cubic (BCC)→HCP→face centered cubic (FCC). Based on this sequence, the stability of the BCC phase has long been predicted to be very high [77]. Therefore, the discovery of a phase transition in vanadium below 70 GPa is very remarkable.

Soon after, Lee et al. [36] confirmed this phase transition with DFT calculations and showed that a metastable RH structure is formed at 73 GPa and becomes the ground state at 84 GPa. This low-pressure RH phase is termed as "RH1", which has an angle of $\alpha = 110.25°$. Furthermore, Lee et al. predicted two other transformations that were not detected in Ding et al.'s experiment: the second transformation to a high-pressure structure "RH2" with an angle $\alpha = 108.14°$ at 120 GPa, and the third transformation back to the high-symmetric BCC structure ($\alpha = 109.47°$) at 280 GPa. As the pressure continues to increase, the BCC phase becomes the only stable structure at 315 GPa. Since the latent heat of BCC→RH transition is much smaller than the thermal fluctuation at room temperature, Lee et al. [36] suggested that this transformation is first-order, which contradicts the second-order transition proposed by Ding et al. [20]. To verify this result, Lee et al. [37] further studied the elastic constants and volume changes associated with two high-pressure RH phase transitions in vanadium. The results shown are that there were small discontinuities in shear modulus and other elastic properties in the phase transitions even at zero temperature, indicating that the phase transitions should be first-order.

The prediction of RH1 and RH2 phases in vanadium was supported by phonon calculations of Luo et al. [35]. They found that the lattice dynamical instability of vanadium starts at 62 GPa and phonon softening leads to a phase transition of BCC→RH1 ($\alpha = 110.5°$). At about 130 GPa, the angle of RH1 phase changes to 108.2°, and the electronic structure changes drastically. At a pressure of 250 GPa, lattice dynamics calculations show that the stability of BCC structure is restored. Luo et al. [35] suggested that the dramatic change in the electronic structures of vanadium under pressure are the driving force behind the structural phase transitions. Later, Verma et al. [38] and Qiu et al. [39] further investigated the structural stability of vanadium under high pressure through first-principles calculations. Both confirmed the existence of the RH1 and RH2 phases, as well as the pressure-induced structural transition sequence of BCC→RH1→RH2→BCC reported by Lee et al. [36]. Meanwhile, detailed electronic structure analysis by Verma et al. showed that the phase transition of BCC→RH1 is caused by the Jahn–Teller mechanism. Although different theoretical studies have reached a consensus on the phase transition sequence of vanadium under pressure, there is still disagreement on the exact value of the phase transition pressure. For example, Qiu et al. reported a BCC→RH1 transition pressure of 32 GPa, while all other theoretical transition pressures were located between 60 and 84 GPa.

To trace the possible origin of the discrepancy between these calculations, two different methods were exploited by Wang et al. [54] to evaluate the phase transition pressures of vanadium. The first method is to choose RH1 and RH2 as the initial structure, and directly optimize them under different pressures without any symmetry constraints. The calculated enthalpy difference with respect to the BCC phase as a function of pressure is shown in Figure 1 (taken from [54]). It can be seen clearly that the BCC→RH1 transition is not at 30 GPa, where RH1 is dynamically instable and spontaneously collapses to BCC phase. According to Wang et al. [54], the relaxation of the RH1 phase to the BCC phase is far from being perfect, and the angle is about 109.51° at 20–40 GPa, which reflects a possibility that non-hydrostatic loading can easily drive vanadium towards RH-like deformations. It is worth noting that another RH phase (RH2) transforms to a similar twisted BCC structure with $\alpha = 109.39°$ when below 110 GPa. In addition, when the pressure is about 98 GPa, the RH1 phase becomes the ground state with an angle of $\alpha = 110.17°$. With the further increase of pressure, the RH1 phase transforms into another RH phase (RH2) at about 128 GPa. At the pressure of 211 GPa, RH2 reaches the maximum stability, with $\alpha = 108.23°$. As compression increases, the RH2 phase eventually collapses to the BCC phase at about 300 GPa.

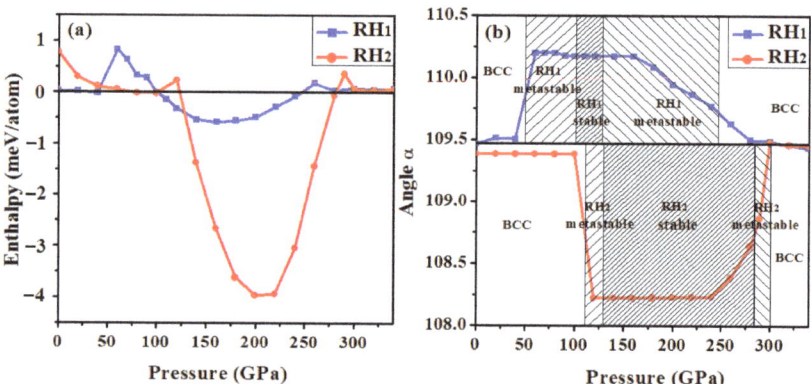

Figure 1. (a) Enthalpy difference of vanadium in RH1 and RH2 structures at zero Kelvin with respect to the BCC phase as a function of pressure. (b) Variation of angle α in RH1 and RH2 structures as a function of pressure at zero Kelvin. Note that $\alpha = 109.47°$ corresponds to the perfect BCC structure. (By the courtesy of Ref. [54]).

In the second method, Wang et al. [54] adopted the same method as Lee et al. [36], that is, twisting the BCC structure along a predetermined path. Note that the unit volume is conserved in this approach. Qiu et al. [39] argued that such a treatment would result in a higher phase transition pressure. According to Lee et al., the error caused by fixed volume can be corrected by the following formula

$$H(\delta, P_0) \approx U(\delta, V_0) + P_0 V_0 - \frac{1}{2B(\delta, V_0)} \Delta P(\delta, V_0)^2 V_0. \tag{1}$$

It is worth noting that only the first term was used in Lee et al.'s calculations [36]. After careful examination, Wang et al. [54] found that the correction of the third term is indeed small, which means that the contribution of volume relaxation can be safely ignored when studying the relative phase stability. This supports the assessment of Lee et al. [36]. In addition, the transition pressures of 'unrelaxed' calculations (method II) by Wang et al. [54]

are in good accordance with the full structural relaxation calculations (method I), therefore, Qiu et al.'s comment on Lee et al.'s results is inappropriate.

Wang et al. [54] further studied the influence of changing the position of Fermi level on the structural stability of vanadium by using a partial jellium model, see Figure 2 (taken from [54]). Here, the Fermi level is shifted by charge transfer (chemical doping) changing the orbital occupations. Since the RH2 phase reaches the maximal stability at 211 GPa, it should have the optimum orbital occupation. Thus, raising (adding electrons) or lowering (removing electrons/adding holes) the Fermi level pushes the system away from the optimum occupation, so the stability of RH2 phase will be weakened in both cases. However, Wang et al.'s calculations showed that moving the Fermi level down further stabilized the RH2 phase, while moving the Fermi level up greatly destabilized the RH2 phase. This is consistent with Landa et al.'s band-filling argument when alloying vanadium with the same transition series [41]; however, it is incompatible with Jahn–Teller mechanism.

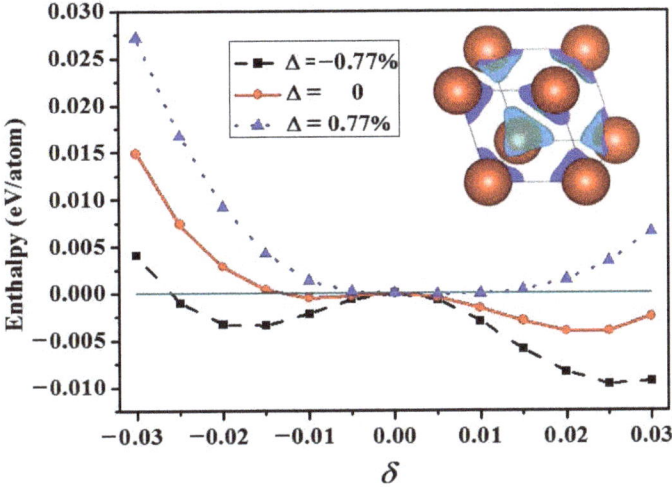

Figure 2. Calculated enthalpy difference with respect to BCC phase as a function of the RH deformation parameter δ when the Fermi level is shifted up or down at a pressure of 211 GPa. Inset: Calculated differential charge density between $\Delta = -0.77\%$ and $\Delta = 0$. (By the courtesy of Ref. [54]).

According to the calculation of Wang et al. [54], RH2 phase reaches the maximal stability at 211 GPa when $\Delta = -2.15\%$ (Δ represents the percentage of total charge added to /removed from the system). Further shifting down the Fermi level reduces the stability of RH phase. When $\Delta < -4.85\%$, BCC becomes stable again. In the inset of Figure 2, Wang et al. plotted the differential charge density between $\Delta = -0.77\%$ and $\Delta = 0$. As shown, the removed electrons/added holes are distributed around the nucleus and mainly exhibit d orbital characteristics. Since the d orbitals' delocalization results in lower electronegativity of the RH2 phase relative to the BCC phase, the localization (or delocalization) of d electrons has an important effect on the stability of the RH phase.

In addition, Wang et al. [54] found that even if the absolute convergence is achieved, the DFT method still has the problem of insufficient accuracy when exploring the phase transition of vanadium. Consider that the semi-local functional PBE may not be able to handle strong electron correlations in narrow-band systems, so they thought that the quality of the exchange-correlation (XC) functional may be an important factor affecting the results. For this reason, an evaluation of the advanced hybrid functionals in vanadium and niobium was further carried out by Wang et al. [66]. The results show that the common HSE06, PBE0, and B3LYP hybrid functionals are complete failures in describing the mechanical properties of these metals. The unexpected failure is due to the very rare localization error in these functionals, which is further supported by a similar failure of the DFT + U method.

To solve this problem, Wang et al. [66] proposed a DFT + J method to promote on-site electron exchange, which well reproduces the experimental shear modulus under ambient conditions. However, Wang et al. found that the PBE + J increases the BCC→RH transition pressure, and the correction of localization error weakens (or even eliminates) the RH phases. They concluded that RH phase could be unstable under more accurate calculation methods, which is a striking prediction and challenges the previously reported structural transition in vanadium.

Following this prediction, an independent XRD experiment was performed by Akahama et al. [73] to study the structural stability of vanadium. The results showed that BCC vanadium is stable up to 189 GPa at room temperature, while the RH phase (α > 109.47°) reported in previous studies should be a metastable phase induced by non-hydrostatic pressure. This supports the prediction of Wang et al. [66] that RH phase may be unstable in vanadium. Furthermore, they observed a new high-pressure phase after annealing at 242 GPa, which was also confirmed from a different experiment at room temperature. Akahama et al. interpreted the phase as RH phase with α < 109.47°. However, the pressure range does not agree with the previous theoretical [35–37,54] and experimental results [21,47,51,55].

At the same time, Stevenson et al. [74] re-performed the XRD experiment at pressures up to 154 GPa using polycrystalline (powder) and single crystal samples with various pressure transfer media (PTM). It was found that only the single-crystal samples reveal two RH phases, and the distortions from cubic symmetry are much smaller than previous results. That is to say, the observed RH phase is far from being perfect, and should be interpreted as a kind of lattice distortion rather than a phase transition. Moreover, Wang et al. [72] measured the sound velocity of vanadium through shock wave experiment recently. They found that when the pressure was above 79 GPa, the sound velocity of the shocked vanadium was closer to the RH phase rather than the BCC phase. The unexpected high-pressure phases along the Hugoniot can be seen as slight distortions of the BCC structure, which may be caused by the dynamic, nonequilibrium, and nonhydrostatic nature of planar shock waves. The above two experimental signatures are compatible with the experimental results of Akahama et al. [73] and further confirm the theoretical assessment of Wang et al. made in Ref. [66].

In other experimental research, Jenei et al. [47] performed DAC experimental studies and found that the BCC structure transformed to RH1 phase at about 30 GPa in non-pressure medium, while it was around 60 GPa when Ne pressure medium was used. In addition, the transition can occur at a much lower pressure if under nonhydrostatic conditions. Nonetheless, in Ding et al.'s experiments [20], the transition pressure of BCC→RH in non-pressure medium and He pressure medium was 69 GPa and 63 GPa, respectively. Thus, the deviation in transition pressure of BCC→RH might not be due to the non-hydrostatic condition. In addition, Antonangeli et al. [55] used inelastic X-ray scattering to detect the phonon dispersion of single crystal vanadium under pressure up to 45 GPa. Their results showed that the transverse acoustic mode has abnormally high-pressure behavior along (100) direction, and the softening of C_{44} causes the RH distortion around 34–39 GPa. It is obvious that the transition pressure is consistent with the diffraction results of 30 GPa in non-hydrostatic conditions by Jenei et al. [47]. It should be noticed that Antonangeli et al. [55] performed the measurements on relatively large single crystals, which are more susceptible to non-hydrostatic stress than powders. Moreover, according to the experimental study of Stevenson et al. [74], an RH high-pressure phase was indeed observed when using single crystal samples; but the high-pressure diffraction profiles from the polycrystalline samples is not suitable for RH lattice, regardless of the PTM used. Why there is such a big difference between the experimental data of powder and single crystal is still an open question that needs further study.

On the other hand, Yu et al. [51] recently measured the sound velocities and yield strength of vanadium through reverse impact experiments. They found an indication of the shock induced BCC→RH transition at about 60.5 GPa by the discontinuity of longitudinal sound velocity against shock pressure, which disagrees with the results of Jenei et al. [47]

and Antonangeli et al. [55] but remains consistent with both DAC measurements by Ding et al. [20]. The aforementioned experiment studies cautioned that the phase transition in group VB metals is elusive, which remains an open question and no general consensus on this issue has been achieved.

2.2. Temperature Effect on Structure Stability

In 2014, Landa et al. [78] explored the phase stability of vanadium at high temperatures and pressures by using the self-consistent ab initio lattice dynamics (SCAILD) approach combined with DFT. In this study, the phonon-phonon interactions at elevated temperatures were considered. In Figure 3 (taken from [78]), Landa et al. showed the calculated phonon frequencies for vanadium under different temperatures at 182 GPa. The maximum stability of RH phase was measured by analyzing the variation in phonon dispersion of BCC phase. Their results showed that temperature promoting the phonon frequencies from being imaginary to being real along the Γ→H and Γ→N lines. When the temperature is above 8000 K, BCC phase becomes stable again. Since this temperature is significantly higher than the shock melting temperature of 6800 ± 800 K at 182 GPa [79], Landa et al. [78] concluded that the BCC phase is actually never stable at this density. Namely, high-pressure RH phases of compressed vanadium should have a very broad pressure-temperature stability region.

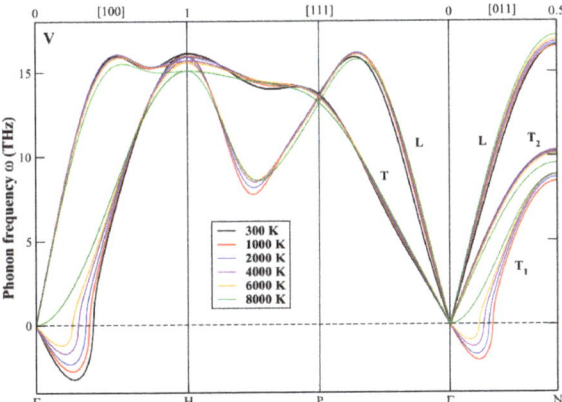

Figure 3. Calculated phonon dispersions for vanadium under different temperatures at a pressure of 182 GPa. (By the courtesy of Ref. [78]).

Nonetheless, lattice dynamics stability is not the unequivocal criterion for the thermodynamic stability of a phase. For the latter purpose, one should resort to free energy difference. To this end, based on Landa et al.'s initial results [78], Wang et al. further investigated the effect of thermo-electrons on the structure stability of vanadium by calculating the free energy using finite temperature DFT method [54]. Since the phase transitions of vanadium are closely related to the changes in the electronic structure, Wang et al. estimated that the contribution of thermo-electrons may be greater than that of lattice dynamics. As shown in Figure 4 (taken from [54]), the electronic temperature significantly reduces the stability of RH phases, and RH1 and RH2 transform back to BCC at around 1440 K (at 140 GPa) and 1915 K (at 211 GPa), respectively. Compared with the results of Landa et al. [78], this new transition temperature is much lower. It clearly demonstrated that compressed vanadium should transition back to BCC structure in the solid state; and the transition temperature is much lower than the melting temperature. Wang et al. analyzed the impact of lattice dynamics by including both thermo-electronic and phonon corrections in their assessment. It revealed that phonon correction will further reduce the transition temperature by about 260 K at 200 GPa [54], comparable to the usual expectation. Further study by Wang et al. indicated that this heating-induced reentrant transition in vanadium was mainly driven by electronic entropy.

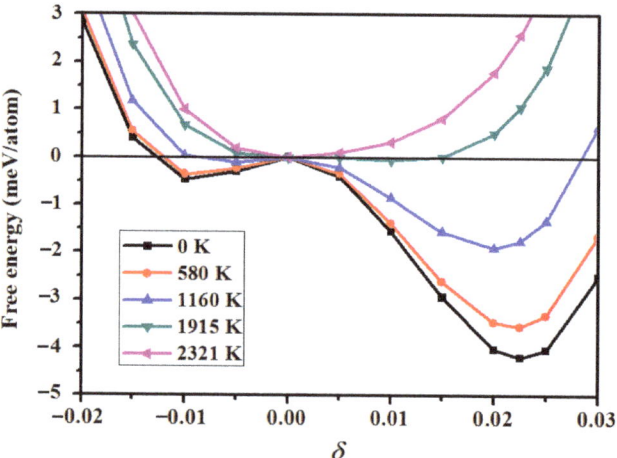

Figure 4. Effect of thermo-electrons on the phase stability of vanadium at 211 GPa. (By the courtesy of Ref. [54]).

Based on these new theoretical data, Wang et al. [54] constructed a comprehensive phase diagram for vanadium under high pressure and temperature for the first time. According to their diagram (see Figure 5 taken from [54]), RH1 phase stabilizes at 100–126 GPa with a maximum transition temperature of 1440 K at 140 GPa, while RH2 phase stabilizes at 126–280 GPa with a maximum transition temperature of 1915 K at 211 GPa. In addition, the stability of RH1 and RH2 phases decreases with increasing temperature, and both transform back to BCC before melting. This picture completely changes our understanding about vanadium which insisted that RH phases could stand up to the melting temperature [79,80]. Meanwhile, Wang et al. identified a triple point at about 1440 K and 140 GPa [54], where there may be spectacular physical properties due to the structural frustration.

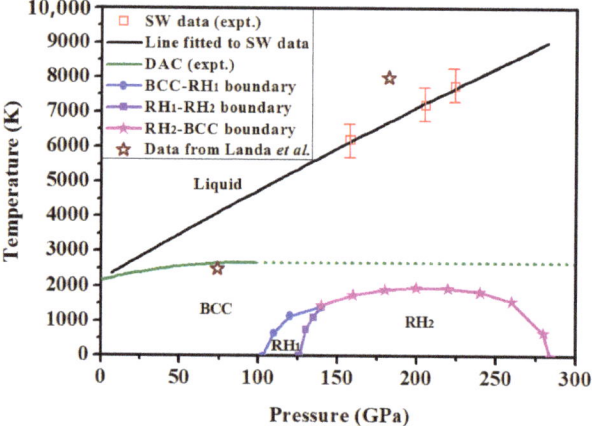

Figure 5. High-pressure and high-temperature phase diagram of vanadium. (By the courtesy of Ref. [54]).

Following Wang et al.'s prediction, Errandonea et al. [62] performed powder XRD experiments on vanadium up to 120 GPa and 4000 K. Under compression, the BCC vanadium was observed up to 53 GPa at room temperature. At higher pressure of 64 GPa, the measured XRD spectra at room temperature belonged to the RH structure of the R-3m space group. This observation supports the previous report by Ding et al. [20]. According to Errandonea et al. [62], the RH phase could be observed at temperatures up to 1560 K at 64 GPa and up to 1700 K at 120 GPa. Under the higher temperature of 1840 K at 64 GPa, the existence of BCC phase can be seen through the measured XRD pattern. This result is in accordance with the prediction of Wang et al. [54] but is less than one quarter of the initial estimate of Landa et al. [78]. Moreover, Errandonea et al. [62] interpreted their observations as the RH lattice distortion in vanadium that is triggered by phonon anomalies at high pressure and can be eliminated by phonon-phonon scattering effects at high temperatures.

According to the experimental results, Errandonea et al. further presented a phase diagram for vanadium under HPHT, as shown in Figure 6 (taken from [62]). The phase boundary of BCC→RH (dashed blue line) is tentatively drawn, which is qualitatively similar to the phase boundary given by Wang et al. [54] and consistent with other available results [19–21,47,78,81]. In addition, Wang et al. [54] also predicted the reentrance of the BCC phase at room temperature around 280 GPa, but this exceeded the pressure limit in Errandonea et al.'s experiments. This ultra-high pressure prediction still requires experimental verification.

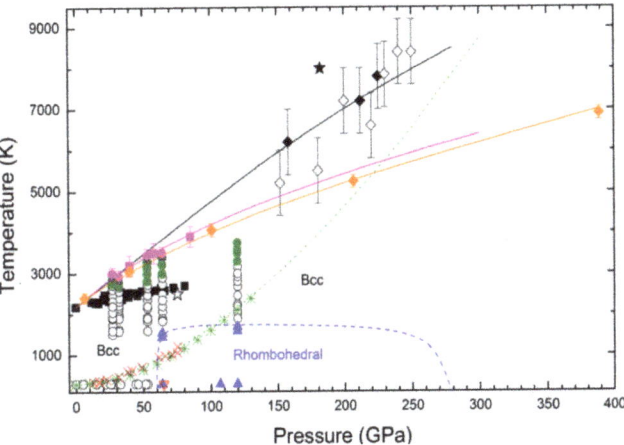

Figure 6. Pressure–temperature phase diagram of vanadium. (By the courtesy of Ref. [62]).

Most recently, Zhang et al. [69] also determined the phase stability of vanadium at 0–4400 K and 20–100 GPa by using synchrotron XRD independently. The results showed that BCC vanadium stabilized below 44 GPa at room temperature. With increasing the pressure above 52 GPa, a BCC→RH phase transition occurs, which is consistent with the observations of Errandonea et al. using NaCl as the pressure medium at 53 GPa [62]. Moreover, Zhang et al.'s experimental results showed that the RH vanadium stabilized between 50 and 100 GPa at room temperature [69]. At the pressure of 52 GPa, the RH phase transformed back to BCC when increasing the temperature to 1881 K. This supports the prediction by Wang et al. [54] that electronic temperatures will reduce the stability of RH phase in vanadium, as well as the experiment of Errandonea et al. [62].

In contrast to vanadium, only few studies [65,82] on the structure stability of niobium and tantalum have been carried out in the literature. In 2018, Haskins et al. [82] examined possible HPHT polymorphism in tantalum with complementary DFT-based model generalized pseudopotential theory (MGPT) multi-ion interatomic potentials. Their results showed that four orthorhombic structures of Pnma, Fddd, Pmma, and α-U are similarly energetically favorable. Moreover, the MGPT-MD simulations of them further revealed possible spontaneous heating-induced Pnma→BCC and Fddd→BCC transitions at modest temperatures. Nevertheless, neither unequivocal experimental (DAC or shock wave) nor direct DFT calculation evidence exists for these proposed phase transitions in tantalum by far.

As for niobium, Errandonea et al. [65] recently reported a result of static laser-heated DAC experiments up to 120 GPa, as well as ab initio quantum molecular dynamics simulations. They found that niobium undergoes a BCC→Pnma phase transition at high temperatures, which can be seen from their experimental XRD data. Errandonea et al.'s finding could provide evidence for the topological similarity of the phase diagrams of niobium and tantalum. All of them may undergo orthogonal phase transitions, and the HPHT phase is Pnma. However, this phase transition in niobium is the only report available so far; and no other relevant studies have been reported. Errandonea et al.'s experiments and theoretical calculations seem to be self-consistent, suggesting that such a type of phase transition may exist. However, more accurate experimental and theoretical studies are still needed for final confirmation.

2.3. Elastic Anomalies in V, Nb, Ta

In addition to the investigations on structural stability, the anomaly in the elastic constants of the group VB transition elements also attracts a lot of attention. Early theoretical and experimental studies [17,21] have shown that the transverse acoustic phonon mode of vanadium and niobium all exhibit softening. Since the shear elastic constant C_{44} is directly related to the transverse acoustic mode in the limit of short q-vector lengths, anomalous softening in this phonon mode implies that vanadium and niobium should have extraordinary elastic moduli. Landa et al. [18,19] confirmed this anomaly in the elastic constants of vanadium and niobium through first-principles calculations. The results show that vanadium is mechanical instability in C_{44} at pressures between 120 and 245 GPa. The results also show a softening in niobium at around 50 GPa. Their further study suggested that the pressure-induced shear instability (softening) in vanadium (niobium) is mainly due to the electronic structure with FSN.

Later, Koči et al. [22] confirmed the mechanical instability of C_{44} in vanadium through first-principles calculations. Meanwhile, they found that the elastic constants of group VB elements (V, Nb, Ta) exhibit anomalous behaviors, while those of group VIB elements (Mo, W) increase monotonically with pressure. According to Koči et al. [22], the calculated C_{44} of both vanadium and niobium is significantly underestimated by comparing to experimental data, while C_{11} and C_{12} are consistent with the experimental results. They further analyzed these metals by Fermi surface calculations and found that the nesting vectors of vanadium, niobium, and tantalum contracted with increasing pressure. This phenomenon, however, was not observed in molybdenum and tungsten. To explore the reason why C_{44} was underestimated in theoretical calculation, Liu et al. [48] further calculated the Fermi surface of these metals. The results suggested that the underestimation of C_{44} is mainly caused by the FSN.

To verify the theoretically predicted shear modulus anomalies, Jing et al. conducted XRD experiments on niobium powders under pressures up to 61 GPa at room temperature using DAC technique [60]. They observed an obvious softening in the yield strength of niobium between 42 and 47 GPa, which unexpectedly follows the trend of abnormal softening in the shear modulus predicted by recent theoretical studies [56]. Therefore, Jing et al. predicted that there should be a close relationship between the abnormal strength softening of niobium and the abnormal shear modulus softening [60]; however, this needs to be confirmed by further experimental evidence.

Furthermore, Li et al. [75] experimentally investigated the sound velocities of niobium up to 69 GPa and 1100 K under shock compression. It was found that both the compressional and shear sound velocities soften significantly between 50 and 60 GPa. Li et al. [75] suggested this anomalous behavior might be due to a pressure-induced ETT at 50–60 GPa. However, their data deviate significantly from all theoretical calculations [18,19,22,48]. Especially, by extrapolating their sound velocity to an impact pressure above 100 GPa, the longitudinal sound velocity of niobium would be much higher than that of vanadium; that is unphysical. In this regard, it is still an open question, and more experimental data with much higher precision are needed to pin down the predicted elastic anomalous softening in niobium.

It should also be mentioned that a similar softening in C_{44} was predicted by calculations for tantalum [83,84]. According to Gülseren et al. [84], the C_{44} of tantalum shows softening at pressures of 100–200 GPa. They also found that the C_{44} softens with temperature at low pressures, but then it becomes rather flat at higher pressures. Soon after, Landa et al. [41] further predicted a similar softening of C_{44} in tantalum between 50 and 80 GPa through first-principles calculations. Later, Antonangeli et al. [43] studied the elasticity of tantalum under pressures over 100 GPa. Their experiments showed that the shear velocity softened between 90 and 100 GPa, and with a pressure dependence above 120 GPa. They suggested that this abnormal behavior may be due to the intraband FSN that causes an ETT and a concomitant transverse acoustic phonon mode softening, which is consistent with the other theoretical predictions [22,48].

Jing et al. thought that there was a potential physical relationship between yield strength and shear modulus, so they further measured the yield strength of tantalum up to 101 GPa at room temperature by XRD-DAC experiment [53]. They detected a yield strength softening at 52–84 GPa, which is in accordance with one of the previous calculations [41] that suggested a significant softening in the shear modulus of tantalum between 50 and 80 GPa, but not others [43,84]. In addition, their measurements showed that the softening trend of the yield stress is roughly the same as that of the shear modulus given by the first-principles calculations [41]. To verify this result, Zhang et al. [85] re-studied the elastic properties of tantalum at high pressures through first-principles calculations. Their calculations showed elastic softening for both the C_{11} and C_{44} at pressures above 100 GPa, rather than at 50 GPa. The softening in C_{44} overall agrees with previous powder tantalum IXS data by Antonangeli et al. [43], but is different from Jing et al.'s results of 52–84 GPa [53].

In order to have a comprehensive understanding about the elastic anomalies of compression-induced softening, Wang et al. [54,56] revisited the elastic properties of the group VB transition metals, by noticing that DFT already correctly predicted many metals that also having Fermi surface nesting or Van Hove singularities. They first calculated the C_{44} and C' of BCC vanadium under different pressures [54], as shown in Figure 7 (taken from [54]). The obtained C_{44} is negative between 125 and 260 GPa, which is in agreement with that of Landa et al.'s full-potential linear muffin-tin orbitals (FP-LMTO) [18]. However, the calculated C_{44} from first-principles by Qiu et al. [39] shows that the first mechanical instability pressure for vanadium occurs at about 60 GPa, which is markedly different from the results of Landa et al. [18,19]. Qiu et al. suggested that this difference is mainly due to the fact that Landa et al.'s work ignored the pressure correction. However, Wang et al. [54] explicitly included the same pressure correction as Qiu et al. [39] when calculating the elastic modulus C_{44}. The obtained results perfectly match with that of Landa et al. [19], demonstrating that the pressure correction is not the reason for the difference between them. By using DFT, Wang et al. [56] further studied the elastic properties of niobium under pressure through first-principles calculations. The results demonstrated that the C_{44} and C' of niobium soften significantly in the range of 20–150 GPa. In addition, a new softening range for C_{44} at 275–400 GPa was also discovered. They suggested that the first anomaly was directly related to the underlying RH distortion, whereas the latter originated in an ETT.

In addition, the shear modulus C_{44} calculated by Wang et al. [66] through GGA (in PBE) is underestimated by about −40% (−30%) for vanadium (niobium) compared with the experimental results [86–89], which is consistent with the evaluation of Liu et al. [48] and Koči et al. [22]. Since the semi-local functional (such as PBE) may not be able to handle strong electron correlations in narrow-band systems, Wang et al. [66] further thoroughly evaluated the accuracy of different XC functionals in describing the C_{44} of vanadium and niobium. The results unexpectedly showed that C_{44} calculated by the common hybrid functionals (PBE0, HSE06, and B3LYP) are negative value at 0 GPa, which means that these functionals incorrectly predict the mechanical stability of these metals. Through systematic analysis, Wang et al. suggested that this unexpected failure is mainly caused by the localization error of these functionals.

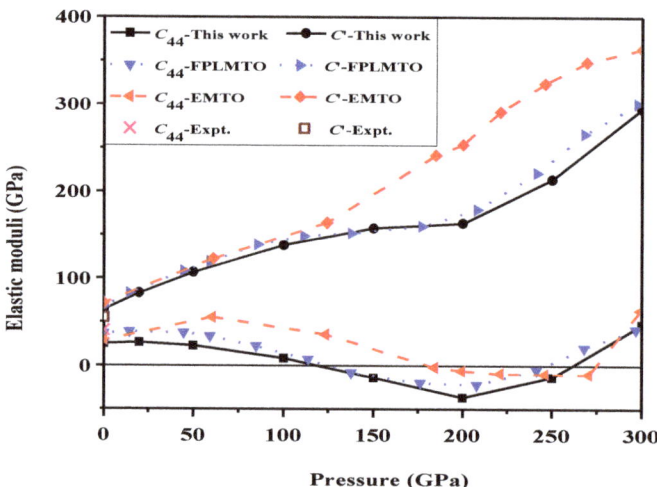

Figure 7. Calculated elastic moduli of BCC vanadium as a function of pressure. (By the courtesy of Ref. [54]).

Subsequently, Wang et al. [66] tentatively proposed a DFT + J method to correct the localization error, which corrected the C_{44} from 25.5 to 37.34 GPa for vanadium, and from 19.77 to 22.07 GPa for niobium at 0 GPa, respectively. To explore the influence of this correction on the high-pressure properties, Wang et al. further used the PBE + J method to study the elastic properties of vanadium under pressure, as shown in Figure 8 (taken from [66]). Obviously, the C_{11} and C_{12} calculated by different methods are consistent with each other, but the C_{44} calculated by different methods as a function of pressure is quite different. Among them, the PBE + J method has the largest correction, especially in the softening pressure range, which means that the correction of localization error could slightly weaken the elastic anomaly of compression-induced softening in group VB transition metals.

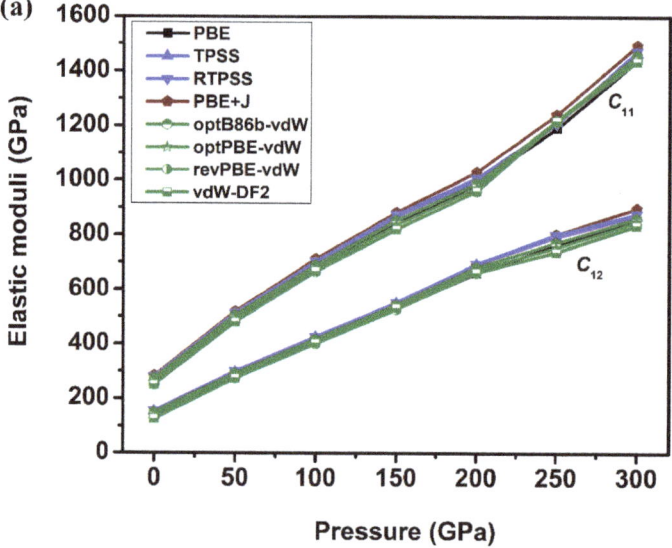

Figure 8. *Cont.*

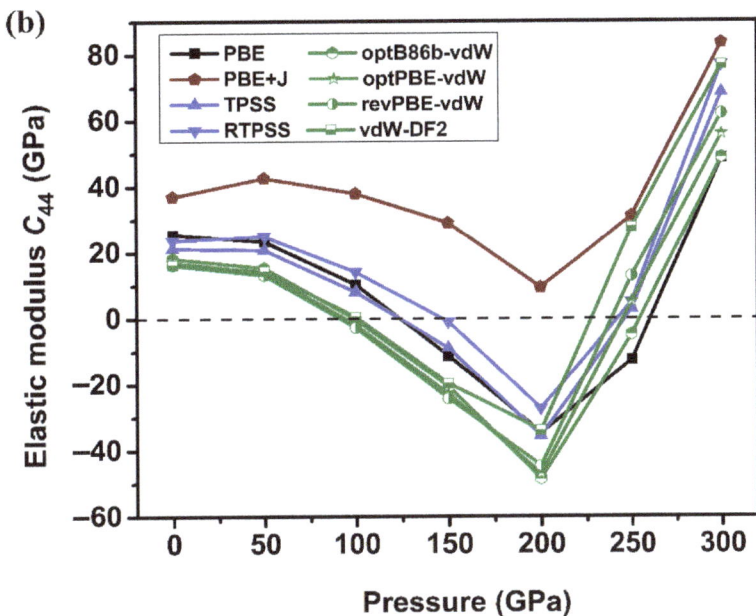

Figure 8. Calculated elastic moduli of BCC vanadium as a function of pressure by using different XC functionals and the PBE + J method (with J = 2 eV): (**a**) C_{11} and C_{12}, and (**b**) C_{44}. (By the courtesy of Ref. [66]).

Although vanadium, niobium, and tantalum have nearly identical valence electronic configuration, their compression-induced softening ranges of C_{44} are not the same. To explore the potential connection between them, Wang et al. [56] carefully calculated and compared the band structures of these metals under pressure of 0–400 GPa (see Figure 9 taken from [56]). They discovered two electronic topological transitions in the Γ→H direction for vanadium, niobium, and tantalum. The first ETT occurs around 300, 110, and 280 GPa in vanadium, niobium, and tantalum, respectively. The second ETT occurs at 300 GPa in niobium and at about 600 GPa in both vanadium and tantalum. According to Wang et al. [56], the second pressure-induced ETT should be the cause of the abnormal softening of C_{44} in niobium at 275–400 GPa, thus they predicted that vanadium and tantalum should have the same elastic modulus C_{44} softening at about 600 GPa. This is a quite interesting prediction that needs to be further confirmed by more precise experimental and theoretical studies.

In addition to the elastic anomaly of compression-induced softening, the heating-induced *hardening* in group VB metals was further predicted by Wang et al. [54]. They evaluated the effect of electronic temperature on the C_{44} of BCC vanadium and found that C_{44} increases with the temperature (see Figure 10 taken from [54]). This means that there is a heating-induced *hardening* in this metal, which is against our empirical intuition. In addition, as shown in the inset of Figure 10, at selected pressures of 50 GPa and 300 GPa, when the temperature increases from 0 to 3000 K, C_{44} increases by about 75% and 53%, respectively. At higher temperatures, however, the thermal motion of the nucleus will inevitably soften the metal. Therefore, as the temperature increases, the strength and shear modulus of vanadium will rise to a maximum and then drop to zero.

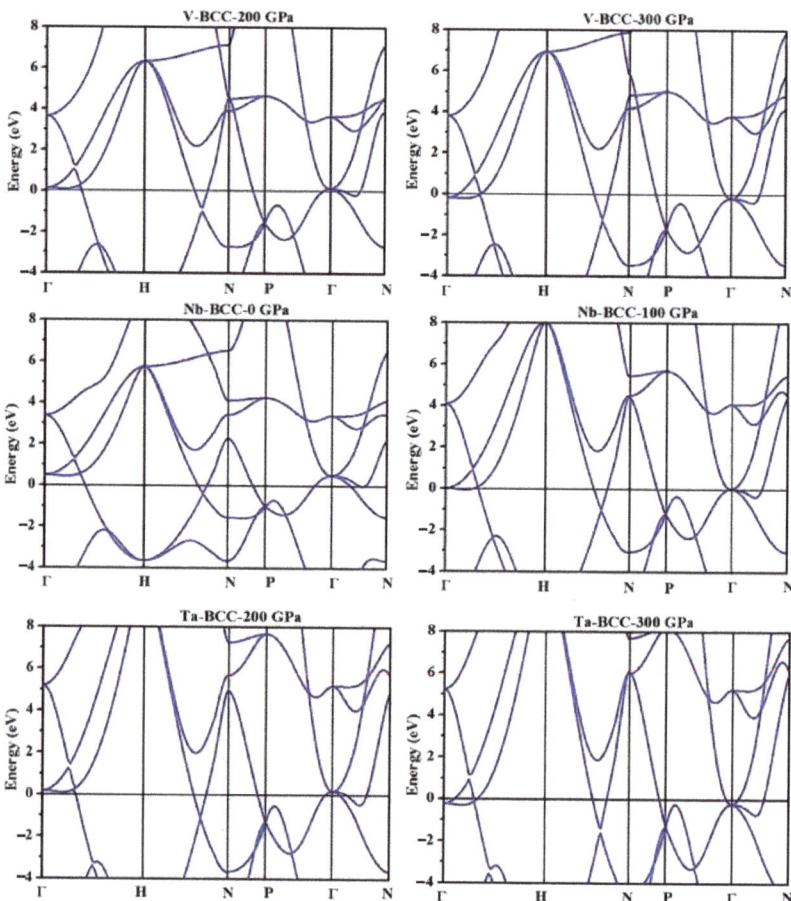

Figure 9. Band structures of vanadium, niobium, and tantalum in BCC phase at various pressures. (By the courtesy of Ref. [56]).

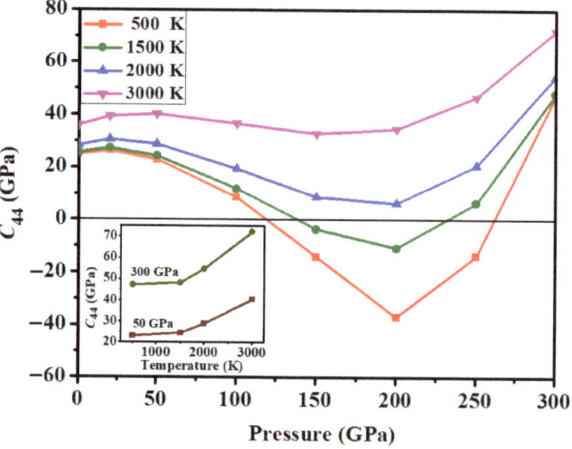

Figure 10. Calculated C_{44} of BCC vanadium as a function of pressure under different electronic temperatures. (By the courtesy of Ref. [54]).

Further investigation by Wang et al. showed that thermo-electrons also have a significant effect on the elastic anomalies of niobium [56]. Like its light neighbor vanadium, the elastic softening in niobium is gradually diminished with increased electronic temperature, effectively causing a heating-induced *hardening* phenomenon. Wang et al. [56] noticed that this thermo-electron effect only presents in the softening pressure ranges for niobium. At the pressure of 75 GPa, the C_{44} of niobium increases by about 135% when the temperature increases from 0 to 2000 K. Further research by Wang et al. [56] showed that the inclusion of phonon contribution could slightly soften the metal; however, it cannot change the conclusion qualitatively.

Subsequently, Keuter et al. [63] studied the anomalous thermoelastic properties of vanadium, niobium, and tantalum based on a DFT model, which could calculate the C_{44} under different temperatures. The results showed that the calculated elastic constants of cuprum and molybdenum decrease monotonically with the increase of temperature, which accords with the general thermoelastic behavior. However, for vanadium and niobium, the C_{44} falls to a minimum at about 500 K and then increases at higher temperatures, which is consistent with the heating-induced *hardening* observed by Wang et al. [56] in these elements.

As a rule of thumb, solids usually harden under compression and soften at high temperatures. Therefore, the theoretical predicted compression-induced softening and heating-induced *hardening* in group VB metals is quite remarkable. Nevertheless, direct experimental evidence has long been lacking. Until recently, Wang et al. [72] measured the HPHT sound velocities at Hugoniot states generated by shock waves and reported the first evidence for the CISHIH counterintuitive phenomenon in group VB metals. They observed that the shock vanadium not only had a significant reduction in sound velocity due to compression, but also had a strong increase in sound velocity due to heating. The former reflects the softening of the shear modulus by compression, while the latter corresponds to the reverse hardening by heat. Their experimental study further highlights the CISHIH dual anomaly behavior in group VB metals and provided inspiration for further theoretical and experimental research on this outstanding problem. The conceptual advancement also might be inspirational in understanding the general exotic behavior of matter, such as electrides [90,91], under extreme conditions.

3. Conclusions

In summary, it is a basic topic in condensed matter physics to reveal and elucidate the trend and mechanism of structural transformation and elastic anomaly of elemental metals. This review covers a large number of theoretical and experimental research on the phase stability and elastic anomalies in group VB transition metals over the last two decades. Two quite different scenarios are emerging from this progress. The first one is focused on phase transition and is represented by the phase diagram shown in Figures 5 and 6. As the basis of this scenario, a tentative theoretical "consensus" has been established on the phase transition sequence of BCC→RH1→RH2→BCC in vanadium under pressure, and RH→BCC under high temperature. Meanwhile, we also showed that some experimental and theoretical facts are incompatible with this scenario. They suggested that the resultant RH is far from being perfect and should be interpreted as lattice distortion rather than a phase transition. The same issue exists for niobium and tantalum. The above discussions cautioned that the phase transition in group VB metals is elusive, and the main problem of this scenario is that it cannot provide a unified and general description for all group VB elements.

The second scenario emphasizes on electronic structure and the resultant elastic anomalies. It viewed the phase transition in vanadium as the natural outcome of the anomaly, rather than the cause of it. In this direction, recent investigations revealed that different from other groups of transition metals, the group VB transition elements exhibit striking anomalous elastic softening under pressure. It is known that vanadium, niobium, and tantalum have nearly identical valence electronic configuration, but the variation of the C_{44} under pressure for them is not the same. According to existing literature reports, the pressure-induced softening of C_{44} in vanadium, niobium, and tantalum originates from the electronic structure, which is closely related to the FSN and ETT. In addition, the elastic softening in vanadium and niobium are gradually diminished with increased temperature, effectively giving rise to a heating-induced *hardening* phenomenon. The most striking feature of this scenario is that there actually could be *no phase transition* in vanadium at all. We have increasing confidence on this picture, especially after the reported facts that accurate DFT + *J* method completely eliminates RH phases from the thermodynamic equilibrium phase diagram [66], the absence/imperfect RH phase in Stevenson et al.'s experiment [74], the stability of BCC and meta-stability of RH as reported in Akahama et al.'s experiment [73], and the unexpected appearance of RH in shock experiment as reported by Wang et al. [72]. All of them cannot be consistently interpreted in the first scenario. Nonetheless, they can be reconciled in the second scenario, in which the "imperfect" RH as reported in various DAC experiments is viewed as lattice distortions caused by deviatoric stress rather than a new phase. These distortions lead to prominent change in modulus and sound velocity, and are modulated by compression and temperature, leading to CISHIH dual anomaly in both vanadium and niobium. We conclude that it seems the second scenario is more promising, but a lot of investigation is still required to achieve a general consensus.

Author Contributions: Y.W. and H.G. conceived the research. H.W. (Hao Wu), Y.L., H.W. (Hao Wang), X.C. and H.G. analyzed the results. Y.W. wrote the manuscript and all the authors commented on it. All authors have read and agreed to the published version of the manuscript.

Funding: This research was funded by National Key R&D Program of China under Grant No. 2021YFB3802300; the NSAF under Grant Nos. U1730248 and U1830101; the National Natural Science Foundation of China under Grant Nos. 11672274, 11602251, 11872056, and 11904282; the CAEP Research Project under Grant No. CX2019002; and the Science Challenge Project Tz2016001.

Conflicts of Interest: The authors declare no conflict of interest.

References

1. Cezairliyan, A. A dynamic technique for measurements of thermophysical properties at high temperatures. *Int. J. Thermophys.* **1984**, *5*, 177–193. [CrossRef]
2. Liu, Z.L.; Cai, L.C.; Chen, X.R.; Wu, Q.; Jing, F.Q. Ab initio refinement of the thermal equation of state for BCC tantalum: The effect of bonding on anharmonicity. *J. Phys. Condens. Matter* **2009**, *21*, 095408. [CrossRef] [PubMed]
3. Mattheiss, L.F. Electronic structure of niobium and tantalum. *Phys. Rev. B* **1970**, *1*, 373. [CrossRef]
4. Halloran, M.H.; Condon, J.H.; Graebner, J.E.; Kunzier, J.E.; Hsu, F.S.L. Experimental study of the fermi surfaces of niobium and tantalum. *Phys. Rev. B* **1970**, *1*, 366. [CrossRef]
5. Papaconstantopoulos, D.A.; Anderson, J.R.; McCaffrey, J.W. Self-consistent energy bands in vanadium at normal and reduced lattice spacings. *Phys. Rev. B* **1972**, *5*, 1214. [CrossRef]
6. Parker, R.D.; Halloran, M.H. Experimental study of the fermi surface of vanadium. *Phys. Rev. B* **1974**, *9*, 4130. [CrossRef]
7. Laurent, D.G.; Wang, C.S.; Callaway, J. Energy bands, Compton profile, and optical conductivity of vanadium. *Phys. Rev. B* **1978**, *17*, 455. [CrossRef]
8. Papaconstantopoulos, D.A.; Klein, B.M. Calculations of the pressure dependence of the superconducting transition temperature of vanadium. *Phys. B* **1981**, *107*, 725–726. [CrossRef]
9. Anderson, J.R.; Papaconstantopoulos, D.A.; Schirber, J.E. Influence of pressure on the Fermi surface of niobium. *Phys. Rev. B* **1981**, *24*, 6790. [CrossRef]
10. Struzhkin, V.V.; Timofeev, Y.A.; Hemley, R.J.; Mao, H.K. Superconducting Tc and electron-phonon coupling in Nb to 132 GPa: Magnetic susceptibility at megabar pressures. *Phys. Rev. Lett.* **1997**, *79*, 4262. [CrossRef]
11. McMillan, W.L. Transition temperature of strong-coupled superconductors. *Phys. Rev.* **1968**, *167*, 331. [CrossRef]
12. Tse, J.S.; Li, Z.Q.; Uehara, K.; Ma, Y.M.; Ahuja, R. Electron-phonon coupling in high-pressure Nb. *Phys. Rev. B* **2004**, *69*, 132101. [CrossRef]
13. Smith, T.F. Pressure dependence of the superconducting transition temperature for vanadium. *J. Phys. F Met. Phys.* **1972**, *2*, 946. [CrossRef]
14. Brandt, N.B.; Zarubina, O.A. Superconductivity of vanadium at pressures up to 250 kbar. *Sov. Phys. Solid State* **1974**, *15*, 3423–3425.
15. Akahama, Y.; Kobayashi, M.; Kawamura, H. Pressure Effect on Superconductivity of V and V-Cr Alloys up to 50 GPa. *J. Phys. Soc. Jpn.* **1995**, *64*, 4049–4050. [CrossRef]
16. Ishizuka, M.; Iketani, M.; Endo, S. Pressure effect on superconductivity of vanadium at megabar pressures. *Phys. Rev. B* **2000**, *61*, 3823. [CrossRef]

17. Suzuki, N.; Otani, M. Theoretical study on the lattice dynamics and electron-phonon interaction of vanadium under high pressures. *J. Phys. Condens. Matter* **2002**, *14*, 10869. [CrossRef]
18. Landa, A.; Klepeis, J.; Söderlind, P.; Naumov, I.; Velikokhatnyi, O.; Vitos, L.; Ruban, A. Ab initio calculations of elastic constants of the bcc V-Nb system at high pressures. *J. Phys. Chem. Solids* **2006**, *67*, 2056–2064. [CrossRef]
19. Landa, A.; Klepeis, J.; Söderlind, P.; Naumov, I.; Velikokhatnyi, O.; Vitos, L.; Ruban, A. Fermi surface nesting and pre-martensitic softening in V and Nb at high pressures. *J. Phys. Condens. Matter* **2006**, *18*, 5079. [CrossRef]
20. Ding, Y.; Ahuja, R.; Shu, J.; Chow, P.; Luo, W.; Mao, H.K. Structural phase transition of vanadium at 69 GPa. *Phys. Rev. Lett.* **2007**, *98*, 085502. [CrossRef]
21. Nakagawa, Y.; Woods, A.D.B. Lattice Dynamics of Niobium. *Phys. Rev. Lett.* **1963**, *11*, 271. [CrossRef]
22. Koči, L.; Ma, Y.; Oganov, A.R.; Souvatzis, P.; Ahuja, R. Elasticity of the superconducting metals V, Nb, Ta, Mo, and W at high pressure. *Phys. Rev. B* **2008**, *77*, 214101. [CrossRef]
23. Cynn, H.; Yoo, C.S. Equation of state of tantalum to 174 GPa. *Phys. Rev. B* **1999**, *59*, 8526. [CrossRef]
24. Ostanin, S.A.; Trubitsin, V.Y.; Savrasov, S.Y.; Alouani, M.; Dreyssé, H. Calculated Nb superconducting transition temperature under hydrostatic pressure. *Comput. Mater. Sci.* **2000**, *17*, 202–205. [CrossRef]
25. Yang, L.H.; Söderlind, P.; Moriarty, J.A. Atomistic simulation of pressure-dependent screw dislocation properties in BCC tantalum. *Mater. Sci. Eng. A* **2001**, *309*, 102–107. [CrossRef]
26. Singh, A.K.; Takemura, K. Measurement and analysis of nonhydrostatic lattice strain component in niobium to 145 GPa under various fluid pressure-transmitting media. *J. Appl. Phys.* **2001**, *90*, 3269–3275. [CrossRef]
27. Louis, C.N.; Iyakutti, K. Electron phase transition and superconductivity of vanadium under high pressures. *Phys. Rev. B* **2003**, *67*, 094509. [CrossRef]
28. Nnolim, N.O.; Tyson, T.A.; Axe, L. Theory of the structural phases of group 5B-6B metals and their transport properties. *J. Appl. Phys.* **2003**, *93*, 4543–4560. [CrossRef]
29. Dewaele, A.; Loubeyre, P.; Mezouar, M. Refinement of the equation of state of tantalum. *Phys. Rev. B* **2004**, *69*, 092106. [CrossRef]
30. Dewaele, A.; Loubeyre, P. Mechanical properties of tantalum under high pressure. *Phys. Rev. B* **2005**, *72*, 134106. [CrossRef]
31. Takemura, K.; Singh, A.K. High-pressure equation of state for Nb with helium-pressure medium: Powder X-ray diffraction experiments. *Phys. Rev. B* **2006**, *73*, 224119.
32. Klepeis, J.E. Electronic topological transitions in high-pressure bcc metals. In *APS March Meeting Abstracts*; American Physical Society: Washington, DC, USA, 2005; p. L11-010. Available online: http://meetings.aps.org/link/BAPS.2005.MAR.L11.10 (accessed on 28 November 2022).
33. Orlikowski, D.; Söderlind, P.; Moriarty, J.A. First-principles thermoelasticity of transition metals at high pressure: Tantalum prototype in the quasiharmonic limit. *Phys. Rev. B* **2006**, *74*, 054109. [CrossRef]
34. Suzuki, N.; Otani, M. The role of the phonon anomaly in the superconductivity of vanadium and selenium under high pressures. *J. Phys. Condens. Matter* **2007**, *19*, 125206. [CrossRef]
35. Luo, W.; Ahuja, R.; Ding, Y.; Mao, H.K. Unusual lattice dynamics of vanadium under high pressure. *Proc. Natl. Acad. Sci. USA* **2007**, *104*, 16428–16431. [CrossRef]
36. Lee, B.; Rudd, R.E.; Klepeis, J.E.; Söderlind, P.; Landa, A. Theoretical confirmation of a high-pressure rhombohedral phase in vanadium metal. *Phys. Rev. B* **2007**, *75*, 180101. [CrossRef]
37. Lee, B.; Rudd, R.E.; Klepeis, J.E.; Becker, R. Elastic constants and volume changes associated with two high-pressure rhombohedral phase transformations in vanadium. *Phys. Rev. B* **2008**, *77*, 134105. [CrossRef]
38. Verma, A.K.; Modak, P. Structural phase transitions in vanadium under high pressure. *Europhys. Lett.* **2008**, *81*, 37003. [CrossRef]
39. Qiu, S.L.; Marcus, P.M. Phases of vanadium under pressure investigated from first principles. *J. Phys. Condens. Matter* **2008**, *20*, 275218. [CrossRef]
40. Bosak, A.; Hoesch, M.; Antonangeli, D.; Farber, D.L.; Fischer, I.; Krisch, M. Lattice dynamics of vanadium: Inelastic X-ray scattering measurements. *Phys. Rev. B* **2008**, *78*, 020301. [CrossRef]
41. Landa, A.; Söderlind, P.; Ruban, A.V.; Peil, A.V.; Vitos, L. Stability in BCC transition metals: Madelung and band-energy effects due to alloying. *Phys. Rev. Lett.* **2009**, *103*, 235501. [CrossRef]
42. Vekilov, Y.K.; Krasil'nikov, O.M. Structural transformations in metals at high compression ratios. *Phys. Usp.* **2009**, *52*, 831–834. [CrossRef]
43. Antonangeli, D.; Farber, D.L.; Said, A.H.; Benedetti, L.R.; Aracne, C.M.; Landa, A.; Söderlind, P.; Klepeis, J.E. Shear softening of tantalum at megabar pressures. *Phys. Rev. B* **2010**, *82*, 132101. [CrossRef]
44. Bondarenko, N.G.; Vekilov, Y.K.; Isaev, E.I.; Krasil'nikov, O.M. Deformation Phase Transition in Vanadium under High Pressure. *JETP Lett.* **2010**, *91*, 611–613. [CrossRef]
45. Klepeis, J.H.P.; Cynn, H.; Evans, W.J.; Rudd, R.E.; Yang, L.H.; Liermann, H.P.; Yang, W. Diamond anvil cell measurement of high-pressure yield strength of vanadium using in situ thickness determination. *Phys. Rev. B* **2010**, *81*, 134107. [CrossRef]
46. Landa, A.; Söderlind, P.; Velikokhatnyi, O.I.; Naumov, I.I.; Ruban, A.V.; Peil, O.E.; Vitos, L. Alloying-driven phase stability in group-VB transition metals under compression. *Phys. Rev. B* **2010**, *82*, 144114. [CrossRef]
47. Jenei, Z.; Liermann, H.P.; Cynn, H.; Klepeis, J.H.P.; Baer, B.J.; Evans, W.J. Structural phase transition in vanadium at high pressure and high temperature: Influence of nonhydrostatic conditions. *Phys. Rev. B* **2011**, *83*, 054101. [CrossRef]

48. Liu, Z.; Shang, J. First principles calculations of electronic properties and mechanical properties of BCC molybdenum and niobium. *Rare Met.* **2011**, *30*, 354–358. [CrossRef]
49. Singh, A.K.; Liermann, H.-P. Strength and elasticity of niobium under high pressure. *J. Appl. Phys.* **2011**, *109*, 113539. [CrossRef]
50. Hu, J.; Dai, C.; Yu, Y.; Liu, Z.; Tan, Y.; Zhou, X.; Tan, H.; Cai, L.; Wu, Q. Sound velocity measurements of tantalum under shock compression in the 10–110 GPa range. *J. Appl. Phys.* **2012**, *111*, 033511. [CrossRef]
51. Yu, Y.; Tan, Y.; Dai, C.; Li, X.; Li, Y.; Wu, Q.; Tan, H. Phase transition and strength of vanadium under shock compression up to 88 GPa. *Appl. Phys. Lett.* **2014**, *105*, 201910. [CrossRef]
52. Krasil'nikov, O.M.; Vekilov, Y.K.; Lugovskoy, A.V.; Mosyagin, I.Y.; Belov, M.P.; Bondarenko, N.G. Structural transformations at high pressure in the refractory metals (Ta, Mo, V). *J. Alloys Compd.* **2014**, *586*, 242–245. [CrossRef]
53. Jing, Q.; Wu, Q.; Xu, J.-A.; Bi, Y.; Liu, L.; Liu, S.; Zhang, Y.; Geng, H. Anomalous softening of yield strength in tantalum at high pressures. *J. Appl. Phys.* **2015**, *117*, 055903. [CrossRef]
54. Wang, Y.X.; Wu, Q.; Xiang, R.; Chen, X.R.; Geng, H.Y. Stability of rhombohedral phases in vanadium at high-pressure and high-temperature: First-principles investigations. *Sci. Rep.* **2016**, *6*, 32419. [CrossRef] [PubMed]
55. Antonangeli, D.; Farber, D.L.; Bosak, A.; Aracne, C.M.; Ruddle, D.G.; Krisch, M. Phonon triggered rhombohedral lattice distortion in vanadium at high pressure. *Sci. Rep.* **2016**, *6*, 31887. [CrossRef] [PubMed]
56. Wang, Y.X.; Geng, H.Y.; Wu, Q.; Chen, X.R.; Sun, Y. First-principles investigation of elastic anomalies in niobium at high pressure and temperature. *J. Appl. Phys.* **2017**, *122*, 235903. [CrossRef]
57. Foster, J.M.; Comley, A.J.; Case, G.S.; Avraam, P.; Rothman, S.D.; Higginbotham, A.; Floyd, E.K.R.; Gumbrell, E.T.; Luis, J.J.D.; McGonegle, D.; et al. X-ray diffraction measurements of plasticity in shock-compressed vanadium in the region of 10-70 GPa. *J. Appl. Phys.* **2017**, *122*, 025117. [CrossRef]
58. Zou, Y.T.; Li, Y.; Chen, H.Y.; Welch, D.; Zhao, Y.S.; Li, B.S. Thermoelasticity and anomalies in the pressure dependence of phonon velocities in niobium. *Appl. Phys. Lett.* **2018**, *112*, 011901. [CrossRef]
59. Xiong, L.; Liu, J. Structural phase transition, strength, and texture in vanadium at high pressure under nonhydrostatic compression. *Chin. Phys. B* **2018**, *27*, 036101. [CrossRef]
60. Jing, Q.M.; He, Q.; Zhang, Y.; Li, S.R.; Liu, L.; Hou, Q.Y.; Geng, H.Y.; Bi, Y.; Yu, Y.Y.; Wu, Q. Unusual softening behavior of yield strength in niobium at high pressures. *Chin. Phys. B* **2018**, *27*, 106201. [CrossRef]
61. Kramynin, S.P.; Akhmedov, E.N. Equation of state and properties of Nb at high temperature and pressure. *J. Phys. Chem. Solids* **2019**, *135*, 109108. [CrossRef]
62. Errandonea, D.; MacLeod, S.G.; Burakovsky, L.; Santamaria-Perez, D.; Proctor, J.E.; Cynn, H.; Mezouar, M. Melting curve and phase diagram of vanadium under high-pressure and high-temperature conditions. *Phys. Rev. B* **2019**, *100*, 094111. [CrossRef]
63. Keuter, P.; Music, D.; Schnabel, V.; Stuer, M.; Schneider, J.M. From qualitative to quantitative description of the anomalous thermoelastic behavior of V, Nb, Ta, Pd and Pt. *J. Phys. Condens. Matter* **2019**, *31*, 225402. [CrossRef] [PubMed]
64. Weck, P.F.; Townsend, J.P.; Cochrane, K.R.; Crockett, S.D.; Moore, N.W. Shock compression of niobium from first-principles. *J. Appl. Phys.* **2019**, *125*, 245905. [CrossRef]
65. Errandonea, D.; Burakovsky, L.; Preston, D.L.; MacLeod, S.G.; Santamaría-Perez, D.; Chen, S.P.; Cynn, H.; Simak, S.I.; McMahon, M.I.; Proctor, J.E.; et al. Experimental and theoretical confirmation of an orthorhombic phase transition in niobium at high pressure and temperature. *Commun. Mater.* **2020**, *1*, 60. [CrossRef]
66. Wang, Y.X.; Geng, H.Y.; Wu, Q.; Chen, X.R. Orbital localization error of density functional theory in shear properties of vanadium and niobium. *J. Chem. Phys.* **2020**, *152*, 024118. [CrossRef]
67. Tidholm, J.; Hellman, O.; Shulumba, N.; Simak, S.I.; Tasnádi, F.; Abrikosov, I.A. Temperature dependence of the Kohn anomaly in BCC Nb from first-principles self-consistent phonon calculations. *Phys. Rev. B* **2020**, *101*, 115119. [CrossRef]
68. Weck, P.F.; Kalita, P.E.; Ao, T.; Crockett, S.D.; Root, S.; Cochrane, K.R. Shock compression of vanadium at extremes: Theory and experiment. *Phys. Rev. B* **2020**, *102*, 184109. [CrossRef]
69. Zhang, Y.J.; Tan, Y.; Geng, H.Y.; Salke, N.P.; Gao, Z.P.; Li, J.; Sekine, T.; Wang, Q.M.; Greenberg, E.; Prakapenka, V.B.; et al. Melting curve of vanadium up to 256 GPa: Consistency between experiments and theory. *Phys. Rev. B* **2020**, *102*, 214104. [CrossRef]
70. Yang, F.C.; Hellman, O.; Fultz, B. Temperature dependence of electron-phonon interactions in vanadium. *Phys. Rev. B* **2020**, *101*, 094305. [CrossRef]
71. Wang, Y.X.; Liu, Y.Y.; Yan, Z.X.; Liu, W.; Geng, H.Y.; Chen, X.R. Ab initio dynamical stability and lattice thermal conductivity of vanadium and niobium at high temperature. *Solid State Commun.* **2021**, *323*, 114130. [CrossRef]
72. Wang, H.; Li, J.; Zhou, X.M.; Tan, Y.; Hao, L.; Yu, Y.Y.; Dai, C.D.; Jin, K.; Wu, Q.; Jing, Q.M.; et al. Evidence for mechanical softening-hardening dual anomaly in transition metals from shock-compressed vanadium. *Phys. Rev. B* **2021**, *104*, 134102. [CrossRef]
73. Akahama, Y.; Kawaguchi, S.; Hirao, N.; Ohishi, Y. High-pressure stability of BCC-vanadium and phase transition to a rhombohedral structure at 200 GPa. *J. Appl. Phys.* **2021**, *129*, 135902. [CrossRef]
74. Stevenson, M.G.; Pace, E.J.; Storm, C.V.; Finnegan, S.E.; Garbarino, G.; Wilson, C.W.; McGonegle, D.; Macleod, S.G.; McMahon, M.I. Pressure-induced BCC-rhombohedral phase transition in vanadium metal. *Phys. Rev. B* **2021**, *103*, 134103. [CrossRef]
75. Li, P.; Huang, Y.F.; Wang, K.; Xiao, S.F.; Wang, L.; Yao, S.L.; Zhu, W.J.; Hu, W.Y. Crystallographic-orientation-dependence plasticity of niobium under shock compressions. *Int. J. Plasticity* **2022**, *150*, 103195. [CrossRef]

76. Li, X.H.; Yang, C.; Gan, B.; Huang, Y.Q.; Wang, Q.M.; Sekine, T.; Hong, J.W.; Jiang, G.; Zhang, Y.J. Sound velocity softening in body-centered cubic niobium under shock compression. *Phys. Rev. B* **2022**, *105*, 104110. [CrossRef]
77. Manghnani, M.H.; Nellis, W.J.; Nicol, M.F. Science and technology of high pressure. In Proceedings of the International Conference on High Pressure Sciene and Technology (AIRAPT-17), Honolulu, HI, USA, 25–30 July 1999; University Press: Hyderabad, India, 2000; Volume 1.
78. Landa, A.; Söderlind, P.; Yang, L.H. Ab initio phase stability at high temperatures and pressures in the V-Cr system. *Phys. Rev. B* **2014**, *89*, 020101. [CrossRef]
79. Dai, C.; Jin, X.; Zhou, X.; Liu, J.; Hu, J. Sound velocity variations and melting of vanadium under shock compression. *J. Phys. D Appl. Phys.* **2001**, *34*, 3064. [CrossRef]
80. Errandonea, D.; Schwager, B.; Ditz, R.; Gessmann, C.; Boehler, R.; Ross, M. Systematics of transition-metal melting. *Phys. Rev. B* **2001**, *63*, 132104. [CrossRef]
81. Landa, A.; Söderlind, P.; Naumov, I.I.; Klepeis, J.E.; Vitos, L. Kohn anomaly and phase stability in group VB transition metals. *Computation* **2018**, *6*, 29. [CrossRef]
82. Haskins, J.B.; Moriarty, J.A. Polymorphism and melt in high-pressure tantalum. II. Orthorhombic phases. *Phys. Rev. B* **2018**, *98*, 144107. [CrossRef]
83. Söderlind, P.; Moriarty, J.A. First-principles theory of Ta up to 10 Mbar pressure: Structural and mechanical properties. *Phys. Rev. B* **1998**, *57*, 10340. [CrossRef]
84. Gülseren, O.; Cohen, R.E. High-pressure thermoelasticity of body-centered-cubic tantalum. *Phys. Rev. B* **2002**, *65*, 064103. [CrossRef]
85. Zhang, Y.J.; Yang, C.; Alatas, A.; Said, A.H.; Salke, N.P.; Hong, J.W.; Lin, J.F. Pressure effect on Kohn anomaly and electronic topological transition in single-crystal tantalum. *Phys. Rev. B* **2019**, *100*, 075145. [CrossRef]
86. Bolef, D.I.; Smith, R.E.; Miller, J.G. Elastic Properties of Vanadium. I. Temperature Dependence of the Elastic Constants and the Thermal Expansion. *Phys. Rev. B* **1971**, *3*, 4100. [CrossRef]
87. Ko, C.R.; Salama, K.; Roberts, J.M. Effect of hydrogen on the temperature dependence of the elastic constants of vanadium single crystals. *J. Appl. Phys.* **1980**, *51*, 1014. [CrossRef]
88. Kojima, H.; Shino, M.; Suzuki, T. Effects of hydrogen and deuterium on the temperature dependence of the shear constants C' of vanadium single crystals. *Acta Metall.* **1987**, *35*, 891. [CrossRef]
89. Trivisonno, J.; Vatanayon, S.; Wilt, M.; Washick, J.; Reifenberger, R. Temperature dependence of the elastic constants of niobium and lead in the normal and superconducting states. *J. Low Temp. Phys.* **1973**, *12*, 153. [CrossRef]
90. Zhang, L.; Geng, H.Y.; Wu, Q. Prediction of anomalous LA-TA splitting in electrides. *Matter Radiat. Extremes* **2021**, *6*, 038403.
91. Zhang, L.; Wu, Q.; Li, S.; Sun, Y.; Yan, X.; Chen, Y.; Geng, H.Y. Interplay of anionic quasi-atoms and interstitial point defects in electrides: Abnormal interstice occupation and colossal charge state of point defects in dense fcc-lithium. *ACS Appl. Mater. Interfaces* **2021**, *13*, 6130–6139. [CrossRef]

Article

An Investigation of the Pressure-Induced Structural Phase Transition of Nanocrystalline α-CuMoO₄

Vinod Panchal [1], Catalin Popescu [2] and Daniel Errandonea [3,*]

[1] Department of Physics, Royal College, Mumbai 401107, India; panchalvinod@yahoo.com
[2] CELLS-ALBA Synchrotron Light Facility, 08290 Barcelona, Spain; cpopescu@cells.es
[3] Departamento de Física Aplicada, Instituto de Ciencias de Materiales, Universidad de Valencia, 46100 Valencia, Spain
* Correspondence: daniel.errandonea@uv.es

Abstract: The structural behavior of nanocrystalline α-CuMoO₄ was studied at ambient temperature up to 2 GPa using in situ synchrotron X-ray powder diffraction. We found that nanocrystalline α-CuMoO₄ undergoes a structural phase transition into γ-CuMoO₄ at 0.5 GPa. The structural sequence is analogous to the behavior of its bulk counterpart, but the transition pressure is doubled. A coexistence of both phases was observed till 1.2 GPa. The phase transition gives rise to a change in the copper coordination from square-pyramidal to octahedral coordination. The transition involves a volume reduction of 13% indicating a first-order nature of the phase transition. This transformation was observed to be irreversible in nature. The pressure dependence of the unit-cell parameters was obtained and is discussed, and the compressibility analyzed.

Keywords: high pressure; phase transition; synchrotron radiation; X-ray diffraction

1. Introduction

Nanomaterials play an important role in catalysis, optics, electricity, and other research fields due to their exclusive and typical characteristics. In the past decades, compared to bulk materials, their nanocrystalline counterparts have gained massive attention due to their novel properties such as thermal, electronic, and magnetic, owing to their specific shape, size, and surface to volume ratio. In addition to that, nanomaterials also find their application in the fields of biology, medicine, and chemical industry [1–4]. Metal molybdates (AMO₄) are important inorganic materials which have gained enormous scientific importance due to their wide range of applicability, such as industrial catalysts, photoluminescence, microwave applications, optical fibers, humidity sensors, scintillator materials and due to their magnetic and electrochemical properties [5–8].

Among molybdates, CuMoO₄ is the compound with a very complex polymorphism [9–12]. Till date, six different polymorphs of CuMoO₄ have been identified in the literature: namely ambient condition α-CuMoO₄ [9], high temperature β-CuMoO₄ [10], low temperature γ-CuMoO₄ [11], high pressure (HP) CuMoO₄-II [12], distorted wolframite CuMoO₄-III [13] and monoclinic ε-CuMoO₄ [14]. Earlier, single-crystal high-pressure and low-temperature X-ray diffraction studies carried by Wiesmann et al. [15] found that phase α-CuMoO₄ undergoes a structural phase transition to γ-CuMoO₄. This occurs during cooling at 190 K at ambient pressure as well as under increasing pressure at 0.2 GPa and room temperature. Congruently, high pressure and low temperature optical-absorption measurements carried out by Rodriguez et al. [16] found that α-CuMoO₄ undergoes a first-order transition at 0.25 GPa to γ-CuMoO₄. They found the same transition by cooling α-CuMoO₄ to 200 K at ambient pressure. The α to γ phase transition involves a change of part of the copper coordination polyhedra, from square-pyramidal (C_{4v}) CuO₅ in α-CuMoO₄ to octahedral elongated (D_{4h}) CuO₆ in γ-CuMoO₄. The transition also involves a volume drop of 13%.

Many of the applications of CuMoO$_4$ involve its use in the form of nanoparticles [17–19]. However, the information available on the structural and mechanical properties of nanocrystalline CuMoO$_4$ is scarce. It is known that the high-pressure (HP) behavior of materials could be different for nanoparticles than for bulk materials [20–23]. In particular, transition pressures and compressibilities could be affected when the particle size is reduced [20–23]. To explore the structural and mechanical properties of nanocrystalline CuMoO$_4$ under compression, the performance of an HP X-ray diffraction (XRD) investigation is needed. However, such studies have not been carried out yet in nanocrystalline α-CuMoO$_4$. In the present work we have investigated nanocrystalline α-CuMoO$_4$ by high pressure powder X-ray diffraction up to a pressure of 2 GPa under hydrostatic conditions.

2. Experiment

CuMoO$_4$ nanoparticles were prepared following the method reported by Hassani et al. [20]. The chemical composition of the obtained CuMoO$_4$ nanoparticles was determined by energy-dispersive X-ray spectrometry (EDXS), which was recorded with a JEOL JEM-6700F device. A EDXS spectrum is shown in Figure 1. The analysis showed that the composition of the nanoparticles was 67 (3) atomic % oxygen, 16 (1) atomic % copper, and 17 (1) atomic % molybdenum, which agrees with the stoichiometry of CuMoO$_4$.

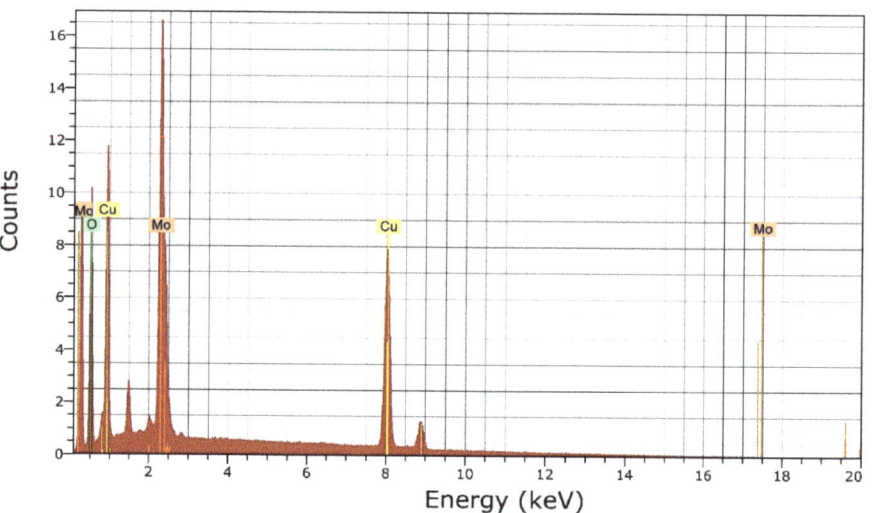

Figure 1. Energy dispersive X-ray analysis (EDXS) spectrum of nanocrystalline CuMoO$_4$.

Ambient conditions powder XRD (λ = 0.4246 Å) performed in the same set up as high-pressure measurements confirmed that CuMoO$_4$ nanoparticles crystallize in the triclinic α polymorph (space group $P\bar{1}$). The average particle size was estimated from powder XRD using the Scherrer equation [24], to be 43 (7) nm. This value was confirmed using MAUD [25], which gave an average particle size of 40 (9) nm. Angle-dispersive powder HP-XRD measurements were performed at the BL04-MSPD beamline of ALBA synchrotron [26]. An incident monochromatic X-ray beam with a wavelength of 0.4246 Å was focused down to a spot size of 20 μm × 20 μm (FWHM). Two-dimensional (2D) XRD data were recorded using a Rayonix SX165 CCD detector. The 2D images were integrated into one-dimensional intensity versus 2θ patterns using Dioptas [27]. The sample-to-detector distance along with detector parameters were determined using a LaB$_6$ calibrant. A pellet of CuMoO$_4$ nanocrystalline powder was loaded into a Boehler-Almax diamond-anvil cell (DAC) with diamond culets of 500 μm using a T301 stainless-steel gasket, pre-indented to a thickness of 50 μm and with a sample chamber of 200 μm diameter in the center. As pressure-transmitting medium, we used a 4:1 methanol–ethanol to allow a direct comparison with previous HP studies performed in microscopic samples [15]. The pressure medium can be considered nearly hydrostatic within the pressure limit of the experiments [28]. Special attention was paid to occupy only a minor fraction on the pressure chamber with sample to avoid sample bridging between diamonds [29]. The pressure was determined using ruby fluorescence [30]. Small ruby spheres with a grain size of ~1 μm and concentration of 3000 ppm Cr^{3+} [31] were used for pressure calibration [31] based on the wavelength shift of the R1 fluorescence band of the trivalent Cr^{3+} [32,33]. The estimated relative error for all measured pressures was better than 1% [34]. The unit-cell parameters were initially determined using the Le Bail analysis incorporated in the GSAS software [35]. Using the same software, Rietveld analysis [36] was also carried out for the initial α-$CuMoO_4$ phase and high pressure γ-$CuMoO_4$ phase.

3. Results and Discussion

At ambient conditions, $CuMoO_4$ crystallizes in a triclinic α-$CuMoO_4$ structure (space group $P\bar{1}$, Z = 6). The α-$CuMoO_4$ structure can be described by MoO_4 tetrahedra, CuO_5 square pyramidal polyhedra, and CuO_6 distorted octahedra, interconnected via common corners and edges forming layers as shown in Figure 2.

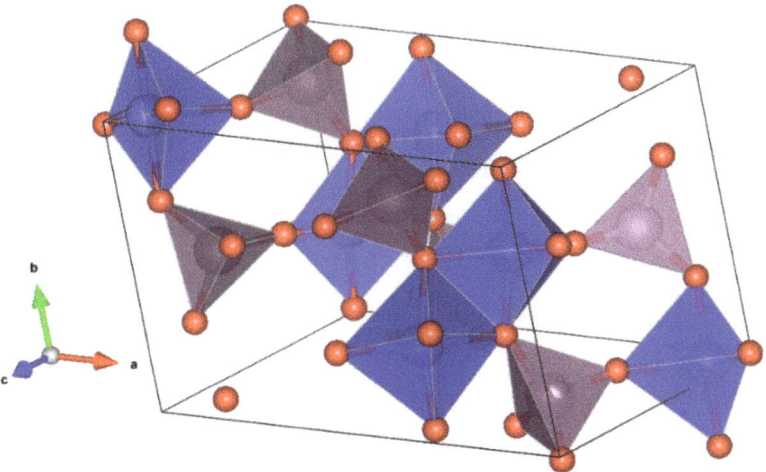

Figure 2. Crystal structure of α-$CuMoO_4$. Blue solid spheres correspond to Cu atoms, gray solid spheres correspond to Mo atoms, and red solid spheres correspond to O atoms. The different coordination polyhedra are illustrated in the figure.

Figure 3 shows a selection of HP powder X-ray diffraction profiles of nanocrystalline α-$CuMoO_4$ at representative pressures. There were no noticeable changes (other than typical peak shifts due to the changes induced by pressure in the unit-cell parameters) in the diffraction patterns up to 0.3 GPa and all the diffraction peaks could be indexed to α-$CuMoO_4$ phase. At 0.5 GPa, the appearance of many weak extra diffraction peaks was observed at 2θ of 4.2, 4.7, 6.8 and 9.5° along with the diffraction peaks of the α-$CuMoO_4$ phase. The red arrows in Figure 3 identify some of the additional peaks observed at 0.5 GPa. The red dashed lines show how these peaks evolve with pressure until the highest pressure corresponding to peaks of the γ-phase. In the pattern at 2.0 GPa (γ-phase) they are identified with asterisks.

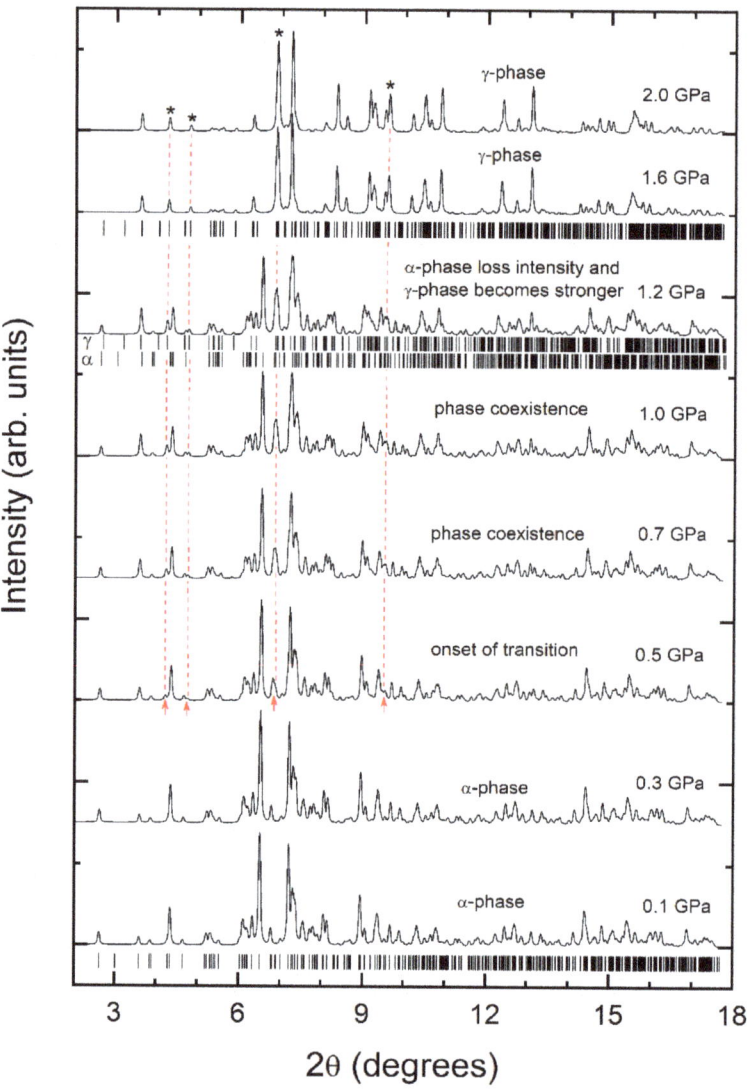

Figure 3. Evolution of the X-ray diffraction patterns of α-CuMoO$_4$ as a function of pressure. Pressures are provided in the figure. Red arrows indicate the appearance of extra diffraction peaks evidencing the onset of the transition to the γ-CuMoO$_4$ phase. Red dashed lines show how these peaks evolve under compression. The asterisks identify the peaks in the γ-phase. Ticks are the position of Bragg peaks for different structures.

These modifications in the diffraction pattern are indicative of the onset of a structural phase transition in α-CuMoO$_4$ at 0.5 GPa. The transition pressure observed in this study is almost double the transition pressure determined from earlier single-crystal high-pressure XRD measurements and optical studies on bulk samples, which identified the phase transition at 0.2 and 0.25 GPa, respectively [15,16]. The increase of the transition pressure could be explained in terms of an increase in surface energy in the newly formed high-pressure phase crystallites [37]. The coexistence of both the phases has been observed till 1.2 GPa. Beyond this pressure, all the diffraction peaks could be assigned to the triclinic γ-CuMoO$_4$ (space group $P\bar{1}$). On further increase of pressure, the γ-CuMoO$_4$ phase has been found to remain stable up to 2 GPa, which is the highest pressure reached in our XRD measurements. On release of pressure, the γ-CuMoO$_4$ phase has been observed at ambient conditions, thus indicating the irreversible nature of the phase transition. In previous studies [15,16] the reversibility of the HP α-γ transition was not evaluated. However, it was shown that the γ-phase has a smaller band gap and different optical properties than the α-phase [16]. In particular, the band gap of γ-CuMoO$_4$ is close to 2 eV, making it ideal for hydrogen production via photocatalytic water splitting [38]. Given the low-pressure requested to obtain γ-CuMoO$_4$ as an ambient-conditions metastable polymorph, HP could be an efficient way to synthesize γ-CuMoO$_4$ for water splitting applications. The crystal structure of the HP γ-CuMoO$_4$ phase is represented in Figure 4. In this structure, all Cu atoms display an octahedral elongated coordination, and all Mo atoms have a distorted octahedral coordination. Rietveld refinements supporting the structural assignments are provided in Figure 5.

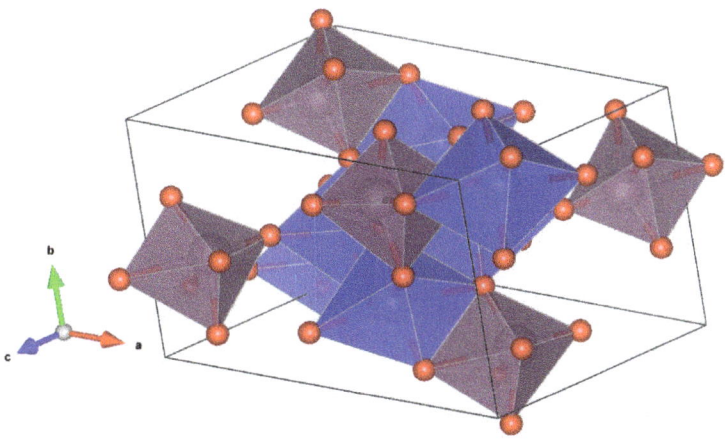

Figure 4. Crystal structure of γ-CuMoO$_4$. Blue solid spheres correspond to Cu atoms, gray solid spheres correspond to Mo atoms, and red solid spheres correspond to O atoms. The different coordination polyhedra are illustrated in the figure.

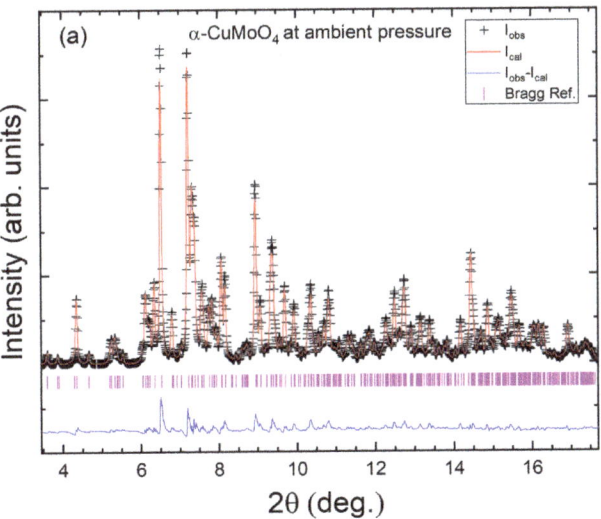

Figure 5. Cont.

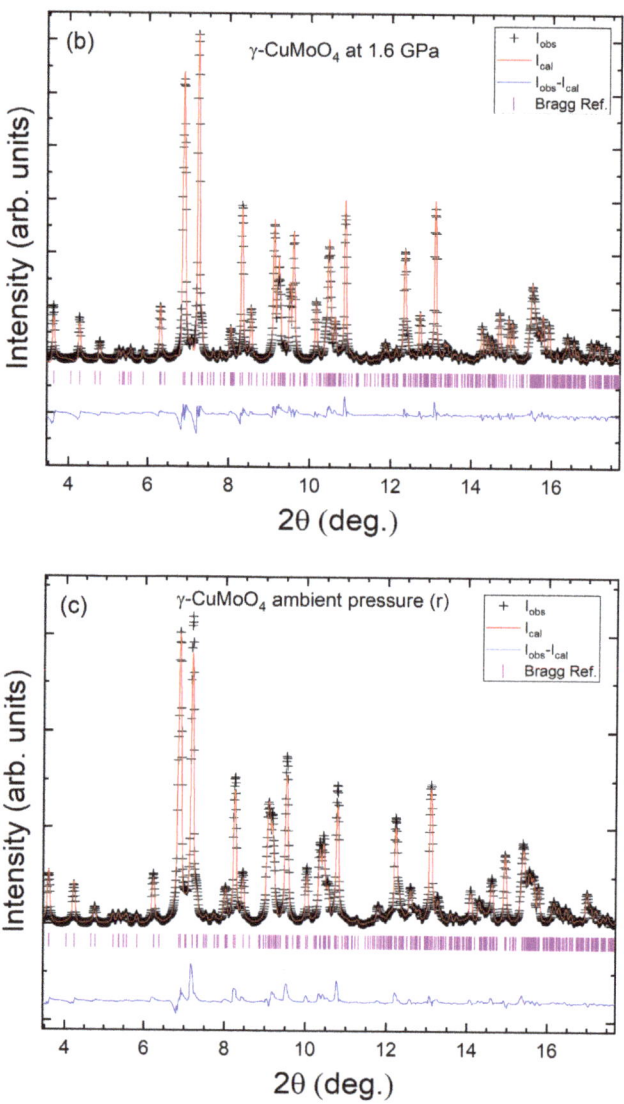

Figure 5. (**a**) XRD pattern and Rietveld refinement of α-CuMoO$_4$ at ambient pressure. (Rwp = 13.4%, Rp = 9.2%). (**b**) XRD pattern and Rietveld refinement of γ-CuMoO$_4$ at 1.6 GPa (Rwp = 11.4%, Rp = 6.8%). (**c**) XRD pattern and Rietveld refinement of γ-CuMoO$_4$ at ambient pressure after pressure release (Rwp = 12.4%, Rp = 8.0%), denoted by (r).

At ambient pressure, the lattice parameters for α-CuMoO$_4$ were refined as a = 9.870 (1) Å, b = 6.764 (1) Å, c = 8.337 (1) Å, α = 101.13(5)°, β = 96.90 (5)°, γ = 107.03 (5)°. The lattice parameters and fractional coordinates of the α-CuMoO$_4$ are given in Table 1. They are in good agreement with those reported in previous single-crystal high-pressure XRD measurements by Wiesmann et al., a = 9.901 (3) Å, b = 6.786 (2) Å, c = 8.369 (3) Å, α = 101.13°, β = 96.88°, γ = 107.01° [15]. On the other hand, the lattice parameters of γ-CuMoO$_4$ phase at 1.6 GPa were found to be a = 9.608 (9) Å, b = 6.219 (7) Å, c = 7.875 (5) Å, α = 94.91 (1)°, β = 103.10 (1)°, γ = 102.48 (9)°. The lattice parameters and fractional coordinates are summarized in Table 2. They are in good agreement with those reported in previous single crystal high-pressure XRD measurements by Wiesmann et al., a = 9.708 (3) Å, b = 6.302 (7) Å,

c = 7.977 (2) Å, α = 94.76°, β = 103.35°, γ = 103.26° [15]. After release of pressure, the lattice parameters for γ-CuMoO$_4$ were found to be a = 9.6605 Å, b = 6.2751 Å, c = 7.9334 Å, α = 94.76°, β = 103.29°, γ = 103.14°. The unit-cell volume of the γ-phase was 11.5% smaller than the volume of the α-phase.

Table 1. Lattice parameters and fractional coordinates of α-CuMoO$_4$ at ambient conditions. Space group $P\bar{1}$, Z = 6, a = 9.870 (1) Å, b = 6.764 (1) Å, c = 8.337 (1) Å, α = 101.13 (5)°, β = 96.90 (5)°, γ = 107.03 (5)°.

Atoms	Sites	x	y	z
Cu1	2i	0.4054 (3)	0.7466 (3)	0.1983 (3)
Cu2	2i	0.9922 (3)	0.0472 (3)	0.2024 (3)
Cu3	2i	0.2365 (3)	0.4626 (3)	0.3853 (3)
Mo1	2i	0.3450 (2)	0.1970 (2)	0.0850 (2)
Mo2	2i	0.1064 (2)	0.4960 (2)	0.7776 (2)
Mo3	2i	0.2500 (2)	0.9927 (2)	0.4648 (2)
O1	2i	0.1425 (12)	0.9355 (12)	0.6056 (12)
O2	2i	0.2751 (12)	0.7459 (12)	0.3590 (12)
O3	2i	0.1839 (12)	0.1582 (12)	0.3414 (12)
O4	2i	0.1644 (12)	0.0642 (12)	0.9873 (12)
O5	2i	0.1846 (12)	0.5020 (12)	0.5974 (12)
O6	2i	0.3669 (12)	0.4518 (12)	0.2238 (12)
O7	2i	0.2405 (12)	0.5940 (12)	0.9484 (12)
O8	2i	0.4122 (12)	0.1567(12)	0.5772 (12)
O9	2i	0.0039 (12)	0.2291 (12)	0.7692 (12)
O10	2i	0.0107 (12)	0.3455 (12)	0.2115 (12)
O11	2i	0.4463 (12)	0.2493 (12)	0.9330 (12)
O12	2i	0.4104 (12)	0.0335 (12)	0.1952 (12)

Table 2. Lattice parameters and fractional coordinates of γ-CuMoO$_4$ at 1.6 GPa. Space group $P\bar{1}$, Z = 6, a = 9.608 (9) Å, b = 6.219 (7) Å, c = 7.875 (5) Å, α = 94.91 (1)°, β = 103.10 (1)°, γ = 102.48 (9)°.

Atoms	Sites	x	y	z
Cu1	2i	0.4339 (3)	0.7158 (3)	0.2360 (3)
Cu2	2i	0.0069 (3)	0.0931 (3)	0.1939 (3)
Cu3	2i	0.3355 (3)	0.4240 (3)	0.5260 (3)
Mo1	2i	0.3481 (2)	0.2165 (2)	0.1269 (2)
Mo2	2i	0.1092 (2)	0.4145 (2)	0.8800 (2)
Mo3	2i	0.2269 (12)	0.9336 (2)	0.4583 (2)
O1	2i	0.1193 (12)	0.8973 (12)	0.6000 (12)
O2	2i	0.2951 (12)	0.6970 (12)	0.4544 (12)
O3	2i	0.1998 (12)	0.2082 (12)	0.3424 (12)
O4	2i	0.1546 (12)	0.1263 (12)	0.9699 (12)
O5	2i	0.1927 (12)	0.4514 (12)	0.7175 (12)
O6	2i	0.4765 (12)	0.4375 (12)	0.3011 (12)
O7	2i	0.2720 (12)	0.5024 (12)	0.0747 (12)
O8	2i	0.3863 (12)	0.1490 (12)	0.5890 (12)
O9	2i	0.9406 (12)	0.1831 (12)	0.7561 (12)
O10	2i	0.9744 (12)	0.3626 (12)	0.1078 (12)
O11	2i	0.4474 (12)	0.2102 (12)	0.9788 (12)
O12	2i	0.3568 (12)	0.9728 (12)	0.2219 (12)

From Rietveld refinements, we obtained the pressure evolution of the lattice parameters and equation of state of α-CuMoO$_4$ and γ-CuMoO$_4$. For the HP phase, they were obtained only for 0.7 GPa and higher pressures because at 0.3 and 0.5 GPa the fraction of γ-CuMoO$_4$ was too small to allow an accurate determination of unit-cell parameters (The peaks of γ-CuMoO$_4$ are too weak at these two pressures). For the low-pressure phase they were obtained up to 1.2 GPa. The pressure dependence of the lattice parameters and angles are plotted in Figure 6. From the figure, it can be seen the behavior under compression is anisotropic.

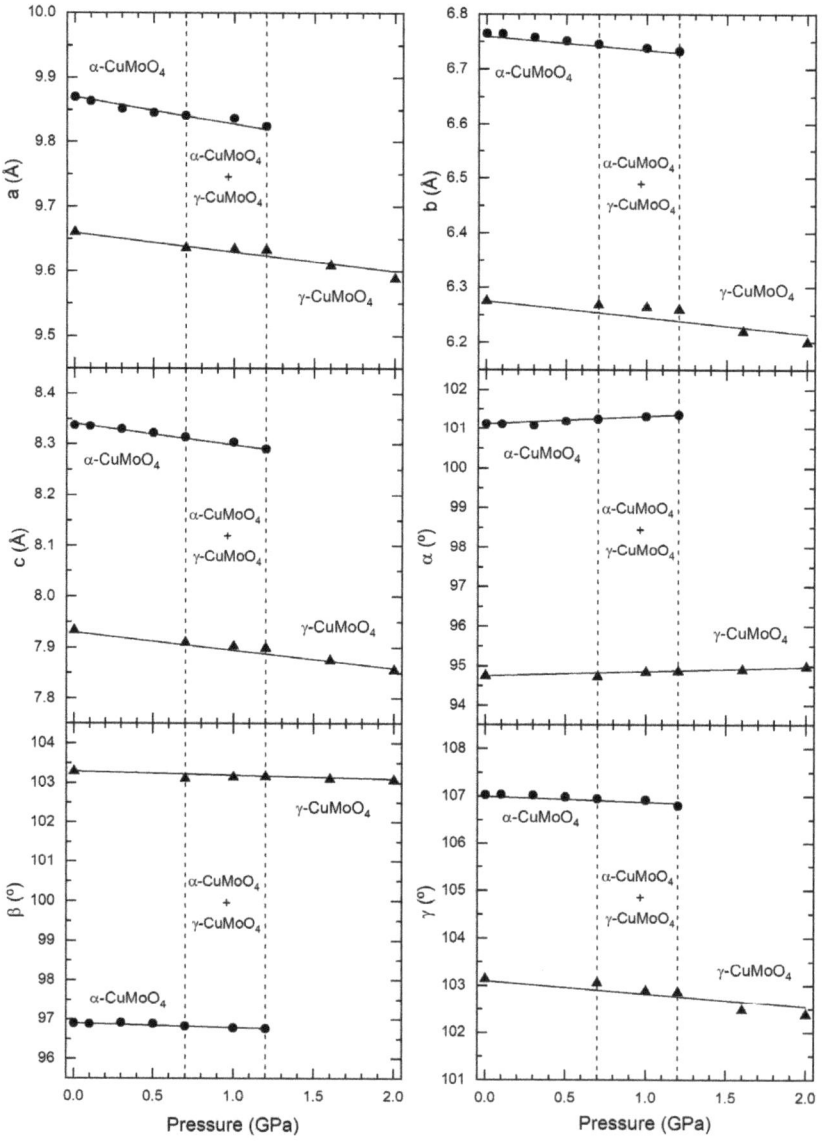

Figure 6. Pressure dependence of the lattice parameters and angles. Filled circles correspond to the α-CuMoO$_4$ phase and solid triangles corresponds to γ-CuMoO$_4$. Errors are smaller than symbol's size. The solid lines in the lattice-parameter plots are linear fits of the data. Vertical dashed lines delimit the region of phase coexistence.

For any triclinic structure, the description of the deformation of the unit cell under compression is provided by the compressibility tensor [39]; in particular, by its eigenvectors ($e_{\lambda i}$) and eigenvalues (λ_i). They give the principal axes of the compressibility and the compressibilities along them. We have obtained $e_{\lambda i}$ and λ_i for α-CuMoO$_4$ using PASCal [40]. The results are given in Table 3.

Table 3. Eigenvalues, λ_i, and Eigenvectors, $e_{\lambda i}$, for the isothermal compressibility tensor of α-CuMoO$_4$.

Eigenvalues	Eigenvectors
$\lambda_1 = 4.2(8) \times 10^{-3}$ GPa^{-1}	$e_{\lambda 1} = (-0.8568, 0.0228, -0.2899)$
$\lambda_2 = 5.8(8) \times 10^{-3}$ GPa^{-1}	$e_{\lambda 2} = (-0.2653, -0.7825, 05632)$
$\lambda_3 = 7.1(8) \times 10^{-3}$ GPa^{-1}	$e_{\lambda 3} = (0.1237, 0.7819, 0.6110)$

The pressure dependence of volume of both the phases is plotted in Figure 7. From the plot, it can be seen that the phase transition from α-CuMoO$_4$ to γ-CuMoO$_4$ exhibits a volume collapse of ~13% at 0.7 GPa which is also in agreement with earlier XRD measurements [15]. The results of Figure 7 were fitted using second- and third-order Birch–Murnaghan equations of state. The fits were carried out using EOSFIT [41]. A fit of the volume vs. pressure data of the α-CuMoO$_4$ phase to a second-order Birch–Murnaghan equation of state ($B'_0 = 4$) gives a bulk modulus of 67 (2) GPa, if we use third order Birch-Murnaghan equation of state it gives bulk modulus of 62 (4) GPa with $B'_0 = 5.1$ (8) which is also in agreement with earlier XRD measurements [15]. A fit of the volume vs. pressure data of the γ-CuMoO$_4$ phase to a second-order Birch-Murnaghan equation of state ($B'_0 = 4$) gives bulk modulus of 62 (6) GPa, if we use third-order order Birch-Murnaghan equation of state it gives bulk modulus of 68 (9) GPa with $B'_0 = 3.1$ (9). In the previous study, $B_0 = 44.0$ (3.4) GPa and $B'_0 = 15.3$ (3.0) were reported for γ-CuMoO$_4$. Such a large pressure derivative for the bulk modulus is unusual in oxides [42]. On the other hand, it is not reasonable to have a smaller bulk modulus for γ-CuMoO$_4$ than for α-CuMoO$_4$, because the first phase is denser than the second one. The mismatch of B_0 and B'_0 in the previous work could be related to the fact that volume at zero pressure (V_0) was unknown, being extrapolated from HP results. In our case, V_0 has been experimentally determined after decompression. V_0, B_0, and B'_0 are correlated parameters [43] and therefore an overestimation of V_0 (466 Å^3 extrapolated in Ref. [15] and 454 Å^3 here measured) could easily lead to inaccurate values of B_0 and B'_0. Interestingly, our results indicate, that despite the volume collapse of the transition, the two phases of CuMoO$_4$ have comparable bulk moduli.

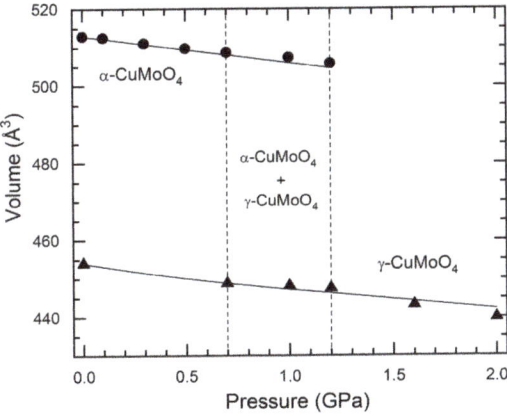

Figure 7. Pressure dependence of the volume. Filled circles correspond to the α-CuMoO$_4$ phase and solid triangles corresponds to γ-CuMoO$_4$. Errors are smaller than symbol size. The solid lines are linear fits of the data.

Interestingly, the two phases of CuMoO$_4$ have considerably smaller bulk moduli than CuWO$_4$ (B_0 = 139 GPa) [44]. This is related to the fact that CuMoO$_4$ has a much more open structure than CuWO$_4$. This is a direct consequence of the way in which polyhedra are linked in CuMoO$_4$, which gives to the structure of the two polymorphs a pseudo-layered characteristic. This feature makes the reduction of the "empty" space between layers easier and consequently leads to a smaller bulk modulus in CuMoO$_4$. The second fact that facilitates the reduction of volume in CuMoO$_4$ is that it can be achieved by tilting of the polyhedra. It should be noted that with respect to its polyhedral constitution, CuMoO$_4$ is more similar to Cu$_3$V$_2$O$_8$ and Cu$_2$V$_2$O$_7$ [45,46] than to typical orthomolybdates [47,48]. This is consistent in that the two vanadates have bulk moduli of 52 and 64 GPa, respectively; values that are comparable to the bulk modulus of α- and γ-CuMoO$_4$.

4. Conclusions

X-ray diffraction studies on nanocrystalline α-CuMoO$_4$ up to 2 GPa suggest that the low pressure α-phase undergoes an irreversible α–γ phase transition. The onset of phase transition occurs at 0.5 GPa, the coexistence of both phases was observed between 0.5 to 1.2 GPa, and the transition completes at 1.6 GPa, indicating a sluggish nature of the phase transition. The α to γ phase transition involves a change of part of the copper coordination polyhedra, from square-pyramidal (C_{4v}) CuO$_5$ in α-CuMoO$_4$ to octahedral elongated (D_{4h}) CuO$_6$ in γ-CuMoO$_4$ along with a reduction in volume by 13%, indicative of first-order phase transition. The transition pressure observed in this case is almost double compared to its bulk counterpart which can be attributed to an increase in surface energy in the newly formed high-pressure phase. In addition, we found that in nanocrystalline CuMoO$_4$ the high-pressure γ-polymorph can be recovered as a metastable phase upon decompression. Since the γ-phase has a smaller band-gap energy than the α-phase, our findings suggest that high pressure can be used to synthesize a metastable polymorph with a tailored band-gap energy for water splitting applications. Finally, we have also discussed the compressibility of both polymorphs and determined their equations of state.

Author Contributions: D.E. conceived the experiments. D.E. and C.P. conducted the experiments. V.P. analyzed the results. The manuscript was written through contributions of all authors. All authors have read and agreed to the published version of the manuscript.

Funding: This work was supported by the Spanish Research Agency (AEI) and Spanish Ministry of Science and Investigation (MCIN) under project PID2019-106383GB-C41 (DOI:10.13039/501100011033), co-financed by EU funds. The authors also acknowledge financial support from the MALTA Consolider Team network, under project RED2018-102612-T.

Institutional Review Board Statement: Not applicable.

Informed Consent Statement: Not applicable.

Data Availability Statement: All relevant data that support the findings of this study are available from the corresponding authors upon request.

Acknowledgments: The authors acknowledge the ALBA synchrotron facilities for provision of beamtime on the beamline MSPD-BL04 (Proposal 2016081779). They also thanks S. Agouram from Servicio Central de Soporte a la Investigación Experimental (SCSIE) at Universidad de Valencia for technical assistance in EXDS measurements.

Conflicts of Interest: The authors declare no conflict of interest.

References

1. Kharisov, B.I.; Dias, H.V.R.; Kharissova, O.V.; Jiménez-Pérez, V.M.; Pérez, B.O.; Flores, B.M. Iron-containing nanomaterials: Synthesis, properties, and environmental applications. *RSC Adv.* **2012**, *2*, 9325–9358. [CrossRef]
2. Tan, X.; Liu, Y.-G.; Gu, Y.-L.; Xu, Y.; Zeng, G.-M.; Hu, X.-J.; Liu, S.-B.; Wang, X.; Liu, S.-M.; Li, J. Biochar-based nano-composites for the decontamination of wastewater: A review. *Bioresour. Technol.* **2016**, *212*, 318–333. [CrossRef] [PubMed]
3. Ullattil, S.G.; Narendranath, S.B.; Pillai, S.C.; Periyat, P. Black TiO$_2$ Nanomaterials: A Review of Recent Advances. *Chem. Eng. J.* **2018**, *343*, 708–736. [CrossRef]

4. Pourmortazavi, S.M.; Hajimirsadeghi, S.S.; Rahimi, N.M.; Zahedi, M.M. Facile and Effective Synthesis of Praseodymium Tungstate Nanoparticles through an Optimized Procedure and Investigation of Photocatalytic Activity. *Mater. Sci. Semicond. Process.* **2013**, *16*, 131–137. [CrossRef]
5. Sundaram, R.; Nagaraja, K.S. Solid state electrical conductivity and humidity sensing studies on metal molybdate–molybdenum trioxide composites (M = Ni^{2+}, Cu^{2+} and Pb^{2+}). *Sens. Actuators B* **2004**, *101*, 353–360. [CrossRef]
6. Brito, J.L.; Barbosa, A.L.; Albornoz, A.; Severino, F.; Laine, J. Nickel molybdate as precursor of HDS catalysts: Effect of phase composition. *Catal. Lett.* **1994**, *26*, 329–337. [CrossRef]
7. Liu, J.P.; Huang, X.T.; Li, Y.Y.; Li, Z.K. A general route to thickness-tunable multilayered sheets of sheelite-type metal molybdate and their self-assembled films. *J. Mater. Chem.* **2007**, *17*, 2754–2758. [CrossRef]
8. Naja, M.; Abbasi, A.; Masteri-Farahani, M.; Rodrigues, V.H. Synthesis, characterization and crystal structure of a copper molybdate coordination polymer as an epoxidation catalyst. *Inorg. Chim. Acta* **2015**, *433*, 21–25.
9. Abrahams, S.C.; Bernstein, J.L.; Jamieson, P.B.; Crystal Structure of the Transition Metal Molybdates and Tungstates. IV. Paramagnetic $CuMoO_4$. *J. Chem. Phys.* **1968**, *48*, 2619–2629. [CrossRef]
10. Kohlmuller, R.; Faurie, J.P. Etude des systemes MoO_3–Ag_2MoO_4 et MoO_3–MO (M–Cu, Zn, Cd). *Bull. Soc. Chim. Fr.* **1968**, *11*, 4379–4382.
11. Ehrenberg, H.; Weitzel, H.; Paulus, H.; Wiesmann, M.; Wltschek, G.; Geselle, M.; Fuess, H. Crystal structure and magnetic properties of $CuMoO_4$ at low temperature (γ-phase). *J. Phys. Chem. Sol.* **1997**, *58*, 153–160. [CrossRef]
12. Sleight, A.W. High Pressure $CuMoO_4$. *Mater. Res. Bull.* **1973**, *8*, 863–866. [CrossRef]
13. Tali, R.; Tabachenko, V.V.; Kovba, L.M.; DemŌyanets, L.N. Kristallicheskaya struktura $CuMoO_4$. *Zhurnal Neorg. Khimii* **1991**, *36*, 1642–1644.
14. Baek, J.; Sefat, A.S.; Mandrus, D.; Halasyamani, P.S. A New Magnetically Ordered Polymorph of $CuMoO_4$: Synthesis and Characterization of ε-$CuMoO_4$. *Chem. Mater.* **2008**, *20*, 3785–3787. [CrossRef]
15. Wiesmann, M.; Ehrenberg, H.; Miehe, G.; Peun, T.; Weitzel, H.; Fuess, H. P-T Phase Diagram of $CuMoO_4$. *J. Solid State Chem.* **1997**, *132*, 88–97. [CrossRef]
16. Rodrıguez, F.; Hernandez, D.; Garcia-Jaca, J.; Ehrenberg, H.; Weitzel, H. Optical study of the piezochromic transition in $CuMoO_4$ by pressure spectroscopy. *Phys. Rev. B* **2000**, *61*, 16497–16501. [CrossRef]
17. Hassani, H.O.; Akouibaa, M.; Rakass, S.; Abboudi, M.; Bali, B.E.; Lachkar, M.; Wadaani, F. A simple and cost-effective new synthesis method of copper molybdate $CuMoO_4$ nanoparticles and their catalytic performance. *J. Sci. Adv. Mater. Dev.* **2021**, *6*, 501–507. [CrossRef]
18. Rahmani, A.; Farsi, H. Nanostructured copper molybdates as promising bifunctional electrocatalysts for overall water splitting and CO_2 reduction. *RSC Adv.* **2020**, *10*, 39037. [CrossRef]
19. Mohamed, B.; El Ouatib, R.; Guillemet, S.; Er-Rakho, L.; Durand, B. Characterization and photoluminescence properties of ultrafine copper molybdate (α-$CuMoO_4$) powders prepared via a combustion-like process. *Int. J. Miner. Metall. Mater.* **2016**, *23*, 1340–1345.
20. Popescu, C.; Sans, J.A.; Errandonea, D.; Segura, A.; Villanueva, R.; Sapiña, F. Compressibility and Structural Stability of Nanocrystalline TiO_2 Anatase Synthesized from Freeze-Dried Precursors. *Inorg. Chem.* **2014**, *53*, 11598–11603. [CrossRef]
21. Wang, J.; Yang, J.; Hu, T.; Chen, X.; Lang, J.; Wu, X.; Zhang, J.; Zhao, H.; Yang, J.; Cui, Q. Structural Phase Transition and Compressibility of CaF_2 Nanocrystals under High Pressure. *Crystals* **2018**, *8*, 199. [CrossRef]
22. Srihari, V.; Verma, A.K.; Pandey, K.K.; Vishwanadh, B.; Panchal, V.; Garg, N.; Errandonea, D. Making $Yb_2Hf_2O_7$ Defect Fluorite Uncompressible by Particle Size Reduction. *J. Phys. Chem. C* **2021**, *125*, 27354–27362. [CrossRef]
23. Yuan, H.; Rodriguez-Hernandez, P.; Muñoz, A.; Errandonea, D. Putting the squeeze on lead chromate nanorods. *J. Phys. Chem. Lett.* **2019**, *10*, 4744–4751. [CrossRef] [PubMed]
24. Patterson, A.L. The Scherrer Formula for X-Ray Particle Size Determination. *Phys. Rev.* **1939**, *56*, 978–982. [CrossRef]
25. Lutterotti, L.; Matthies, S.; Wenk, H.R. MAUD: A friendly Java program for material analysis using diffraction. *IUCr Newsl. CPD* **1999**, *21*, 14–15.
26. Fauth, F.; Peral, I.; Popescu, C.; Knapp, M. The new Material Science Powder Diffraction beamline at ALBA Synchrotron. *Powder Diffr.* **2013**, *28*, S360–S370. [CrossRef]
27. Prescher, C.; Prakapenka, V.B. DIOPTAS: A program for reduction of two-dimensional X-ray diffraction data and data exploration. *High Press. Res.* **2015**, *35*, 223–230. [CrossRef]
28. Klotz, S.; Chervin, J.C.; Munsch, P.; Le Marchand, G. Hydrostatic limits of 11 pressure transmitting media. *J. Phys. D Appl. Phys.* **2009**, *42*, 075413. [CrossRef]
29. Errandonea, D. Exploring the properties of MTO_4 compounds using high-pressure powder x-ray diffraction. *Cryst. Res. Technol.* **2015**, *50*, 729–736. [CrossRef]
30. Mao, H.K.; Xu, J.; Bell, P.M. Calibration of the ruby pressure gauge to 800 kbar under quasi-hydrostatic conditions. *J. Geophys. Res.* **1986**, *91*, 4673–4676. [CrossRef]
31. Chervin, J.C.; Canny, B.; Mancinelli, M. Ruby-spheres as pressure gauge for optically transparent high-pressure cells. *High Press. Res.* **2001**, *21*, 305–314. [CrossRef]
32. Dai, L.; Liu, K.; Li, H.; Wu, L.; Hu, H.; Zhang, Y.; Yang, L.; Pu, C.; Liu, P. Pressure-induced irreversible metallization with phase transitions of Sb_2S_3. *Phys. Rev. B* **2018**, *97*, 024103.

33. Yang, L.; Jiang, J.; Dai, L.; Hu, H.; Hong, M.; Zhang, X.; Li, H.; Liu, P. High-pressure structural phase transition and metallization in Ga_2S_3 under non-hydrostatic and hydrostatic conditions up to 36.4 GPa. *J. Mater. Chem. C* **2021**, *9*, 2912–2918.
34. Chijioke, A.D.; Nellis, W.J.; Soldatov, A.; Silvera, I.F. The ruby pressure standard to 150 GPa. *J. Appl. Phys.* **2005**, *98*, 114905. [CrossRef]
35. Brian, H.T. EXPGUI, a graphical user interface for GSAS. *J. Appl. Crystallogr.* **2001**, *34*, 210–213.
36. Rietveld, H.M. A profile refinement method for nuclear and magnetic structures. *J. Appl. Crystallogr.* **1969**, *2*, 65–71. [CrossRef]
37. Tolbert, S.H.; Alivisatos, A.P. High-pressure structural transformations in semiconductor nanocrystals. *Annu. Rev. Phys. Chem.* **1995**, *46*, 595–625. [CrossRef]
38. Razek, S.A.; Popeil, M.R.; Wangoh, L.; Rana, J.; Suwandaratne, N.; Andrews, J.L.; Watson, D.F.; Banerjee, S.; Piper, L.F.J. Designing catalysts for water splitting based on electronic structure considerations. *Electron. Struct.* **2020**, *2*, 023001. [CrossRef]
39. Curetti, N.; Sochalski-Kolbus, L.M.; AngeL, R.J.; Benna, P.; Nestola, F.; Bruno, E. High-pressure structural evolution and equation of state of albite. *Am. Mineral.* **2011**, *96*, 383–392. [CrossRef]
40. Cliffe, M.J.; Goodwin, A.L. PASCal: A principal axis strain calculator for thermal expansion and compressibility determination. *J. Appl. Crystallogr.* **2012**, *45*, 1321–1329. [CrossRef]
41. Gonzalez-Plattas, J.; Alvaro, M.; Nestola, F.; Angel, R. EosFit7-GUI: A new graphical user interface for equation of state calculations, analyses and teaching. *J. Appl. Crystallogr.* **2016**, *49*, 1377–1382. [CrossRef]
42. Errandonea, D.; Manjon, F.J. Pressure effects on the structural and electronic properties of ABX_4 scintillating crystals. *Prog. Mater. Sci.* **2008**, *53*, 711–773. [CrossRef]
43. Anzellini, S.; Errandonea, D.; MacLeod, S.G.; Botella, P.; Daisenberger, D.; De'Ath, J.M.; Gonzalez-Platas, J.; Ibáñez, J.; McMahon, M.I.; Munro, K.A.; et al. Phase diagram of calcium at high pressure and high temperature. *Phys. Rev. Mater.* **2018**, *2*, 083608. [CrossRef]
44. Ruiz-Fuertes, J.; Errandonea, D.; Lacomba-Perales, R.; Segura, A.; González, J.; Rodríguez, F.; Manjón, F.J.; Ray, S.; Rodríguez-Hernández, P.; Muñoz, A.; et al. High-pressure structural phase transitions in $CuWO_4$. *Phys. Rev. B* **2010**, *81*, 224115. [CrossRef]
45. Díaz-Anichtchenko, D.; Turnbull, R.; Bandiello, E.; Anzellini, S.; Achary, S.N.; Errandonea, D. Pressure-induced chemical decomposition of copper orthovanadate (α-$Cu_3V_2O_8$). *J. Mater. Chem. C* **2021**, *9*, 13402–13409. [CrossRef]
46. Turnbull, R.; Gonzáles-Platas, J.; Rodríguezc, F.; Liang, A.; Popescu, C.; Santamaría-Pérez, Z.D.; Rodríguez-Hernandez, P.; Muñoz, A.; Errandonea, D. Pressure-induced phase transition and band-gap collapse in semiconducting β-$Cu_2V_2O_7$. *Inorg. Chem.* **2022**, *61*, 3697–3707. [CrossRef]
47. Errandonea, D.; Gracia, L.; Lacomba-Perales, R.; Polian, A.; Chervin, J.C. Compression of scheelite-type $SrMoO_4$ under quasi-hydrostatic conditions: Redefining the high-pressure structural sequence. *J. Appl. Phys.* **2013**, *113*, 123510. [CrossRef]
48. Errandonea, D.; Ruiz-Fuertes, J. A Brief Review of the Effects of Pressure on Wolframite-Type Oxides. *Crystals* **2018**, *8*, 71. [CrossRef]

Article

Computational Investigation of the Stability of Di-*p*-Tolyl Disulfide "Hidden" and "Conventional" Polymorphs at High Pressures

Valeriya Yu. Smirnova [1,2], Anna A. Iurchenkova [1,3] and Denis A. Rychkov [1,2,*]

[1] Laboratory of Mechanochemistry, Institute of Solid State Chemistry and Mechanochemistry, SB RAS, Kutateladze 18, 630090 Novosibirsk, Russia
[2] Laboratory of Physicochemical Fundamentals of Pharmaceutical Materials, Novosibirsk State University, Pirogova 2, 630090 Novosibirsk, Russia
[3] Ångstrom Laboratory, Nanotechnology and Functional Materials, Department of Materials Science and Engineering, Faculty of Science and Technology, Uppsala University, 75121 Uppsala, Sweden
* Correspondence: rychkov.dennis@gmail.com

Abstract: The investigation of molecular crystals at high pressure is a sought-after trend in crystallography, pharmaceutics, solid state chemistry, and materials sciences. The di-*p*-tolyl disulfide $(CH_3-C_6H_4-S-)_2$ system is a bright example of high-pressure polymorphism. It contains "conventional" solid–solid transition and a "hidden" form which may be obtained only from solution at elevated pressure. In this work, we apply force field and periodic DFT computational techniques to evaluate the thermodynamic stability of three di-*p*-tolyl disulfide polymorphs as a function of pressure. Theoretical pressures and driving forces for polymorphic transitions are defined, showing that the compressibility of the γ phase is the key point for higher stability at elevated pressures. Transition state energies are also estimated for α → β and α → γ transitions from thermodynamic characteristics of crystal structures, not exceeding 5 kJ/mol. The β → γ transition does not occur experimentally in the 0.0–2.8 GPa pressure range because transition state energy is greater than 18 kJ/mol. Relations between free Gibbs energy (in assumption of enthalpy) of phases α, β, and γ, as a function of pressure, are suggested to supplement and refine experimental data. A brief discussion of the computational techniques used for high-pressure phase transitions is provided.

Keywords: high-pressure polymorph; relative stability; di-*p*-tolyl disulfide; high-pressure DFT; molecular crystals; hidden polymorph

1. Introduction

High-pressure research of molecular crystals is an important direction of modern crystallography [1–11]. A lot of organic systems have polymorphic modifications at ambient pressure, while other organic crystals undergo phase transitions at extreme conditions [12,13]. These new high-pressure structures may be the result of direct solid-solid transformation and may be called "conventional". Nevertheless, the formation of new polymorphs may also be hindered by the high transition state energies and crystalize only from liquid phase. These solid–liquid–solid phase transitions (which can also undergo through mobile fluid phase, e.g., partial dissolution) are called "hidden" polymorphs [14–16]. The comparison of the thermodynamic stability of both "hidden" and "conventional" phases is a complicated question, especially if several phases exist at the same conditions. High-pressure experiments could only show the structural stability of such phases, but not the thermodynamic stability. If there is no direct phase transition between polymorphs, it becomes impossible to investigate their thermodynamic stability. Both computational and experimental approaches should be used in such cases [13,17].

Di-p-tolyl disulfide (p-Tol$_2$S$_2$) is a demonstrative organic system that contains a "conventional" form and a "hidden" high-pressure form (Scheme 1). The "hidden" form recrystallizes from solution at elevated pressures, while the "conventional" form is obtained via solid–solid transition [18].

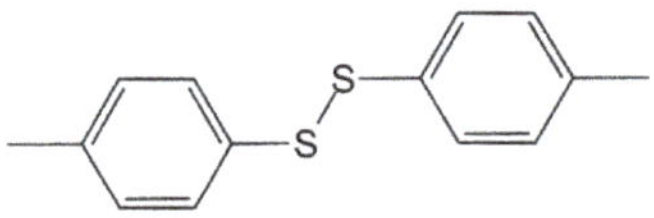

Scheme 1. Molecular structure of di-p-tolyl disulfide (p-Tol$_2$S$_2$).

Szymon Sobczak and Andrzej Katrusiak previously found that the α phase is the most stable phase at ambient conditions, while, at higher pressures, two phase transitions occur with the formation of β and γ polymorphs [18]. On the one hand, the reversible $\alpha \to \beta$ phase transition occurs at a pressure near 1.6 GPa and the β phase remains stable up to 2.8 GPa (glycerin was used as a pressure transmitting media). On the other hand, α transforms to the γ-phase form near 0.45 GPa via recrystallization (methanol, ethanol, and isopropanol solvents were simultaneously used as hydrostatic fluids). The γ-phase form is also stable up to 2.8 GPa, similarly to the β one. Based on those experiments, the authors suggested a scheme of phase transitions and schematic relations between the free Gibbs energy of the α, β, and γ phases as a function of pressure for p-Tol$_2$S$_2$ (Figure 1).

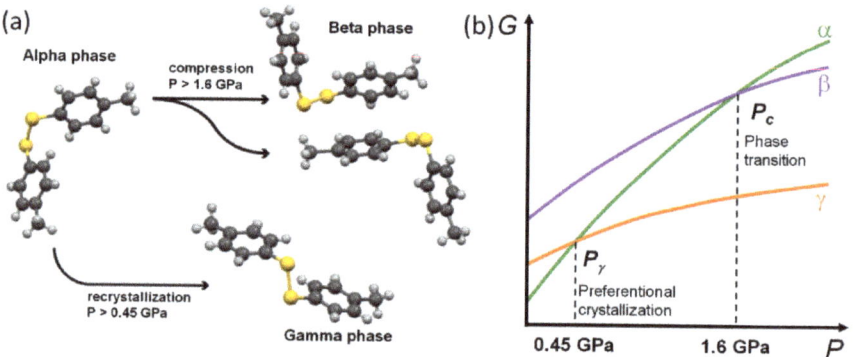

Figure 1. (a) Principal scheme of the experimentally observed phase transitions in a di-p-tolyl disulfide system. The $\alpha \to \beta$ phase transition is given by two arrows due to formation of two different molecular conformations. (b) Suggested relations between the free Gibbs energy of phases α, β, and γ as a function of pressure based on experimental study (adapted from [18]).

It is impossible to measure the relative stability of β and γ polymorphs in different pressure regions due to the absence of a direct phase transition between them. Thus, the aim of this study is to calculate the thermodynamic stability of all three polymorphs of the p-Tol$_2$S$_2$ system and explain the driving force for experimentally observed structural reorganization at pressures of 0.45 GPa and 1.6 GPa. Based on several previous works [19–24] Force Field (FF) and periodic DFT methods in conjunction with equations of states (EoS), the enthalpies of all polymorphs are calculated as a first rough approximation of the Gibbs energy for each.

2. Computational Details

Lattice energies were calculated using different approaches. The force field approach, as implemented in the CrystalExplorer 21.5 (CE21) software package, calculates pair-wise energies, summing up the energy of all intermolecular energies in the experimental crystal structure in terms of kJ/mol [25–27]. CE21 uses data calculated via DFT (e.g., electron density mapped on a Hirshfeld surface at B3LYP/6-31G (d,p)) and should be strictly named as an FF-based method, but, following major principles, it is further referred to as an FF method. Periodic DFT, as implemented in VASP 5.4.4 [28–31], calculates an overall (electronic) energy of the optimized system, which may be presented as a sum of inter- and intra-molecular interactions (which are of course interrelated from molecular geometry). Thus, periodic DFT (apart from obvious differences in principles in comparison to FF methods) considers conformational changes in crystal structures, whereas the CE21 algorithm does not. Enthalpies were calculated by manual addition of the PV term at pressure points, where volume was known experimentally (for FF calculations) or from EoS calculations (for DFT calculations). Zero-point energy was not considered, as well as temperature (T = 0 K). An overall workflow is presented in Scheme S1.

2.1. FF Method Calculations

CrystalExplorer 21.5 [25] was used with a Gaussian09 [32] backend at D2-B3LYP/6-31G (d,p) level of theory (LOT) with four further terms (electrostatic, polarization, dispersion, and exchange-repulsion) separation. Corresponding default scaling factors for this LOT were used to calculate total energy, which were further interpreted as lattice energy.

2.2. Periodic DFT Calculations

All periodic DFT calculations were performed using the VASP 5.4.4 package using a PBE functional [33] plane-wave basis set with a kinetic energy cut-off of 500 eV and PAW atomic pseudopotentials [34,35]. The integrals in reciprocal space were calculated on a k-point Monkhorst-Pack mesh of $4 \times 4 \times 2$ [36]. Grimme D3 dispersion correction with Becke–Johnson damping was used for better simulation of van der Waals (VdW) interactions in crystal structures [37]. Convergence criteria were applied, where the maximum change in system energy was 10^{-5} eV and the norms of all forces were smaller than 10^{-4}. The volumes of unit cells in all periodic DFT calculations were fixed at calculated values from corresponding EoS (which are respectively calculated from experimental data), while cell shapes and atom positions were fully relaxed during the optimization procedure (ISIF = 4). Crystal structures for all calculations were prepared manually by editing the experimental lowest pressure structure (cif) file cell parameters. This editing considered the V_p/V_o ratio (preserving same angles as in the initial structures), where V_p is the calculated volume of the crystal structure at pressure P from EoS. All unit cell parameters including angles were fully relaxed further during DFT optimization procedure, keeping volume fixed only. Experimental structure (.cif) files were obtained from the CCDC database [38].

2.3. Equations of States

Experimentally obtained pressure dependencies of unit cell volumes of the polymorphs (Table S1 in Supplementary Materials) were fitted by the EoSFit7-GUI software [39,40], using the third order Birch–Murnaghan equation of state [40,41]. Standard deviations of the volumes and pressures were considered.

3. Results and Discussion

The thermodynamic stability of different polymorphs is described as the enthalpy of these forms in the current study. The authors are aware of possible significant changes in the entropy term due to change in lattice vibrations under pressure. Nevertheless, we suggest such simplification reasonable for this system while taking into account the isothermal conditions of the experiments, minor changes in the molecular arrangement of the α and β forms, which results in very small lattice changes, and the preserved symmetry (P)2 point group of the α and γ polymorphs (Figure S1 and Table S1). Thus, the entropy difference should be less significant than that which is expected for major materials and may be neglected as a rough approximation of the Gibbs energies to save computational resources. Enthalpy may be divided into several terms to find a driving force for phase transitions (Equation (1)).

$$H = U_{crystal} + P \times V = U_{inter} + U_{intra} + P \times V \quad (1)$$

Lattice energy is presented as an $U_{crystal}$ and presents a sum of inter- and intramolecular interactions. Considering the computational techniques, Equation (1) transforms to Equation (2) in the case of FF calculations as implemented in CrystalExplorer 21.5 software (where only intermolecular interaction energies are parametrized).

$$H = U_{crystal} + P \times V = U_{inter} + P \times V \quad (2)$$

U_{intra}, being conformational energy, was calculated in the work [18] separately using several gas-phase DFT methods, and is reproduced in Figure S2, showing γ-phase conformations are 6–8 kJ/mol less favorable than the α and γ forms at the same pressures. Thus, U_{intra} may be added to U_{inter} manually, which in fact does not change the overall picture for used FF method (Figure S3).

Following this logic, an overall behavior of p-Tol$_2$S$_2$ system may be assessed regarding $U_{crystal}$ lattice energies (mostly intermolecular interactions) and PV term (compressibility) to find a driving force for phase transitions and evaluate relative stability of polymorphs using different computational techniques.

3.1. Force Field Calculations

Lattice energies of all experimentally founded crystal structures were calculated using the CrystalExplorer 21.5 software package at corresponding experimental pressure values (Figure 2). The function of lattice energies on the pressure shows relative stability of all three phases in the regions of their structural stability. A significant decrease in the lattice energy of α phase in the interval 0.0–1.5 GPa is probably an artifact of CE21 parametrization, which was already discussed in the work [24]. It was shown that decrease of lattice energies with an initial increase in pressure is similar for all phases/interactions and does not affect results of polymorphs relative energies.

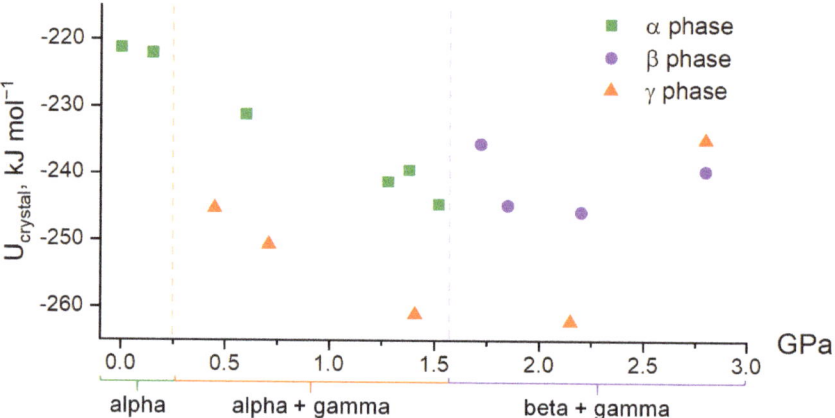

Figure 2. Lattice energies ($U_{crystal}$ = U_{inter}) of phases α, β, and γ as a function of pressure calculated based on experimental data. Regions of structural stability according to experimental data [18] are shown with dashed lines.

According to the lower lattice energy of the γ phase (Figure 2), it has higher stability in the pressure range of 0.45–2.5 GPa with relatively reversed stability for the β and γ polymorphs at 2.8 GPa. The addition of conformational energies does not change this tendency (Figure S3). Thus, based only on the lattice energies, one can expect a γ to β phase transition at higher pressures. Nevertheless, compressibility of different phases should be considered, which may affect PV term and consequently H drastically [42].

The enthalpies of all phases were calculated at experimental pressures by manual addition of a PV term (Figure S4). Higher lattice energy of the γ polymorph was fully compensated by PV term (compressibility of γ phase) and the γ phase preserved relative stability at 2.8 GPa. It is impossible to compare the enthalpies of all phases at the same pressures in a whole pressure interval (0–3 GPa) due to the limitation of lacking a region where the phases are all structurally stable. This is one of the key limitations for the FF method, because atom coordinates for other pressures cannot be obtained without expensive DFT calculations or new labor-intensive experiments. Linearization of enthalpy data may be successfully used in such cases [17] and be applied for a p-Tol$_2$S$_2$ system (Figure 3).

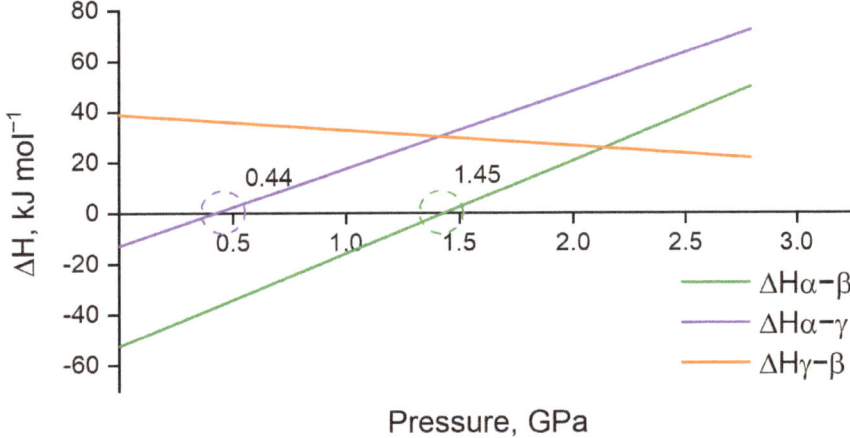

Figure 3. Calculated enthalpy differences between p-Tol$_2$S$_2$ polymorph α and γ (green), α and β (purple), and β and γ (orange) as a function of pressure (i.e., ΔHα − β = Hα − Hβ, ΔHα − γ = Hα − Hγ, and ΔHα − β = Hα − Hβ). Phase transition between the involved polymorphs can be expected when their enthalpy difference is equal to zero, marked with circles.

The phase transitions from α → γ and α → β can be expected when the free energies (in the approximation of the calculations used here—the enthalpies) of the corresponding phases become equal. These calculations predict this to happen at 0.44 GPa and 1.45 GPa, respectively (Figure 3). The obtained theoretical values are extremely close to the experimental values of 0.45 and 1.6 GPa for the α → γ and α → β phase transitions, respectively (Table S2). Nevertheless, the applied approximation of linear enthalpy change (due to high PV energies) at higher pressures is a significant simplification and is recommended for use only with short pressure intervals. The periodic DFT calculations, in conjunction with EoS, were applied to estimate the possible margins of the used approach.

3.2. Periodic DFT Calculations with EoS

The prediction of structure change at finite pressure points is mandatory to overcome the limitations of experimental data at different pressures. A previously reported [21] technique of fully automated structure change prediction at any pressure using the PSTRESS tag in VASP turned out to be inaccurate for small pressure ranges, at least for a p-Tol$_2$S$_2$ system. Thus, a concept similar to those reported in the work [43] was applied. Equations of state were calculated based on experimental data for all three polymorphs to predict the cell volume of each phase at the 0.0–3.0 GPa pressure range with a 0.5 GPa step size (Figure S5 and Table S3). At each pressure point, the full electronic energy was calculated using DFT optimization with a fixed cell volume (Figure 4).

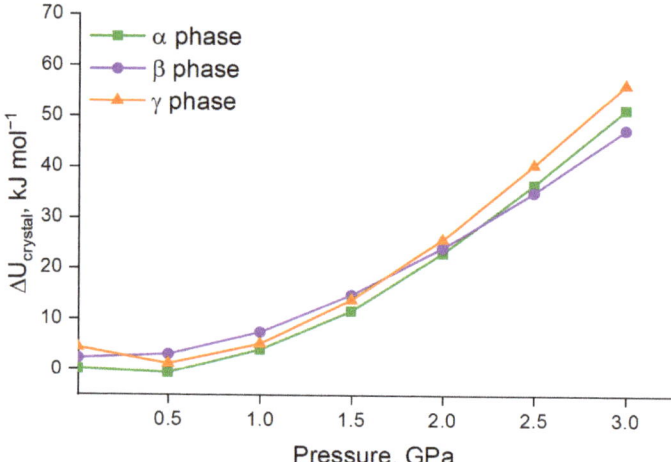

Figure 4. Relative lattice energies of phases α, β, and γ as a function of pressure calculated based on EoS predicted data. Full electronic structure energy contains both inter and intramolecular interaction energies. The α phase energy at ambient pressure is taken as zero.

Lattice energies show a similar trend as calculated using the FF approach, predicting that the relative stability of β to γ changes at elevated pressures. The γ phase is slightly less stable than the α phase in the whole pressure range, which may be explained by less favorable conformations. What is more important, however, is that these calculations predict γ and β phase instability at ambient pressure, which fully coincides with the experimental study.

It is not only possible to estimate phase stability, but also to compare separate enthalpy terms and find a prevailing term for phase transition. Pairwise comparisons of lattice energies with PV in terms of polymorphs are presented in Figure 5a,b.

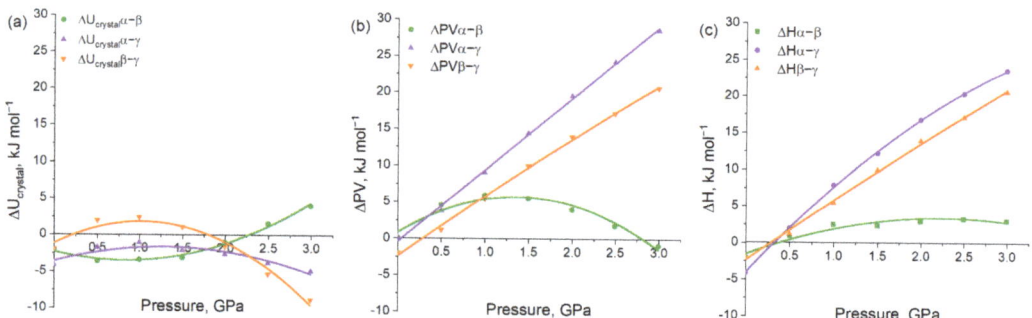

Figure 5. Calculated (**a**) lattice energy $\Delta U_{crystal}$, (**b**) ΔPV term and (**c**) ΔH differences between p-Tol$_2$S$_2$ polymorphs α and γ (green), α and β (purple), β and γ (orange) as a function of pressure (i.e., $\Delta U\alpha - \beta = U\alpha - U\beta$, $\Delta PV\alpha - \beta = PV\alpha - PV\beta$, $\Delta H\alpha - \beta = H\alpha - H\beta$, etc.). Phase transition between the involved polymorphs can be expected when their enthalpy difference is equal to zero.

Phase transition α → γ is justified by higher compressibility (the PV component of the enthalpy, rather than the specific pairwise energies) of γ polymorph (what also shown with K′ coefficients from EoS data (Table S3)), whereas α → β transition cannot be unequivocally defined by any term. This also coincides well with structure data, showing β and α phases being very similar in terms of geometry (Figure S1).

Calculated cell volumes from EoS helps to exclude simplification of enthalpy linearization—it is possible to calculate enthalpies at each point by direct sum of $U_{crystal}$ ($U_{inter} + U_{intra}$) and PV term (Figure S6). Enthalpy difference is plotted in Figure 5c (and Figure S7 and Table S5 with linearization).

Based on this more sophisticated approach, one can predict phase transitions at 0.34 GPa and 0.36 GPa for $\alpha \to \gamma$ and $\alpha \to \beta$ transitions in comparison to experimental 0.45 GPa and 1.6 GPa. Such deviations may be explained by several factors, apart from the absence of calculated thermal effects. It is mandatory to understand that experimental techniques show a range of pressures where transition occurs, due to features of the high-pressure devices (e.g., a diamond anvil cell and the limitations of a single crystal diffraction technique at elevated pressure). The experimental data are limited, so the exact pressure of phase transition is not reported: $\alpha \to \beta$ transition occurred in the interval 1.52–1.60 GPa, whereas the γ phase was crystallized at 0.45 GPa and other pressures were not checked. On the other hand, transition state energy during phase change may be significant and lead to hysteresis and higher pressures of experimental phase transition in comparison to thermodynamic data [21,44,45]. Finally, the entropy term may also affect transitions, changing the Gibbs energies of different polymorphs. Nevertheless, it is possible to estimate energy limits of the transition state from the suggested enthalpy functions.

The differences in polymorph enthalpies, when at pressure of experimental phase transition, denote the upper limit for the transition state energy. This is because the initial phase (α in this case) stores "excessive" energy in relation to the final phase. Following this logic, the formation of the "hidden" γ polymorph has an energy barrier less than 2 kJ/mol, which is reasonable for solid–liquid–solid phase transition. The "conventional" β polymorph also has very small transition state energy less than 3 kJ/mol, which may be explained by extremely small structure changes during this transition. The $\beta \to \gamma$ phase transition has a TS energy barrier of no less than 18 kJ/mol. So, the "excessive" energy of the β phase is not enough to result in solid–solid crystal structure rearrangement.

Finally, it is possible to refine the phase diagram for a p-Tol$_2$S$_2$ system based on the enthalpy calculations, where the α form is the most stable in the pressure range of 0.0–0.34 GPa, the γ phase is the most stable at a pressure range from 0.34 GPa to 3.0 GPa, and the β phase is metastable in the whole pressure range from 0.0 GPa to 3.0 GPa (nevertheless, being more stable than the γ phase in the pressure range of 0.0–0.32 GPa) (Table S4). Figure 6 illustrates the thermodynamic relation between p-Tol$_2$S$_2$ polymorphs, which turned out to be more complicated than suggested from experimental data (Figure 1b, Table S6).

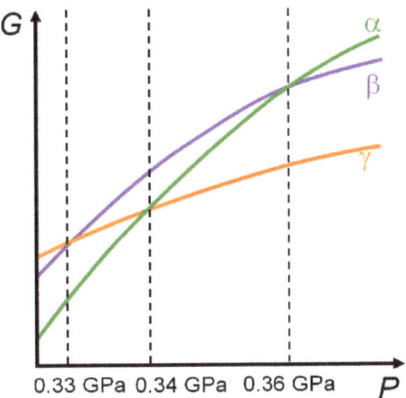

Figure 6. Suggested relations between the free Gibbs energy of phases α, β, and γ as a function of pressure based on a periodic DFT study where only enthalpies were calculated. Table S4 shows the relative phase stability according to calculated enthalpies in different pressure ranges.

4. Conclusions

In this study, we show how different computational techniques supplement previous experimental studies to evaluate thermodynamic stability of p-Tol$_2$S$_2$ polymorphs in the approximation of enthalpy changes. Force field methods (as implemented in CE21) show reliable stability ranking of polymorphs at pressure ranges where phases coexist, if enthalpies are calculated. The "hidden" γ form was predicted to be more stable than the α form in the 0.45–1.52 GPa experimental pressure range, and more stable than the "conventional" β phase in the interval of 1.60–2.8 GPa. Periodic DFT methods, in conjunction with EoS, help to answer questions regarding polymorph stability in the whole pressure (not only experimental) range, even if a specific form does not exist experimentally. It was shown that the α form is the most stable form when in the pressure range 0.0–0.34 GPa, while the γ phase is the most stable from 0.34–3.0 GPa due to higher compressibility (PV term). Upper limits for transition state energies for both experimental phase transitions were estimated from the suggested relations between the free Gibbs energy values of phases α, β, and γ as a function of pressure. Both methods proved to reveal valuable information about relative polymorph stability (FF for experimental and periodic DFT for the whole pressure range). A refined phase diagram (in a rough assumption of calculated enthalpies) over pressure shows the necessity of computational methods for high-pressure crystallography.

Supplementary Materials: The following supporting information can be downloaded at: https://www.mdpi.com/article/10.3390/cryst12081157/s1. **Scheme S1**: An overall workflow of computational procedures. **Table S1**: Summarized structural data for p-Tol$_2$S$_2$ polymorphs used in CE21 and EoSFit7 calculations. **Figure S1**: Crystal structures of (**a**) alpha, (**b**) beta and (**c**) gamma polymorphs of p-Tol$_2$S$_2$. Colors according to symmetry equivalence in corresponding crystal structures. **Figure S2**: Potential energy changes (ΔE_p) calculated by Gaussian (Experimental Section of [14]) for the isolated molecule in its conformation, experimentally determined in the p-Tol$_2$S$_2$ structures of phases α, β, and γ. **Figure S3**: Lattice energies (the sum of U_{inter} and U_{intra}) of phases α, β, and γ as a function of pressure calculated based on experimental data. Regions of structural stability according to experimental data are shown with dashed lines. U_{intra} is reproduced from work [14]. **Figure S4**: Enthalpies of phases α, β, and γ as a function of pressure calculated based on experimental data using CE21. **Table S2**: Predicted by FF and experimental phase transition pressures. **Figure S5**: p-Tol$_2$S$_2$ volume-pressure dependence based on calculated EoS. **Table S3**: EoS Birch-Murnaghan 3rd order coefficients. V_0—reference pressure volume at ambient pressure, K_0—bulk modulus, K_p—derivative of Bulk Modulus (dK/dP). **Figure S6**: Enthalpies of phases α, β, and γ as a function of pressure calculated using periodic DFT and EoS. **Table S4**: Relative stability of p-Tol$_2$S$_2$ polymorphs in different pressure ranges calculated by periodic DFT. **Figure S7**: ΔH differences between p-Tol$_2$S$_2$ polymorph α and γ (green), α and β (purple), β and γ (orange) as a function of pressure (i.e., $\Delta U\alpha - \beta = U\alpha - U\beta$, $\Delta PV\alpha - \beta = PV\alpha - PV\beta$, $\Delta H\alpha - \beta = H\alpha - H\beta$, etc.). **Table S6**: Phase transitions of p-Tol$_2$S$_2$ polymorphs under pressure according to various experimental and computational techniques.

Author Contributions: Conceptualization, D.A.R.; methodology, D.A.R.; software, V.Y.S. and A.A.I.; validation, A.A.I. and D.A.R.; formal analysis, V.Y.S. and A.A.I.; investigation, V.Y.S., A.A.I. and D.A.R.; resources, D.A.R.; data curation, V.Y.S. and A.A.I.; writing—original draft preparation, D.A.R.; writing—review and editing, V.Y.S., A.A.I. and D.A.R.; visualization, V.Y.S., A.A.I. and D.A.R.; supervision, D.A.R.; project administration, D.A.R.; funding acquisition, D.A.R. All authors have read and agreed to the published version of the manuscript.

Funding: This research was funded by Ministry of Higher Education and Science, grant number FWUS-2021-0005 (FF and periodic DFT calculations, data processing and analysis) and the Russian Science Foundation, grant number 18-73-00154 (system choice, database search, preliminary calculations).

Institutional Review Board Statement: Not applicable.

Informed Consent Statement: Not applicable.

Data Availability Statement: Data available on request from corresponding author.

Acknowledgments: The Siberian Branch of the Russian Academy of Sciences (SB RAS) Siberian Supercomputer Center (http://www.sscc.icmmg.nsc.ru/, accessed on 12 August 2022) is gratefully acknowledged for providing access to their supercomputer facilities. The authors also acknowledge the Supercomputing Center of the Novosibirsk State University (http://nusc.nsu.ru, accessed on 12 August 2022) for providing computational resources.

Conflicts of Interest: The authors declare no conflict of interest. The funders had no role in the design of the study; in the collection, analyses, or interpretation of data; in the writing of the manuscript; or in the decision to publish the results.

References

1. Errandonea, D. Pressure-Induced Phase Transformations. *Crystals* **2020**, *10*, 595. [CrossRef]
2. Thiel, A.M.; Damgaard-Møller, E.; Overgaard, J. High-Pressure Crystallography as a Guide in the Design of Single-Molecule Magnets. *Inorg. Chem.* **2020**, *59*, 1682–1691. [CrossRef] [PubMed]
3. Moggach, S.A.; Oswald, I.D.H. Crystallography Under High Pressures. In *21st Century Challenges in Chemical Crystallography I. Structure and Bonding*; Springer: Cham, Switzerland, 2020; pp. 141–198, ISBN 9783030647421.
4. Görbitz, C.H. Crystal structures of amino acids: From bond lengths in glycine to metal complexes and high-pressure polymorphs. *Crystallogr. Rev.* **2015**, *21*, 160–212. [CrossRef]
5. Skakunova, K.D.; Rychkov, D.A. Low Temperature and High-Pressure Study of Bending L-Leucinium Hydrogen Maleate Crystals. *Crystals* **2021**, *11*, 1575. [CrossRef]
6. Sharma, S.M.; Garg, N. Material Studies at High Pressure. In *Materials under Extreme Conditions*; Elsevier: Amsterdam, The Netherlands, 2017; pp. 1–47, ISBN 9780128014424.
7. Colmenero, F. Organic acids under pressure: Elastic properties, negative mechanical phenomena and pressure induced phase transitions in the lactic, maleic, succinic and citric acids. *Mater. Adv.* **2020**, *1*, 1399–1426. [CrossRef]
8. Colmenero, F.; Lunelli, B. Fluorine-substituted cyclobutenes in the solid state: Crystal structures, vibrational spectra and mechanical and thermodynamic properties. *J. Phys. Chem. Solids* **2022**, *160*, 110337. [CrossRef]
9. Ratajczyk, P.; Katrusiak, A.; Bogdanowicz, K.A.; Przybył, W.; Krysiak, P.; Kwak, A.; Iwan, A. Mechanical strain, thermal and pressure effects on the absorption edge of an organic charge-transfer polymer for flexible photovoltaics and sensors. *Mater. Adv.* **2022**, *3*, 2697–2705. [CrossRef]
10. Konar, S.; Hobday, C.L.; Bull, C.L.; Funnell, N.P.; Chan, Q.F.; Fong, A.; Atceken, N.; Pulham, C.R. High-Pressure Structural Behavior of para-Xylene. *Cryst. Growth Des.* **2022**, *22*, 3862–3869. [CrossRef]
11. Wilson, C.J.G.; Cervenka, T.; Wood, P.A.; Parsons, S. Behavior of Occupied and Void Space in Molecular Crystal Structures at High Pressure. *Cryst. Growth Des.* **2022**, *22*, 2328–2341. [CrossRef]
12. Neumann, M.A.; van de Streek, J.; Fabbiani, F.P.A.; Hidber, P.; Grassmann, O. Combined crystal structure prediction and high-pressure crystallization in rational pharmaceutical polymorph screening. *Nat. Commun.* **2015**, *6*, 7793. [CrossRef]
13. Mazurek, A.; Szeleszczuk, Ł.; Pisklak, D.M. Can We Predict the Pressure Induced Phase Transition of Urea? Application of Quantum Molecular Dynamics. *Molecules* **2020**, *25*, 1584. [CrossRef] [PubMed]
14. Paliwoda, D.; Dziubek, K.F.; Katrusiak, A. Imidazole Hidden Polar Phase. *Cryst. Growth Des.* **2012**, *12*, 4302–4305. [CrossRef]
15. Anioła, M.; Katrusiak, A. Conformational Conversion of 4,4′-Bipyridinium in a Hidden High-Pressure Phase. *Cryst. Growth Des.* **2015**, *15*, 764–770. [CrossRef]
16. Patyk-Kaźmierczak, E.; Kaźmierczak, M. A new high-pressure benzocaine polymorph—Towards understanding the molecular aggregation in crystals of an important active pharmaceutical ingredient (API). *Acta Crystallogr. Sect. B Struct. Sci. Cryst. Eng. Mater.* **2020**, *76*, 56–64. [CrossRef]
17. Fedorov, A.Y.; Rychkov, D.A.; Losev, E.A.; Zakharov, B.A.; Stare, J.; Boldyreva, E.V. Effect of pressure on two polymorphs of tolazamide: Why no interconversion? *CrystEngComm* **2017**, *19*, 2243–2252. [CrossRef]
18. Sobczak, S.; Katrusiak, A. Colossal Strain Release by Conformational Energy Up-Conversion in a Compressed Molecular Crystal. *J. Phys. Chem. C* **2017**, *121*, 2539–2545. [CrossRef]
19. Rychkov, D.A. A Short Review of Current Computational Concepts for High-Pressure Phase Transition Studies in Molecular Crystals. *Crystals* **2020**, *10*, 81. [CrossRef]
20. Munday, L.B.; Chung, P.W.; Rice, B.M.; Solares, S.D. Simulations of High-Pressure Phases in RDX. *J. Phys. Chem. B* **2011**, *115*, 4378–4386. [CrossRef]
21. Rychkov, D.A.; Stare, J.; Boldyreva, E.V. Pressure-driven phase transition mechanisms revealed by quantum chemistry: L-serine polymorphs. *Phys. Chem. Chem. Phys.* **2017**, *19*, 6671–6676. [CrossRef]
22. Wood, P.A.; Francis, D.; Marshall, W.G.; Moggach, S.A.; Parsons, S.; Pidcock, E.; Rohl, A.L. A study of the high-pressure polymorphs of L-serine using ab initio structures and PIXEL calculations. *CrystEngComm* **2008**, *10*, 1154. [CrossRef]
23. Thomas, S.P.; Spackman, M.A. The Polymorphs of ROY: A Computational Study of Lattice Energies and Conformational Energy Differences. *Aust. J. Chem.* **2018**, *71*, 279. [CrossRef]
24. Fedorov, A.Y.; Rychkov, D.A. Comparison of different computational approaches for unveiling the high-pressure behavior of organic crystals at a molecular level. Case study of tolazamide polymorphs. *J. Struct. Chem.* **2020**, *61*, 1356–1366. [CrossRef]

25. Spackman, P.R.; Turner, M.J.; McKinnon, J.J.; Wolff, S.K.; Grimwood, D.J.; Jayatilaka, D.; Spackman, M.A. CrystalExplorer: A program for Hirshfeld surface analysis, visualization and quantitative analysis of molecular crystals. *J. Appl. Crystallogr.* **2021**, *54*, 1006–1011. [CrossRef] [PubMed]
26. Thomas, S.P.; Spackman, P.R.; Jayatilaka, D.; Spackman, M.A. Accurate Lattice Energies for Molecular Crystals from Experimental Crystal Structures. *J. Chem. Theory Comput.* **2018**, *14*, 1614–1623. [CrossRef]
27. Mackenzie, C.F.; Spackman, P.R.; Jayatilaka, D.; Spackman, M.A. CrystalExplorer model energies and energy frameworks: Extension to metal coordination compounds, organic salts, solvates and open-shell systems. *IUCrJ* **2017**, *4*, 575–587. [CrossRef]
28. Kresse, G.; Hafner, J. Ab initio molecular dynamics for liquid metals. *Phys. Rev. B* **1993**, *47*, 558–561. [CrossRef]
29. Kresse, G.; Furthmüller, J. Efficient iterative schemes for ab initio total-energy calculations using a plane-wave basis set. *Phys. Rev. B* **1996**, *54*, 11169–11186. [CrossRef]
30. Kresse, G.; Hafner, J. Ab initio molecular-dynamics simulation of the liquid-metal–amorphous-semiconductor transition in germanium. *Phys. Rev. B* **1994**, *49*, 14251–14269. [CrossRef]
31. Kresse, G.; Furthmüller, J. Efficiency of ab-initio total energy calculations for metals and semiconductors using a plane-wave basis set. *Comput. Mater. Sci.* **1996**, *6*, 15–50. [CrossRef]
32. Frisch, M.J.; Trucks, G.W.; Schlegel, H.B.; Scuseria, G.E.; Robb, M.A.; Cheeseman, J.R.; Scalmani, G.; Barone, V.; Mennucci, B. *Gaussian 09, Revision D.01*; Gaussian, Inc.: Wallingford, CT, USA, 2009.
33. Perdew, J.P.; Burke, K.; Ernzerhof, M. Generalized Gradient Approximation Made Simple. *Phys. Rev. Lett.* **1996**, *77*, 3865–3868. [CrossRef]
34. Kresse, G.; Joubert, D. From ultrasoft pseudopotentials to the projector augmented-wave method. *Phys. Rev. B* **1999**, *59*, 1758–1775. [CrossRef]
35. Blöchl, P.E. Projector augmented-wave method. *Phys. Rev. B* **1994**, *50*, 17953–17979. [CrossRef] [PubMed]
36. Monkhorst, H.J.; Pack, J.D. Special points for Brillouin-zone integrations. *Phys. Rev. B* **1976**, *13*, 5188–5192. [CrossRef]
37. Grimme, S.; Ehrlich, S.; Goerigk, L. Effect of the damping function in dispersion corrected density functional theory. *J. Comput. Chem.* **2011**, *32*, 1456–1465. [CrossRef]
38. Groom, C.R.; Bruno, I.J.; Lightfoot, M.P.; Ward, S.C. The Cambridge Structural Database. *Acta Crystallogr. Sect. B Struct. Sci. Cryst. Eng. Mater.* **2016**, *72*, 171–179. [CrossRef]
39. Gonzalez-Platas, J.; Alvaro, M.; Nestola, F.; Angel, R. EosFit7-GUI: A new graphical user interface for equation of state calculations, analyses and teaching. *J. Appl. Crystallogr.* **2016**, *49*, 1377–1382. [CrossRef]
40. Angel, R.J.; Alvaro, M.; Gonzalez-Platas, J. EosFit7c and a Fortran module (library) for equation of state calculations. *Z. für Krist.-Cryst. Mater.* **2014**, *229*, 405–419. [CrossRef]
41. Birch, F. Finite elastic strain of cubic crystals. *Phys. Rev.* **1947**, *71*, 809–824. [CrossRef]
42. Bernstein, J. *Polymorphism in Molecular Crystals*; Oxford University Press: New York, NY, USA, 2002; Volume 14, ISBN 978-0-19-923656-5.
43. Hunter, S.; Sutinen, T.; Parker, S.F.; Morrison, C.A.; Williamson, D.M.; Thompson, S.; Gould, P.J.; Pulham, C.R. Experimental and DFT-D Studies of the Molecular Organic Energetic Material RDX. *J. Phys. Chem. C* **2013**, *117*, 8062–8071. [CrossRef]
44. Kolesnik, E.N.; Goryainov, S.V.; Boldyreva, E.V. Different behavior of L- and DL-serine crystals at high pressures: Phase transitions in L-serine and stability of the DL-serine structure. *Dokl. Phys. Chem.* **2005**, *404*, 169–172. [CrossRef]
45. Moggach, S.A.; Allan, D.R.; Morrison, C.A.; Parsons, S.; Sawyer, L. Effect of pressure on the crystal structure of L-serine-I and the crystal structure of L-serine-II at 5.4 GPa. *Acta Crystallogr. Sect. B Struct. Sci.* **2005**, *61*, 58–68. [CrossRef] [PubMed]

Article

On the Definition of Phase Diagram

Kamil Filip Dziubek

European Laboratory for Non-Linear Spectroscopy (LENS), Via Nello Carrara 1, 50019 Sesto Fiorentino, Italy; dziubek@lens.unifi.it

Abstract: A phase diagram, which is understood as a graphical representation of the physical states of materials under varied temperature and pressure conditions, is one of the basic concepts employed in high-pressure research. Its general definition refers to the equilibrium state and stability limits of particular phases, which set the stage for its terms of use. In the literature, however, a phase diagram often appears as an umbrella category for any pressure–temperature chart that presents not only equilibrium phases, but also metastable states. The current situation is confusing and may lead to severe misunderstandings. This opinion paper reviews the use of the "phase diagram" term in many aspects of scientific research and suggests some further clarifications. Moreover, this article can serve as a starting point for a discussion on the refined definition of the phase diagram, which is required in view of the paradigm shift driven by recent results obtained using emerging experimental techniques.

Keywords: phase diagram; phase transitions; high pressure; metastability; phase boundaries; kinetic lines; nonequilibrium conditions

Citation: Dziubek, K.F. On the Definition of Phase Diagram. *Crystals* 2022, 12, 1186. https://doi.org/10.3390/cryst12091186

Academic Editor: Daniel Errandonea

Received: 25 July 2022
Accepted: 20 August 2022
Published: 23 August 2022

Publisher's Note: MDPI stays neutral with regard to jurisdictional claims in published maps and institutional affiliations.

Copyright: © 2022 by the author. Licensee MDPI, Basel, Switzerland. This article is an open access article distributed under the terms and conditions of the Creative Commons Attribution (CC BY) license (https://creativecommons.org/licenses/by/4.0/).

1. Introduction

In their pioneering work on semantics, semiotics, and the science of symbolism, published almost a century ago, Ogden and Richards recounted: *"In all discussions we shall find that what is said is only in part determined by the things to which the speaker is referring. Often without a clear consciousness of the fact, people have preoccupations which determine their use of words. Unless we are aware of their purposes and interests at the moment, we shall not know what they are talking about and whether their referents are the same as ours or not"* [1]. Today, this statement is still very true, even in the scientific language, where precise definitions are to be generally agreed upon. One of the frequently used and, at the same time, elusive physico-chemical terms that defies a single accepted definition is *phase diagram*. This expression is not reported in the IUPAC Gold Book [2], nor in the IUCr Online Dictionary of Crystallography [3]. In a recent review on pressure-induced phase transitions, Grochala noted that *"the phase diagram is in its most common meaning a graph in which a property or a state of matter is shown in the function of pressure and temperature,"* adding a caveat about the described system being in equilibrium [4]. Therefore, I decided to use the following general Oxford English Dictionary (OED) definition [5] as a reference point:

> **phase diagram** *n*. Chemistry a diagram which represents the limits of stability of the various phases of a chemical system at equilibrium, with respect to two or more variables (commonly composition and temperature); an equilibrium diagram.

In the high-pressure research, phase diagrams are usually constructed for systems of constant composition (empirical formula), with two variables: temperature (T) and pressure (P). Hence, the OED definition applied to the P–T space assumes that: (1) the limits of the stability of the phases are represented in a P–T chart as separate fields that correspond to each phase and are delimited by phase-boundary lines, also known as coexistence curves (representing the P–T conditions at which phase transitions occur), or the abscissa and ordinate axes (representing the $P = 0$ isobar and $T = 0$ isotherm, respectively); (2) phase diagrams pertain only to the equilibrium state. Note that the stability conditions may be ambiguous, as there are various definitions of stability (this will be discussed in more detail later). The OED definition therefore describes an *ideal phase diagram,* and it will be referred to as such from here onwards.

The phase diagrams published in the literature are determined in experimental studies or are predicted using various theoretical approaches. Both methods have their intrinsic limitations. In predicted phase diagrams, the boundary lines between the crystalline phases and melting curves are usually constructed by comparing the Gibbs free energies of individual phases. Because they can be computed using different models and approximations that also include temperature effects, the plots reported by various authors may differ significantly. On the contrary, the disagreements between experimental phase diagrams depend not only on the laboratory protocols, but also on the character of the phase transition.

To aid the reader and for clarity, a glossary of key terms related to the concept of the phase diagram is provided below.

2. Glossary of the Key Terms Related to Phase Diagrams

- An *ideal phase diagram* is a *P–T* plot of thermodynamically stable phases in the equilibrium state. There is only one ideal phase diagram for any stoichiometric assembly of atomic species. It is impossible to experimentally determine an *ideal phase diagram* because all experiments are dynamic, hysteresis in first-order phase transitions cannot be completely eliminated, and nonhydrostatic stresses and *P,T* gradients are inevitable;
- A *phase diagram* is the best estimate of the *ideal phase diagram* based on experimental and theoretical constraints;
- A *dynamic P–T diagram* represents observed or predicted phases that can be produced during the course of dynamic compression or decompression (or cooling/heating). A *dynamic P–T diagram* can include metastable or transitional states and must include descriptions of the necessary conditions: the compression or decompression rate, cooling or heating rate, stress–strain conditions, etc. Note that some authors refer to *dynamic P–T diagrams* as "*dynamic phase diagrams*". The reason for this discrepancy is related to the definition of a phase, and the question of whether the thermodynamic metastable state can be regarded a phase. While a full discussion of this issue is beyond this opinion paper's scope, I am leaning towards using the first option (i.e., a *dynamic P–T diagram*);
- A *transitional P–T diagram* (sometimes referred to as a *transitional phase diagram*) represents the *P–T* diagram that includes metastable states (i.e., the states outside of the stability region in *the ideal phase diagram*). Contrary to an *ideal phase diagram,* a *transitional P–T diagram* depicts a system that is far from thermodynamic equilibrium;
- The difference between *dynamic* and *transitional diagrams* can be characterized as semantic rather than phenomenological. The term *dynamic P–T diagram* is used mainly in dynamic-loading studies, where fast pressure variation is the intrinsic feature of the experimental methodology (shock compression, ramp compression, piezo-electrically driven dynamic diamond-anvil cells), and the observed states often have very short lifetimes. A *transitional P–T diagram* applies mostly to quasistatic experiments, where metastable states are "quenched" from the initial thermodynamic equilibrium and can be stabilized for a relatively long time (from minutes to years). This type of *P–T* diagram is largely related to multicomponent materials, but it can also describe chemical elements. For example, a *transitional P–T diagram* of buckminsterfullerene (C_{60}) reports the forms of carbon in which the molecular integrity of a C_{60} molecule is preserved. It is different than the *ideal phase diagram* of carbon, as all the states of carbon consisting of C_{60} molecules are metastable under any *P–T* condition. Note that there are also other terms used to convey similar notions, such as *phase evolution diagram* or *kinetic phase diagram* [6]. The term *metastable phase diagram,* which sometimes occurs in the literature, is ambiguous and should be abandoned;
- By using terms such as *reaction P–T diagram* [7] or *transformation P–T diagram* [8], some authors intend to emphasize that the transitions between different phases or

states displayed on a diagram involve chemical reactions and the rearrangement of the bonding pattern (e.g., transformations from a molecular to an extended solid), although this additional specification seems redundant given the previously defined types of P–T diagrams.

3. Classification of Phase Transitions and Definitions of Phase Stability

In the most basic terms, phase transitions are usually related to the transformations of the physical properties of matter, which can be quantified by a change in the state parameters. The concept of a phase-transition order was introduced by Paul Ehrenfest in 1933 [9–11]. Ehrenfest's approach is purely thermodynamic: Assuming that the thermodynamic potential (usually the Gibbs free energy (G) or chemical potential (μ)) is continuous across the phase boundary of any equilibrium phase transition, one can investigate its particular derivatives concerning an external thermodynamic variable, such as temperature or pressure. Hence, in the general case, for an nth-order phase transition:

$$\Delta\left(\frac{\partial^n G}{\partial T^n}\right)_P \neq 0; \quad \Delta\left(\frac{\partial^{n-1} G}{\partial T^{n-1}}\right)_P = 0 \quad \text{at } T = T_{trans}, \quad (1)$$

$$\Delta\left(\frac{\partial^n G}{\partial P^n}\right)_T \neq 0; \quad \Delta\left(\frac{\partial^{n-1} G}{\partial P^{n-1}}\right)_T = 0 \quad \text{at } P = P_{trans}, \quad (2)$$

where G represents the Gibbs free energy, and T_{trans} and P_{trans} are the equilibrium transition temperature and pressure, respectively. Thus, the order of the transition is determined by the lowest nonzero derivative of the thermodynamic potential. At a first-order transition, the first derivatives of the G (namely, entropy (S) and volume (V)) are discontinuous; at a second-order transition, the S and V are continuous, but the second derivatives of the G (for instance, the constant-pressure heat capacity (C_P), the isothermal compressibility (β_T), and the isobaric thermal expansion coefficient (α)) are discontinuous, and so on. This simple view, however, is often inadequate for the characterization of some more complex transitions beyond the first order, which cannot be effectively described by a jump in any nth derivative of the thermodynamic potential, but by the nonanalytic behavior of these functions. In the modern classification scheme, which is essentially a generalization of the Ehrenfest one, phase transitions are classified into two broad categories: (1) "discontinuous" first-order transitions, which are distinguished by the presence of the latent heat and discontinuity in the order parameter defined on the grounds of the phenomenological theory formulated by Lev Landau [12]; (2) "continuous" second-order transitions (also comprising higher-order transitions, as defined in the Ehrenfest scheme), which are marked by a steady evolution of the order parameters, which are fully reversible across the phase boundary. This claim has significant consequences for the experimental determination of phase diagrams.

Consider a multidimensional space where the thermodynamic potential is a function of different variables. In a system of constant composition, each phase can be represented as a potential-energy surface in terms of pressure and temperature. The equilibrium-transformation conditions between two phases are defined by the intersection of surfaces, corresponding to the phase-boundary lines. In the example shown in Figure 1, if an isothermal compression is considered, then one can expect that the thermodynamic potential of the system should vary along the path 1–2–3–4–5 upon compression (5–4–3–2–1 upon decompression), with the sharp transformation of the entire sample occurring at Point 3, corresponding to the pressure (P_{trans}) at which the potentials of Phases I and II become equal. In practice, however, during the course of first-order transitions, the potential often follows the path 1–2–3–7–4–5 upon compression (5–4–3–8–2–1 upon decompression). Phase I, which exists for $P_{trans} < P < P^{**}$ during compression, is not in equilibrium because it is not the minimum potential configuration but a "superpressurized" metastable state. Likewise, Phase II, which persists upon decompression for $P^* < P < P_{trans}$, corresponds to a "superdepressurized" state (named by an analogy to superheated and supercooled states). Hence, the experimental onset of the transition upon the pressure increase can be regarded as an upper bound of the metastability range, while the onset of the reverse transition marks a lower bound. The pressure range $\Delta P = P^{**} - P^*$ of this hysteresis (and the values of P^* and P^{**}) depend on many factors, such as the pressure-change rate and the purity of the substance. Moreover, note that the intervals (P^{**}, P_{trans}) and (P_{trans}, P^*) are not necessarily equal, and so the center of the hysteresis loop does not need to coincide with the transition pressure. Therefore, the actual value of the P_{trans} is difficult, if not virtually impossible, to locate from the experiment.

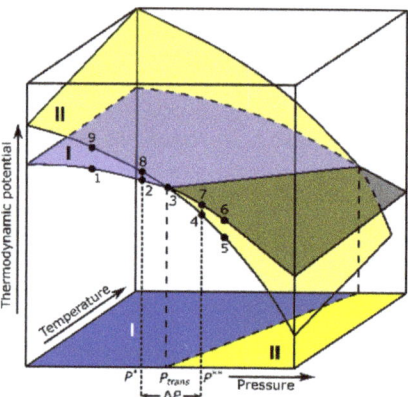

Figure 1. Thermodynamic-potential surfaces for two phases, I and II, as functions of pressure and temperature. The projection onto the P–T plane represents the phase diagram, whereas the line of intersection of the two surfaces corresponds to the I–II-phase-transition boundary. Black dots represent the stages of experimental isothermal compression or decompression routes. P^* and P^{**} define the limits of the metastability range, while P_{trans} denotes the equilibrium transition pressure.

One of the most common examples is the first-order solid–liquid phase transitions and the determination of the melting line. In the isothermal runs, crystallization often occurs from the superpressurized liquid, while in standard experimental settings, virtually all the solids melt immediately under the equilibrium solid–liquid coexistence conditions without entering the underpressurized (superdepressurized) state. The reason behind this is that, in the case of heterogeneous melting, the crystal surfaces and grain boundaries suppress the energy barrier required for melt nucleation [13–15].

Furthermore, it has to be stressed that the solid-state first-order transitions proceed through the nucleation and growth of interfaces. If the kinetics of nucleation is much slower than the rate of the pressure (or temperature) change, then one can observe the phase coexistence, where the mass fraction between the two phases changes across the transformation [16]. If the crystal structures are not similar, and the change from one structure to the other involves the significant reorganization of the bonding scheme at the atomic level, then the energy barrier of nucleation may be considerable and the transformation rate sluggish. Moreover, many other factors (such as the pressure- or temperature-variation rate, sample microstructure, or sample environment) may affect this process. Figure 2 demonstrates the fraction of the transformed phase as a function of pressure in a pressure-induced first-order transition for different kinetic rates.

Moreover, many of the reported high-pressure datasets are collected on compression only. The reasons are varied but are often quite mundane: limited beamtime or technical obstacles (jammed diamond-anvil cell, plastic deformation of the gasket, failure of the diamonds). Therefore, the observed conditions at which the new phase begins to appear may be far away from the point where the Gibbs free energies of the two phases are equal. This discrepancy is even more severe in the ultrahigh-pressure regime.

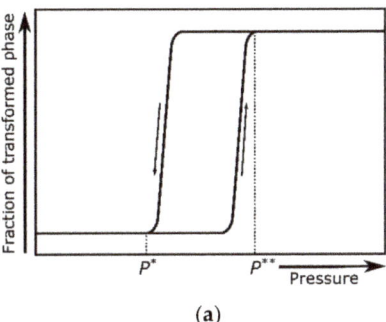 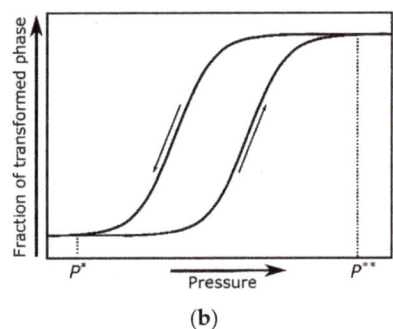

Figure 2. The fraction of the transformed phase as a function of pressure representing a relatively fast (**a**) and slow (**b**) rate of interconversion. P^* and P^{**} mark the lower and upper bounds of the phase coexistence range.

To sum up, various generic features are associated with first-order transitions, such as metastability, phase coexistence, hysteresis, and their dependence on the path followed in the pressure–temperature space. All these factors can affect the shape of the experimentally determined phase-boundary lines and, consequently, the phase diagram.

Finally, one should be aware that the phase stability has different definitions. The phase is *thermodynamically stable* if it corresponds to the minimum of the thermodynamic potential at a given pressure and temperature with respect to the other phases. It is roughly consistent with the IUPAC definition of "stable" regarding the chemical stability: "*As applied to chemical species, the term expresses a thermodynamic property, which is quantitatively measured by relative molar standard Gibbs energies. A chemical species A is more stable than its isomer B if $\Delta_r G° > 0$ for the (real or hypothetical) reaction A $\rightarrow$ B under standard conditions*" [17]. An immediate conclusion from this definition is that, at the given pressure and temperature, there is only one thermodynamically stable phase. At the same time, the other structures are *metastable* (i.e., thermodynamically unstable states that can be identified with the local minima on the potential-energy hypersurface and persist due to kinetic hindrance). A good general definition of *metastability* was provided by Tschoegl [18]: "*When a thermodynamic system is in* stable equilibrium, *a perturbation will result only in small (virtual) departures from its original conditions and these will be restored upon removal of the cause of the perturbation. A thermodynamic system is in* unstable equilibrium *if even a small perturbation will result in large, irreversible changes in its conditions. A system in* metastable equilibrium *will act as one in stable equilibrium if perturbed by a small perturbation but will not return to its initial conditions upon a large perturbation.*"

Under equilibrium conditions, a crystal is *dynamically stable* (or *vibrationally stable*; these terms are synonymous) if its potential energy always increases for any combination of atomic displacements; hence, all of the atoms are trapped within the local energy minima. In the harmonic approximation, this corresponds to the real and positive phonon frequencies of all the phonon modes [19]. The presence of imaginary (negative) frequencies of vibration at reciprocal lattice vectors in the Brillouin zone testifies to the dynamic instability.

A crystal is *mechanically stable* (or *elastically stable*; synonymous terms) if it fulfills Born's stability criteria, which were formulated for the first time in his 1940 paper [20]. In the harmonic approximation (i.e., for an unstressed material at zero pressure), the second-order elastic stiffness tensor (**C**) is defined as:

$$C_{ij} = \frac{1}{V_0}\left(\frac{\partial^2 E}{\partial \varepsilon_i \partial \varepsilon_j}\right), \tag{3}$$

where E is the elastic energy of the crystal, V_0 is its equilibrium volume, ε_i, and ε_j are the components of the strain tensor, and C_{ij} are the elastic constants (i.e., the elements of the elastic tensor). The elastic energy of a crystal is quadratic with applied strains, as follows:

$$E = E_0 + \frac{V_0}{2} \sum_{i=1}^{6} \sum_{j=1}^{6} C_{ij} \varepsilon_i \varepsilon_j. \tag{4}$$

According to the Born *mechanical (elastic)-stability* criterion, the crystal is *mechanically (elastically) stable* if the elastic energy is always positive. This corresponds to the requirement that the tensor (**C**) is positive definite, or equivalently, that all the eigenvalues of the **C** are positive. In a general case, the symmetric 6 × 6 s-order elastic stiffness tensor (**C**) contains 21 independent elements. While a more detailed analysis is beyond the scope of this paper, the interested reader is referred to the comprehensive article of Mouhat and Coudert [21], who generalized the necessary and sufficient Born mechanical-stability conditions for all crystal classes. It should be emphasized that the *mechanical (elastic)-stability* conditions can be generalized to systems under an arbitrary external load (σ) by introducing an elastic stiffness tensor (**B**) under load. The applied external stress is often a game-changing factor for the phase stability. If a system is mechanically unstable but dynamically stable under given conditions, then this may indicate a thermodynamically metastable state.

The definitions of *dynamic* and *mechanical* stability refer to different physical variables and are therefore independent of each other. For example, scandium carbide (ScC) was found to be *mechanically stable*, but *dynamically unstable* [22], while the bcc titanium structure is *mechanically unstable*, but *dynamically stable* [23].

4. Case Studies

The following examples will illustrate a variety of issues and concerns regarding the literature phase diagrams. Questions about their conformity to the universally accepted *ideal-phase-diagram* definition will be discussed.

4.1. Stable or Metastable, That Is the Question
4.1.1. Chemical Elements

Chemical elements have been extensively studied under pressure since the pioneering works of Percy Bridgman at the beginning of the last century. Several monographs are entirely devoted to the phase diagrams of elements [24,25]. The systematic behavior of chemical elements over a range of pressures and temperatures was also illustrated graphically in a specific form of the periodic table [26]. Despite the expected simplicity, designing phase diagrams of elements can pose a real challenge.

Carbon is arguably the element with the highest known number of chemical compounds due to its ability to bind covalently with other carbon atoms and form extended structures. It is also a chemical element with an overwhelming number of predicted crystal forms; the most recent release of the comprehensive Samara Carbon Allotrope Database (SACADA) contains 524 allotropes [27,28]. This complexity is not reflected in the experimental phase diagram of carbon [8,29–32], which contains only two experimentally determined solid phases, graphite and diamond, and other predicted phases, still not confirmed in experiments (see Figure 3). The first-order graphite ⇌ diamond phase transition has a high energy barrier, and the room-temperature compression of graphite does not yield diamond, but another three-dimensional metastable structure, in which chemical bonds between the neighboring graphene sheets are formed: the monoclinic M-carbon [33]. As expected, the kinetic barrier between the phases is lowered with the increase in temperature. On the one hand, synthetic laboratory-grown diamonds are produced at the pressure of ~5 GPa and a temperature of ~1800 K. On the other hand, the back transformation of diamond to graphite is kinetically hindered at room temperature and atmospheric pressure; hence, "diamonds are forever." Fullerenes and nanotubes, which are famous families of carbon allotropes, are not found in the phase diagram of carbon because they are metastable under any *P,T* conditions with respect to graphite or diamond. Another well-known carbon allotrope, lonsdaleite, or the so-called "hexagonal diamond," is not a separate phase but actually adopts a stacking-disordered diamond structure [34,35]. Hypothetical superdense carbon allotropes, some of them calculated to be thermodynamically stable in the terapascal regime, have been included in predicted phase diagrams of carbon [31,36,37]. However, a recent study on the ramp compression of diamond to 2 TPa demonstrates that the transition to stable dense allotropes is kinetically arrested due to a high energy barrier, similar to the hindered graphite ⇌ diamond transition in the modest-pressure range [32].

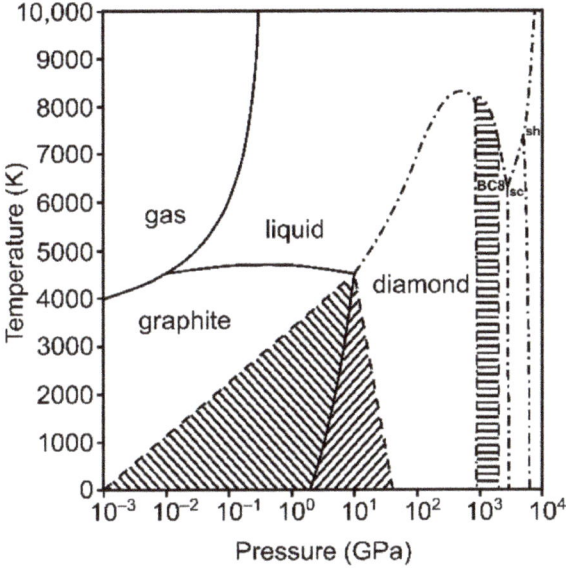

Figure 3. The simplified *transitional P–T diagram* of carbon, after [8,29–32]. Note the logarithmic scale for pressure. The left and right hashed areas correspond to the regions of metastability of graphite and diamond, with metastability boundaries marked as dashed lines. Dash-dotted contours delineate the calculated melting curve of diamond and stability fields of the predicted BC8, simple cubic (sc), and simple hexagonal (sh) phases [31], which have been never confirmed experimentally. The horizontal hashed area demonstrates the persistence of diamond in a ramp-compression experiment up to 2 TPa [32] within the predicted thermodynamic stability field of the BC8 phase. While the kinetics of the graphite ⇌ diamond transition can be very slow (at room temperature and ambient pressure, the diamond → graphite conversion timescale is longer than the age of the universe), the proposed metastability of diamond above 1 TPa is reported in the order of magnitude of nanoseconds.

As mentioned above, the experimental determination of phase boundaries may be affected by numerous factors. A striking example is the phase transitions in lithium, where quantum and isotope effects play a crucial role in shaping its phase diagram [38]. It was known from earlier studies [39–41] that, upon cooling below ~77 K at atmospheric pressure, bcc ^{7}Li transforms to a rhombohedral close-packed α-Sm-like 9R structure (space group $R\bar{3}m$), which, for a long time, was regarded as the ground state of lithium [40,41]. Careful selection of the pressure–temperature pathway (pressure-induced bcc → fcc transition at room temperature, followed by cooling to 20 K and subsequent decompression along the T = 20 K isotherm) helped to circumvent the bcc → 9R transition and facilitated the establishment of the actual ground state of the element, which is the fcc phase. By a prudent choice of words, the authors prevented ambiguities: while presenting the metastable 9R state on *P–T* charts, they entitled the figure "*Observed stable and metastable crystal structures of ^{6}Li and ^{7}Li measured along the identified P–T paths*", avoiding the term "phase diagram" [38].

Another interesting case is phosphorous, where two textbook examples of allotropes, white and red phosphorous, are thermodynamically metastable under any P–T conditions and are missing in the phase diagram [42]. Black phosphorous, a ground-state structure built of corrugated layers of covalently bonded atoms, was synthesized for the first time by Percy Bridgman in 1914 by heating the white modification at high pressure [43]. It was one of the first spectacular achievements of high-pressure research. This form exhibits an interesting phenomenon of layer destabilization upon heating [44], although no other stable crystal phase was revealed in this experiment.

As already mentioned, the mechanical stability of the crystal can be tuned by the external stress. For example, first-principles calculations and stability analysis suggest that diamond becomes mechanically unstable when the difference between the radial- and axial-stress components exceeds 200 GPa [45]. In his review article, Levitas distinguished three types of phase transitions at high pressure: pressure-induced, stress-induced, and strain-induced transformations [46], and he described the differences between them in the following manner: "*In most cases nucleation of the product phase occurs heterogeneously at some defects (dislocations, grain and twin boundaries), which produce a concentration of the stress tensor and/or provide some initial surface energy. Temperature-induced transformations nucleate predominantly at pre-existing defects without stresses at the specimen surface. Similarly, pressure-induced transformations occur mostly by nucleation at the same pre-existing defects under action of external hydrostatic pressure.* Stress-induced transformations *occur at the same defects when external nonhydrostatic stresses do not exceed the macroscopic yield strength in compression σ_y. If the phase transformations take place during plastic deformations they are classified as* strain-induced transformations. *They occur by nucleation at new defects generated during plastic deformation*". The effects of uniaxial stress on the phase-transition onset, hysteresis, and phase-coexistence range were demonstrated experimentally for the $\alpha \rightleftarrows \varepsilon$ transition of iron [47] and the $\alpha \rightleftarrows \omega$ transition of titanium [48], employing various pressure-transmitting media.

4.1.2. Chemical Compounds

Chemical compounds exhibit a higher degree of complexity. The P–T phase diagrams are determined for all the phases with the same net stoichiometry (constant empirical formula). Even for the simplest cases, there are many contradictory results, which have stirred major controversies. One of the examples is carbon dioxide (CO_2), which, despite the simple stoichiometry, has quite a complex phase diagram that consists of several crystalline molecular polymorphs below ~35 GPa. Above this pressure, in laser-heated diamond-anvil-cell experiments, they transform to a polymeric tetragonal crystal phase, which is named CO_2-V, which can be described as a network of fourfold coordinated carbon atoms interconnected by oxygen bridges [49,50]. The substantial kinetic barrier that results from the reconstruction of the bonding scheme may preclude the formation of this phase, leading to the quenching of metastable states in the stability field of CO_2-V. Indeed, other extended covalent (CO_2-VI [51]) and even ionic (i-CO_2 [52]) structures of CO_2 have been suggested in experimental studies of CO_2 in the nonmolecular P–T regime. Recent reports that extend the pressure range beyond 100 GPa [53] and that employ in situ high-pressure–high-temperature (HP–HT) X-ray diffraction [54] confirmed that CO_2-V is the only stable modification of carbon dioxide at $P > 35$ GPa. Moreover, a previously reported dissociation of CO_2 to elements [55–57] was not observed, even in long laser-heating cycles up to a pressure above 100 GPa and T = 6200 K [54]. The P–T diagram of CO_2 showing the relationships between the equilibrium crystal phases and nonequilibrium states is illustrated in Figure 4. Please note that the position of the so-called kinetic transition lines may depend strongly on the pressure- and temperature-change rate and the P–T pathway.

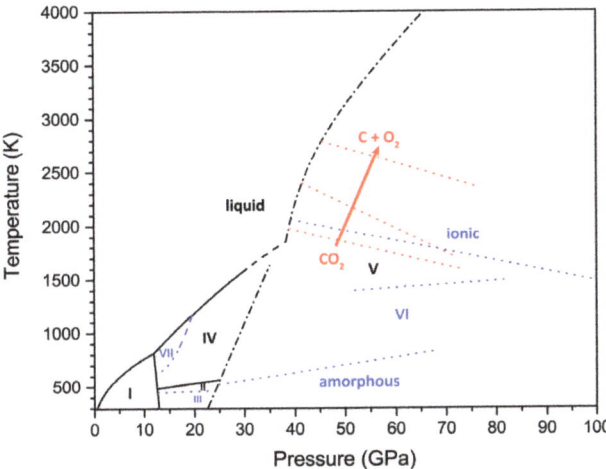

Figure 4. The P–T diagram of carbon dioxide compiled from literature data. Black solid lines represent the phase boundaries of thermodynamically stable phases delineating the stability range of CO_2-I and the melting curves of molecular phases [58,59], as well as the coexistence line between CO_2-II and CO_2-IV [60]. The black dashed line is the extrapolated CO_2-IV melting curve. Black dash-dotted lines correspond to the computed boundary line between molecular phases and CO_2-V [61] and the calculated CO_2-V melting curve [62], used in the absence of reliable experimental data. Metastable states are highlighted in blue: CO_2-III is most probably a metastable manifestation of CO_2-VII, and both have the same structure [61], while CO_2-V appears to be the only thermodynamically stable crystal phase of carbon dioxide above ~35 GPa [53,54,61,63]. Blue dotted traces are the kinetic transition lines involving the metastable states: between CO_2-IV and CO_2-VII (the extrapolation indicated by the dashed line) [59], between CO_2-II and CO_2-III [51], between amorphous CO_2 and CO_2-VI [51], between CO_2-V and CO_2-VI [51], and between CO_2-V and ionic i-CO_2 [52]. Red dotted lines denote the thresholds of the dissociation of CO_2 into chemical elements (indicated by an arrow), which was reported in three independent studies [55–57]. This reaction, however, was not confirmed by the subsequent in situ HP–HT study [54].

The above example refers to a simple system that consists of two chemical elements. In principle, the vast majority of the materials are metastable, including almost all the organic compounds, as reflected, for example, in the description of the methodology of the constrained-evolutionary-algorithm calculations by Zhu et al. [64]: "*Most of the molecular compounds are thermodynamically less stable than the simpler molecular compounds from which they can be obtained (such as H_2O, CO_2, CH_4, NH_3). This means that a fully unconstrained global optimization approach in many cases will produce a mixture of these simple molecules*". This statement falls squarely into the category of chemical stability and supports the expectation that the decomposition of complex molecules into a mixture of simpler structures is thermodynamically preferred; hence, such compounds are chemically metastable. Because the rate of transition is very often so slow (the energetic barrier is so high) that the transition to the thermodynamic ground state is practically not observed, and the shelf lives of many molecular compounds are in a timescale of years, the P–T diagrams of most of them (such as, e.g., paracetamol [65]) do not conform to the *ideal-phase-diagram* definition. Indeed, as Brazhkin noted in his viewpoint article [66]: "*the objects of the physics of condensed media are primarily the equilibrium states of substances with metastable phases viewed as an exception, while in chemistry the overwhelming majority of organic substances under investigation are metastable. It turns out that at normal pressure many simple molecular compounds based on light elements (. . .) are metastable substances too, i.e., they do not match the Gibbs free energy minimum for a given atomic chemical composition. (. . .) Actually almost all so-called 'phase diagrams' of molecular substances in the reference books are in fact transitional diagrams (. . .)*". Such transitional *diagrams* are useful tools for analyzing molecular systems. In many organic compounds, the upper P–T limit of the metastability range of the molecular state is not a melting curve but a chemical-stability boundary, above which decomposition, polymerization, or pyrolysis is observed [67,68].

It is worth mentioning that the transformations between constrained metastable states have most of the features of phase transitions. In particular, the transition point in molecular solids can depend on factors such as the pressure-transmitting medium [69], nonhydrostatic conditions [70], and so on, affecting the shape of the associated transitional diagrams.

Lastly, it should be noted that, in the computational structure predictions of chemical compounds, the idea of the so-called convex hull is widely employed to predict stable phases. In the simplest case, for a binary AB system at zero temperature and zero pressure, it shows the enthalpy of formation (ΔH_F), plotted against the A:B molar ratio, represented on a linear segment between the pure A and B elements (unary phases) as the endmembers (composition–energy hull) [71]. Convex hulls contain only thermodynamically stable phases with the lowest possible formation energy for a given composition. Most frequently, these phases correspond to stoichiometric compounds that are characterized by the simplest whole-number A:B ratios. This idea can be extended to more complex systems and more dimensions: ternary systems are represented graphically in an equilateral triangle, quaternary systems in a tetrahedron, and so forth. Pressure can be included by adding volume as a parameter (composition–volume–energy hull) [71]. This concept can be helpful in the quest for new materials under high pressure [72], and even ones as complicated as the H–C–N–O quaternary system [73]. While the convex hulls cannot be directly rendered into P–T phase diagrams (mostly due to the difficulties related to accounting for temperature effects and creating a composition–volume–temperature–energy hull), a quick glimpse at a hull allows us to answer the question of whether the phase is stable towards decomposition into the constituent elements or other compounds with different stoichiometry by comparing their ΔH_F under the given thermodynamic conditions.

4.2. Pressure-Induced Amorphization, Glassy States, and Liquid–Liquid Transitions

Pressure-induced amorphization is a ubiquitous phenomenon that is observed as crystalline solids are compressed beyond their stability range under specific conditions (fast compression rate, low temperature) that hamper the expected crystal-to-crystal phase transition [74,75]. Amorphization can also occur upon the decompression of high-density phases. It can be associated with a chemical reaction; for example, at a moderate temperature (up to 680 K), molecular CO_2 starts to polymerize, forming amorphous carbonia glass [76], which is a chemically distinct species that contains carbon in threefold and fourfold coordination (see Figure 4). This transformation occurs when the temperature is too low to initiate and complete the phase transition to the thermodynamically stable CO_2-V crystal phase.

As McMillan and Wilding noted [75]: "Polymorphic transformations between different amorphous "phases" that appear to mimic crystalline phase transitions also occur as a function of variables such as pressure (P) or temperature (T). Because glasses and other amorphous materials are not in internal thermodynamic equilibrium, such analogies must be made with care, however". Indeed, glasses are not phases but are kinetically trapped metastable states (even if one can plot boundaries between the glassy states), and they shall not be present in the phase diagrams. For instance, the formation of amorphous ice results from the sluggish transition kinetics and can be overcome if the compression rate is very slow [77]. In this regard, one can plot a phase diagram of ice reporting the crystalline phases and a transitional P–T diagram revealing amorphous states (with boundaries between low-density amorphous, high-density amorphous, and very-high-density amorphous ice). Presenting "mixed" plots that comprise both stable and metastable states is also possible [78], but they have to be presented with appropriate comments to avoid any ambiguities.

Phase transitions in liquids are a different subject because they pertain to the state, which corresponds to the energy minimum in the *P–T* range beyond the melting curve. First-order liquid–liquid phase transitions usually occur between the states that are characterized by different local structures and short-range bonding schemes [75]. Examples are transitions in phosphorus between the low-density-liquid (LDL) phase consisting of discrete P_4 molecules and the polymeric high-density liquid (HDL) [79], or between the LDL and HDL phases of sulfur [80]. In the latter case, the phase-boundary line between the LDL and HDL ends with a critical point, which was a first experimental observation of a critical point in liquid–liquid phase transitions.

4.3. Critical Points and Supercritical States

As per the definition, in a phase diagram of a pure substance, liquid and gas are undistinguishable beyond the liquid–vapor critical point and constitute a single fluid phase. However, boundary lines that delimit the gas-like and liquid-like states of supercritical fluid can be noted in some instances. These boundaries do not correspond to classical phase transitions but can be revealed in experimental studies (e.g., investigating sound velocity in supercritical fluid argon reveals the so-called Widom line) [81].

Critical points can also be observed at the end of the phase-boundary lines between the liquid states (e.g., in sulfur [80]). In solid phases, they are rare, primarily due to symmetry requirements. Usually a solid–solid phase-boundary line is intercepted by other phase boundaries or runs to infinity. One of the notable exceptions is the $\alpha \rightleftarrows \gamma$ phase transition in cerium discovered by Percy Bridgman in 1927 [82], which has been intensively studied ever since [83]. Both phases share the same space-group symmetry (*Fm-3m*), and the observed first-order transition of electronic origin [84] is associated with a substantial volume collapse, which decreases with the pressure and temperature rise along the phase-boundary line, and eventually vanishes at the critical point. Interestingly, upon the extension of the α–γ phase boundary, no evidence of a second-order phase transition is found, but a minimum in the isothermal bulk modulus in the function of pressure is observed to the highest experimentally achieved temperature [85]. The *P–V* compression data revealed in the radiography study indicated that, at the extrapolation of the α–γ coexistence line to the liquid-state regime, the *P(V)* isotherm shows an inflection point, which indicates a plausible liquid–liquid phase transition [86] (see Figure 5).

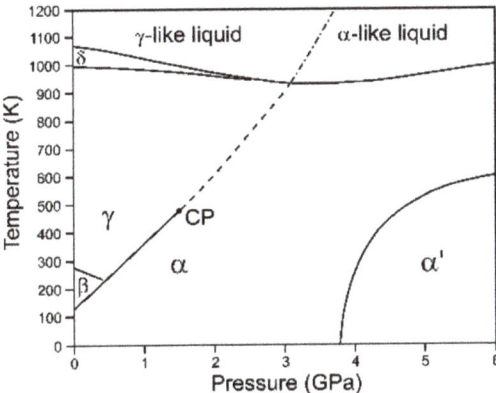

Figure 5. The phase diagram of cerium adapted from the literature data [86,87]. Solid traces denote the phase-boundary lines, and CP marks the critical point. The dashed line corresponds to the minimum observed in the isothermal-bulk-modulus plots [85]. The dash-dotted line refers to the liquid–liquid transition [86]. The phase denoted in the literature as α'' (mC4 in Pearson notation) and claimed to coexist with α' in part of its *P–T* field was not confirmed in the recent experimental study [87]. The stability of the α' and α'' structures is a matter of ongoing debate.

4.4. Nonequilibrium Studies and Dynamic P–T Diagrams

The advent of dynamic-compression techniques has enabled experimental studies of matter under *P–T* conditions that are unattainable in static studies, and that allow for a reinvestigation of the previous reports with a new method. In shock-compressed silicon, the observed onset of phase transitions is lower than in similar static experiments [88]. In shock-compressed antimony, a similar lowering of the transition pressure is observed and an additional structure, not existing under "static" (i.e., very slow) compression, is revealed [89]. Some authors report the *"dynamic phase diagrams"* and speculate about the "modification of the phase diagram" of studied materials. However, it has to be emphasized in the strongest possible terms that dynamic *P–T* charts correspond to nonequilibrium conditions and cannot be regarded as phase diagrams in light of the *ideal-phase-diagram* definition. A perfect illustration of this fact is the accurate determination of phase-transition pressures in bismuth in a broad compression-rate range (10^{-2}–10^3 GPa/s) using a dynamic diamond-anvil cell [90]. The variation in the transition pressure of the Bi-III $\rightleftarrows$ Bi-V transition can actually be determined as a function of the compression rate. Similar relations can be observed in compression-rate-dependent transitional diagrams of organic compounds (e.g., L-serine) [91]. Such phenomena are therefore relatively frequent, and there is no single *"dynamic phase diagram"* for a given system, but a continuum of *P–T* diagrams, where the compression rate is an additional variable. One should also consider that, in dynamic experimental studies, many other factors can effectively tune the observed phase boundaries (such as the sample microstructure and environment (see, e.g., the differences between the runs on powder and foil bismuth, or the experiments on bismuth surrounded by neon versus those performed without a pressure-transmitting medium [90])). Finally, it must be taken into account that nonequilibrium processes can lead to the kinetic stabilization of metastable states. This refers to the definition of a *transitional P–T diagram* and the difference between *dynamic* and *transitional diagrams*, as formulated in the glossary of terms (Section 2 of this article). As the timescale is the principal and perhaps even the only basis for distinguishing these two terms, with no quantitative threshold, the criterion of discrimination is therefore purely discretionary. Hence, I am inclined to believe that only one of these terms should be officially recommended. In this regard, I would prefer to use the term *transitional* rather than *dynamic*, as virtually any variable-pressure study is de facto *dynamic*, although some experiments can appear as quasistatic.

5. Conclusions and Call for Actions

Phase diagrams are simple visual representations that communicate key data related to the state of matter under varied thermodynamic conditions. Their focus and clarity make them understandable and accessible, even at the undergraduate level [92]. However, the notion of a phase diagram needs to be better defined. After running through all these examples, it is time to go back to the opening quotation from Ogden and Richards' "The Meaning of Meaning" and reflect on the status quo. I think the community should settle the terms and find a common language to avoid further misunderstandings. For many, "phase diagram" is a buzzword that automatically springs to mind when referring to any pressure–temperature chart. Do not worry, there is no scientific police that hounds for the misuse of the *phase-diagram* definition. Let me conclude with some closing remarks, which may serve as a starting point for a further discussion. It would be perfect if this debate resulted in the reaching of a consensus on the most significant working definitions and the issuing of official recommendations by international scientific organizations, such as IUPAC, IUCr, and AIRAPT [93]. To catalyze and facilitate the exchange of ideas, in Table 1, I present some preliminary suggestions and measures to tackle the fundamental problems in the current presentation of *phase diagrams*, providing insight into issues and potential solutions.

Table 1. The issues associated with the presentation of phase diagrams and suggestions for improvements.

Issues	Potential Solutions
The definition of an *ideal phase diagram* is quite limited to the elements and simplest chemical compounds. P–T plots of more complex systems, often reported as phase diagrams, are transitional P–T diagrams or "metastable phase diagrams." Even when regarding only chemical elements, one should be extremely careful with wording [1].	It should be acknowledged that a presented *phase diagram* is the best estimate of the *ideal phase diagram* based on experimental and theoretical constraints. At the base of the wording *phase diagram* lies the assumption that: (1) the presented states are *thermodynamically stable* phases; (2) the described situation is as close to the thermodynamic equilibrium as possible.
There is no harm in presenting nonequilibrium P–T *diagrams* if only experimental conditions are described in detail. Reporting nonequilibrium states (in particular, amorphous glasses) incorporated within an equilibrium *phase diagram* may be utterly confusing when not described appropriately.	*Transitional P–T diagrams* that include nonequilibrium states can be reported only if additional information is indicated and commented on in the legend and caption. The careful description is substantial, especially to distinguish between the *thermodynamically stable* phases and kinetically trapped metastable states. In such plots, the position of the kinetic transition lines may strongly depend on the experimental conditions and the P–T pathway. Such P–T diagrams should not be referred to as *phase diagrams*.
Undoubtedly, discovering new stable phases is of paramount importance in condensed-matter chemistry and physics. However, some literature P–T plots report numerous "phases" that appear to be metastable states after a more thorough investigation. Articles containing the catchphrase "we revised the *phase diagram* of ..." make little sense if the authors of previous studies understand the term differently. To paraphrase Occam's razor, supposedly stable phases should not be multiplied beyond necessity.	The reported phases should conform to the definition of a *thermodynamically stable* phase. Under no circumstances can a claim of the revision of a *phase diagram* be accepted if the reported states are *metastable*.
Some specific boundary lines (such as liquid–liquid-transition curves and Widom lines) and critical points may appear in a *phase diagram* or *transitional P–T diagram*. It is essential, however, to characterize them well in the plot or figure caption.	All the elements of the reported diagrams should be communicated in a clear and unambiguous fashion.
Dynamic P–T diagrams (based on the results of compression experiments) do not comply with the *phase-diagram* definition, as they do not correspond to an equilibrium state. The compression rate can shift phase boundaries and, as such, may be regarded as an additional variable, but one has to bear in mind that many other factors can also influence the boundary lines under nonequilibrium conditions. Having said that, an ideally static P–T diagram does not exist, as any compression and decompression is a time-dependent process. The crucial factor is, therefore, the ratio between the rate of transformation and the rate of pressurization or depressurization.	The terms *dynamic P–T diagram* and *transitional P–T diagrams* are almost synonymous; however, the second one is used more often to report the results of quasistatic compression experiments. As there is no universally accepted compression-rate threshold that unequivocally distinguishes these two notions, the term *dynamic P–T diagram* appears to be redundant.

[1] For example, a *transitional P–T diagram* of buckminsterfullerene (C_{60}) is not a phase diagram of carbon, and a *transitional P–T diagram* of ozone (O_3) is not a phase diagram of oxygen, and so on.

Funding: This research received no external funding.

Data Availability Statement: Not applicable.

Acknowledgments: Some of the thoughts included in this article were presented by the author in the 56th course of the International School of Crystallography in Erice, "Crystallography under Extreme Conditions: The Future is Bright and very Compressed", 3–11 June 2022. The author gratefully acknowledges Agnès Dewaele for the valuable comments on the manuscript and fruitful discussions. The help of Demetrio Scelta with designing the carbon dioxide diagram is greatly appreciated. The anonymous reviewer is kindly thanked for the constructive comments, which helped to improve the quality and presentation of this article.

Conflicts of Interest: The author declares no conflict of interest.

References

1. Ogden, C.K.; Richards, I.A. *The Meaning of Meaning: A Study of the Influence of Language upon Thought and of the Science of Symbolism*, 1st ed.; Routledge & Kegan Paul: London, UK, 1923.
2. IUPAC Gold Book. Available online: https://goldbook.iupac.org (accessed on 25 July 2022).
3. IUCr Online Dictionary of Crystallography. Available online: https://dictionary.iucr.org (accessed on 25 July 2022).
4. Grochala, W. Pressure-induced phase transitions. In *Chemical Reactivity in Confined Systems: Theory, Modelling and Applications*, 1st ed.; Chattaraj, P.K., Chakraborty, A., Eds.; Wiley: Hoboken, NJ, USA, 2021; pp. 25–47.
5. Oxford English Dictionary Online. Available online: https://www.oed.com (accessed on 25 July 2022).
6. Luo, K.; Liu, B.; Hu, W.; Dong, X.; Wang, Y.; Huang, Q.; Gao, Y.; Sun, L.; Zhao, Z.; Wu, Y.; et al. Coherent interfaces with mixed hybridization govern direct transformation from graphite to diamond. *Nature* **2022**, *607*, 486–491. [CrossRef] [PubMed]
7. Turnbull, R.; Hanfland, M.; Binns, J.; Martinez-Canales, M.; Frost, M.; Marqués, M.; Howie, R.T.; Gregoryanz, E. Unusually complex phase of dense nitrogen at extreme conditions. *Nat. Commun.* **2018**, *9*, 4717. [CrossRef] [PubMed]
8. Bundy, F.P.; Bassett, W.A.; Weathers, M.S.; Hemley, R.J.; Mao, H.K.; Goncharov, A.F. The pressure-temperature phase and transformation diagram for carbon; updated through 1994. *Carbon* **1996**, *34*, 141–153. [CrossRef]
9. Ehrenfest, P. *Phasenumwandlungen im ueblichen und erweiterten Sinn, Classifiziert nach dem Entsprechenden Singularitaeten des Thermodynamischen Potentiales*; N. V. Noord-Hollandsche Uitgevers Maatschappij: Amsterdam, The Netherlands, 1933; Volume 36, pp. 153–157.
10. Jaeger, G. The Ehrenfest classification of phase transitions: Introduction and evolution. *Arch. Hist. Exact Sci.* **1998**, *53*, 51–81. [CrossRef]
11. Sauer, T. A look back at the Ehrenfest classification. Translation and commentary of Ehrenfest's 1933 paper introducing the notion of phase transitions of different order. *Eur. Phys. J. Spec. Topics* **2017**, *226*, 539–549. [CrossRef]
12. Landau, L.D.; Lifshitz, E.M. *Statistical Physics*, 3rd ed.; Pergamon Press: Oxford, UK, 1980.
13. Frenkel, J. *Kinetic Theory of Liquids*, 1st ed.; Clarendon Press: Oxford, UK, 1946.
14. Turnbull, D. Phase changes. In *Solid State Physics*, 1st ed.; Seitz, F., Turnbull, D., Eds.; Academic Press: New York, NY, USA, 1956; Volume 3, pp. 225–306.
15. Dash, J.G. Melting from one to two to three dimensions. *Contemp. Phys.* **2002**, *43*, 427–436. [CrossRef]
16. Mnyuk, Y. Mechanism and kinetics of phase transitions and other reactions in solids. *Am. J. Condens. Matter Phys.* **2013**, *3*, 89–103.
17. Muller, P. Glossary of terms used in physical organic chemistry (IUPAC Recommendations 1994). *Pure Appl. Chem.* **1994**, *66*, 1077–1184. [CrossRef]
18. Tschoegl, N.W. *Fundamentals of Equilibrium and Steady-State Thermodynamics*, 1st ed.; Elsevier: Amsterdam, The Netherlands, 2000.
19. Wallace, D.C. *Thermodynamics of Crystals*, 1st ed.; Wiley: New York, NY, USA, 1972.
20. Born, M. On the stability of crystal lattices. I. *Math. Proc. Cambridge Philos. Soc.* **1940**, *36*, 160–172. [CrossRef]
21. Mouhat, F.; Coudert, F.-X. Necessary and sufficient elastic stability conditions in various crystal systems. *Phys. Rev. B* **2014**, *90*, 224104. [CrossRef]
22. Abu-Jafar, M.S.; Leonhardi, V.; Jaradat, R.; Mousa, A.A.; Al-Qaisi, S.; Mahmoud, N.T.; Bassalat, A.; Khenata, R.; Bouhemadou, A. Structural, electronic, mechanical, and dynamical properties of scandium carbide. *Results Phys.* **2021**, *21*, 103804. [CrossRef]
23. Kadkhodaei, S.; Hong, Q.-J.; van de Walle, A. Free energy calculation of mechanically unstable but dynamically stabilized bcc titanium. *Phys. Rev. B* **2017**, *95*, 064101. [CrossRef]
24. Young, D.A. *Phase Diagrams of the Elements*, 1st ed.; University of California Press: Berkeley, CA, USA, 1991.
25. Tonkov, E.Y.; Ponyatovsky, E.G. *Phase Transformations of the Elements Under High Pressure*, 1st ed.; CRC Press: Boca Raton, FL, USA, 2005.
26. Holzapfel, W.B. Structures of the elements—Crystallography and art. *Acta Crystallogr. B* **2014**, *70*, 429–435. [CrossRef] [PubMed]
27. SACADA—Samara Carbon Allotrope Database. Available online: https://www.sacada.info (accessed on 25 July 2022).
28. Hoffmann, R.; Kabanov, A.A.; Golov, A.A.; Proserpio, D.M. Homo Citans and Carbon Allotropes: For an Ethics of Citation. *Angew. Chem. Int. Ed. Engl.* **2016**, *55*, 10962–10976. [CrossRef] [PubMed]
29. Bundy, F.P. Pressure-temperature phase diagram of elemental carbon. *Phys. A Stat. Mech. Its Appl.* **1989**, *156*, 169–178. [CrossRef]
30. Zazula, J.M. *On Graphite Transformations at High Temperature and Pressure Induced by Absorption of the LHC Beam*; CERN-LHC-Project-Note-78/97; CERN: Geneva, Switzerland, 1997.
31. Benedict, L.X.; Driver, K.P.; Hamel, S.; Militzer, B.; Qi, T.; Correa, A.A.; Saul, A.; Schwegler, E. Multiphase equation of state for carbon addressing high pressures and temperatures. *Phys. Rev. B* **2014**, *89*, 224109. [CrossRef]
32. Lazicki, A.; McGonegle, D.; Rygg, J.R.; Braun, D.; Swift, D.C.; Gorman, M.G.; Smith, R.F.; Heighway, P.; Higginbotham, A.; Suggit, M.J.; et al. Metastability of diamond ramp-compressed to 2 terapascals. *Nature* **2021**, *589*, 532–535. [CrossRef]
33. Wang, Y.; Panzik, J.E.; Kiefer, B.; Lee, K.K.M. Crystal structure of graphite under room-temperature compression and decompression. *Sci. Rep.* **2012**, *2*, 520. [CrossRef]
34. Németh, P.; Garvie, L.A.J.; Aoki, T.; Natalia, D.; Dubrovinsky, L.; Buseck, P.R. Lonsdaleite is faulted and twinned cubic diamond and does not exist as a discrete material. *Nat. Commun.* **2014**, *5*, 5447. [CrossRef]
35. Salzmann, C.G.; Murray, B.J.; Shephard, J.J. Extent of stacking disorder in diamond. *Diam. Relat. Mater.* **2015**, *59*, 69–72. [CrossRef]
36. Correa, A.A.; Bonev, S.A.; Galli, G. Carbon under extreme conditions: Phase boundaries and electronic properties from first-principles theory. *Proc. Natl. Acad. Sci. USA* **2006**, *103*, 1204–1208. [CrossRef] [PubMed]

37. Sun, J.; Klug, D.D.; Martoňák, R. Structural transformations in carbon under extreme pressure: Beyond diamond. *J. Chem. Phys.* **2009**, *130*, 194512. [CrossRef] [PubMed]
38. Ackland, G.J.; Dunuwille, M.; Martinez-Canales, M.; Loa, I.; Zhang, R.; Sinogeikin, S.; Cai, W.; Deemyad, S. Quantum and isotope effects in lithium metal. *Science* **2017**, *356*, 1254–1259. [CrossRef] [PubMed]
39. Barrett, C.S. A low temperature transformation in lithium. *Phys. Rev.* **1947**, *72*, 245. [CrossRef]
40. Overhauser, A.W. Crystal structure of lithium at 4.2 K. *Phys. Rev. Lett.* **1984**, *53*, 64–65. [CrossRef]
41. Smith, H.G. Martensitic phase transformation of single-crystal lithium from bcc to a 9R-related structure. *Phys. Rev. Lett.* **1987**, *58*, 1228–1231. [CrossRef]
42. Scelta, D.; Baldassarre, A.; Serrano-Ruiz, M.; Dziubek, K.; Cairns, A.B.; Peruzzini, M.; Bini, R.; Ceppatelli, M. Interlayer bond formation in black phosphorus at high pressure. *Angew. Chem. Int. Ed. Engl.* **2017**, *56*, 14135–14140. [CrossRef]
43. Bridgman, P.W. Two new modifications of phosphorus. *J. Am. Chem. Soc.* **1914**, *36*, 1344–1363. [CrossRef]
44. Henry, L.; Svitlyk, V.; Mezouar, M.; Sifré, D.; Garbarino, G.; Ceppatelli, M.; Serrano-Ruiz, M.; Peruzzini, M.; Datchi, F. Anisotropic thermal expansion of black phosphorus from nanoscale dynamics of phosphorene layers. *Nanoscale* **2020**, *12*, 4491–4497. [CrossRef]
45. Zhao, J.-J.; Scandolo, S.; Kohanoff, J.; Chiarotti, G.L.; Tosatti, E. Elasticity and mechanical instabilities of diamond at megabar stresses: Implications for diamond-anvil-cell research. *Appl. Phys. Lett.* **1999**, *75*, 487–488. [CrossRef]
46. Levitas, V.I. High pressure phase transformations revisited. *J. Phys. Condens. Matter* **2018**, *30*, 163001. [CrossRef] [PubMed]
47. Boehler, R.; von Bargen, N.; Chopelas, A. Melting, thermal expansion, and phase transitions of iron at high pressures. *J. Geophys. Res.* **1990**, *95*, 21731–21736. [CrossRef]
48. Errandonea, D.; Meng, Y.; Somayazulu, M.; Häusermann, D. Pressure-induced α→ω transition in titanium metal: A systematic study of the effects of uniaxial stress. *Phys. B Condens. Matter* **2005**, *355*, 116–125. [CrossRef]
49. Santoro, M.; Gorelli, F.A.; Bini, R.; Haines, J.; Cambon, O.; Levelut, C.; Montoya, J.A.; Scandolo, S. Partially collapsed cristobalite structure in the non molecular phase V in CO_2. *Proc. Natl. Acad. Sci. USA* **2012**, *109*, 5176–5179. [CrossRef]
50. Datchi, F.; Mallick, B.; Salamat, A.; Ninet, S. Structure of polymeric carbon dioxide CO_2-V. *Phys. Rev. Lett.* **2012**, *108*, 125701. [CrossRef]
51. Iota, V.; Yoo, C.-S.; Klepeis, J.-H.; Jenei, Z.; Evans, W.; Cynn, H. Six-fold coordinated carbon dioxide VI. *Nat. Mater.* **2007**, *6*, 34–38. [CrossRef]
52. Yoo, C.-S.; Sengupta, A.; Kim, M. Carbon dioxide carbonates in the Earth's mantle: Implications to the deep carbon cycle. *Angew. Chem. Int. Ed.* **2011**, *50*, 11219–11222. [CrossRef]
53. Dziubek, K.F.; Ende, M.; Scelta, D.; Bini, R.; Mezouar, M.; Garbarino, G.; Miletich, R. Crystalline polymeric carbon dioxide stable at megabar pressures. *Nat. Comm.* **2018**, *9*, 3148. [CrossRef]
54. Scelta, D.; Dziubek, K.F.; Ende, M.; Miletich, R.; Mezouar, M.; Garbarino, G.; Bini, R. Extending the stability field of polymeric carbon dioxide phase V beyond the Earth's geotherm. *Phys. Rev. Lett.* **2021**, *126*, 065701. [CrossRef]
55. Tschauner, O.; Mao, H.-k.; Hemley, R.J. New transformations of CO_2 at high pressures and temperatures. *Phys. Rev. Lett.* **2001**, *87*, 075701. [CrossRef]
56. Seto, Y.; Hamane, D.; Nagai, T.; Fujino, K. Fate of carbonates within oceanic plates subducted to the lower mantle, and a possible mechanism of diamond formation. *Phys. Chem. Miner.* **2008**, *35*, 223–229. [CrossRef]
57. Litasov, K.D.; Goncharov, A.F.; Hemley, R.J. Crossover from melting to dissociation of CO_2 under pressure: Implications for the lower mantle. *Earth Planet. Sci. Lett.* **2011**, *309*, 318–323. [CrossRef]
58. Giordano, V.M.; Datchi, F.; Dewaele, A. Melting curve and fluid equation of state of carbon dioxide at high pressure and high temperature. *J. Chem. Phys.* **2006**, *125*, 054504. [CrossRef] [PubMed]
59. Giordano, V.M.; Datchi, F. Molecular carbon dioxide at high pressure and high temperature. *EPL* **2007**, *77*, 46002. [CrossRef]
60. Iota, V.; Yoo, C.S. Phase diagram of carbon dioxide: Evidence for a new associated phase. *Phys. Rev. Lett.* **2001**, *86*, 5922–5925. [CrossRef]
61. Cogollo-Olivo, B.H.; Biswas, S.; Scandolo, S.; Montoya, J.A. Ab initio determination of the phase diagram of CO_2 at high pressures and temperatures. *Phys. Rev. Lett.* **2020**, *124*, 095701. [CrossRef]
62. Teweldeberhan, A.M.; Boates, B.; Bonev, S.A. CO_2 in the mantle: Melting and solid-solid phase boundaries. *Earth Planet. Sci. Lett.* **2013**, *373*, 228–232. [CrossRef]
63. Datchi, F.; Weck, G. X-ray crystallography of simple molecular solids up to megabar pressures: Application to solid oxygen and carbon dioxide. *Z. Kristallogr.* **2014**, *229*, 135–157. [CrossRef]
64. Zhu, Q.; Oganov, A.R.; Glass, C.W.; Stokes, H.T. Constrained evolutionary algorithm for structure prediction of molecular crystals: Methodology and applications. *Acta Crystallogr. B* **2012**, *68*, 215–226. [CrossRef]
65. Smith, S.J.; Montgomery, J.M.; Vohra, Y.K. High-pressure high-temperature phase diagram of organic crystal paracetamol. *J. Phys. Condens. Matter* **2016**, *28*, 035101. [CrossRef]
66. Brazhkin, V.V. Metastable phases and 'metastable' phase diagrams. *J. Phys. Condens. Matter* **2006**, *18*, 9643–9650. [CrossRef]
67. Ciabini, L.; Santoro, M.; Gorelli, F.A.; Bini, R.; Schettino, V.; Raugei, S. Triggering dynamics of the high-pressure benzene amorphization. *Nat. Mater.* **2006**, *6*, 39–43. [CrossRef] [PubMed]
68. Dziubek, K.; Citroni, M.; Fanetti, S.; Cairns, A.B.; Bini, R. Synthesis of high-quality crystalline carbon nitride oxide by selectively driving the high-temperature instability of urea with pressure. *J. Phys. Chem. C* **2017**, *121*, 19872–19879. [CrossRef]

69. Zakharov, B.A.; Seryotkin, Y.V.; Tumanov, N.A.; Paliwoda, D.; Hanfland, M.; Kurnosov, A.V.; Boldyreva, E.V. The role of fluids in high-pressure polymorphism of drugs: Different behaviour of β-chlorpropamide in different inert gas and liquid media. *RSC Adv.* **2016**, *6*, 92629–92637. [CrossRef]
70. Guńka, P.A.; Olejniczak, A.; Fanetti, S.; Bini, R.; Collings, I.E.; Svitlyk, V.; Dziubek, K.F. Crystal structure and non-hydrostatic stress-induced phase transition of urotropine under high pressure. *Chem. Eur. J.* **2021**, *27*, 1094–1102. [CrossRef]
71. Amsler, M.; Hegde, V.I.; Jacobsen, S.D.; Wolverton, C. Exploring the high-pressure materials genome. *Phys. Rev. X* **2018**, *8*, 041021. [CrossRef]
72. Hilleke, K.P.; Bi, T.; Zurek, E. Materials under high pressure: A chemical perspective. *Appl. Phys. A* **2022**, *128*, 441. [CrossRef]
73. Conway, L.J.; Pickard, C.J.; Hermann, A. Rules of formation of H–C–N–O compounds at high pressure and the fates of planetary ices. *Proc. Natl. Acad. Sci. USA* **2021**, *118*, e2026360118. [CrossRef]
74. Machon, D.; Meersman, F.; Wilding, M.C.; Wilson, M.; McMillan, P.F. Pressure-induced amorphization and polyamorphism: Inorganic and biochemical systems. *Prog. Mater. Sci.* **2014**, *61*, 216–282. [CrossRef]
75. McMillan, P.F.; Wilding, M.C. Polyamorphism and liquid–liquid phase transitions. In *Encyclopedia of Glass Science, Technology, History, and Culture*, 1st ed.; Richet, P., Ed.; Wiley: Hoboken, NJ, USA, 2021; Volume 1, pp. 359–370.
76. Santoro, M.; Gorelli, F.A.; Bini, R.; Ruocco, G.; Scandolo, S.; Crichton, W.A. Amorphous silica-like carbon dioxide. *Nature* **2006**, *441*, 857–860. [CrossRef]
77. Tulk, C.A.; Molaison, J.J.; Makhluf, A.R.; Manning, C.E.; Klug, D.D. Absence of amorphous forms when ice is compressed at low temperature. *Nature* **2019**, *569*, 542–545. [CrossRef] [PubMed]
78. Amann-Winkel, K.; Böhmer, R.; Fujara, F.; Gainaru, C.; Geil, B.; Loerting, T. Colloquium: Water's controversial glass transitions. *Rev. Mod. Phys.* **2016**, *88*, 011002. [CrossRef]
79. Katayama, Y.; Inamura, Y.; Mizutani, T.; Yamakata, M.; Utsumi, W.; Shimomura, O. Macroscopic separation of dense fluid phase and liquid phase of phosphorus. *Science* **2004**, *306*, 848–851. [CrossRef]
80. Henry, L.; Mezouar, M.; Garbarino, G.; Sifré, D.; Weck, G.; Datchi, F. Liquid–liquid transition and critical point in sulfur. *Nature* **2020**, *584*, 382–386. [CrossRef] [PubMed]
81. Simeoni, G.G.; Bryk, T.; Gorelli, F.A.; Krisch, M.; Ruocco, G.; Santoro, M.; Scopigno, T. The Widom line as the crossover between liquid-like and gas-like behaviour in supercritical fluids. *Nat. Phys.* **2010**, *6*, 503–507. [CrossRef]
82. Bridgman, P.W. The compressibility and pressure coefficient of resistance of ten elements. *Proc. Am. Acad. Arts. Sci.* **1927**, *62*, 207–226. [CrossRef]
83. Decremps, F.; Belhadi, L.; Farber, D.L.; Moore, K.T.; Occelli, F.; Gauthier, M.; Polian, A.; Antonangeli, D.; Aracne-Ruddle, C.M.; Amadon, B. Diffusionless $\gamma \rightleftarrows \alpha$ phase transition in polycrystalline and single-crystal cerium. *Phys. Rev. Lett.* **2011**, *106*, 065701. [CrossRef]
84. Devaux, N.; Casula, M.; Decremps, F.; Sorella, S. Electronic origin of the volume collapse in cerium. *Phys. Rev. B* **2015**, *91*, 081101. [CrossRef]
85. Lipp, M.J.; Jackson, D.; Cynn, H.; Aracne, C.; Evans, W.J.; McMahan, A.K. Thermal signatures of the Kondo volume collapse in cerium. *Phys. Rev. Lett.* **2008**, *101*, 165703. [CrossRef]
86. Lipp, M.J.; Jenei, Z.; Ruddle, D.; Aracne-Ruddle, C.; Cynn, H.; Evans, W.J.; Kono, Y.; Kenney-Benson, C.; Park, C. Equation of state measurements by radiography provide evidence for a liquid-liquid phase transition in cerium. *J. Phys. Conf. Ser.* **2014**, *500*, 032011. [CrossRef]
87. Munro, K.A.; Daisenberger, D.; MacLeod, S.G.; McGuire, S.; Loa, I.; Popescu, C.; Botella, P.; Errandonea, D.; McMahon, M.I. The high-pressure, high-temperature phase diagram of cerium. *J. Phys. Condens. Matter* **2020**, *32*, 335401. [CrossRef] [PubMed]
88. McBride, E.E.; Krygier, A.; Ehnes, A.; Galtier, E.; Harmand, M.; Konôpková, Z.; Lee, H.J.; Liermann, H.-P.; Nagler, B.; Pelka, A.; et al. Phase transition lowering in dynamically compressed silicon. *Nat. Phys.* **2019**, *15*, 89–94. [CrossRef]
89. Coleman, A.L.; Gorman, M.G.; Briggs, R.; McWilliams, R.S.; McGonegle, D.; Bolme, C.A.; Gleason, A.E.; Fratanduono, D.E.; Smith, R.F.; Galtier, E.; et al. Identification of phase transitions and metastability in dynamically compressed antimony using ultrafast X-ray diffraction. *Phys. Rev. Lett.* **2019**, *122*, 255704. [CrossRef] [PubMed]
90. Husband, R.J.; O'Bannon, E.F.; Liermann, H.-P.; Lipp, M.J.; Méndez, A.S.J.; Konôpková, Z.; McBride, E.E.; Evans, W.J.; Jenei, Z. Compression-rate dependence of pressure-induced phase transitions in Bi. *Sci. Rep.* **2021**, *11*, 14859. [CrossRef] [PubMed]
91. Fisch, M.; Lanza, A.; Boldyreva, E.; Macchi, P.; Casati, N. Kinetic control of high-pressure solid-state phase transitions: A Case Study on L-serine. *J. Phys. Chem. C* **2015**, *119*, 18611–18617. [CrossRef]
92. Gramsch, S.A. A closer look at phase diagrams for the general chemistry course. *J. Chem. Ed.* **2000**, *77*, 718–723. [CrossRef]
93. International Association for the Advancement of High Pressure Science and Technology (Association Internationale pour l'Avancement de la Recherche et de la Technologie aux Hautes Pressions). Available online: https://www.airapt.org (accessed on 20 July 2022).

Article

Chinese Colorless HPHT Synthetic Diamond Inclusion Features and Identification

Ying Ma [1], Zhili Qiu [2,*], Xiaoqin Deng [2], Ting Ding [1], Huihuang Li [1], Taijin Lu [3], Zhonghua Song [3], Wenfang Zhu [1] and Jinlin Wu [1]

1. National Gemstone Testing Center Shenzhen Lab. Company Ltd., ShenZhen 518000, China
2. School of Earth Sciences and Engineering, Sun Yat-sen University, GuangZhou 510275, China
3. Jewelry Technology Administrative Center, Ministry of Natural Resources, Beijing 100013, China
* Correspondence: qiuzhili@mail.sysu.edu.cn; Tel.: +86-138-2600-8304

Abstract: Chinese HPHT diamonds have improved dramatically in recent years. However, this brings a challenge in identifying type IIa colorless diamonds. In this study, eleven HPHT and three natural, colorless, gem-quality IIa diamonds were analyzed using magnified observation, Raman, PL and chemical element analysis. The results show that only HPHT samples possessed kite-like inclusions and lichenoid inclusions, as verified by their complex Raman spectra (100–750 cm^{-1}). Through PL mapping, HPHT and natural IIa diamonds were distinguished by their growth environments, which were reflected by PL peaks at 503, 505, 575, 637, 693, 694 and 737 nm. The chemical components of HPHT IIa diamond carbide inclusions are mainly Fe, Co, Ni and Mn, but those of Natural IIa are mainly Fe and Ni. As a result, the chemical components can be used to distinguish a natural colorless IIa diamond from a synthetic diamond.

Keywords: HPHT diamond; inclusion; PL mapping; chemical components

1. Introduction

HPHT synthetic diamonds are fundamental for high-pressure research (Crystals 2021, 11, 1185). Thanks to advancements in experimental techniques and computer simulations, the rate of new, important discoveries has significantly increased. Chinese HPHT synthetic diamonds have improved dramatically since the first was produced in China in 1963 [1–3]. After 2016, manufacturers began to produce high-quality IIa diamonds which have been made available in the Chinese jewelry market in large quantities [2–5] (see Figure 1). Nevertheless, HPHT synthetic diamonds are often sold as natural diamonds, or natural diamond parcels are intentionally mixed with HPHT synthetics [4]. There have been reports of cases of fraud in the global diamond business [2–4]. Therefore, there is a demand for distinguishing between natural and synthetic diamonds in the jewelry market.

Luminescence is currently used to identify IIa colorless diamonds. DeBeers company developed the DiamondView fluorescence imaging microscope using an ultra-shortwave excitation of 225 nm in the 1990s to distinguish HPHT-growth diamonds from natural diamonds [6]. They have developed a variety of methods since then, including cathodoluminescence (CL) images, X-ray images, PL spectrums, IR spectrums and inclusion tests [2,3,7,8]. DiamondView fluorescence imaging, PL spectrums and IR spectrums are the most commonly used methods for identifying diamonds. However, under DiamondView fluorescence imaging, many synthetic diamonds have a similar fluorescence color or IR spectrum to natural diamonds, especially IIa diamonds. There are also difficulties distinguishing between natural and synthetic diamonds using only fluorescence imaging, PL and IR spectrums with the progress of synthetic diamond technology. Gemologists face a challenge in identifying IIa colorless diamonds.

Inclusion characteristics, which reflect the diamond's formation environment, are also necessary for identifying diamonds [3,5]. This article discusses eleven HPHT-growth IIa colorless diamonds from three producers (HH-HuangHe, J-HuaJing and Z-ZhongNan) and three natural ones from natural primary diamond deposits (Guizhou). With the addition of previous corresponding research results, this article reports that special inclusions, their Raman features and chemical components are useful for identifying natural diamonds and HPHT synthetic colorless diamonds.

Figure 1. Jewelry made from Chinese HPHT synthetic diamonds (weights ranging from 0.01 to 5.00 ct), 2019. Images courtesy of two companies (Zhengzhou Sino-Crystal Diamond Co., Ltd. Shenzhen Multi-Win Business Co., Ltd.-VCK growth Diamond, China).

2. Materials and methods

2.1. Materials

In this study, 14 samples were tested, including 11 HPHT type IIa colorless diamonds from three Chinese factories (HH, J, Z) and three natural type IIa colorless diamonds selected from the primary diamond mine in Guizhou (China), which are presented in Table 1 and Figure 2. Eleven HPHT type IIa colorless diamonds were cut and polished to expose inclusions near the surface, with the cuts oriented by the {111}, {100} and {110} planes. We also cut the three rough Guizhou diamonds with irregular shapes into plates. We cleaned all of the plates without inclusions on the diamond face in acetic acid for 1 h at 100 °C. The detailed inclusions and PL peak characteristics of these samples are described in Table 1.

Figure 2. Diamond samples.

Table 1. Characteristic of Chinese HPHT synthetic diamonds in recent years.

Sample NO.	Color/Type	Weight/ct	Shape	Inclusion Properties	Inclusion Assemblage	Emission Peaks	Origin
HH-1	Colorless/IIa	0.23	Hextubbiness	Tiny point-like inclusions, less than 1 μm	-	-	HuangHe
HH-5	Colorless/IIa	0.35	Hexoctahedron	Block-shaped, rod-like, greater than 150 μm	Carbide, magnetite, hematite, sulfide, graphite, methane	416, 509, 533, 538, 693, 694, 883, 884 nm	HuangHe
HH-6	Colorless/IIa	0.20	Hexoctahedron	Water-drop, tiny point-like inclusions, less than 1 μm	Carbide, graphite	693, 694, 883, 884 nm	HuangHe
HH-20	Colorless/IIa	0.38	Hexoctahedron	Tiny point-like inclusion distributed along the edge of a crystal, less than 1 μm	-	883, 884 nm	HuangHe
J-2	Colorless/IIa	0.07	Hexoctahedron	Kite-like, tubular (cone like) inclusions (1~10 μm)	-	505, 737, 883 nm	HuaJing
J-3	Colorless/IIa	0.06	Hexoctahedron	Disc shape, point-like inclusions (1~90 μm)	Carbide, graphite	505, 737, 877 nm	HuaJing
J-20	Colorless/IIa	0.05	Hexoctahedron	Point-like inclusions (1~50 μm)	-	575, 637, 737 nm	HuaJing
J-10	Colorless/IIa	0.15	Hexoctahedron	Tiny point-like inclusions, less than 1 μm, Tubular (cone like) inclusions	-	737 nm	HuaJing
Z-2	Colorless/IIa	1.02	Hexoctahedron	Rod like inclusions (1~5 μm)	Carbide, graphite, methane	484, 489, 491, 883, 884 nm	ZhongNan
Z-3	Colorless/IIa	1.00	Hexoctahedron	Lichenoid (tree-like) inclusions, point-like inclusions (1~80 μm)	Metal alloy, methane	484, 489, 507, 883, 884 nm	ZhongNan
Z-4	Colorless/IIa	0.50	Hexoctahedron	Water-drop, point-like inclusions (1~50 μm)	Metal alloy, carbide, graphite	693, 694, 883, 884 nm	ZhongNan
GZ-18	Colorless/IIa	0.12	Irregular	Irregular inclusions	Graphite	491, 496, 503, 505, 536, 575, 579, 637, 612, 676, 710, 741 nm	GuiZhou
GZ-25	Colorless/IIa	0.17	Irregular		-	406, 415, 491, 496, 741 nm	GuiZhou
GZ-42	Colorless/IIa	0.10	Irregular			406, 415, 491, 496, 741, 945 nm	GuiZhou

2.2. Methods

Photomicrographs of samples were tested at the School of Earth Sciences and Engineering (SESE), Sun Yat-sen University, using digital (VHX-5000, Keyence, Japan) and polarizing (BX51, Olympus, Japan) microscopic image technology.

A Renishaw inVia Raman micro-spectrometer was used to collect Raman and photoluminescence (PL) spectra with four laser excitations (325, 473, 532 and 785 nm) at room temperature and liquid nitrogen temperature (77 K), at 1 cm^{-1} resolution, for which the laser power was 20 mw × 24%–20 mw × 100%.

We performed PL mapping with a Thermo Scientific DXR 2xi Raman imaging microscope equipped with an Olympus optical microscope and an EMCCD detector at Sun Yat-sen University's School of Earth Sciences and Engineering, with an accuracy of 100 nm. We conducted mapping with 455 and 532 nm laser excitation wavelengths. An Olympus 10 × 0.25 numerical aperture objective lens was used, 1–3 scans were conducted at 2–4 cm^{-1} resolution and the laser power was 10 mw × 20%–10 mw × 100%. In addition, the sample was mounted on a liquid-nitrogen-cooled Linkam cold stage for analysis at 123 K. The data were processed using Thermo Scientific's OMNIC 2xi analysis software package and baseline-corrected peak area profiles were used to produce the observed PL maps.

LA–ICP–MS analyses were performed at the Metallurgical Geology Bureau (MGB) Shandong Bureau, China. The ThermoFisher ThermoX2 laser ablation–inductively coupled plasma–mass spectrometer, coupled with 193 nm LA system was used to analyze trace elements of inclusions in nine samples employed a 10 Hz pulse rate, 30 µm spot size and ablation time of 30–45 s. The US National Institute of Standards and Technology (NIST) standard reference materials 610 and 612 were used for internal calibration. The uncertainty was 1σ.

3. Results and Discussion

3.1. Inclusion Feature

Using microphotographs, inclusions in Chinese type IIa colorless HPHT diamonds appear transparent to opaque, silver or black in color, and range in size from a few hundred to a few microns. Our samples contained rod-like, fine point-like, lichenoid (tree-like), tubular (cone-like), kite-like, pear-shaped and water-drop-shaped inclusions, as shown in Figure 3. All diamonds from the three factories displayed black point-like, rod-like and irregular clusters. Inclusions of a lichenoid (tree-like) nature were mostly found in diamonds from factories Z and HH, see Figure 3. Tubular or kite-like inclusions were only found in factory J, and these were transparent or filled with black materials, reflecting the differences in production methods, see Figure 3J-2. Rod-like, tree-like, pear-shaped and water drop inclusions were metal alloys or carbides, with minor sulfides and phosphides (see Sections 3.2 and 3.3). Some contained graphite coatings and methane jackets (in Raman). As the tubular and kite-like inclusions were too small to analyze, their chemical composition is unknown.

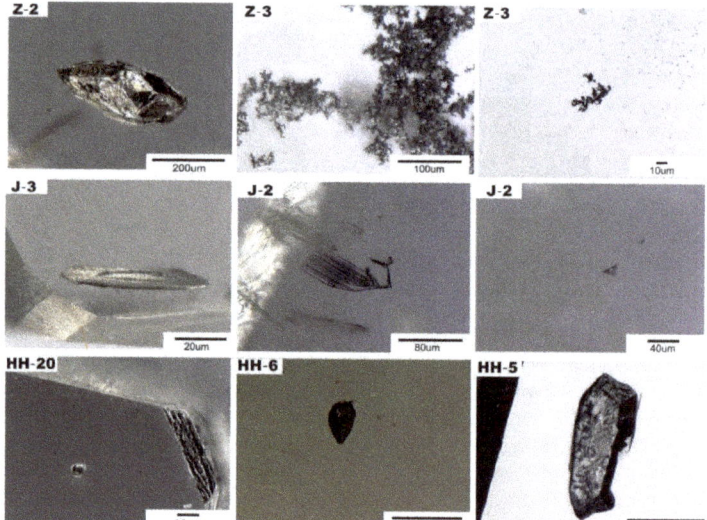

Figure 3. The inclusions in Chinese HPHT synthetic diamonds collected from factories Z, J and HH. **The first row**: Rod-like, lichenoid (tree-like) and fine-point-like inclusions from factory Z, {100}; **The second row**: Disc shape, tubular and kite-like inclusions, {111}, from factory J; **The third row**: Pear or water-drop inclusions, fine-point-like and rod-like inclusions, {111}, from factory HH. These samples were produced in 2017–2020. Photographs by Ying Ma and Xu Ye.

The tubular inclusions (Figure 3J-2) are similar to earlier HPHT diamond cone-like inclusions (the temperature was from 1290 to 1250 °C, the pressure was 6 Gpa) from Frumkin Institute of Physical Chemistry and Electrochemistry RAS [9]. Differing from cone-like inclusions, our samples contained opaque black materials. According to previous research, small-angle grain boundaries (GBs) and associated dislocation bundles decorated with microscopic oxide inclusions are the most plausible explanation for the cone-like defects. Black materials fill the kite-like inclusions in HPHT IIa diamonds (Figure 3J-2). Although there appeared to be networks of planar cracks in natural pink diamond [10], the networks of planar cracks in the natural pink diamonds had multiple planar due to the slip of the crystal plane. However, HPHT synthetic diamonds do not have linear slips in the same way as natural diamonds. The cause of kite-like inclusions appearing in HPHT diamonds is still unclear; however, they can help to distinguish HPHT IIa diamonds from natural diamonds.

3.2. Inclusion Raman Spectroscopy

The Raman method was used to determine inclusions in diamonds using non-destructive testing. In this study, all samples displayed approximately a 1332 cm^{-1} diamond intrinsic Raman line. The three HPHT manufacturers differed. Factory Z samples had a wider HFWT, while J factory samples had a peak at 1400 cm^{-1}, representing graphite D bands. The Raman line of the Guizhou sample showed graphite inclusion and contained the sample's intrinsic peak. Sample differences are not representative of the specific production quality of the three manufacturers and may be caused by sampling. We observed Raman peaks around 2912 cm^{-1} (Figure 4), which corresponds to CH_4 [11,12], see Figure 4J-5. Some HPHT samples contain cohenite inclusions, pear-shaped inclusions and were opaque, silver or black in color, with a graphite coating on the surface, as can be seen in Figures 3 and 4 for HH-20. Furthermore, the Raman spectrum contained many impurity emission peaks, including 4382 cm^{-1} and 4382 cm^{-1}, equal to 692 nm and 693 nm, respectively, which corresponded to Ni (see Figure 4Z-3, right side). Our HPHT sample Z-3 lichenoid inclusions (tree-like inclusions) displayed a complex Raman feature (see Figure 4Z-3, left side). For lichenoid inclusions in HPHT synthetic diamonds, the fit analysis ranged from 100 cm^{-1} to 750 cm^{-1} with software PeakFit V4.12 and Origin 2015, R^2 > 94%. By fitting and dividing the original Raman peak position, Raman emission peaks (100–750 cm^{-1}) of metal oxide were observed. Ni emission peaks can be seen in the lichenoid inclusion complex's Raman feature at 535–554 nm. The 547 nm (546.6 nm), 540nm (539.4 nm) and 535 nm (535.2 nm) emission peaks are related to Ni (see Figure 4Z-3, left side). More detailed electronic activities remain to be studied in the future. To the best of the authors' knowledge, Raman spectroscopy of lichenoid inclusions in HPHT IIa diamonds has yet to be found in natural diamonds.

Our research results were compared with those of previous studies to reveal the differences between type IIa natural diamonds and synthetic diamond inclusions (see Table 2). Perovskites and other mineral crystals are common inclusions in natural type IIa diamonds, but amorphous metal compounds, carbides and others can also occur. Inclusions in natural IIa diamonds and HPHT share various similarities, as shown in Table 2. There are inclusions such as metal alloys, carbides, H_2 and CH_4 in both natural and HPHT IIa diamonds, with some inclusions grouped in <111> chains [13–15]. Since HPHT samples do not undergo as many geological processes as natural diamonds, they do not contain inclusions associated with healed cracks. Crystal minerals appear only in natural IIa diamonds, but not in every diamond. It is important to distinguish metal carbides in natural and HPHT synthetic IIa diamonds when there are no crystalline mineral inclusions. Thus, the morphology and Raman spectroscopy of inclusions in IIa diamonds are useful indicators.

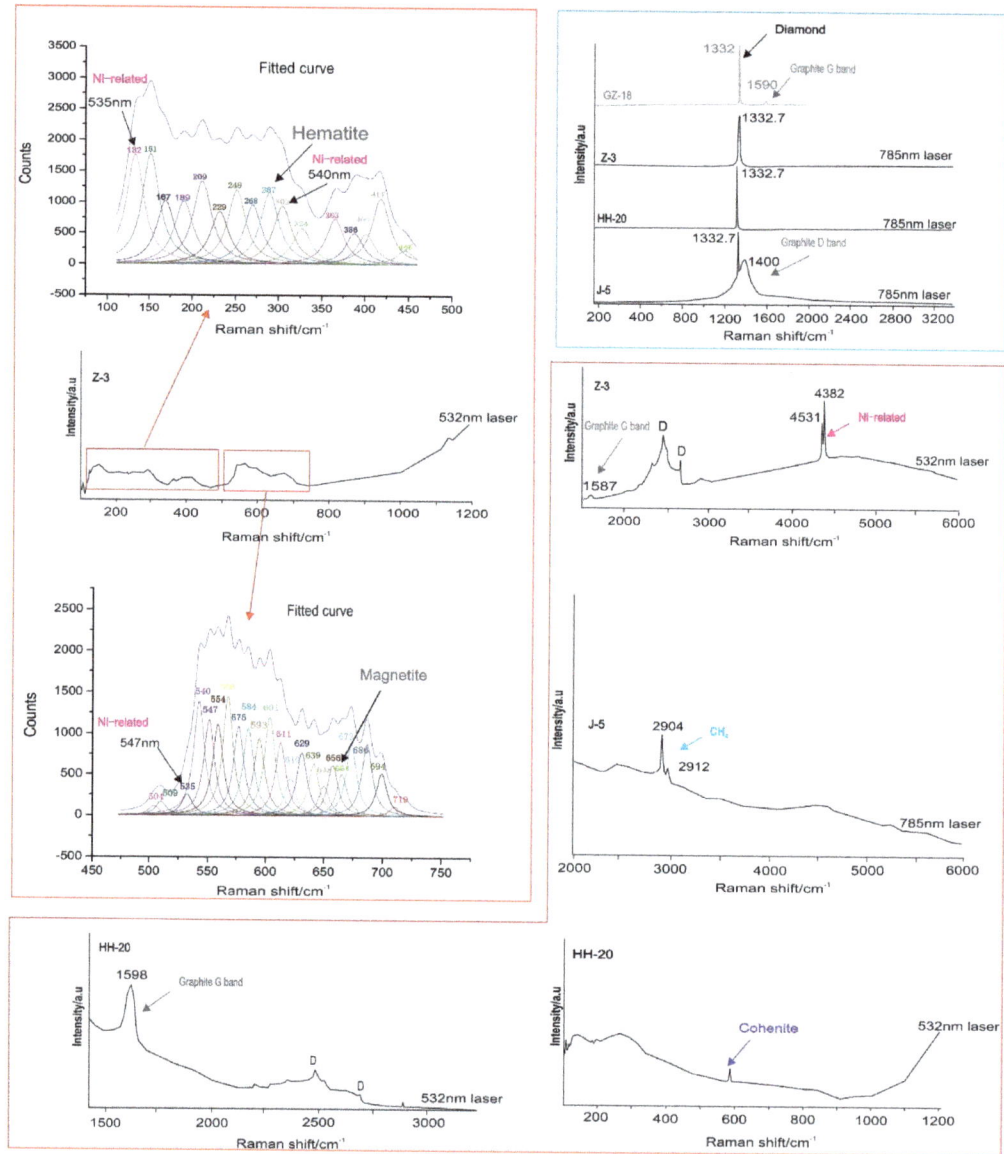

Figure 4. Raman spectroscopy of diamonds and inclusions in Chinese HPHT synthetic diamonds. **The red frame:** Raman Spectroscopy of inclusions in Chinese HPHT synthetic diamond. In sample Z-3, lichenoid inclusions demonstrated complex Raman and electron emission peaks; two red arrows point to the fit and division of the sample Z-3 original Raman lichenoid inclusions. The Cohenite in sample HH-20 are coated with graphite, which is often observed in our HPHT samples. Sample J-5 contains methane gas inclusions encircling metal alloys. Diamond Raman features are marked "D". **The blue frame:** Raman spectroscopy of samples.

Table 2. Summary of inclusions observed in type IIa colorless diamonds.

No. of Diamonds	Color/Style	Largest in Group	Inclusion Assemblage	Inclusion Properties	Origin	Delivery Time/Reference
11	Colorless/IIa	1.02ct	Undetected, metal alloy, carbide, and containing complicated emission peaks.	Tiny point-like inclusions, silver/black color transparent to opaque, less than 1 μm. Rod-like, pear, water-drop inclusions along with four angles at the end of <111> chains, 4 also had graphite + CH_4 jacket (did not have detectable H_2 in Raman). Lichenoid (tree-like) inclusion tubular (cone like), kite-like inclusion along face {100} margin.	HH,J,Z	2018–2020
3	Colorless/IIa	0.17ct	Undetected.	Graphite.	Guizhou of China	2020
60	Colorless/IIa	30.13ct	Crystal minerals, metallic Fe-Ni-C-S (inferred to be primary melt inclusions).	Perovskite, walstromite, majoritic garnet, titanite, larnite, magnetic, silver/black color, opaque, grouped in <111> chains, CH_4 fluid jacket (22 also had detectable H_2 in Raman); associated with healed cracks; altered to red-brown (hematite).	South Africa	Smith et al., 2016, 2017
1	Colorless/IaB	Almost 0.80ct	Metal alloy, metallic compound.	-	Brazil	Kaminsky et al., 2011

3.3. PL Spectroscopy and Mapping

For testing purposes, we directionally sectioned the synthetic samples in this study. There was a smooth plane thrown out for the sample without a crystal row in the Guizhou diamond samples. Diamond inclusions are invisible microscopic defects that can reflect the conditions under which diamonds are formed. The position of PL peaks is often used to identify diamonds. HH, J and Z factory HPHT diamonds (total eight grains) had peaks at 693, 694, 883 and 884 nm, which were attributed to an interstitial Ni^+ atom that was distorted [16,17]. As shown in Figure 5, sample Z-3 exhibited an emission multiplet with lines at 484 nm (2.56 eV), which was Ni-related. The negatively charged silicon-vacancy doublet (SiV^-, 737 nm, 1.68 eV) was only detected for J factory in four of the samples. We did not observe some PL peaks features which were previously observed in products of ZhongNan (peaks at 494, 503, 658 nm) [18], Power (489 nm) [2] and ZhongWu (peaks at 450, 659, 670, 707 and 714 nm) [19] which were HPHT diamonds factories of China. The peaks at 406, 612, 676, 710, 741, 745 and 945 nm were only observed in the Guizhou natural type IIa diamonds. We found that the greater the number of inclusions, the more impurity defect peaks there were. For more details, see Table 1 and Figure 5.

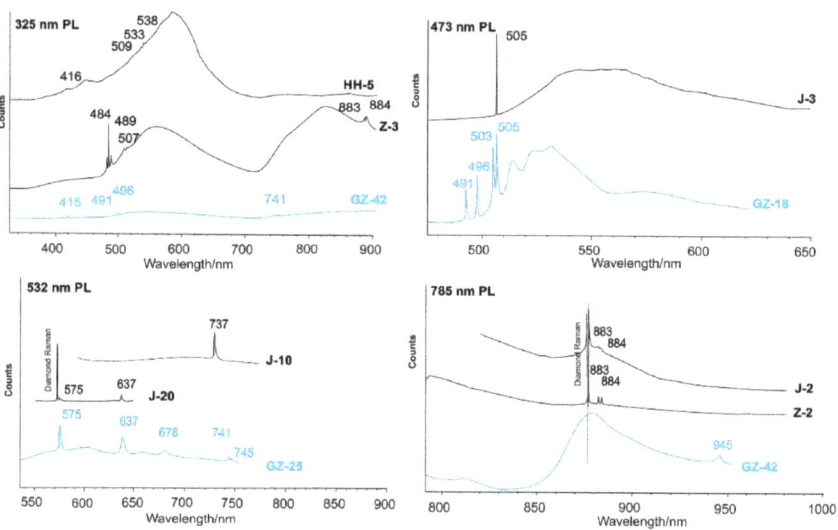

Figure 5. PL Spectroscopy of typical samples (GZ samples in blue color, HPHT samples in black color).

Previous studies indicated which PL peaks are unique to natural diamonds and which are unique to synthetic diamonds [16]. However, with the improvement of HPHT type IIa synthesis technology, impurity defects are decreasing. Some only appeared at 505 and 637 peak positions, or only showed the same PL peaks (693, 694, 737, 883 and 884 nm) as natural diamonds. In this study, both the HPHT and the Guizhou IIa diamonds possessed the H_3, NV^0 and NV^- centers with peaks at 505, 575 and 637 nm, respectively. Additionally, a 737 nm peak (SiV^-) was found in both the HPHT and natural samples [2,20]. Additionally, PL peaks at 693 and 694 nm were found in synthetic diamonds from IIa HPHT in AOTC [17], as well as natural ultra-deep mantle diamonds [21]. Prior studies distinguished between natural and synthetic colorless diamonds by different types (Ia, Ib, IIa) of diamonds, mineral inclusions, fluorescence and spectral characteristics caused by defects from impurities entering the crystal lattice [22]. Similar impurity defects (fluorescence and spectral characteristics) make it increasingly difficult to distinguish between colorless HPHT synthetic and natural diamonds.

We performed PL mapping at the same peak position in HPHT and natural diamonds, with the aim of distinguishing between them. We mapped each sample using a smooth plane's PL, with peak area distribution peaks at 637, 575 and 496 nm. Figure 6 demonstrates the typical samples. HPHT synthetics have an interior morphological structure quite different from that of naturally grown IIa diamonds according to PL mapping (637 nm). An HPHT IIa diamond revealed growth sector distribution, whereas natural diamonds had multiple lines and a cloud-like growth morphology. This underlying mechanism was also useful in studying the difference between HPHT-grown and Natural IIa diamonds during PL mapping (Figure 6A,C). The PL peak intensity is essential for this method; otherwise it is ineffective.

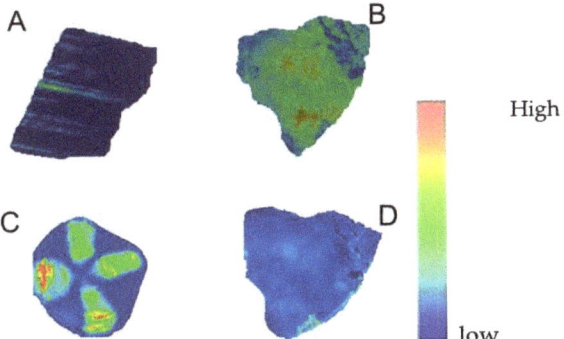

Figure 6. PL mapping images of samples. The baseline-corrected peak area intensity for each laser excitation wavelength. (**A**) Sample GZ-25, the peak area distribution of the 637 nm peak (532 nm excitation). (**B**) Sample GZ-42, the peak area distribution of the 575 nm peak (455 nm excitation). (**C**) Sample J-2, The peak area distribution of the 637 nm peak (532 nm excitation). (**D**) Sample GZ-42, The peak area distribution of the 496 nm peak (455 nm excitation).

3.4. Chemical Composition of Inclusions

Solvents and catalysts were added to decrease the pressure and temperature of Chinese synthetic HPHT diamonds. Among the Chinese catalysts used are the Ni-Mn-Co, Fe-Al Fe-Ni, Fe-Ni-Si, Fe, Fe-S, Fe-Ni–Co, Ni–Mn–Co–Si, and Fe–Ni–B systems that include metal -Fe, Co, Mn, Al, Ni, Pt, Ru, Rh, Pd, Ir, Os, Ta and Cr- and non-metal-N, and B [1,3]. Catalyst materials can cause inclusions in samples. Aside from catalysts, pressure transmitting media and solvents also contribute to HPHT diamond inclusions [1]. As previously found, the inclusions in HPHT diamonds have mainly solid mineral and gas inclusions including graphite, magnetite, carbide Fe_3C, $(FeNi)_{23}C_6$, NiC, $FeSi_2$, FeNi, FeS, SiC, native metals (e.g., F, Co, Ni, Mn); trace elements (e.g., Ca, Al, Si, S, Cr, Cu, Na, Cl, K, Na, Ti, Mg, P, B, and N) and CH_4 and H_2 (gases) [3,11,23,24]. There is evidence that various elements (N, B, H, O, Ni, Co, S, Ti, P, Si, Ge and Sn, etc) can enter the diamond crystal lattice. Nevertheless, Fe3C and (Fe, Ni)3C carbide inclusions were also found in type IIa super-deep diamonds from Juina Rio Soriso, Brazil; Fe–Ni alloy, Fe-alloy and pyrrhotite also occur in these diamonds. [13–15]. Metallic Fe–Ni–C–S melt systems in natural diamonds have been described previously [14].

For inclusions protruding from diamond faces, we performed a chemical analysis with LA–ICP–MS. Using four colorless typical samples, we tested their inclusions, resulting in 26 points of data. The chemical compositions are shown in the Appendix A, with major elements of Fe-Co-Ni and Co-Mn-Ni distributions (see Figure 7), and different products shown as blue, green and gray balls, respectively, in Chinese gem-quality HPHT IIa diamonds. Trace elements in inclusions in HPHT diamonds include B, Ti, Cu, Zn, Ca, S, P, Cr, etc. (see Appendix A).

A previous study demonstrated that Ni, Mn and Co are the predominant elements in Chinese small abrasive diamond inclusion samples (50–250 m) [24]. The Ni content in inclusions of type Ib small abrasive diamonds has obvious advantages. Iron-carbide inclusions in deep-mantle IIa natural diamonds mainly contain Fe elements, as well as small amounts of Ni and lack Co and Mn [15]. The elements B, Ti and Cu are found in colorless HPHT diamonds at a higher quantity than in yellow diamonds. In recent years, Fe-Co and Fe-Co-Ni systems have been mainly used in Chinese colorless HPHT diamonds; B, Ti and Cu may be nitrogen collectors in HPHT diamonds. We compared the composition of the carbide inclusions in natural diamonds and synthetic diamonds. The major difference between natural and synthetic diamonds is the absence of Co and Mn elements in the natural carbide inclusions. Carbide inclusions in natural diamonds were rare and little information was available about them. Ideally, more trace elements from natural IIa diamond carbide inclusions should be examined in the future to compare the differences between HPHT diamonds and natural IIa diamonds.

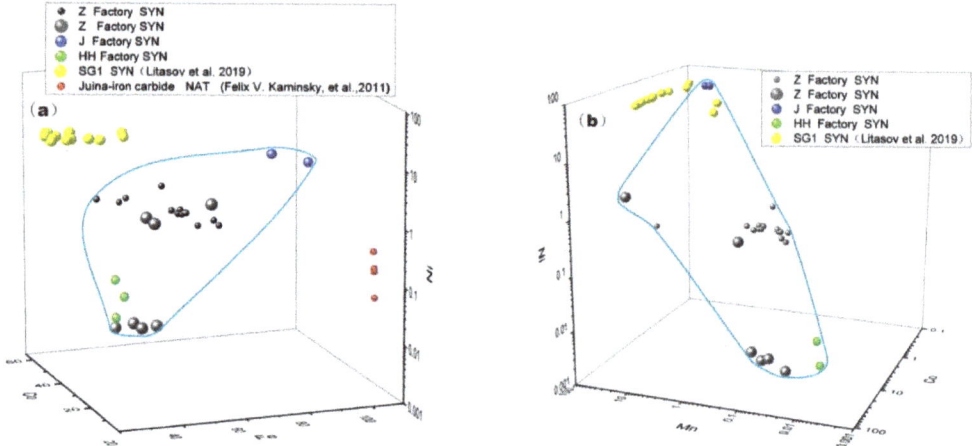

Figure 7. (a) Fe–Co–Ni, (b) Co–Mn–Ni elements plots in inclusions of diamonds, data did normalization operation. Sample numbers are HH-5, HH-20, Z-1, Z-3, J-3 in this study; SG1 are yellow abrasive diamonds, data from Litasov et al. [24]; Julia-iron carbide is a natural super deep diamond, data from Kaminsky et al. [15]; SYN- Synthetic diamond, NAT- Natural diamond; colorless HPHT diamonds characteristics in the blue circle.

4. Conclusions

This study indicates that it is difficult to identify colorless type IIa diamonds, but some details can provide support. The main conclusions are as follows:

(1) Finding characteristic inclusions and testing their Raman spectra to identify IIa diamonds is useful. Different manufacturers have different inclusions in their products due to their different techniques. The kite-like and lichenoid inclusion were not found in natural diamonds, and the abnormal electronic activity characteristics of the lichenoid (tree-like) inclusions in the range of the Raman shift wave-number of 100 to 750 cm^{-1} were clearly different from those of natural diamond inclusions. Furthermore, the HPHT samples did not have healed cracks around the inclusions. Stress differences were also observed around inclusions.

(2) There were no IIa diamonds with characteristic inclusions or Raman spectra of inclusions. PL spectroscopy and mapping can provide evidence for identification. We observed peaks only in Guizhou natural type IIa diamonds at 406 nm, 612 nm, 676 nm, 710 nm, 741 nm, 745 nm and 945 nm. These peaks are missing in HPHT diamonds. The IIa diamond growth environments can be determined by performing particular peak position PL mapping, such as at 503, 505, 694 and 737 nm. These peaks were found in both natural and synthetic HPHT IIa diamonds. HPHT IIa colorless diamond PL mapping revealed the outlines of the growth sectors, whereas natural IIa colorless diamonds had a line and cloud-like growth morphology.

(3) The chemical composition of the iron-carbide inclusion in samples where the inclusion was exposed helped to effectively distinguish HPHT from natural IIa colorless diamonds. The iron carbide inclusions of natural IIa diamonds are not dominated by Co and Mn elements. Trace elements in inclusions in HPHT IIa diamonds include B, Ti, Cu, Zn, Ga, Se, S, P and Zr, etc. We still need more natural samples for comparison of the trace elements in inclusions in natural IIa diamonds.

Thus, the morphology of inclusions and complex Raman spectroscopy of lichenoid (tree-like) inclusion, PL perks, particular peak position PL mapping and the chemical composition of the iron-carbide inclusion can help to distinguish natural diamonds from HPHT IIa colorless diamonds. These tests provided more information than can be determined from a DiamondView image, IR and PL alone. However, for diamonds with no characteristic inclusions and no PL peaks, we suggest that other images such as CL images and X-ray images (XRT) are needed for further analysis.

Author Contributions: Conceptualization, Y.M. and Z.Q.; data curation, Z.S. and T.L.; formal analysis, Y.M.; funding acquisition, X.D. and Z.Q.; investigation, Y.M., T.D., Z.S. and H.L.; methodology, Y.M. and Z.Q.; software, Y.M.; supervision, T.D., H.L. and Z.S.; validation, Y.M., J.W. and W.Z.; visualization, Y.M.; writing—original draft, Y.M.; writing—review and editing, Y.M. and Z.Q. All authors have read and agreed to the published version of the manuscript.

Funding: This research was funded by NSFC (Nos. 41473030 and 42073008).

Institutional Review Board Statement: Not applicable.

Informed Consent Statement: Not applicable.

Data Availability Statement: Not applicable.

Acknowledgments: Thanks to General Manager Weizhang Liang of the Guangzhou Diamond Exchange; and Jianguo Liu, Jingru Shao and Dufu Wang for arranging our factory visits during 2017–2020. The authors also thank anonymous contributors for generously donating diamond samples, and the analysts at the Shandong Bureau of China Metallurgical Geology Bureau. We also would like to thank Thiering Gerg and J.W. Harris for their assistance during the writing process.

Conflicts of Interest: The authors declare no conflict of interest.

Appendix A

Table A1. Trace element compositions of the inclusions in HPHT diamonds.

sample No.	Spot	Fe	Co	Ni	Cu	Zn	B	Na	Mg	Al	Si	P	S	K	Ca	Ti	Cr	Mn
ZN-1	1.00	18722.95	9207.83	108.49	26.49	0.00	394.11	381.85	1966.25	15671.47	0.00	0.00	0.00	1.35	0.00	13488.73	79.33	120.71
ZN-1	2.00	425245.97	286763.11	4462.69	20.33	788.07	290.37	2420.00	251.68	3602.65	2810.52	411.33	899.71	92.24	1289.13	10142.57	112.75	153.15
ZN-1	3.00	341126.35	377834.59	5927.05	38.86	1.87	25.48	19.08	5.00	4584.04	288.80	85.46	8303.31	0.00	0.00	6204.30	44.40	135.64
ZN-1	4.00	291955.62	424632.93	4562.00	28.97	5.16	23.45	157.12	6.52	3330.97	1070.61	77.55	1755.68	74.89	343.42	2248.54	69.26	95.17
ZN-1	5.00	330289.86	390259.89	4744.13	56.04	2.78	48.67	93.10	8.55	3453.61	142.42	34.71	5031.86	2669.60	15012.94	114.91	79.94	200.56
ZN-1	6.00	409077.40	280794.83	4556.43	104.72	4.79	36.90	56.30	7.69	4459.46	2166.45	0.00	0.00	99.95	0.00	2268.67	71.89	66.16
ZN-1	7.00	382435.64	267411.61	4029.91	80.42	30.10	25.11	0.00	4.17	4440.86	29777.87	860.26	736.29	951.93	2758.90	152.41	29.57	35.83
ZN-1	8.00	391707.73	289360.44	4642.30	44.61	19.01	0.00	207.25	0.00	5016.74	10301.09	0.00	0.00	487.78	3691.14	56.73	83.79	89.96
ZN-1	9.00	481099.98	241767.23	4394.33	145.88	12.88	24.14	33.97	2.97	3198.08	0.00	139.19	0.00	0.00	0.00	2466.27	63.16	131.61
ZN-1	10.00	410618.62	266586.04	4561.86	1601.21	1411.06	621.82	4226.40	474.80	6262.01	7084.22	889.36	18713.37	3010.27	6400.09	678.71	72.46	117.90
ZN-1	11.00	434109.29	285879.80	4851.82	30.98	29.43	7.60	0.00	3.62	5306.09	0.00	0.00	13579.38	1026.45	3227.97	52104.71	243.05	120.62
ZN-1	12.00	490450.56	234721.96	3678.41	6.69	619.85	290.80	1805.92	123.16	2383.57	0.00	374.05	1128.02	1369.91	7710.33	4815.41	380.84	177.24
ZN-1	13.00	391844.93	312419.35	12586.81	0.58	620.44	341.90	1333.56	74.66	4571.32	3532.20	127.67	12730.19	1017.74	8485.03	2862.41	575.06	127.13
ZN-1	14.00	410179.80	280336.94	5130.10	78.26	16.46	67.23	186.02	0.00	3143.57	19452.95	0.00	2964.44	223.43	0.00	20678.02	115.93	59.14
ZN-3	1.00	45.37	457987.59	82.56	41447.08	1523.03	9.22	9932.21	405.97	839.12	0.00	258.25	6953.02	368.32	11470.27	15764.10	110.60	76.28
ZN-3	2.00	45.42	387428.42	47.74	17763.66	361.22	26.02	3131.49	127.86	227.71	8746.18	176.40	1434.60	84.70	0.00	49078.02	115.25	38.96
ZN-3	3.00	456165.76	386714.29	44.72	8026.88	22.06	0.00	6.67	6.85	33.41	0.00	6.30	0.00	145.28	0.00	8970.77	189.59	231.03
ZN-3	4.00	399980.98	416146.11	40.30	16723.89	119.14	0.00	176.86	21.09	83.83	0.00	0.00	0.00	133.98	521.82	25865.94	170.90	126.48
ZN-3	5.00	9233.82	7923.87	59.51	22.75	170.02	151.06	404.78	506.06	67122.66	58993.98	1585.44	66812.65	521.35	15324.74	324081.21	47.50	0.00
ZN-3	6.00	2459.78	1252.57	42.94	14.12	72.01	0.00	245.86	803.56	44053.01	55168.43	0.00	172330.18	0.00	0.00	314811.22	26.95	155.53
ZN-3	7.00	29238.30	26965.70	236.74	33.95	134.33	270.57	602.76	3947.10	99587.17	69705.19	2614.04	139507.44	542.01	0.00	359955.68	71.55	28.15
J-3	1.00	1.74	62.18	3197.96	198.82	217.77	82.84	17799.33	13361.19	926027.11	390.98	583.16	16541.57	697.62	77.01	41315.53	178.52	89.63
J-3	2.00	0.56	38.87	2039.09	126.91	519.46	133.02	7544.77	8564.33	939312.41	3336.38	866.11	0.00	0.00	0.00	9.89	55.04	401.05
HH-5	1.00	50.13	424225.41	76.74	22537.87	55.22	16.20	374.55	19.49	279.83	0.00	183.52	0.00	129.67	0.00	0.00	51.84	444.76
HH-5	2.00	56.85	356434.84	88.11	26616.94	195.71	11.94	597.58	223.01	276.97	0.00	93.25	780.53	25.88	0.00	0.00	41.84	419.09
HH-20	1.00	44.54	438244.96	96.49	14054.54	77.61	13.53	293.56	34.53	144.00	338.60	0.00	12908.53	2285.82	15461.07	90.48	37.04	366.48
HH-20	2.00	52.17	366766.51	55.21	16184.89	284.54	0.00	2081.87	171.15	817.33	316.45	0.00	2740.80	3044.31	1998.53	353.62	49.51	408.98

References

1. Luo, X.Y.; Xu, Y.J.; Liu, Y.B. The development and key technological improvement of synthetic diamond in China. *Powder Metall. Ind.* **2016**, *26*, 1–13.
2. Song, Z.H.; Lu, T.J.; Ke, J.; Su, J.; Tang, S.; Gao, B.; Hu, N.; Zhang, J.; Wang, D.F. Identification characteristic of large near-colourless HPHT synthetic diamond from China. *J. Gems Gemmol.* **2016**, *18*, 1–8.
3. Lu, T.; Ke, J.; Lan, Y.; Song, Z.; Zhang, J.; Tang, S.; Su, J.; Dai, H.; Wu, X. Current Status of Chinese Synthetic Diamonds. *J. Gemmol.* **2019**, *36*, 748–757. [CrossRef]
4. Choi, H.; Kim, Y.; Seok, J. Recent trends of gem-quality colorless synthetic diamonds. *J. Korean Cryst. Growth Cryst. Technol.* **2017**, *27*, 149–153.
5. He, X.; Du, M.; Zhang, Y.; Chu, P.K.; Guo, Q. Gemologic and Spectroscopy Properties of Chinese High-Pressure High-Temperature Synthetic Diamond. *JOM* **2019**, *71*, 2531–2540. [CrossRef]

6. Welbourn, C.M.; Cooper, M.; Spear, P.M. De Beers Natural versus Synthetic Diamond Verification Instruments. *Gems Gemol.* **1996**, *32*, 156–169. [CrossRef]
7. Breeding, C.M.; Shen, A.H.; Eaton-Magaña, S.; Rossman, G.R.; Shigley, J.E.; Gilbertson, A. Developments in Gemstone Analysis Techniques and Instrumentation During the 2000s. *Gems Gemol.* **2010**, *46*, 241–257. [CrossRef]
8. Meng, Y.F.; Peng, M.S.; Wang, J.Y.; Huang, W.X.; Yuan, Q.X.; Zhu, P.P. Monochromatic X-ray topography of diamonds with syn-chrotron radiation. *Acta Mineral. Sin.* **2005**, *25*, 353–356.
9. Shiryaev, A.A.; Zolotov, D.A.; Suprun, O.M.; Ivakhnenko, S.A.; Averin, A.A.; Buzmakov, A.V.; Lysakovskyi, V.V.; Dyachkova, I.G.; Asadchikov, V.E. Unusual types of extended defects in synthetic high pressure—High temperature diamonds. *Cryst. Eng. Comm.* **2018**, *20*, 7700–7705. [CrossRef]
10. Muyal, J. Lab Notes: Cleavage System in Pink Diamond. *Gems Gemol.* **2015**, *51*, 324–325.
11. Frezzotti, M.L.; Tecce, F.; Casagli, A. Raman spectroscopy for fluid inclusion analysis. *J. Geochem. Explor.* **2012**, *112*, 1–20. [CrossRef]
12. Smith, E.M.; Wang, W. Fluid CH_4 and H_2 trapped around metallic inclusions in HPHT synthetic diamond. *Diam. Relat. Mater.* **2016**, *68*, 10–12. [CrossRef]
13. Smith, E.M.; Shirey, S.B.; Wang, W. The Very Deep Origin of the World's Biggest Diamonds. *Gems Gemol.* **2017**, *53*, 388–403. [CrossRef]
14. Smith, E.M.; Shirey, S.B.; Nestola, F.; Bullock, E.S.; Wang, J.; Richardson, S.H.; Wang, W. Large gem diamonds from metallic liquid in Earth's deep mantle. *Science* **2016**, *354*, 1403–1405. [CrossRef] [PubMed]
15. Kaminsky, F.V.; Wirth, R. Iron Carbide Inclusions In Lower-Mantle Diamond From Juina, Brazil. *Can. Mineral.* **2011**, *49*, 555–572. [CrossRef]
16. Zaitsev, A.M. *Optical Properties of Diamond: A Data Handbook*; Springer: Berlin/Heidelberg, Germany; New York, NY, USA, 2001; pp. 48–275.
17. D'Haenens-Johansson, U.F.; Moe, K.S.; Johnson, P.; Wong, S.Y.; Lu, R.; Wang, W. Near-Colorless HPHT Synthetic Diamonds from AOTC Group. *Gems Gemol.* **2014**, *50*, 30–45. [CrossRef]
18. Tang, S.; Lu, T.J.; Song, Z.H.; Ke, J.; Zhang, J.; Zhang, J.; Liu, Q.K. Gemological properties and identification of ZhongNan Gem-quality HPHT diamonds. *China Gems* **2017**, *1*, 212–215.
19. Liang, R.; Lan, Y.; Zhang, T.Y.; Lu, T.J.; Chen, M.Y.; Wang, X.Q.; Zhang, X.H. Multi-spectroscopy studies on large grained HPHT synthetic diamonds from Shandong, China. *Spectrosc. Spectr. Anal.* **2019**, *39*, 1840–1845.
20. Breeding, C.M.; Wang, W. Occurrence of the Si-V defect center in natural colorless gem diamonds. *Diam. Relat. Mater.* **2008**, *17*, 1335–1344. [CrossRef]
21. Ma, Y.; Li, H.H.; Zhu, X.X.; Ding, T.; Lu, T.J.; Qiu, Z.L. Lab Notes: D color natural IIa diamond with the walstromite inclusion. *Gems Gemol.* **2018**, *4*, 241–243.
22. Breeding, C.M.; Shigley, J.E. The "Type" Classification System of Diamonds and Its Importance in Gemology. *Gems Gemol.* **2009**, *45*, 96–111. [CrossRef]
23. Guo, M.-M.; Li, S.-S.; Feng, L.; Hu, M.-H.; Su, T.-C.; Gao, G.-J.; Wang, J.-Z.; You, Y.; Nie, Y. The effect of adding Cu on the nitrogen removal efficiency of Ti for the synthesis of a large type IIa diamond under high temperature and high pressure. *New Carbon Mater.* **2020**, *35*, 559–566. [CrossRef]
24. Litasov, K.D.; Kagi, H.; Bekker, T.B.; Hirata, T.; Makino, Y. Cuboctahedral type Ib diamonds in ophiolitic chromitites and perido-tites: The evidence for anthropogenic contamination. *High Press. Res.* **2019**, *39*, 480–488. [CrossRef]

Article

Understanding the Semiconducting-to-Metallic Transition in the CF$_2$Si Monolayer under Shear Tensile Strain

Tarik Ouahrani [1,2,*] and Reda M. Boufatah [1]

1 Laboratoire de Physique Théorique, Université de Tlemcen, Tlemcen 13000, Algeria
2 École Supérieure des Sciences Appliquées, B.P. 165, Tlemcen 13000, Algeria
* tarik.ouahrani@univ-tlemcen.dz

Abstract: With the ever-increasing interest in low-dimensional materials, it is urgent to understand the effect of strain on these kinds of structures. In this study, taking the CF$_2$Si monolayer as an example, a computational study was carried out to investigate the effect of tensile shear strain on this compound. The structure was dynamically and thermodynamically stable under ambient conditions. By applying tensile shear, the structure showed a strain-driven transition from a semiconducting to a metallic behavior. This electronic transition's nature was studied by means of the electron localization function index and an analysis of the noncovalent interactions. The result showed that the elongation of covalent bonds was not responsible for this metallization but rather noncovalent interactions governing the nonbonded bonds of the structure. This strain-tuned behavior might be capable of developing new devices with multiple properties involving the change in the nature of chemical bonding in low-dimensional structures.

Keywords: ab initio calculations; tensile strain; electronic transition; topological analysis of bonds

Citation: Ouahrani, T.; Boufatah, R.M. Understanding the Semiconducting-to-Metallic Transition in the CF$_2$Si Monolayer under Shear Tensile Strain. *Crystals* **2022**, *12*, 1476. https://doi.org/10.3390/cryst12101476

Academic Editor: Dmitri Donetski

Received: 17 September 2022
Accepted: 16 October 2022
Published: 18 October 2022

Publisher's Note: MDPI stays neutral with regard to jurisdictional claims in published maps and institutional affiliations.

Copyright: © 2022 by the authors. Licensee MDPI, Basel, Switzerland. This article is an open access article distributed under the terms and conditions of the Creative Commons Attribution (CC BY) license (https://creativecommons.org/licenses/by/4.0/).

1. Introduction

Strain is ubiquitous in solid-state materials; it can be caused by the melt growth of the material [1], due to the quantum effect driven by the structure [2], or simply by an external application. One fundamental transformation induced by this strain is the change in bonding and structural pattern [3]. Strain applied to a versatile low-dimensional material and also a bulk counterpart could combine several intriguing mechanisms that are still in their infancy. For example, we could obtain the conjunction of metallic (a compound with a vanishing electronic band gap) and semiconducting regions on a monolithic catalytic MX$_2$ nanosheet that could be used to make electronic devices [4,5]. Furthermore, as is the way with pressure, the application of external strains on such structures can become radically different from what they are at ambient pressure, as shown in recent examples with the giant piezoelectricity induced by the mean of pressure in monolayer tellurene [6] and the experimental study showing evidence of an electronic phase transition in molybdenum disulfide [7].

Two-dimensional silicon carbide, a famous member of the 2D family, has been studied extensively owing to its distinctive electronic, optical, and mechanical properties. This quality pushed an increase in the synthesis of new structures based on this material. For instance, graphitic SiC [8], planar SiC$_2$ silagraphene with tetracoordinate Si [9], planar graphitic SiC$_2$ [10], carbon-rich SiC$_3$ [11], g-SiC$_2$ [10], pt-SiC$_2$ [9], SiC$_6$-SW, SiC$_2$-b, SiC$_2$-p [12], quasi-planar tetragonal SiC and SiC$_2$ [13], penta-SiC$_2$ [14], a series of silagraphyne [13], silicon-rich Si$_3$C [12], and the recently reported tetrahex SiC [15] have been experimentally synthesized. All these structures show covalent bonding with strong C–Si bonds. In this contribution, we are aiming to analyze the effect of strain on a buckled geometry, namely the CF$_2$Si structure. Its network also has fluorine atoms on its surface, which functionalizes the C–F bonds and improves structural integrity, surface activity, and processability, opening up new opportunities for catalytic applications. Such a functionalization not only inherits the C–F bonds but also brings about a promise to alter the planar SiC structure, transforming the hybridization from sp^2 to sp^3, and enhancing dispersion, orientation, interaction, and electronic properties [16].

An in-depth understanding of the relationship between chemical structure and macroscopic behavior holds the key to rationalizing the design of new synthetic routes addressing a certain property [17–19]. One of the best possible strategies to unravel these transformations is to look at the bonding patterns, especially in the electronic population [20,21]. Such a link can be analyzed by the so-called localization index as the localization tensor (LT), the localized-electrons detector (LED) [22,23], and the electron localization function (ELF) [24]. With this in mind, in this paper, we look at how the bonding pattern changes from the ambient condition to the highly strained structure. The main goal is to see if the buckled structure of CF_2Si behaves like its planar or bulk SiC counterpart when subjected to no hydrostatic strain/stress deformations. In this manner, the band gap of the title structure can be effectively manipulated through various strategies, including mechanical strain application. The task requires, however, access to the more stable ambient structure. For this purpose, a brief analysis of dynamical, thermal, and thermodynamic stability is done by the density functional theory method. The effect of tensile strain is also analyzed within this strategy. The result gives us insight into whether the modulation vanishing of the electronic band gap is caused by the change in the structure or also a result of the change in the bonding trend.

2. Computational Details

The study of the bonding pattern of ground or strained material required first an analysis of its stability. This task was carried out by means of the density functional theory (DFT), as implemented in the Vienna ab initio simulation package (VASP) [20,25,26]. The projector augmented-wave (PAW) [27] method was employed. Furthermore, the kinetic energy cutoff for the plane-wave basis was converged at 500 eV. We used the exchange potential of Perdew–Burke–Ernzerhof (PBE) [28] weighted by the DFT-D3/BJ approach of the Becke–Jonson damping [29,30], which simulates the van der Waals (vdW) interactions between the adjacent layers. Here, we used a $2 \times 2 \times 1$ supercell with a 20 Å vacuum along the c-direction. A Monkhorst–Pack grid of $12 \times 12 \times 1$ k-points was considered. Structural optimizations were deemed to be converged when all the forces reached their convergence below $F_{tol} = 10^{-6}$ eV/Å. The mechanical properties were estimated within the stress–strain methodology carried through the VASPKIT toolkit [31].

For the calculations of the phonon dispersion of a system, we utilized the direct method as implemented in VASP and interfaced in the Phonopy open-source package [32]. The phonon frequencies were investigated in a $3 \times 3 \times 1$ supercell within a self-consistent way by alternating between the DFT calculations of the Hellmann–Feynman forces acting on atoms displaced from their equilibrium positions and the calculations of improved phonon frequencies and atomic displacement vectors. To achieve calculation convergence, we selected displacements of 0.05 Å as appropriate values. All the topological analyses were carried out with the TOPCHEM2 package [33] using very fine grids of size $300 \times 300 \times 100$.

3. Results and Discussion

3.1. Optimized Structure and Its Stability

The structure under investigation has a space group P3m1 with a trigonal shape (T1), see Figure 1, which means that the lattice parameters a and b are equal. As a result, the optimized lattice parameter was limited to the a parameter, which was calculated to be 3.16 Å. Unfortunately, no experimental data were available to compare with it. As a result, it was necessary to examine the structure's stability as well as its growth conditions. In terms of thermodynamic stability, the chemical potentials of the constituent atoms of the CF$_2$Si low-dimensional structure obeyed several restrictions based on the energetic equilibrium as follows:

$$\begin{aligned}
\Delta\mu_C + 2\Delta\mu_F + \Delta\mu_{Si} &= \Delta E(CF_2Si) = -1.44 \text{ eV} \\
2\Delta\mu_{Si} &\leq \Delta E(Si_2) = 0.64 \text{ eV} \\
2\Delta\mu_C &\leq \Delta E(C_2) = -0.01 \text{ eV} \\
2\Delta\mu_C + 2\Delta\mu_F &\leq \Delta E(C_2F_2) = -0.91 \text{ eV} \\
\Delta\mu_C + 2\Delta\mu_F &\leq \Delta E(CF_2) = -1.39 \text{ eV} \\
2\Delta\mu_F + 2\Delta\mu_{Si} &\leq \Delta E(F_2Si_2) = -1.77 \text{ eV}
\end{aligned} \quad (1)$$

where $\Delta\mu_C$, $\Delta\mu_F$, $\Delta\mu_{Si}$, $\Delta E(Si_2)$, $\Delta E(C_2)$, $\Delta E(CF_2)$, $\Delta E(F_2Si_2)$, and $\Delta E(C_2F_2)$ are the chemical potentials for the computing elements, and $\Delta E(CF_2Si)$ corresponds to the investigated low-dimensional structure

We present the resulting limiting conditions and the intersection points bounding the stability region in Table 1. Figure 2 shows its corresponding diagram. According to this diagram, the title structure competes mainly with the F$_2$Si$_2$ and C$_2$F$_2$ monolayers. We can stipulate this conclusion by taking a number of equations into consideration, considering the constraints to building this structure.

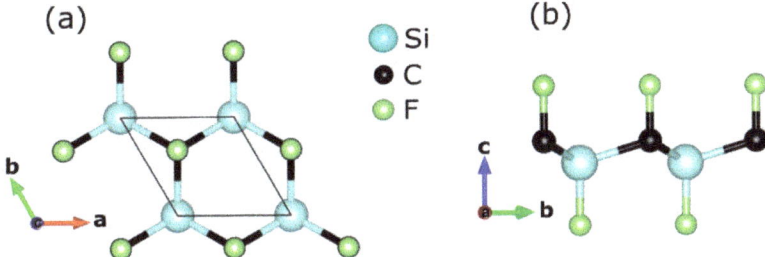

Figure 1. (**a**) The top and (**b**) the side views of CF$_2$Si in its low-dimensional structure.

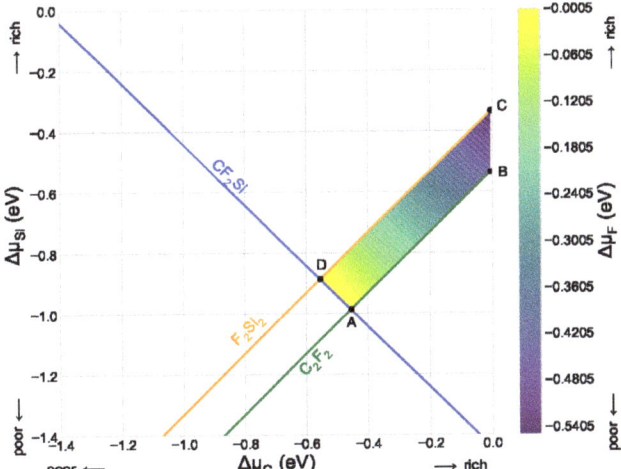

Figure 2. Chemical potential's phase diagram, showing the region of stability of CF$_2$Si in terms of the excess chemical potentials. (Color online) Variation in $\Delta\mu_F$ formation enthalpy as a function of chemical potential, shown within the stability region for the formation of low dimensional CF$_2$Si. Points A, B, C, and D are shown only in order to delimit the zone of stability.

Table 1. The limiting inequalities applied to the independent variables $\Delta\mu_C$ and $\Delta\mu_{Si}$. The point column lists the intersection points bonding the stability region. The corresponding values of the dependent variable $\Delta\mu_F$ and the relevant competing phases are also given. All energies are in eV.

Limiting Inequalities	Point ($\Delta\mu_C, \Delta\mu_{Si}, \Delta\mu_F$)	Competing Phases
$\Delta\mu_C + \Delta\mu_{Si} > -1.4405$	A($-0.4550, -0.9855, 0.0000$)	CF_2Si, C_2F_2
$-\Delta\mu_C + \Delta\mu_{Si} < -0.3295$	B($-0.0050, -0.5355, -0.4500$)	C_2F_2
$2\Delta\mu_C + \Delta\mu_{Si} < -0.0100$	C($-0.0050, -0.3345, -0.5505$)	F_2Si_2
$\Delta\mu_C - \Delta\mu_{Si} < 0.5305$	D($-0.5555, -0.8850, 0.0000$)	CF_2Si, F_2Si_2
$\Delta\mu_C > -1.4405$		
$\Delta\mu_C < 0.0000$		
$\Delta\mu_{Si} < 0.0000$		
$\Delta\mu_{Si} > -1.4405$		

According to all of the above equations and the diagram in Figure 2, the CF_2Si growth is preferred for intersection points bounding along points A (Si-poor, C-rich, F-rich) where the CF_2Si competes. We also analyzed the dynamical stability of the low-dimensional structure in the title. This task was done via the calculation of the phonon dispersion spectrum displayed in Figure 3a. Due to the absence of an imaginary frequency, we can clearly see that the structure is mechanically stable. The shape of the plot seems to share the general scheme of 2D chalcogenide structures. We can note some anomalies in the form of longitudinal acoustic branches, the signature of weak interplanar interactions. According to [34], these acoustic modes have a linear dependence in the **q**-space near the Γ point. The modes at the center of the Γ point, according to the crystal point group, obey the formula $\Gamma = 4A_1 + 4E$, where $\Gamma_{acoustic} = A_1 + E$ and $\Gamma_{optic} = 3A_1 + 3E$. The E mode denotes the double degenerated one. The analysis of the activity of each mode shows that they are both Raman and infrared ones. The calculated wavenumbers for each mode are gathered in Table 2.

Table 2. Calculated modes of 2D CF_2Si. Raman (R) and infrared (IR) modes are indicated.

Symmetry	ω (cm^{-1})	Activity
A_1	494.5	IR/R
A_1	939.2	IR/R
A_1	970.0	IR/R
E	137.2	IR/R
E	171.7	IR/R
E	797.0	IR/R

At the same time, the phonon density of state gathered in Figure 3 shows that the acoustic branches were mainly formed from the lighter atoms, namely the F one, and the branches at high frequency originated from the carbon one. To assess more information on the mechanical properties of the studied structure, we also evaluated the elastic constants of CF_2Si. The results are shown in Table 3. The mechanical stability of the 2D CF_2Si monolayer can be examined by using the elastic constants C_{ij}. For symmetry reasons, we had only two elastic stiffness components, C_{11} and C_{12}, the $C_{66} = (C_{11} - C_{12})/2$. Because CF_2Si is bidimensional, the Born criteria of mechanical stability should satisfy the conditions $C_{11} > 0$ and $C_{66} > 0$ [35–37]. According to the calculated results of C_{ij}, the mechanical stability was satisfied, implying that the investigated structure would eventually be mechanically stable in its low-dimensional structure.

Table 3. Calculated lattice parameters, elastic components, and elastic modulus of CF_2Si monolayer.

a (Å)	C_{11} (N/m)	C_{12} (N/m)	E (N/m)	G_{xy} (N/m)	K (N/m)	ν
3.16086	118.543	22.410	114.306	48.067	70.477	0.189

According to the results of this table, and due to its buckled nature, the compound was rather rigid and had comparable properties to its SiC counterpart in the 2D honeycomb structure [21]. The application of strain on the 2D-SiC honeycomb structure showed a stable structure until 17% of compression. CF_2Si in fact manifested comparable dynamical properties as TH-SiC_2 and TH-SiC structures. C_{11} (119.7 N/m) and C_{12} (26.5 N/m) had values that were close to TH-SiC_2 ((T) and (H) stand, respectively, for tetragonal and hexagonal structures). The main small difference was mainly due to the F atom bonded in the out-plane direction of the Si–C bonds. As a result, the structure was less covalent than a TH-SiC_2, TH-SiC, or h-SiC monolayer and then more ductile along the shear direction. Furthermore, the shear modulus of CF_2Si was 17% lower than that of TH-SiC_2 [38]. This allowed a precise application of strain along the shear direction.

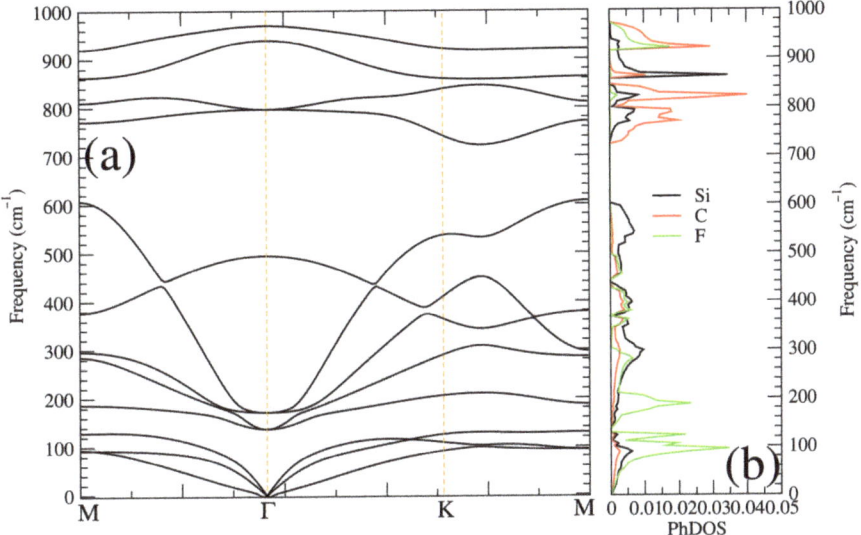

Figure 3. (a) Phonon dispersion plot and its corresponding (b) phonon density of states of CF_2Si monolayer in cm^{-1}.

3.2. Low-Dimensional CF_2Si Structure under Shear Tensile Strain

Starting with the fully relaxed 2D crystal structure CF_2Si, we simulated the shear tensile strain by decreasing the value of the γ angle from 120° to 98° (see Figure 4a,b). The tensile strain was defined as $\varepsilon = (\theta - \theta_0)/\theta_0 \times 100\%$, where θ and θ_0 are the lattice constants of the strained and relaxed structure, respectively. After the application of such a strain, the structure was fully relaxed, keeping the volume and lattice parameters constant, but allowing the relaxation of the internal atomic coordinates. The shear strain was applied step by step until ε reached a value of 98° of the initial $\gamma = 120°$ angle.

To gain a better understanding of the structure's trend under shear deformation, we plotted the evolution of the C–Si bonds as a function of θ in three directions in Figure 4b. The linear compressibility of the d_1 and d_2 axial bonds was correlated. They gave a value of $\kappa = 1.6 \times 10^{-3}$ GPa^{-1}, where the lateral direction d_3 had a linear compressibility of $\kappa = 3.7 \times 10^{-3}$ GPa^{-1}. According to this result, we believe that the enlargement of the d_2 C–Si bonds under shear compression was responsible for the increase of most high-frequency modes and was a consequence of the change in the band-gap nature. Qualitatively, under the harmonic approximation, the stretching mode was proportional to frequency $\omega^{-2/3}$ [39]. Thus, according to our mode assignment, the strongest Raman active mode related to the carbon atoms, and located at a wavenumber of 970.1 cm^{-1}, increased to 1354.86 cm^{-1} under a shear deformation of 18%. This result implied that the structure was rather compressible under shear deformation.

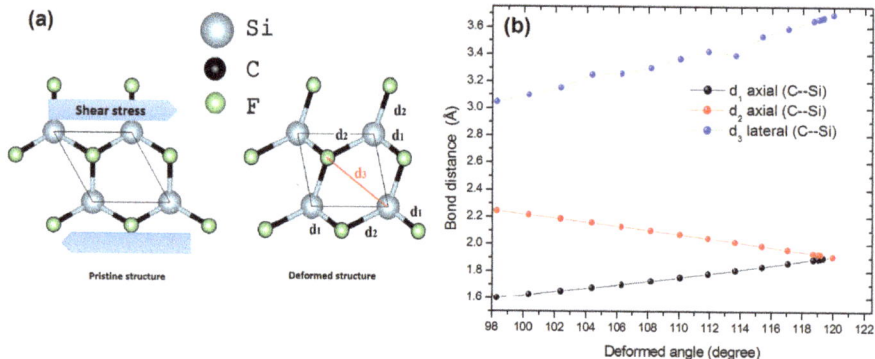

Figure 4. (a) Schematic representation of the pristine and strained (deformed) structure of CF$_2$Si monolayer. (b) Calculated strain dependence of C–Si bond distances. d_1, d_2 represent the axial bonds and d_3 the lateral bonds.

In a subsequent step, we analyzed the effect of an imposed strain on the electronic band gap of the investigated structure. The result is gathered in Figure 5. The calculation of the band gap of the unstrained structure within the Heyd–Scuseria–Ernzerhof hybrid density functional (HSE06) [40] gave a value of 3.31 eV, whereas the PBE gave a band gap of 1.94 eV. The plot shows that with the increase of tensile strain, the band gap decreased and vanished for a strain of 18% on the structure. The new Si–C–Si reached a value of 98°, and this behavior (semiconductor-to-metal transition) was also shown in the application of a zigzag strain on the 2D–SiC [21]. As a result, we suspect a corresponding electronic transformation on the CF$_2$Si structure.

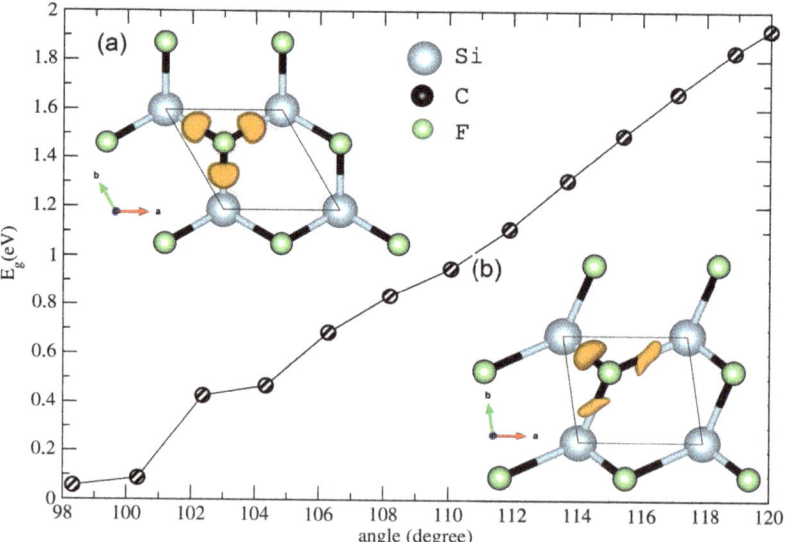

Figure 5. Evolution of the band gap as a function of gamma angle. The inset shows the 3D-ELF isosurface of the structure at (a) (unstrained) 0% and (b) (strained) 18% of the tensile strain.

In order to understand this electronic transformation, we analyzed the nature of both Si–C and C–F bonds for strained and pristine structures. For this task, we performed the electron localization function (ELF) designed by Becke and Edgecombe [24] according to the method proposed by Savin et al. [41]. The ELF topology gives a partition into localized electronic domains known as basins. They are used to rationalize the bonding schemes. The synaptic order of a valence ELF basin is determined by the number of core basins with which they share a common boundary. The basins' spatial locations are very close to the valence-shell electron-pair repulsion domains [42]. According to Savin et al. [41], in the DFT framework, the ELF can be understood as a local measure of the excess of local kinetic energy of electrons, $t_p(\vec{r})$. This quantity is computed by subtracting the bosonic contribution, $|\nabla \rho(\vec{r})|^2/8$, from the kinetic energy density of the system, $t(\vec{r})$. A rescaling of it with respect to the homogeneous electrons gas provides the core of the ELF, $\chi(\vec{r})$:

$$\chi(\vec{r}) = \frac{t_p(\vec{r})}{t_w(\vec{r})} = \frac{t(\vec{r}) - \frac{1}{8}\frac{|\nabla \rho(\vec{r})|^2}{\rho}}{c_F \rho^{5/3}(\vec{r})} \qquad (2)$$

where c_F is the Fermi constant. Accordingly, the regions of electron pairing would have a small χ value. In order to inverse this relationship and map it in a closed interval, the final function was defined as follows:

$$\eta(\vec{r}) = \frac{1}{1 + \chi^2(\vec{r})} \qquad (3)$$

The ELF partition shown in the inset of Figure 5 depicts two cases: the top for pristine structure, which depicts three disynaptic basins ELF–V(Si, C) built at isosurface = 0.95.

These basins had similar shapes and populations (see Table 4), while both basins V(Si,C) and V(C,Si) had $\eta(\vec{r}) = 0.96$ and their population was approximately equal to two electrons, indicating their covalent nature. Table 4 also indicates the existence of a monosynaptic basin V(F) with an ionic nature ($\eta(\vec{r}) = 0.79$), plus a small disynaptic V(C, F) basin ($\eta(\vec{r}) = 0.85$), with a weak electronic population. This behavior was a result of the strong polarization between the in-plane Si–C and C–F bonds. While the charge density (ρ) in the covalent bonds was weak, the ionic one in V(F) was strong. On the other hand, the strained structure showed deformed electronic basins. In the strained structure, the covalent disynaptic V(C, Si) basins also appeared, as well as new monosynaptic ones, for instance, a V(Si) basin with a small $\eta(\vec{r}) = 0.5$, as well as three polarized ionic V(F) basins, where we could show an increase in the electron population. We could also show an increase in the metallicity of the bonds and an extension of the charge density. The asymmetry in the bond population and the electronic volume of the emerged basins were attributed to the strong distortion of the strained structure.

A deeper analysis can untangle this behavior with both the 1D ELF profile and the NCI index. In Figure 6a,b, we displayed the ELF profile of Si–C–F, of unstrained and strained structures in black, the deformed Si–C–F bond of a strained structure in red, and also the lateral direction of Si–C–F of both structures in blue. The 1D ELF profile confirmed that the Si–C bond was covalent and C–F ionic, but it also confirmed the existence of a metallic plateau in the lateral direction of Si–C–F showing a region of very low ELF values with a small hill around $\eta(\vec{r}) = 0.5$. The same trend was noticed whilst studying the bulk SiC under strain [20]. In fact, in the application of tensile strain, the Si–C bond length built from the unit cell of the CF_2Si structure became different. This distortion enhanced the noncovalent interactions in the nonbonded Si and C direction. Given the relevance of this result, we used the same tool as in [20]. The noncovalent interactions' (NCI) isosurface is presented in Figure 6c,d. The noncovalent domain seemed to increase from the unstrained to the strained structure in the nonbonded Si–C–F direction. This region, in fact, did not play any meaningful role in the cohesion of low-dimensional CF_2Si, but rather in the stability and the enhancement of the polarization of bonds. This could be shown in the region of blue surrounding the distorted Si–C bonds of the strained structure, which explained the increase in ionicity and metallicity.

Table 4. Electron localization analysis of a basin of bonding of the CF$_2$Si monolayer. V(Ω), q, bond metallicity ($\rho/\nabla(\rho)$), and ρ stand, respectively, for basin volumes of disynaptic V(X,Y) and monosynaptic V(X) bonds, electronic charges of the bond, calculated bond metallicity and electronic density charge (ρ), and ELF (3, −3) basin.

Basin	V(Ω) (Bohr³)	q (Electrons)	ELF	Bond Metallicity ($\rho/\nabla(\rho)$)	ρ (Atomic Units)
pristine					
V(C,Si)	67.784	−2.1654	0.9617	−0.4906	0.12872651
V(Si,C)	71.167	−2.2023	0.9617	−0.4906	0.12872651
V(F)	57.249	−2.4110	0.7945	−1.2574	2.07097897
V(F,C)	3.137	−0.7520	0.8466	0.2796	0.37723410
strained					
V(C,Si)	19.022	−1.9622	0.9608	−0.3428	0.23750961
V(C,Si)	35.711	−0.6928	0.9252	−0.5682	0.07178674
V(Si,C)	33.542	−1.4756	0.9236	−0.5340	0.09279469
V(C,Si)	33.642	−1.4252	0.9175	−2.4263	0.10376149
V(Si)	52.592	−2.2627	0.4977	−0.3529	2.07237211
V(F)	73.435	−3.3421	0.8837	−0.5505	0.06132580
V(F)	42.653	−3.3972	0.8746	−0.5486	0.05989116
V(F)	48.171	−1.9589	0.8696	0.2769	0.37582815
V(F,C)	3.124	−0.7088	0.8478	0.2791	0.37814979

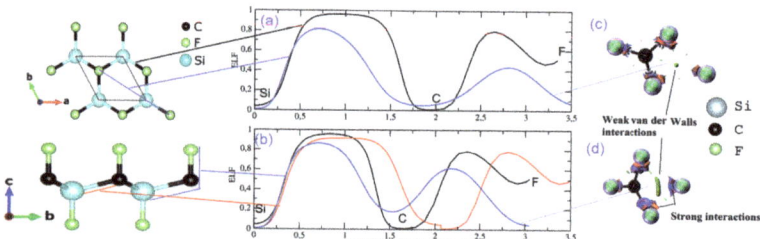

Figure 6. *Left*, one-dimensional ELF profile of bonds for (**a**) pristine and (**b**) strained CF$_2$Si monolayer. The black, blue, and red correspond, respectively, to Si–C–F, deformed Si–C–Si for strained structure, and Si–C–F lateral direction. *Right*, the noncovalent interactions index NCI = 0.3 isosurface of (**c**) pristine and (**d**) strained structures.

Finally, to investigate the relationship between conductivity and shear effect on the low-dimensional CF$_2$Si structure, we used the semiclassical Boltzmann transport theory with a fixed relaxation time approximation to predict the electrical conductivity (σ). This task was done by the use of the BoltzTraP code [43]. We note that, as the procedure was related to the band structure calculation, we took care to increase the grid of the calculation by the use of 37 × 37 × 1 k-points. Figure 7 gives the σ/τ (τ being the relaxation time) at 300 K as a function of the carrier concentration for both pristine (0%) and strained (18%) structures. We can see that the tensile shear strain increased in both the **p**- and **n**-type doped systems. In fact, the increase in σ coincided with that of the band gap. Therefore, the electrical conductivity was efficiently tuned by the strain via band structure engineering. It was noted that the p-type doped region was more affected due to the closing of the band gap.

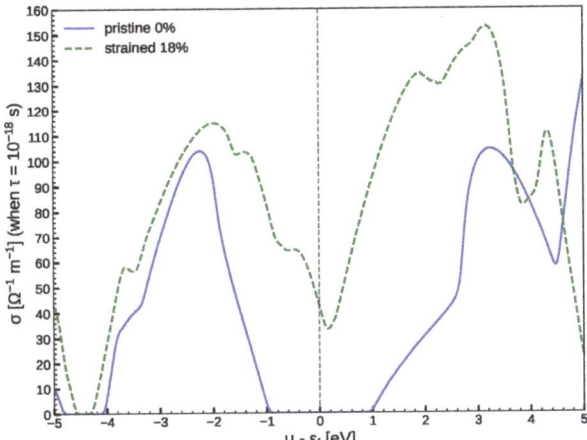

Figure 7. Calculated electrical conductivity of pristine (0%) and strained (18%) structures at ambient temperature. The electrical conductivity was calculated with respect to the relaxation time τ. The transporting directions are not distinguished here.

4. Summary

We showed in this contribution that when the CF_2Si low-dimensional structure was submitted to the application of tensile strain along the shear direction, an electronic transition occurred from the semiconducting phase with a wide electronic band gap to a metallic one with a vanishing band gap. A first attempt to understand this transformation was made within the density functional method calculation, by calculating the stability and mechanical properties of the investigated structure. Furthermore, to understand this mechanism, we used the inherent topological bonding. This task allowed us to understand that this transition was a response to the new repartition of the electronic population and an enhancement of weak noncovalent interactions along the nonbonded directions of the unit cell. Overall, such interactions could be responsible for the relative stability of the structure and also play an important role in controlling the nature of the electronic wave function describing the computed structure. The increase of electric conductivity as a function of strain stipulated the poor performance of the CF_2Si low-dimensional structure as a thermoelectric system.

Author Contributions: Conceptualization, methodology, validation, formal analysis, investigation, writing—original draft preparation, writing—review and editing, supervision, T.O.; DFT calculation, review and editing, R.M.B. All authors have read and agreed to the published version of the manuscript.

Funding: This research received no external funding.

Institutional Review Board Statement: Not applicable.

Informed Consent Statement: Not applicable.

Data Availability Statement: Data are available upon request from the corresponding author.

Acknowledgments: T.O. acknowledges the financial support given by the PRFU, no. B00L02EP13022-0230001.

Conflicts of Interest: The authors declare no conflict of interest.

References

1. Barreteau, C.; Michon, B.; Besnard, C.; Giannini, E. High-pressure melt growth and transport properties of SiP, SiAs, GeP, and GeAs 2D layered semiconductors. *J. Cryst. Growth* **2016**, *443*, 75–80. [CrossRef]
2. Schaack, S.; Depondt, P.; Huppert, S.; Finocc, H.F. Quantum driven proton diffusion in brucite-like minerals under high pressure. *Sci. Rep.* **2020**, *10*, 8123. [CrossRef] [PubMed]
3. Li, W.; Qian, X.; Li, J. Phase transitions in 2D materials. *Nat. Rev. Mater.* **2021**, *6*, 829–846. [CrossRef]
4. Voiry, D.; Yamaguchi, H.; Li, J.; Silva, R.; Alves, D.C.B.; Fujita, T.; Chen, M.; Asefa, T.; Shenoy, V.B.; Eda, G.; et al. Enhanced catalytic activity in strained chemically exfoliated WS_2 nanosheets for hydrogen evolution. *Nat. Mater.* **2013**, *12*, 850–855. [CrossRef] [PubMed]
5. Eda, G.; Fujita, T.; Yamaguchi, H.; Voiry, D.; Chen, M.; Chhowalla, M. Coherent atomic and electronic heterostructures of single-layer MoS_2. *ACS Nano* **2012**, *6*, 7311–7317. [CrossRef]
6. Cai, X.; Ren, Y.; Wu, M.; Xu, D.; Luo, X. Strain-induced phase transition and giant piezoelectricity in monolayer tellurene. *Nanoscale* **2020**, *12*, 167. [CrossRef]
7. Nayak, A.P.; Bhattacharyya, S.; Zhu, J.; Liu, J.; Wu, X.; Pandey, T.N.; Jin, C.; Singh, A.K.; Akinwande, D.; Lin, J.-F. Pressure-induced semiconducting to metallic transition in multilayered molybdenum disulphide. *Nat. Commun.* **2014**, *5*, 3731. [CrossRef]
8. Fleurence, A.; Friedlein, R.; Ozaki, T.; Kawai, H.; Wang, Y.; Yamada-Takamura, Y. Experimental evidence for epitaxial silicene on diboride thin films. *Phys. Rev. Lett.* **2012**, *108*, 245501. [CrossRef]
9. Li, Y.; Li, F.; Zhou, Z.; Chen, Z. SiC_2 silagraphene and its one-dimensional derivatives: Where planar tetracoordinate silicon happens. *J. Am. Chem. Soc.* **2011**, *133*, 900–908. [CrossRef]
10. Zhou, L.J.; Zhang, Y.F.; Wu, L.M. SiC_2 siligraphene and nanotubes: Novel donor materials in excitonic solar cells. *Nano Lett.* **2013**, *13*, 5431–5436. [CrossRef]
11. Ding, Y.; Wang, Y. Geometric and electronic structures of two-dimensional SiC_3 compound. *J. Phys. Chem. C* **2014** *118*, 4509–4515. [CrossRef]
12. Borlido, P.; Huran, A.W.; Marques, M.A.L.; Botti, S. Novel two-dimensional silicon-carbon binaries by crystal structure prediction. *Phys. Chem. Chem. Phys.* **2020**, *22*, 8442–8449. [CrossRef] [PubMed]
13. Fan, D.; Lu, S.; Guo, Y.; Hu, X. Novel bonding patterns and optoelectronic properties of the two-dimensional Si_xC_y monolayers. *J. Mater. Chem. C* **2017**, *5*, 3561–3567. [CrossRef]
14. Xu, Y.; Ning, Z.; Zhang, H.; Ni, G.; Shao, H.; Peng, B.; Zhang, X.; He, X. Anisotropic ultrahigh hole mobility in two-dimensional penta-SiC_2 by strain-engineering: electronic structure and chemical bonding analysis. *RSC Adv.* **2017**, *7*, 45705–45713. [CrossRef]
15. Kilic, M.E.; Lee, K.-R. Tetrahex Carbides: Two-Dimensional Group-IV Materials for Nanoelectronics and Photocatalytic Water Splitting. *Carbon* **2021**, *174*, 15. [CrossRef]
16. Feng, W.; Long, P.; Feng, Y.; Li, Y. Two-Dimensional Fluorinated Graphene: Synthesis, Structures, Properties and Applications *Adv. Sci.* **2016**, *3*, 1500413. [CrossRef]
17. Glass, C.W.; Oganov, A.R.; Hansen, N. USPEX—Evolutionary crystal structure prediction. *Comput. Phys. Commun.* **2006**, *175*, 713. [CrossRef]
18. Martonak, R.; Laio A.; Bernasconi, M.; Ceriani, C.; Raiteri, P.; Parrinello, M. Simulation of structural phase transitions by metadynamics. *Z. Krist.* **2005**, *220*, 489
19. Ouahrani, T.; Merad-Boudia, I.; Baltache, H.; Khenata, R.; Bentalha, Z. Effect of pressure on the global and local properties of cubic perovskite crystals. *Phys. Scr.* **2011**, *84*, 025704 [CrossRef]
20. Guedda, H.Z.; Ouahrani, T.; Morales-García, A.; Franco, R.; Salvado, M.A.; Pertierra, P.; Recio, J.M. Computer simulations of 3C-SiC under hydrostatic and non-hydrostatic stresses . *Chem. Phys. Chem. Phys.* **2016**, *18*, 8132–8139. [CrossRef]
21. Belarouci, S.; Ouahrani, T.; Benabdallah, N.; Morales-García, A.; Belabbas, I. Two-dimensional silicon carbide structure under uniaxial strains, electronic and bonding analysis. *Comp. Mater. Sci.* **2018**, *151*, 288–295. [CrossRef]
22. Bohorquez, H.; Boyd, R.J. A localized electrons detector for atomic and molecular systems. *Theor. Chem. Acc.* **2010**, *127*, 393. [CrossRef]
23. Bohorquez, H.J.; Matta, C.F.; Boyd, R.J. The localized electrons detector as an ab initio representation of molecular structures. *Int. J. Quantum Chem.* **2010**, *110*, 2418. [CrossRef]
24. Becke, A.D.; Edgecombe, K.E. A simple measure of electron localization in atomic and molecular systems. *J. Chem. Phys.* **1990**, *92*, 5397–5404. [CrossRef]
25. Kresse, G.; Furthmüller, J. Efficiency of ab initio total energy calculations for metals and semiconductors using a plane-wave basis set. *Comput. Mat. Sci.* **1996**, *6*, 15–50. [CrossRef]
26. Kresse, G.; Hafner, J. Ab initio molecular dynamics for liquid metals. *Phys. Rev. B* **1993**, *47*, 558–561. [CrossRef]
27. Kresse, G.; Joubert, D. From ultrasoft pseudopotentials to the projector augmented-wave method. *Phys. Rev. B* **1999**, *59*, 1758–1775. [CrossRef]
28. Perdew, J.P.; Burke, K.; Ernzerhof, M. Generalized Gradient Approximation Made Simple. *Phys. Rev. Lett.* **1996**, *77*, 3865–3868. [CrossRef]
29. Grimme, S.; Antony, J.; Ehrlich, S.; Krieg, S. A consistent and accurate ab initio parametrization of density functional dispersion correction $(DFT-D)$ for the 94 elements $H-Pu$. *J. Chem. Phys.* **2010**, *132*, 154104. [CrossRef]

30. Grimme, S.; Ehrlich, S.; Goerigk, L. Effect of the damping function in dispersion corrected density functional theory. *J. Comp. Chem.* **2011**, *32*, 1456–1465. [CrossRef]
31. Wang, V.; Xu, N.; Liu, J.C.; Tang, G.; Geng, W.T. VASPKIT: A User-Friendly Interface Facilitating High-Throughput Computing and Analysis Using VASP Code. *Comput. Phys. Commun.* **2021**, *267*, 108033. [CrossRef]
32. Togo, A.; Tanaka, I. First principles phonon calculations in materials science. *Scr. Mater.* **2015**, *108*, 1–5. [CrossRef]
33. Kozlowski, D.; Pilmé, J. New Insights in Quantum Chemical Topology Studies Using Numerical Grid-based Analyses. *J. Comput. Chem.* **2011**, *32*, 3207. [CrossRef] [PubMed]
34. Ougherb, C.; Ouahrani, T.; Badawi, M.; Morales-García, A. Effect of the sulfur termination on the properties of Hf_2CO_2 MXene. *Phys. Chem. Chem. Phys.* **2022**, *24*, 7243–7252. [CrossRef] [PubMed]
35. Born, M.; Huang, H. *Dynamical Theory of Crystal Lattices*; Oxford University Press: Clarendon, UK, 1954; 432p, ISBN 9780198503699.
36. Mouhat, F.; Coudert, F.X. Necessary and sufficient elastic stability conditions in various crystal systems. *Phys. Rev. B.* **2014**, *90*, 224104. [CrossRef]
37. Sekkal, A.; Benzair, A.; Ouahrani, T.; Faraoun, H.I.; Merad, G.; Aourag, H. Mechanical properties and bonding feature of the YAg, CeAg, HoCu, LaAg, LaZn, and LaMg rare-earth intermetallic compounds: An ab initio study. *Intermetallics* **2014**, *45*, 65–70. [CrossRef]
38. Wei, Q.; Yang, Y.; Yang, G.; Peng, V. New stable two dimensional silicon carbide nanosheets. *J. Alloys Compd.* **2021**, *868*, 159201. [CrossRef]
39. Chen, G.; Haire, R.G.; Peterson, J.R. Compressibilities of $TbVO_4$ and $DyVO_4$ Calculated from Spectroscopic Data. *Appl. Spectrosc.* **1992**, *46* 1495–1497. [CrossRef]
40. Heyd, J.; Scuseria, G.E.; Ernzerhof, M. Hybrid functionals based on a screened Coulomb potential. *J. Chem. Phys.* **2003**, *118*, 8207–8215 . [CrossRef]
41. Savin, A.; Jepsen, O.; Flad, J.; Anderson, L.K.; Preuss, H.; von Schnering, H.G. Electron localization in solid-state structures of the elements: The diamond structure. *Angew. Chem. Int. Ed. Engl.* **1992**, *32*, 187–188. [CrossRef]
42. Gillespie, R.J.; Nyholm, R.S. Inorganic stereochemistry. *Quart. Rev.* **1957**, *11*, 339-380. [CrossRef]
43. Madsen, G.K.H.; Singh, D.J. BoltzTraP. A code for calculating band-structure dependent quantities. *Comput. Phys. Commun.* **2006**, *175*, 67–71. [CrossRef]

Article

Effects of Physical and Chemical Pressure on Charge Density Wave Transitions in LaAg$_{1-x}$Au$_x$Sb$_2$ Single Crystals

Li Xiang [1,†], Dominic H. Ryan [1,2], Paul C. Canfield [1] and Sergey L. Bud'ko [1,*]

[1] Ames National Laboratory, Department of Physics and Astronomy, Iowa State University, Ames, IA 50011, USA
[2] Physics Department and Centre for the Physics of Materials, McGill University, 3600 University Street, Montreal, QC H3A 2T8, Canada
[*] Correspondence: budko@ameslab.gov
[†] Current address: National High Magnetic Field Laboratory, Florida State University, Tallahassee, FL 32310, USA.

Abstract: The structural characterization and electrical transport measurements at ambient and applied pressures of the compounds of the LaAg$_{1-x}$Au$_x$Sb$_2$ family are presented. Up to two charge density wave (CDW) transitions could be detected upon cooling from room temperature and an equivalence of the effects of chemical and physical pressure on the CDW ordering temperatures was observed with the unit cell volume being a salient structural parameter. As such LaAg$_{1-x}$Au$_x$Sb$_2$ is a rare example of a non-cubic system that exhibits good agreement between the effects of applied, physical, pressure and changes in unit cell volume from steric changes induced by isovalent substitution. Additionally, for LaAg$_{0.54}$Au$_{0.46}$Sb$_2$ anomalies in low temperature electrical transport were observed in the pressure range where the lower charge density wave is completely suppressed.

Keywords: charge density wave; pressure; chemical substitution; resistivity; LaAgSb$_2$; LaAuSb$_2$

1. Introduction

Physical (hydrostatic) pressure and chemical substitution are two common ways to tune the physical properties of materials. Whereas hydrostatic pressure is considered to be a clean parameter that does not introduce additional disorder, as well as changes in band filling in many cases, the experimental techniques available under pressure are more limited. Chemical substitution necessarily involves some additional disorder. In the case of aliovalent substitutions the corresponding electron- or hole- doping effects are often dominant. For isovalent substitution the primary effect is thought to be steric and the comparison with physical pressure can be more relevant. Whereas such isovalent substitutions can be referred to as "chemical pressure" differences in how pressure and substitution affect a compound, especially a non-cubic one, can be greater than similarities in some cases. It is of particular importance when there is a desire to stabilize a particular high pressure phase/state (like high temperature superconductivity [1]) using chemical and/or physical pressure. Additionally, observation of an apparent equivalence [2–4] or non-equivalence [5] of chemical and physical pressure can help in understanding of structure—property relations and in recognizing relevant structural motifs.

The members of the family of compounds chosen for this study, LaAg$_{1-x}$Au$_x$Sb$_2$, demonstrate charge density wave (CDW) transitions, or spontaneous superstructures formed by electrons [6]. Decades ago Peierls showed the instability of a (one-dimensional) metal interacting with the lattice towards a lattice distortion and the opening of a gap in the electronic spectrum [7]. This concept is often applied to CDW formation in low-dimensional materials, although alternatives are widely discussed [8–10]. Studies of CDW phenomena in solids and competition of CDW with other collective phenomena remain one of the active subfields of quantum materials research.

In this work, we study effects of pressure on single crystals of selected members of the LaAg$_{1-x}$Au$_x$Sb$_2$ family which are then compared to those of chemical pressure implemented via Ag ↔ Au substitution. The end-compounds, LaAgSb$_2$ and LaAuSb$_2$ were first synthesized almost three decades ago [11,12] and were reported to crystallize in the same tetragonal ZrCuSi$_2$-type structure (P4/nmm, No. 129). CDW transitions were observed in electrical transport at ~210 K and ~100 K for LaAgSb$_2$ and LaAuSb$_2$, respectively, refs. [13,14]. Synchrotron X-ray scattering study [15] and further thermodynamic measurements [16] identified second, lower temperature, CDW transition at ~185 K in LaAgSb$_2$. Similarly, in addition to T_{CDW1} ≈ 110 K, a second CDW transition at T_{CDW2} ~90 K was detected by electrical transport measurements in near stoichiometric LaAuSb$_2$ [17]. The synthesis and evolution of the higher temperature, T_{CDW1} in the LaAg$_{1-x}$Au$_x$Sb$_2$ series was reported in Ref. [18] but without any measurements of T_{CDW2} or companion applied pressure studies.

For LaAgSb$_2$, pressure reportedly suppressed T_{CDW1} [19–23] as well as T_{CDW2} [21,23] with the results being consistent in majority of publications [19–21]. Qualitatively similar behavior under pressure was also observed for LaAuSb$_2$ [17,24,25]. Additionally, in LaAgSb$_2$ and LaAuSb$_2$, low temperature superconductivity was discovered and studied under pressure [23,24]. It has to be noted that the exact Au stoichiometry in LaAu$_x$Sb$_2$ depends on details of the synthesis and affects both the ambient pressure values of T_{CDW1} and T_{CDW2} and their pressure derivatives [17].

Earlier comparison of the effects of pressure and chemical substitution in this family of materials [19] was based on a study of the La$_{1-x}$R$_x$AgSb$_2$ series (R = Y, Ce, Nd, Gd) and a significant contribution of disorder prevailing in the case of substitution was found, thus resulting in a significant difference between physical and chemical pressure. In this work, we address the same question in the different, transition metal site, substitution series.

2. Materials and Methods

Single crystals of LaAg$_{1-x}$Au$_x$Sb$_2$ were grown from an antimony-rich self-flux following the method described in Refs. [13,17,26]. Pure elements were loaded into an alumina Canfield crucible set [27] which was placed into an amorphous silica tube and sealed in partial atmosphere of argon. The sealed tubes were heated to 1050 °C over 10 h, held for 8 h, then cooled to 800 °C over a period of 10 h prior to starting the crystal growth. Crystal growth occurred during the 100 h cooling from 800 °C to 670 °C, after which the excess flux was decanted with the aid of a centrifuge.

In this work, crystals of LaAg$_{1-x}$Au$_x$Sb$_2$ with nominal compositions x = 0, 0.25, 0.5, 0.75 were grown with the initial La:T:Sb (T = Ag$_{1-x}$Au$_x$) growth compositions: 1:2:20 (T2). To investigate whether reported Au deficiency [14,17,18] is relevant and can be tuned for the intermediate Au concentrations in LaAg$_{1-x}$Au$_x$Sb$_2$, for x = 0.25, 0.75 the growth composition of 1:6:20 (T6) was used as well. For the end compound, LaAu$_x$Sb$_2$, the data from the recent Ref. [17] are used when appropriate.

Cu-K$_\alpha$ X-ray diffraction patterns were taken using a Rigaku Miniflex-II diffractometer. The crystals were ground and the powder was mounted on a low-background single-crystal silicon plate using a trace amount of Dow Corning silicone vacuum grease. The mount was spun during data collection to reduce possible effects of texture. Data taken for Rietveld refinement were collected in two overlapping blocks: 10° ≤ 2θ ≤ 48° and 38° ≤ 2θ ≤ 100°, with the second block counted for 4–5 times longer than the first to compensate for the loss of scattered intensity at higher angles due to the X-ray form factors. The two data blocks for each sample were co-refined within GSAS [28,29] using a single set of structural and instrumental parameters but with independent scale factors to allow for the different counting times used. Parameters for both the primary phase and any impurity were refined. We found that the materials were easy to grind into a random powder and no texture or preferential orientation effects were observed in the residuals. The diffractometer and analysis procedures were checked using Al$_2$O$_3$ (SRM 676a [30]); our fitted values of a = 4.7586(2) Å and c = 12.9903(7) Å were both 1.6(4) × 10^{-4} Å smaller than the values on the certificate [30], suggesting a small but statistically significant mis-calibration of the instrument. The fitted lattice parameters given in the analysis that follows do not include this correction.

Chemical analysis of the crystals was performed using an Oxford Instruments energy-dispersive X-ray spectroscopy (EDS) system on a Thermo Scientific Teneo scanning electron microscope. The measurements were performed on polished ab surfaces of single crystals with four to eight points taken for every sample.

Standard, linear 4-probe ac resistivity was measured on bar—shaped samples of LaAg$_{1-x}$Au$_x$Sb$_2$ in two arrangements: $I||ab$ and, when needed, $I||c$. The size of the samples was 1.5–2 mm length, 0.2–0.4 mm width and about 0.1 mm thickness. The frequency used was 17 Hz, typical current values were 3 mA for in-plane electrical transport and 5 mA for the c-axis measurements. The contact resistances between the leads and the samples were below 1 Ω. Based on our experience with the LaAu$_x$Sb$_2$ samples with similar size and contact resistance [17], we do not expect heating effects to be observed either at ambient pressure or in the pressure cell environment. The measurements were performed using the ACT option of a Quantum Design Physical Property Measurement System (PPMS).

For selected samples, resistivity measurements under pressure were performed in a hybrid, BeCu/NiCrAl piston-cylinder pressure cell (modified version of the one used in Ref. [31]) in the temperature environment provided by a PPMS instrument. A 40:60 mixture of light mineral oil and n-pentane was used as a pressure-transmitting medium. This medium solidifies at room temperature in the pressure range of 30–40 kbar, [4,31,32] which is above the maximum pressure in this work. Elemental Pb was used as a low temperature pressure gauge [33]. It has been shown that in piston-cylinder pressure cells the value of pressure depends on temperature (see Ref. [34] for mineral oil:n-pentane pressure medium and this particular design of the cell). Below we use the Pb gauge pressure value. Given that the upper transition for LaAgSb$_2$, highest in the series, is at ambient pressure at ~200 K, this may give rise to pressure differences with the values determined by Pb gauge by at most 2 kbar.

3. Results

3.1. Structure and Substitution

The X-ray diffraction patterns for all LaAg$_{1-x}$Au$_x$Sb$_2$ samples were fitted using the GSAS/EXPGUI packages [28,29]. Small amounts of residual flux were generally observed as impurity phases and were included in the fits as necessary. Figure 1 shows a typical X-ray diffraction data set for the T2 growth of LaAg$_{0.75}$Au$_{0.25}$Sb$_2$ with ~1 wt.% Sb as impuritiy. In the fit, the occupations of the La, Sb1, and Sb2 sites as well as the total occupation of the T = Ag/Au 2b site were fixed as 1, whereas the Ag/Au ratio was allowed to vary. As the parameter that actually contributes to the scattering from a given site in the structure is the average scattering length for that site, it was not meaningful to refine both the Au/Ag ratio and a possible vacancy level using a single measurement (our X-ray diffraction patterns) of the average scattering length. The same (reduced) average scattering could be constructed from Au-only + some level of vacancy, a fully occupied site with Au + some Ag, or some appropriate, and continuously variable combination of Au + Ag + vacancy. The results from Rietveld analysis of the powder X-ray data for LaAgSb$_2$ and five LaAg$_{1-x}$Au$_x$Sb$_2$ samples are listed in Tables A1 and A2 in the Appendix A. The EDS results for LaAgSb$_2$ and four LaAg$_{1-x}$Au$_x$Sb$_2$ samples are presented in Table A3 in the Appendix A. The values in the table are the average of the measurements taken at between four and eight different places on the samples' surfaces, standard deviations are listed in the parentheses.

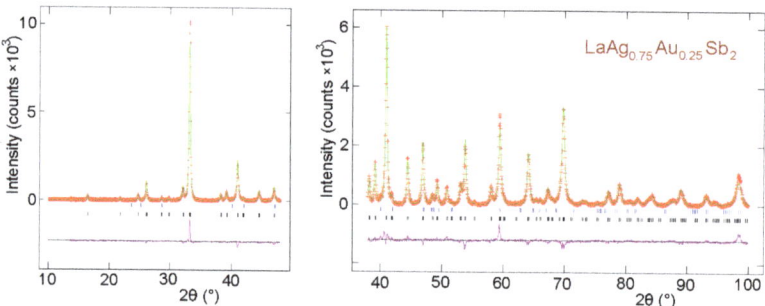

Figure 1. Cu-K$_\alpha$ X-ray diffraction patterns for the T2 growth of LaAg$_{0.75}$Au$_{0.25}$Sb$_2$ showing the two overlapping data blocks that were co-fitted using the GSAS/EXPGUI packages [28,29]. The red points are the data and the green lines show the fits with the residuals shown below each fitted pattern. The Bragg markers show the positions of the reflections from (top) Sb, and (bottom) LaAg$_{1-x}$Au$_x$Sb$_2$.

Analysis of the X-ray diffraction as well as EDS results show (Figure 2) that the measured Ag/Au ratio deviates from the nominal with the experimental points for x_{meas} being slightly below the $x_{meas} = x_{nom}$ line with $x_{meas}/x_{nom} = 0.88 \pm 0.02$ and 0.90 ± 0.03 for X-ray diffraction and EDS results, respectively. In the rest of the text we will use x-values determined from the X-ray diffraction.

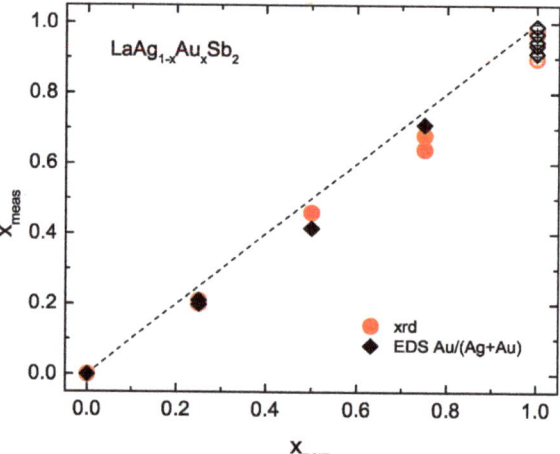

Figure 2. Measured vs. nominal values of Au concentration x in LaAg$_{1-x}$Au$_x$Sb$_2$ (filled symbols). Double filled red circles for $x_{nom} = 0.25$ (not clearly discerned on the plot) and 0.75, as well as double filled black rhombi for $x_{nom} = 0.25$ correspond to T2 and T6 growths, see Section 2. These data are presented in the Appendix A in a tabular form (Tables A2 and A3). For example, of two red circles at $x_{nom} = 0.75$, higher corresponds to T2 and lower to T6 growth. Data for LaAu$_x$Sb$_2$ [17] (open symbols) are added for the reference, here again multiple symbols correspond to different initial La:Au:Sb growth compositions [17]. Dashed line corresponds to $x_{meas} = x_{nom}$.

The lattice parameters, unit cell volume and the c/a ratio as a function of Au substitution are presented in Figure 3. All these quantities have an approximately linear dependence of x, in good agreement with Ref. [18].

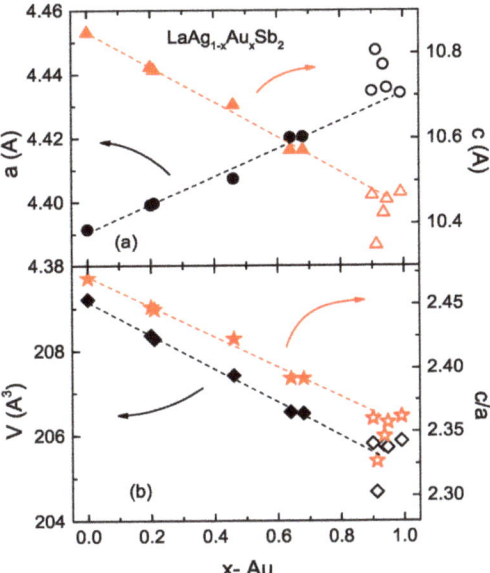

Figure 3. (a) Lattice parameters, (b) unit cell volume and c/a ratio vs. Au concentration in LaAg$_{1-x}$Au$_x$Sb$_2$ determined from Rietveld refinement (filled symbols). Data for LaAu$_x$Sb$_2$ [17] (open symbols) are added for the reference. Dashed lines are guide to the eye.

3.2. CDW at Ambient Pressure

Whereas in the case of LaAu$_x$Sb$_2$ the CDW temperatures were significantly affected by initial growth compositions [17], this appears to be not so critical for LaAg$_{1-x}$Au$_x$Sb$_2$ with the intermediate Au compositions. For the nominal LaAg$_{0.75}$Au$_{0.25}$Sb$_2$ and LaAg$_{0.25}$Au$_{0.75}$Sb$_2$ samples the difference between T2 and T6 initial compositions in the XRD-refined Au concentrations is 0.01–0.04 (5–6%) (Table A2) and in the CDW ordering temperatures 3–6 K (2–5%), with the difference, not surprisingly, being larger for the latter samples with higher Au concentration. The normalized in-plane resistivity and the CDW ordering temperatures for T2 and T6 samples of LaAg$_{0.75}$Au$_{0.25}$Sb$_2$ and LaAg$_{0.25}$Au$_{0.75}$Sb$_2$ are shown in Figure 4a,b, respectively.

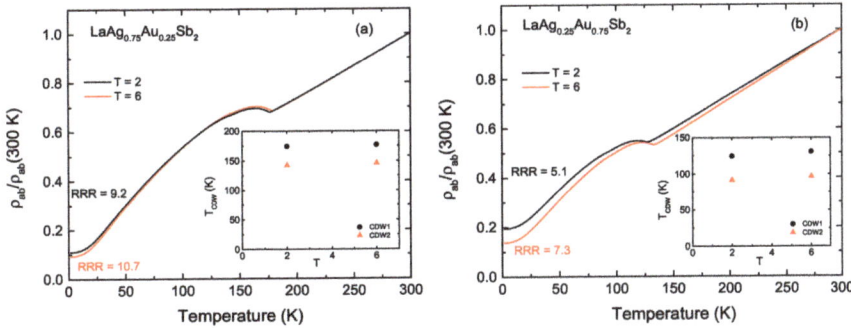

Figure 4. Normalized in-plane resistivity, $\rho_{ab}/\rho_{ab}(300\ K)$ and the ordering temperatures CDW1 and CDW2 for T2 and T6 samples with nominal compositions (a) LaAg$_{0.75}$Au$_{0.25}$Sb$_2$ and (b) LaAg$_{0.25}$Au$_{0.75}$Sb$_2$.

The overall evolution of the in-plane resistivity of LaAg$_{1-x}$Au$_x$Sb$_2$ is shown in Figure 5. CDW transition temperatures decrease with Au substitution. The suppression of T_{CDW1} is in fair agreement with the prior results [18]. The ambient pressure $x - T$ phase diagram based on the data of Figures 4 and 5 is presented in Figure 6. The T_{CDW} values were determined from extrema in the $d\rho_{ab}/dT$ data; an example of which is shown in the inset to Figure 5.

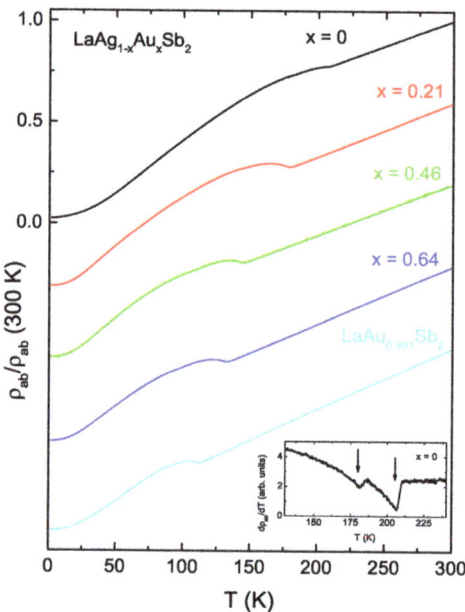

Figure 5. Normalized in-plane resistivity for LaAg$_{1-x}$Au$_x$Sb$_2$. Data are vertically shifted for clarity. The curve for LaAu$_{0.991}$Sb$_2$ [17] is added for the reference. The inset shows an example of $d\rho_{ab}/dT$ for LaAgSb$_2$ with two CDW transitions marked.

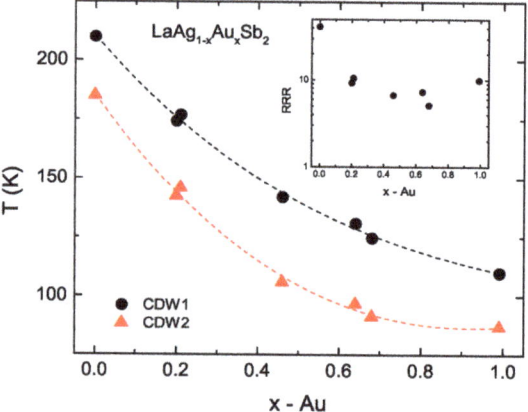

Figure 6. Transition temperatures, CDW1 and CDW2, as a function of x in LaAg$_{1-x}$Au$_x$Sb$_2$. Data are vertically shifted for clarity. Dashed lines are guide for the eye. The inset shows residual resistivity ratio, $RRR = \rho_{ab}(300\ K)/\rho_{ab}(1.8\ K)$ as a function of x-Au.

On going from LaAgSb$_2$ to LaAuSb$_2$, based on electrical transport data, both CDW transitions persist, but both of them are suppressed by ∼100 K without a significant change of the value of $T_{CDW1} - T_{CDW2}$. The $T_{CDW1}(x)$ and $T_{CDW2}(x)$ behavior has an upward curvature. Most likely the disorder induced by substitution contributes to additional suppression of CDW transition temperatures, although to the extent significantly smaller than, e.g., in 2H—TaSe$_{2-x}$S$_x$ [35]. The presence of substitutional disorder is, expectedly, seen in the evolution of residual resistivity ratio ($RRR = \rho_{ab}(300\ \text{K})/\rho_{ab}(1.8\ \text{K})$) with Au substitution (Figure 6, inset), which shows a broad local minimum for intermediate substitution values. Similar moderate but visible effect of substitutional disorder was observed in studies of superconducting transition temperature in Y$_x$Lu$_{1-x}$Ni$_2$B$_2$C [36]. A clearer example can be found in a similar, isoelectronic substitution in the Mn(Pt$_{1-x}$Pd+x)$_5$P series [37].

3.3. CDW under Pressure

Note that often (see above) both CDW1 and CDW2 are reasonably well discerned in the derivatives of in-plane resistivity, with the feature associated with CDW2 being less pronounced. Having in mind that (at least in LaAgSb$_2$ [15]) the CDW2 wavevector is along the c-axis, the ρ_c measurements provide better identification of the T_{CDW2} with T_{CDW1} still being strong. Therefore the measurements for LaAgSb$_2$ under pressure were performed in $I\|c, H\|ab$ geometry.

Main panel of Figure 7 presents c-axis resistivity data taken for LaAgSb$_2$ at different pressures. The overall resistivity is suppressed as pressure increases. The CDW transitions are moving down in temperature. The insets help to quantify above statement. The relative change of the c-axis resisitivity under pressure, is similar for both temperatures presented, 300 K and 50 K, $1/\rho_c(0) \cdot d\rho_c/dP = -(0.012-0.014)$ kbar^{-1}. Both CDW temperatures decrease under in a close to linear fashion, with the derivatives $dT_{CDW1}/dP = -4.6 \pm 0.2$ K/kbar and $dT_{CDW2}/dP = -7.0 \pm 0.3$ K/kbar. Simple, linear, extrapolation suggests that CDW2 will be suppressed to 0 K at ~25 kbar and CDW1 at ~43 kbar. The observed T_{CDW} derivatives are consistent with the published values of $-(4.3-5.1)$ K/kbar for CDW1 [19–21] and -8.0 K/kbar for CDW2 [21].

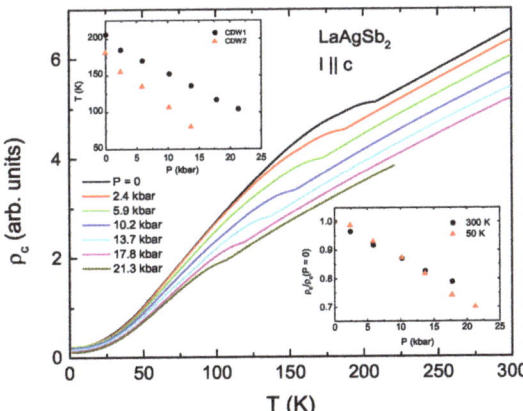

Figure 7. Temperature-dependent c-axis resistivity of LaAgSb$_2$ measured at different applied pressures. Upper inset: CDW transition temperatures, T_{CDW1} and T_{CDW2}, as a function of pressure. Lower inset: normalized resistivity at 300 K and 50 K as a function of pressure.

In order to extend the pressure dependence of T_{CDW1} and T_{CDW2} across the substitutional series, a similar data set (but for ρ_{ab}) for LaAg$_{0.54}$Au$_{0.46}$Sb$_2$ (nominal LaAg$_{0.5}$Au$_{0.5}$Sb$_2$) is presented in Figure 8. The relative change in the in-plane resistivity at 300 K and 200 K is $1/\rho_{ab}(0) \cdot d\rho_{ab}/dP = -(0.009-0.01)$ kbar^{-1}, the same as that in LaAgSb$_2$ [20]. The initial pressure derivatives of the CDW transitions are $dT_{CDW1}/dP = -4.9 \pm 0.1$ K/kbar and $dT_{CDW2}/dP = -9.4 \pm 0.1$ K/kbar. For CDW2 simple, linear, extrapolation yields ~12.5 kbar as a critical pressure of complete suppression of CDW2. Since we cannot observe any distinguishable feature in $d\rho_{ab}/dT$ data at 9.3 kbar below 50 K (see Figure A1 in the Appendix B), it is possible that the $T_{CDW2}(P)$ behavior is super-linear and the critical pressure for CDW2 is lower than ~12.5 kbar obtained from the linear extrapolation. Alternatively, the feature associated with CDW2 could be suppressed so much, that it cannot be detectable within our signal-to-noise ratio and digital differentiation. The data in Figure 9 potentially favors the former possibility. $T_{CDW1}(P)$ dependence has some curvature, the data extrapolate to the value of the critical pressure of ~26 kbar.

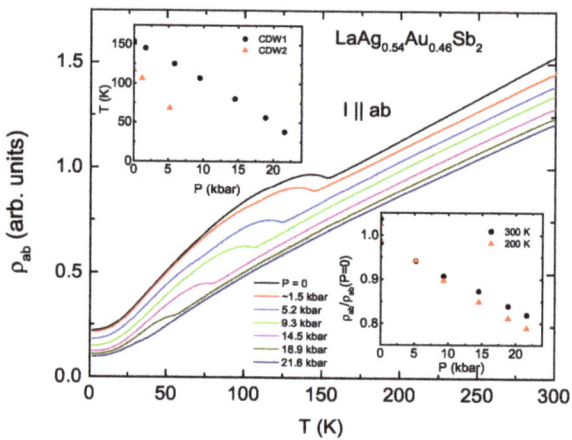

Figure 8. Temperature-dependent in-plane resistivity of LaAg$_{0.54}$Au$_{0.46}$Sb$_2$ measured at different applied pressures. Upper inset: CDW transition temperatures, T_{CDW1} and T_{CDW2}, as a function of pressure. Lower inset: normalized resistivity at 300 K and 200 K as a function of pressure. Note, some minor corrections of pressure values, following Ref. [17] were applied.

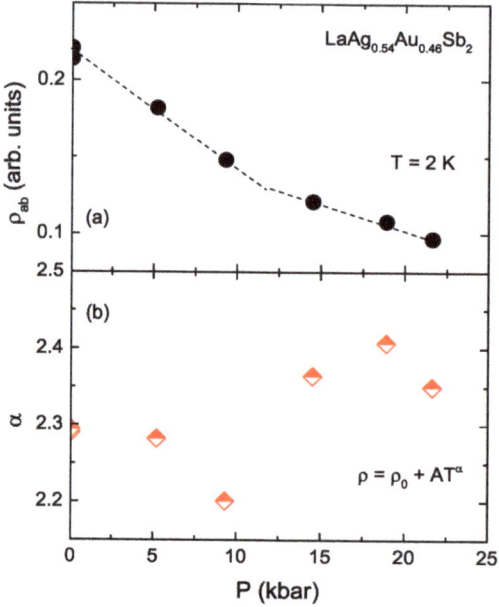

Figure 9. Pressure dependence of (**a**) in-plane resistivity at 2 K (dashed lines are guide to the eye); (**b**) exponent α in $\rho = \rho_0 + AT^\alpha$ fit of low temperature resistivity (fit performed between 1.8 K and 20 K) for LaAg$_{0.54}$Au$_{0.46}$Sb$_2$.

Since for LaAg$_{0.54}$Au$_{0.46}$Sb$_2$ CDW2 appears to be suppressed to 0 K within our pressure range, we examine if this suppression has any bearing on the low temperature electrical transport. Indeed, the zero applied field data in Figure 9a,b show changes in behavior near 9.3 kbar with the power law exponent, α, having the clearest signature of a possible transition near the 9.3 kbar. So most probably the value of the critical pressure for CDW2 is around 9.3 kbar. The changes observed are rather subtle, however the features associated with CDW suppression in LaAgSb$_2$ [21] and LaAu$_x$Sb$_2$ [17] were subtle as well.

4. Discussion

The pressure dependence of the CDW transitions of LaAgSb$_2$, LaAg$_{0.54}$Au$_{0.46}$Sb$_2$ and LaAu$_x$Sb$_2$ samples is shown in Figure 10. For different members of the family the behavior is very similar. It is noteworthy that the CDW2 suppression rates are almost a factor of 2 higher than those for CDW1. This is possibly due to different effect of pressure on the nesting features along a- and c-axes (note that for LaAgSb$_2$ the CDW wave-vectors were found to be (0.026 0 0) and (0 0 0.16) for CDW1 and CDW2, respectively, [15]).

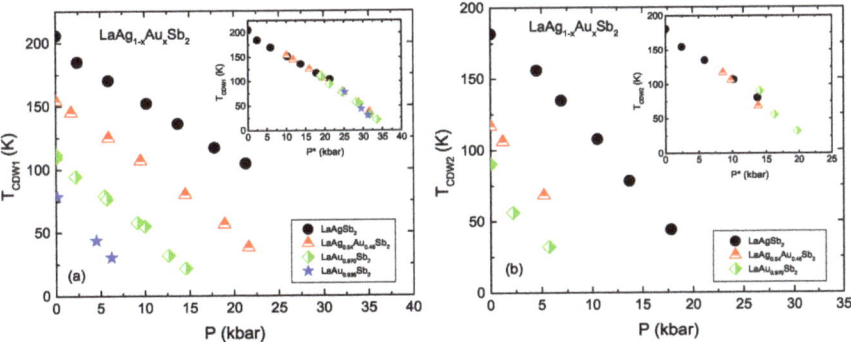

Figure 10. Pressure dependence of (**a**) T_{CDW1} and (**b**) T_{CDW2} for different LaAg$_{1-x}$Au$_x$Sb$_2$ shown on the same plot. Data for LaAu$_x$Sb$_2$ are taken from Ref. [17]. Insets: the same data positioned on universal lines by horizontal shifts, ΔP.

The $T_{CDW}(P)$ data could be combined on the same universal line by horizontal shift of the data (as shown in the insets to Figure 10). This universal behavior suggest equivalence of the chemical and physical pressure. The approximate scaling is 2 kbar~0.1 x-Au for CDW1 and slightly smaller pressure shift per 0.1 x-Au for CDW2. We recall that in a similar way the $P - T$ phase diagrams for Ba(Fe$_{1-x}$Ru$_x$)$_2$As$_2$ with different value of x were combined to form a universal phase diagram by ΔP shifts with 30 kbar~0.1 x-Ru [4]. In contrast, the pressure and substitution data in the La$_{1-x}$R$_x$AgSb$_2$ (R = Ce, Nd) series [20] cannot be combined on the same line by ΔP shifts. Apparently rare earth and transition metal substitutions in LaAgSb$_2$ affect the pressure derivatives of CDW transition temperatures in different manner, with R-substituted compounds having higher (and R-dependent) suppression rates. Of course, whereas both Ag/Au and La/R substitutions are isoelectronic, substitution of Ce or Nd for La brings local moment magnetism that subsitution of Au for Ag does not.

To gain some further insight on which structural parameter could be of importance for change of the CDW temperature under pressure and with Au substitution we plot T_{CDW1} as a function of basic structural parameters, a, c, c/a and V in Figure 11. For ambient pressure data the structural parameters obtained from the Rietveld refinement are used. For the high pressure data the structural parameters were obtained from the $P = 0$ values using LaAgSb$_2$ elastic constants from Ref. [19] and assuming that their change within the LaAg$_{1-x}$Au$_x$Sb$_2$ series is insignificant.

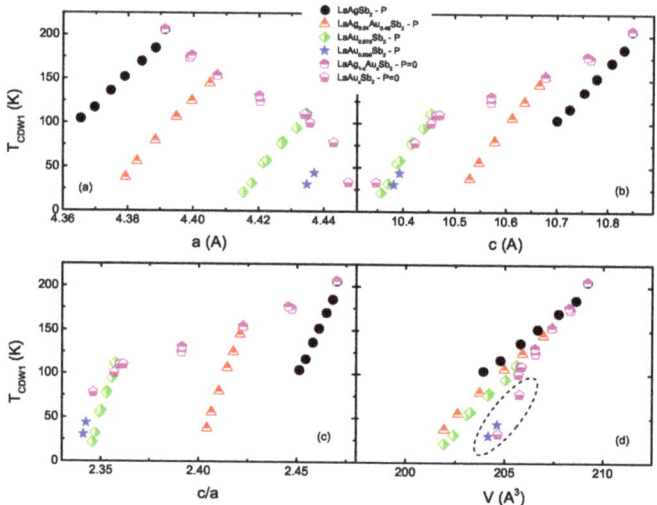

Figure 11. CDW1 transition temperature for LaAg$_{1-x}$Au$_x$Sb$_2$ and LaAu$_x$Sb$_2$ [17] at ambient and high pressure as a function of (**a**) a; (**b**) c; (**c**) c/a; and (**d**) V structural parameters. Encircled points discussed in the text.

The data in Figure 11 clearly show that whereas using a, c, and c/a as a structural parameter results in distinctly different trends for chemical and physical pressure, all the data on T_{CDW1} vs. V plot fall fairly well on the same line. To check if this holds for CDW2 as well, we plotted T_{CDW2} vs. V in Figure 12 as well. The T_{CDW2} data also scale with the unit cell volume well. We have few outlier points encircled in Figures 11d and 12. These point correspond to measurably *off-stoichiometric* LaAu$_x$Sb$_2$ [17], whereas the rest of the data are for the compounds with stiochiometry very close to 1:1:2. That is possibly the reason for these few point being outliers. The T_{CDW} vs. V scaling could be even better if the elastic constants measured for each compound were used, however to address this, further elastic properties measurements should be performed. It is of a surprise, that despite LaAg$_{1-x}$Au$_x$Sb$_2$ being tetragonal, anisotropic compounds, the chemical and physical pressure appear to be equivalent with a salient structural parameter being the unit cell volume (that lacks any anisotropic information). Hopefully further band structure calculations will be able to address this issue.

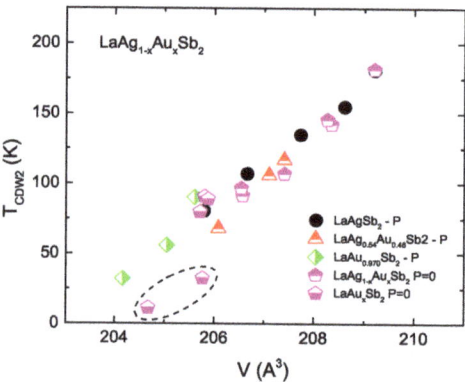

Figure 12. CDW2 transition temperature for LaAg$_{1-x}$Au$_x$Sb$_2$ and LaAu$_x$Sb$_2$ [17] at ambient and high pressure as a function of the unit cell volume, V. Encircled points discussed in the text.

5. Summary

Study of compounds of the LaAg$_{1-x}$Au$_x$Sb$_2$ family at ambient and high pressure show that both CDW transitions are suppressed with Au substitution and under pressure in a manner that indicate equivalence of chemical and physical pressure in this series with the unit cell volume being a suitable structural control parameter and with suppression rates being different for CDW1 and CDW2. Such equivalence of physical and chemical pressure was not observed in the La$_{1-x}$R$_x$AgSb$_2$ series [20]. Different CDW suppression rates probably reflect the fact that (at least for LaAgSb$_2$) the CDW wave-vectors are orthogonal, along a- and c-axis for CDW1 and CDW2, respectively.

Additionally, for LaAg$_{0.54}$Au$_{0.46}$Sb$_2$ anomalies in low temperature electrical transport were observed in the pressure range where CDW2 is completely suppressed.

Author Contributions: Conceptualization, S.L.B.; formal analysis, L.X., D.H.R. and S.L.B.; investigation, L.X. and D.H.R.; data curation, L.X. and D.H.R.; writing—original draft preparation, S.L.B.; writing—review and editing, L.X., D.H.R. and P.C.C.; visualization, L.X., D.H.R. and S.L.B.; supervision, P.C.C. and S.L.B.; funding acquisition, P.C.C. All authors have read and agreed to the published version of the manuscript.

Funding: Work at the Ames National Laboratory was supported by the U.S. Department of Energy, Office of Science, Basic Energy Sciences, Materials Sciences and Engineering Division. The Ames National Laboratory is operated for the U.S. Department of Energy by Iowa State University under contract No. DE-AC02-07CH11358. L.X. was supported, in part, by the W. M. Keck Foundation. Much of this work was carried out while D.H.R. was on sabbatical at Iowa State University and Ames Laboratory and their generous support (again under under contract No. DE-AC02-07CH11358) during this visit is gratefully acknowledged. D.H.R. was supported as well by Fonds Québécois de la Recherche sur la Nature et les Technologies.

Data Availability Statement: The data presented in this study are available on reasonable request from the corresponding author.

Acknowledgments: We thank Warren E. Straszheim for help with the EDS measurements, and Elena Gati and Raquel A. Ribeiro for useful discussions. PCC would like to acknowledge Paul Delvaux for ongoing inspiration.

Conflicts of Interest: The authors declare no conflict of interest.

Appendix A. Rietveld Refinement and EDS Results

This Appendix A contains tables with the results of Rietveld refinements and EDS chemical analysis of the LaAg$_{1-x}$Au$_x$Sb$_2$ samples. Data for LaAu$_x$Sb$_2$ [17] in Table A1 are added for comparison.

Table A1. Lattice parameters of LaAg$_{1-x}$Au$_x$Sb$_2$ samples (labels in parentheses indicate initial growth compositions, see Experimental details section for more details).

Sample	a (Å)	c (Å)	V (Å^3)
LaAgSb$_2$ (T2)	4.3915(1)	10.8485(4)	209.21(1)
LaAg$_{0.75}$Au$_{0.25}$Sb$_2$ (T2)	4.3991(1)	10.7669(4)	208.36(1)
LaAg$_{0.75}$Au$_{0.25}$Sb$_2$ (T6)	4.3996(1)	10.7601(4)	208.28(1)
LaAg$_{0.5}$Au$_{0.5}$Sb$_2$ (T2)	4.4074(2)	10.6777(6)	207.42(2)
LaAg$_{0.25}$Au$_{0.75}$Sb$_2$ (T2)	4.4205(2)	10.5715(5)	206.52(2)
LaAg$_{0.25}$Au$_{0.75}$Sb$_2$ (T6)	4.4202(1)	10.5716(4)	206.55(1)
LaAu$_x$Sb$_2$ (T2) [17]	4.4430(2)	10.4237(4)	205.77(1)
LaAu$_x$Sb$_2$ (T6) [17]	4.4347(1)	10.4653(3)	205.88(1)

Table A2. Atomic coordinates, occupancy, and isotropic displacement parameters of LaAg$_{1-x}$Au$_x$Sb$_2$ samples (labels in parentheses indicate initial growth compositions, see Experimental details section for more details).

Sample	Atom	Site	x	y	z	Occupancy	U_{eq}
LaAgSb$_2$ (T2)	La	2c	0.25	0.25	0.2397(1)	1	0.0260(5)
	Ag	2b	0.75	0.25	0.5	1	0.0297(6)
	Sb1	2a	0.75	0.25	0	1	0.0275(5)
	Sb2	2c	0.25	0.25	0.6691(2)	1	0.0275(5)
LaAg$_{0.75}$Au$_{0.25}$Sb$_2$ (T2)	La	2c	0.25	0.25	0.2424(2)	1	0.0335(6)
	Ag	2b	0.75	0.25	0.5	0.80(1)	0.033(1)
	Au	2b	0.75	0.25	0.5	0.20(1)	0.033(1)
	Sb1	2a	0.75	0.25	0	1	0.0318(7)
	Sb2	2c	0.25	0.25	0.6696(2)	1	0.0318(7)
LaAg$_{0.75}$Au$_{0.25}$Sb$_2$ (T6)	La	2c	0.25	0.25	0.2418(2)	1	0.0281(7)
	Ag	2b	0.75	0.25	0.5	0.79(2)	0.027(1)
	Au	2b	0.75	0.25	0.5	0.21(2)	0.027(1)
	Sb1	2a	0.75	0.25	0	1	0.0275(7)
	Sb2	2c	0.25	0.25	0.6700(2)	1	0.0275(7)
LaAg$_{0.5}$Au$_{0.5}$Sb$_2$ (T2)	La	2c	0.25	0.25	0.2448(2)	1	0.0366(7)
	Ag	2b	0.75	0.25	0.5	0.54(2)	0.042(1)
	Au	2b	0.75	0.25	0.5	0.46(1)	0.042(1)
	Sb1	2a	0.75	0.25	0	1	0.0357(8)
	Sb2	2c	0.25	0.25	0.6700(2)	1	0.0357(8)
LaAg$_{0.25}$Au$_{0.75}$Sb$_2$ (T2)	La	2c	0.25	0.25	0.2455(3)	1	0.0159(8)
	Ag	2b	0.75	0.25	0.5	0.32(2)	0.021(1)
	Au	2b	0.75	0.25	0.5	0.68(2)	0.021(1)
	Sb1	2a	0.75	0.25	0	1	0.0177(8)
	Sb2	2c	0.25	0.25	0.6999(3)	1	0.0177(8)
LaAg$_{0.25}$Au$_{0.75}$Sb$_2$ (T6)	La	2c	0.25	0.25	0.2453(2)	1	0.0298(6)
	Ag	2b	0.75	0.25	0.5	0.36(2)	0.0320(8)
	Au	2b	0.75	0.25	0.5	0.64(2)	0.0320(8)
	Sb1	2a	0.75	0.25	0	1	0.0331(6)
	Sb2	2c	0.25	0.25	0.6998(2)	1	0.0331(6)

Table A3. EDS results for LaAg$_{1-x}$Au$_x$Sb$_2$ samples.

Sample	La at.%	Ag at. %	Au at.%	Sb at. %	Au/(Ag + Au)	3(Ag + Au)/(La + Sb)
LaAgSb$_2$ (T2)	25.4(2)	25.9(1)	0	48.7(2)	0	1.05(2)
LaAg$_{0.75}$Au$_{0.25}$Sb$_2$ (T2)	25.6(1)	20.4(2)	5.08(7)	48.9(1)	0.199(5)	1.03(6)
LaAg$_{0.75}$Au$_{0.25}$Sb$_2$ (T6)	25.5(2)	20.2(2)	5.40(4)	48.9(1)	0.211(4)	1.03(3)
LaAg$_{0.5}$Au$_{0.5}$Sb$_2$ (T2)	25.6(1)	14.8(2)	10.50(8)	49.2(1)	0.415(9)	1.01(3)
LaAg$_{0.25}$Au$_{0.75}$Sb$_2$ (T6)	25.7(1)	7.4(1)	17.7(1)	49.28(8)	0.71(1)	1.00(2)

Appendix B. LaAg$_{0.54}$Au$_{0.46}$Sb$_2$ under Pressure

Figure A1 presents the derivatives of the resistivity data taken at 5.2 kbar and 9.3 kbar for LaAg$_{0.54}$Au$_{0.46}$Sb$_2$ sample.

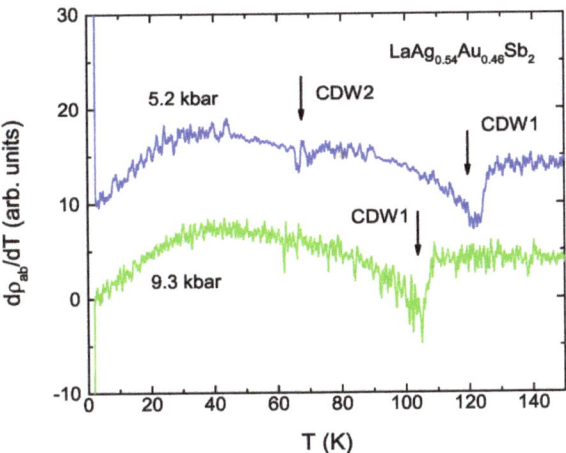

Figure A1. Resistivity derivatives, $d\rho_{ab}/dT$, data at 5.2 kbar and 9.3 kbar for $LaAg_{0.54}Au_{0.46}Sb_2$. Arrows point to CDW transition temperatures. The 5.5 kbar data are shifted vertically by 10 for clarity.

References

1. Wang, D.; Ding, Y.; Mao, H.K. Future Study of Dense Superconducting Hydrides at High Pressure. *Materials* **2021**, *14*, 7563. [CrossRef]
2. Klintberg, L.E.; Goh, S.; Kasahara, S.; Nakai, Y.; Ishida, K.; Sutherland, M.; Shibauchi, T.; Matsuda, Y.; Terashima, T. Chemical Pressure and Physical Pressure in $BaFe_2(As_{1-x}P_x)_2$. *J. Phys. Soc. Jpn.* **2010**, *79*, 123706. [CrossRef]
3. Paglione, J.; Greene, R.L. High-temperature superconductivity in iron-based materials. *Nat. Phys.* **2010**, *6*, 645. [CrossRef]
4. Kim, S.K.; Torikachvili, M.S.; Colombier, E.; Thaler, A.; Bud'ko, S.L.; Canfield, P.C. Combined effects of pressure and Ru substitution on $BaFe_2As_2$. *Phys. Rev. B* **2011**, *84*, 134525. [CrossRef]
5. Fernandes, A.A.R.; Santamaria, J.; Bud'ko, S.L.; Nakamura, O.; Guimpel, J.; Schuller, I.K. Effect of physical and chemical pressure on the superconductivity of high-temperature oxide superconductors. *Phys. Rev. B* **1991**, *44*, 7601. [CrossRef]
6. Monceau, P. Electronic crystals: An experimental overview. *Adv. Phys.* **2012**, *61*, 325.
7. Peierls, R.E. *Quantum Theory of Solids*; Oxford University: Oxford, UK, 1955.
8. Johannes, M.D.; Mazin, I.I. Fermi surface nesting and the origin of charge density waves in metals. *Phys. Rev. B* **2008**, *77*, 165135. [CrossRef]
9. Eiter, H.M.; Lavagnini, M.; Hackl, R.; Nowadnick, E.A.; Kemper, A.F.; Devereaux, T.P.; Chu, J.H.; Analytis, J.G.; Fisher, I.R.; Degiorgi, L. Alternative route to charge density wave formation in multiband systems. *Proc. Natl. Acad. Sci. USA* **2013**, *110*, 64. [CrossRef]
10. Zhu, X.; Cao, Y.; Zhang, J.; Plummer, E.W.; Guo, J. Classification of charge density waves based on their nature. *Proc. Natl. Acad. Sci. USA* **2015**, *112*, 2367. [CrossRef]
11. Brylak, M.; Möller, M.H.; Jeitschko, W. Ternary Arsenides $ACuAs_2$ and Ternary Antimonides $AAgSb_2$ (A = Rare-Earth Elements and Uranium) with $HfCuSi_2$-Type Structure. *J. Solid State Chem.* **1995**, *115*, 305. [CrossRef]
12. Sologub, O.; Hiebl, K.; Rogl, P.; Noël, H.; Bodak, O. On the crystal structure and magnetic properties of the ternary rare earth compounds $RETSb_2$ with RE = rare earth and T = Ni, Pd, Cu and Au. *J. Alloys Compd.* **1994**, *210*, 153. [CrossRef]
13. Myers, K.D.; Bud'ko, S.L.; Fisher, I.R.; Islam, Z.; Kleinke, H.; Lacerda, A.H.; Canfield, P.C. Systematic study of anisotropic transport and magnetic properties of $RAgSb_2$ (R = Y, La – Nd, Sm, Gd – Tm). *J. Magn. Magn. Mater.* **1999**, *205*, 27. [CrossRef]
14. Seo, S.; Sidorov, V.A.; Lee, H.; Jang, D.; Fisk, Z.; Thompson, J.G.; Park, T. Pressure effects on the heavy-fermion antiferromagnet $CeAuSb_2$. *Phys. Rev. B* **2012**, *85*, 205145. [CrossRef]
15. Song, C.; Park, J.; Koo, J.; Lee, K.-B.; Rhee, J.Y.; Bud'ko, S.L.; Canfield, P.C.; Harmon, B.N.; Goldman, A.I. Charge-density-wave orderings in $LaAgSb_2$: An X-ray scattering study. *Phys. Rev. B* **2003**, *68*, 035113. [CrossRef]
16. Bud'ko, S.L.; Law, S.A.; Canfield, P.C.; Samolyuk, G.D.; Torikachvili, M.S.; Schmiedeshoff, G.M. Thermal expansion and magnetostriction of pure and doped $RAgSb_2$ (R = Y, Sm, La) single crystals. *J. Phys. Condens. Matter* **2008**, *20*, 115210. [CrossRef]
17. Xiang, L.; Ryan, D.H.; Straszheim, W.E.; Canfield, P.C.; Bud'ko, S.L. Tuning of charge density wave transitions in $LaAu_xSb_2$ by pressure and Au stoichiometry. *Phys. Rev. B* **2020**, *102*, 125110. [CrossRef]
18. Masubuchi, S.; Ishii, Y.; Ooiwa, K.; Fukuhara, T.; Shimizu, F.; Sato, H. Chemical Substitution Effect on CDW State in $LaAgSb_2$. *JPS Conf. Proc.* **2014**, *3*, 011053.
19. Bud'ko, S.L.; Wiener, T.A.; Ribeiro, R.A.; Canfield, P.C.; Lee, Y.; Vogt, T.; Lacerda, A.H. Effect of pressure and chemical substitutions on the charge-density-wave in $LaAgSb_2$. *Phys. Rev. B* **2006**, *73*, 184111. [CrossRef]

20. Torikachvili, M.S.; Bud'ko, S.L.; Law, S.A.; Tillman, M.E.; Mun, E.D.; Canfield, P.C. Hydrostatic pressure study of pure and doped La$_{1-x}$R$_x$AgSb$_2$ (R = Ce, Nd) charge-density-wave compounds. *Phys. Rev. B* **2007**, *76*, 235110. [CrossRef]
21. Akiba, K.; Nishimori, H.; Umeshita, N.; Kobayashi, T.C. Successive destruction of charge density wave states by pressure in LaAgSb$_2$. *Phys. Rev. B* **2021**, *103*, 085134. [CrossRef]
22. Zhang, B.; An, C.; Chen, X.; Zhou, Y.; Zhou, Y.; Yuan, Y.; Chen, Y.; Zhang, C.; Yang, L.; Yang, Z. Structural and electrical transport properties of charge density wave material LaAgSb$_2$ under high pressure. *Chin. Phys. B* **2021**, *30*, 076201. [CrossRef]
23. Akiba, K.; Umeshita, N.; Kobayashi, T.C. Observation of superconductivity and its enhancement at the charge density wave critical point in LaAgSb$_2$. *Phys. Rev. B* **2022**, *106*, L161113. [CrossRef]
24. Du, F.; Su, H.; Luo, S.S.; Shen, B.; Nie, Z.Y.; Yin, L.C.; Chen, Y.; Li, R.; Smidman, M.; Yuan, H.Q. Interplay between charge density wave order and superconductivity in LaAuSb$_2$ under pressure. *Phys. Rev. B* **2020**, *102*, 144510. [CrossRef]
25. Lingannan, G.; Joseph, B.; Vajeeston, P.; Kuo, C.N.; Lue, C.S.; Kalaiselvan, G.; Rajak, P.; Arumugam, S. Pressure-dependent modifications in the LaAuSb$_2$ charge density wave system. *Phys. Rev. B* **2021**, *103*, 195126. [CrossRef]
26. Zhao, L.; Yelland, E.A.; Bruin, J.A.; Sheikin, I.; Canfield, P.C.; Fritsch, V.; Sakai, H.; Mackenzie, A.P.; Hicks, C.W. Field-temperature phase diagram and entropy landscape of CeAuSb$_2$. *Phys. Rev. B* **2016**, *93*, 195124. [CrossRef]
27. Canfield, P.C.; Kong, T.; Kaluarachchi, U.S.; Jo, N.H. Use of frit-disc crucibles for routine and exploratory solution growth of single crystalline samples. *Philos. Mag.* **2016**, *96*, 84. [CrossRef]
28. Larson, A.C.; Von Dreele, R.B. *General Structure Analysis System (GSAS)*; Los Alamos National Laboratory Report LAUR 86-748; Los Alamos National Laboratory: Los Alamos, NM, USA, 2004.
29. Toby, B.H. EXPGUI, a graphical user interface for GSAS. *J. Appl. Cryst.* **2001**, *34*, 210. [CrossRef]
30. *Standard Reference Material 676a*; NIST: Gaithersburg, MD, USA, 2015.
31. Bud'ko, S.L.; Voronovskii, A.N.; Gapotchenko, A.G.; Itskevich, E.S. The Fermi surface of cadmium at an electron-topological phase transition under pressure. *Zh. Eksp. Teor. Fiz.* **1984**, *86*, 778; [English translation: *Sov. Phys. JETP* **1984**, *59*, 454].
32. Torikachvili, M.S.; Kim, S.K.; Colombier, E.; Bud'ko, S.L.; Canfield, P.C. Solidification and loss of hydrostaticity in liquid media used for pressure measurements. *Rev. Sci. Instrum.* **2015**, *86*, 123904. [CrossRef]
33. Eiling, A.; Schilling, J.S. Pressure and temperature dependence of electrical resistivity of Pb and Sn from 1 to 300 K and 0–10 GPa-use as continuous resistive pressure monitor accurate over wide temperature range; superconductivity under pressure in Pb, Sn and In. *J. Phys. F Met. Phys.* **1981**, *11*, 623. [CrossRef]
34. Xiang, L.; Gati, E.; Bud'ko, S.L.; Ribeiro, R.A.; Ata, A.; Tutsch, U.; Lang, M.; Canfield, P.C. Characterization of the pressure coefficient of manganin and temperature evolution of pressure in piston-cylinder cells. *Rev. Sci. Instrum.* **2020**, *91*, 095103. [CrossRef] [PubMed]
35. Li, L.; Deng, X.; Wang, Z.; Liu, Y.; Abeykoon, M.; Dooryhee, E.; Tomic, A.; Huang, Y.; Warren, J.B.; Bozin, E.S.; et al. Superconducting order from disorder in 2H-TaSe$_{2-x}$S$_x$. *npj Quant. Mater.* **2017**, *2*, 11. [CrossRef]
36. Fuchs, G.; Müller, K.-H.; Freudenberger, J.; Nenkov, K.; Drechsler, S.-L.; Shulga, S.V.; Lipp, D.; Gladun, A.; Cichorek, T.; Gegenwart, P. Influence of disorder on superconductivity in non-magnetic rare-earth nickel borocarbides. *Pramana* **2002**, *58*, 791. [CrossRef]
37. Slade, T.J.; Mudiyanselage, R.S.D.; Furukawa, N.; Smith, T.R.; Schmidt, J.; Wang, L.L.; Kang, C.-J.; Wei, K.; Shu, Z.; Kong, T.; et al. Mn(Pt$_{1-x}$Pd$_x$)$_5$P: Isovalent Tuning of Mn Sublattice Magnetic Order. *arXiv* **2022**, arXiv:2211.01818.

Article

Corresponding States for Volumes of Elemental Solids at Their Pressures of Polymorphic Transformations

Oliver Tschauner

Department of Geoscience, University of Nevada Las Vegas, Las Vegas, NV 89154, USA; oliver.tschauner@unlv.edu

Abstract: Many non-molecular elemental solids exhibit common features in their structures over the range of 0 to 0.5 TPa that have been correlated with equivalent valence electron configurations. Here, it is shown that the pressures and volumes at polymorphic transitions obey corresponding states given by a single, empirical universal step-function $V_{tr}/L = -0.0208(3) \cdot P_{tr} + N_i$, where V_{tr} is the atomic volume in Å^3 at a given transformation pressure P_{tr} in GPa, and L is the principal quantum number. N_i assumes discrete values of approximately 20, 30, 40, etc. times the cube of the Bohr radius, thus separating all 113 examined polymorphic elements into five discrete sets. The separation into these sets is not along L. Instead, strongly contractive polymorphic transformations of a given elemental solid involve changes to different sets. The rule of corresponding states allows for predicting atomic volumes of elemental polymorphs of hitherto unknown structures and the transitions from molecular into non-molecular phases such as for hydrogen. Though not an equation of state, this relation establishes a basic principle ruling over a vast range of simple and complex solid structures that confirms that effective single-electron-based calculations are good approximations for these materials and pressures The relation between transformation pressures and volumes paves the way to a quantitative assessment of the state of very dense matter intermediate between the terrestrial pressure regime and stellar matter.

Keywords: corresponding states; high pressure; Wigner–Seitz radii; elemental solids; phase transformation

1. Introduction

In difference to gases whose states can be described by the ideal gas- or the van der Waals equations of state, solids seem to defy the concept of a general equation of state. The range of structures and properties as well as their changes upon compression appear too large to be comprised within a single formula. Over a range of pressure of 0 to 500 GPa, many elements that are molecular solids at ambient conditions transform into atomic metals [1], while others that are normal metals at ambient conditions become semiconductors [2–4], and some monatomic elemental solids assume complex structures under compression that do not exist at ambient pressure [5,6].

These changes are not arbitrary but reflect general trends in the effect of pressure on the valence electron structure of solid matter over large pressure intervals [1–12]. The changes in the valence electron configuration reflect the different response of different orbital states to the increase in electron density upon compression [2,7,10–15]. The atomic volume often is markedly reduced upon such transitions. The transition from the bcc to the hcp structure in many transition metal elements is correlated with an increase in the hybridization of filled d-states with p-states of the next higher principal quantum number [16,17]. Similarly, the pressure-induced transformations of lanthanides from hcp to the Sm type to dhcp to fcc is explained through the increased occupancy of valence d-states [11–13], whereas the marked volume collapse and transitions to low symmetric phases in lanthanides are assigned to delocalization of the 4f states [12,13,15]. A different type of delocalization occurs in the low-Z elemental alkaline metals, which at high pressure transform into electrides in which inner shell electron states overlap with valence s-states and non-bonding p-states are occupied on expenses of the bonding s- and p-hybrid states [2,4,14]. As a consequence, elements such as Li and Na, which are normal metals at ambient conditions, become semiconductors at pressures between 100 and 200 GPa [2–4]. Changes in the electronic valence states have been studied quantitatively for some alkaline metals, transition group elements, and lanthanides through X-ray spectroscopy [10,14–17]. These changes are correlated with structural transformations. It has been noted that these pressure-induced valence changes follow general correlations of the nuclear charge number with the pressure, volume, and structure type [5,8–10]. In addition, there are correlations between the atomic volumes of equivalent polymorphs of different elements [8,18].

Here, it is shown that all known non-molecular elemental solids follow a rule of corresponding states if only the volumes and pressures at the polymorphic transitions are considered. This observation extends to high-pressure atomic polymorphs of elements that are molecular at ambient conditions, such as O, N, S, P, Cl, Br, and I. Only experimental data are considered here (see Section 2), hence the relation presented is without any kind of simplifying assumptions or theory.

2. Materials and Methods

Only published volumes and transformation pressures of polymorphs of elemental solids are used in this paper. Recently published data and work conducted under hydrostatic or nearly hydrostatic conditions are given preference. This includes data obtained from samples embedded in He or Ne as pressure media or data from crystals grown from melt at a high pressure. Data from compression experiments without pressure media are only included if the average pressure has been below 10 GPa. However, the induced transformations in Ir, Pb, and Th under non-hydrostatic stresses are included for reference. Similarly, compression data of some lanthanides are included due to their principal interest, although most of these data were not and possibly could not be acquired with the use of hydrostatic media. The plotted volumes V_{tr} and pressures P_{tr} at the polymorphic transitions are generally those of the first observation of a high-pressure polymorph upon compression at 300 K. In a few cases in which the transformations exhibits hystereses larger than 5 GPa, the arithmetic middle pressure has been taken as P_{tr}. In all other cases, the hysteresis has been added to the uncertainty. For some elemental high-pressure polymorphs, the structures have been redetermined in more recent studies but without reassessment of the 300 K isotherms. In such cases, the atomic volume of the correct structure has been used to reassess the atomic volume at the pressure of the transformation that has been reported in the earlier studies. The complete data are given in Table 1 at the end of the paper.

3. Results

For all of the examined 113 elemental solids, the volume at the phase transition $V_{tr} = V(P_{tr})$ (P_{tr} = transition pressure) divided by the principal quantum number L exhibits a universal linear pressure dependence of $-0.0208(3) \cdot P_{tr}$ (P_{tr} in GPa; atomic volume in Å^3; the adjusted R^2 of the linear fit was 0.9985; see Supplemental Figure S1). This is shown in Figure 1. Furthermore, within the uncertainties, the constant term of this equation provides a volume that for all of the examined elemental solids is nearly equal to $n \cdot 10 \cdot r_B^3$, where n is an integer number between 2 and 6, and r_B is the Bohr radius in Å. This is shown in Figure 2. In other words:

$$V_{tr}/L + 0.0208(3) \cdot P_{tr} = N_i \approx n \cdot 10 \cdot r_B^3 \tag{1}$$

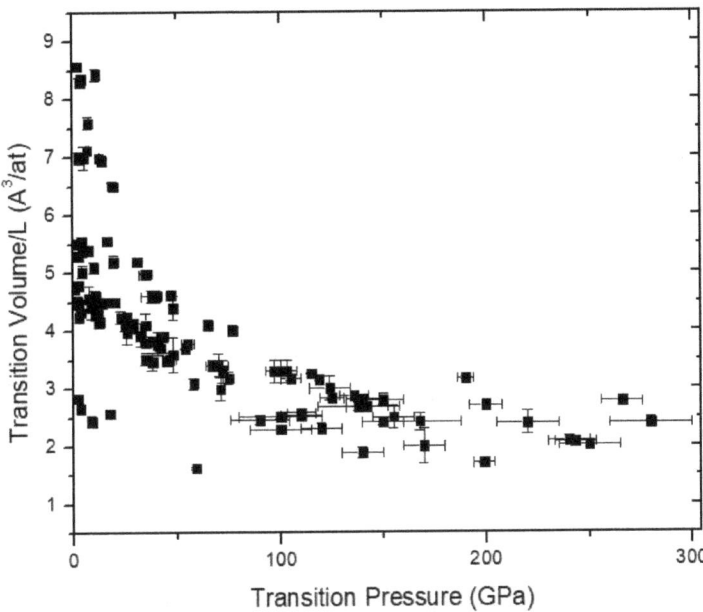

Figure 1. Volume of elemental polymorphs in Å³/at at 300 K divided by the principal quantum number L as a function of pressure in GPa. The volumes are those that are observed at the pressure of their formation through direct phase transformation from lower pressure polymorphs; the plotted pressures are the pressures of the transformation. The linear relation between these volumes and pressures and the existence of five distinct sets of pressure–volume relations is clearly visible.

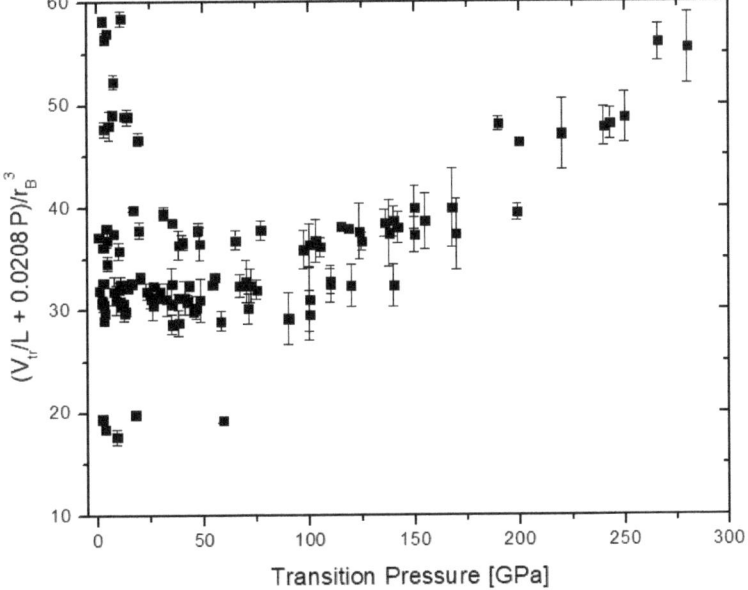

Figure 2. This figure shows the data from Figure 1 with the volume corrected for the linear pressure dependence from Equation (1). The remaining constant term of each set is approximately equal to integer multiples of 2, 3, 4, 5, and 6 of 10 times the cube of the Bohr radius.

The actual values of the five distinct volumes for N_i are 2.88(8), 4.58(4), 5.51(5), 7.14(5), 8.53(6) Å^3/at, which is 19.5(4), 30.9(4), 37.2(8), 48.2(1), and 57.6(5) times the cube of the Bohr radius r_B^3 in Å^3, respectively. Therefore, Figures 1 and 2 show that the volumes of elemental solids at their polymorphic transitions assume corresponding states. It is important to recall that this universal relation of corresponding states only holds for the pressures and volumes at the phase transitions: Within the stability field of each polymorph, the pressure–volume relation is controlled by its specific compressibility that itself depends on pressure and temperature and that, in general, is far from being linear, universal, and discrete. Equation (1) is therefore not an equation of state.

4. Discussion

The 'collective quantization' shown in Figure 1 may, upon first glance, be assigned to the factorization by the principal quantum number L. However, the series of corresponding states N_i does not follow the rows of the periodic table. Rather, many elements with subsequent polymorphic transitions change between states N_i. For instance, the high-pressure polymorphs fcc, cI16, and oc120 of Na reside in $N_4 = 5.51(5)$ Å^3/at $\approx 40 \cdot r_B^3$ along with K-IV, Rb-IV, -V, Cs-IV, and Ca-III to -IV; whereas the h4p-type Na resides in $N_5 = 7.14(5)$ Å^3/at $\approx 50 \cdot r_B^3$ along with Rb-II and -III. In some cases, for instance Si, all polymorphs remain within one set N_i; however, for most elements at least some polymorphs fall into different sets. For instance, Cs-II and Cs-III are in N_6 but Cs-IV and -V belong to N_4 and Cs-VI to N_3. A sequence of stepwise transitions to sets with a lower state N_i along with an increasing transformation pressure is found for many elements: the high-pressure polymorphs of Rb fall into the sets N_5 (Rb-II and -III), N_4 (Rb-IV, -V, -VIII), and N_3 (Rb-VI, -VII); Li into N_6 (Li-II), N_4 (Li-III, -IV,-VI), and N_3 (Li-V); and the various polymorphs of the alkaline earths Ca, Sr, and Ba are found in the sets N_3 to N_6 (Figures 1 and 2), where the higher pressure polymorphs belong to sets with a smaller N_i. Lower sets N_i have markedly smaller volumes at a given pressure than the corresponding higher sets N_{i+1} (Figure 1), and the polymorphic transitions that involve changes $N_i \rightarrow N_{i-1}$ are accompanied by marked volume contractions $(V^{LP}_{tr} - V^{HP}_{tr})/V_0$ where V^{LP}_{tr} and V^{HP}_{tr} are the atomic volumes of the low- and the high-pressure polymorphs, respectively, at the pressure of the phase transitions; and V_0 is the volume of the ambient pressure phase at standard conditions. A smaller number of elements also exhibit transitions into higher sets N_{i+1} at higher pressure. In particular, this holds for the polymorphic transitions to Sc-V [19], bcc-Ti [20,21], hp4-Na [3], and Rb-VIII (Figure 1), but also for the elastic anomaly of Os around 290 GPa [22], which is taken here for an isostructural transition. There is a simple reason for these 'upward transitions' $N_i \rightarrow N_{i+1}$: at the given transformation pressure, the corresponding volumes of the next lower set N_{i-1} would be negative. It is to be expected that for all elements at sufficiently high pressures, the cascade of successive, markedly contractive transitions $N_i \rightarrow N_{i-1}$ is followed by transitions $N_i \rightarrow N_{i+1}$ to polymorphs in higher sets with a less pronounced contraction once $N_i \rightarrow N_{i-1}$ reaches the limit of negative volumes. Transitions to higher sets come with smaller contractions $(V^{LP}_{tr} - V^{HP}_{tr})/V_0$ than those into lower sets.

4.1. Theoretical Explanation

'Upward' and 'downward' transitions between different states N_i can be correlated to changes in the valence electron configuration wherever such information has been obtained experimentally [10,15,17] or through calculation [14]. The fact that the volumes N_i are nearly equal to integer multiples of r_B^3 has a straightforward theoretical basis: It implies that the volumes that occur at polymorphic transitions of non-molecular elements are controlled by the number of valence electrons rather than by their interactions. Besides the limited experimental accuracy the electron exchange- and Coulomb interactions account for part of the deviations from Equation (1) and from integer N_i values; thus, the interaction terms are of the second order with respect to the volumes. In other words, Equation (1) is the equivalent to an ideal equation of state for elemental solids, only that it does not describe continuous pressure–volume changes but discrete changes that occur at the phase transitions. This finding can be further quantified by considering the effect of the different valence orbital states as based on the experimentally determined pressure-induced hybridization of d-states of the alkaline metals Rb, Cs, and K (the latter element exhibits no measurable d-state occupancy up to 40 GPa) [10]. Figure 1 may be re-parametrized in terms of pressure-induced changes in the d-state occupancy at the valence level. In particular, the d-state occupancy of the elemental polymorphs can be correlated with the five sets N_2 to N_6 shown in Figure 1 through their 0 GPa intercepts. This is shown in Figure 3a. For instance, the d-state occupancy of Cs increases for phase Cs-II to -VI from ~0.0 to ~0.6, and for Rb from ~0.2 to 0.4 electrons between Rb-II and -VI [10]. The high-P polymorphs of Li and Na have little or no occupied d-states in this diagram and according to earlier computational work [2,4,14]. The correlation between $\Delta n(d)$ and N_i is linear, at least for the polymorphs K-II, Rb-II,-IV, Cs-II, and Cs-III (black squares in Figure 3a): Within uncertainty, the slopes are equal for all sets. In terms of multiples of the electron elemental charge, this linear correlation between d-state occupancy, V_{tr}, and L is:

$$N_i = N_i^0 - 6.11(14) \cdot 10^{-11} \cdot n(d) \cdot e \qquad (2)$$

where $N_i^0 = N_i$ at $n(d) = 0$, the constant factor is given in $m^3/(C \cdot at)$, and e is the electron elemental charge. However, the correlation between N_i and $n(d)$ is not continuous: Equations (1) and (2) can be combined to give:

$$\frac{V_{tr}}{L} = -2.08(3) \cdot 10^{-32} \cdot P - 6.11(14) \cdot 10^{-11} \cdot \sum_{i=0}^{m-1} i \cdot n(d) \cdot e + N_m \qquad (3)$$

where N_m is the value assumed at $n(d) = 0$, P is in GPa, V is in m^3/at, and e is the electron elemental charge. The effect of the d → p transfer on shifts into sets with a higher N_i has been discussed above. Overall, there are fewer data on pressure-induced changes in the valence p-state occupancy (summarized in Supplementary Figure S2, see refs. [10,16,17]), but within uncertainties, the relation between V_{tr}/L and $n(p)$ has the same slope as in (3) but with an opposed sign (consistent with the transfer into higher sets N_i, whereas changes in the d-state occupancy generally result in changes to a lower N_i). Implementing $n(p)$ into (3), dividing through r_B^3, and using the approximation $N_i\, r_B^3 \sim 10 \cdot m$ gives:

$$\frac{V_{tr}}{L \cdot r_B^3} = -0.1405(20) \cdot P + 10 \cdot m - 66.1(15) \cdot \sum_{i=0}^{m-1} i \cdot 0.15(4) + 66(2) \cdot \sum_{j=p}^{m} j \cdot$$
$$0.09(21) = -0.1405(20) \cdot P + 10 \cdot m - 10.0(6) \cdot \sum_{i=0}^{m-1} i + 6(1) \cdot \sum_{j=p}^{m} j \qquad (4)$$

where the pressure is given in GPa, the volumes N_i are taken as integer multiples $10 \cdot i \in \mathbb{N}$ (the set of natural numbers) of r_B^3 per atom, $p \leq n(d)$, and the increments $\Delta n(d)$ are taken as 0.151; hence: $66.1 \cdot 0.151 = 10$ d-electrons and accordingly $66(2) \cdot 0.09(1) = 6(1)$ for the p-electrons. Thus, within uncertainties, the factors of the sum terms equal the number of p- and d-states and an additional term for f-electrons with factor -14 may be added. The actual values N_i may be substituted for 10m.

The volumes of elemental atomic solids at ambient conditions are not bound to Equation (1) because they do not represent volumes at phase transitions. However, the correlation $V = N_i \cdot L$ is within ±10% of the observed volumes at ambient conditions for 28 and within 20% for 38 out of the 42 non-molecular solid elements (Figure 3b). Considering that the atomic volumes of elemental metals are equal to the cube of the Wigner–Seitz radii, their approximate relation to $N_i \cdot L$ is equivalent to a relation recently found for the ionic radii [23]. However, in the present case of elemental solids the relation does not rely on any approximate model of bonding.

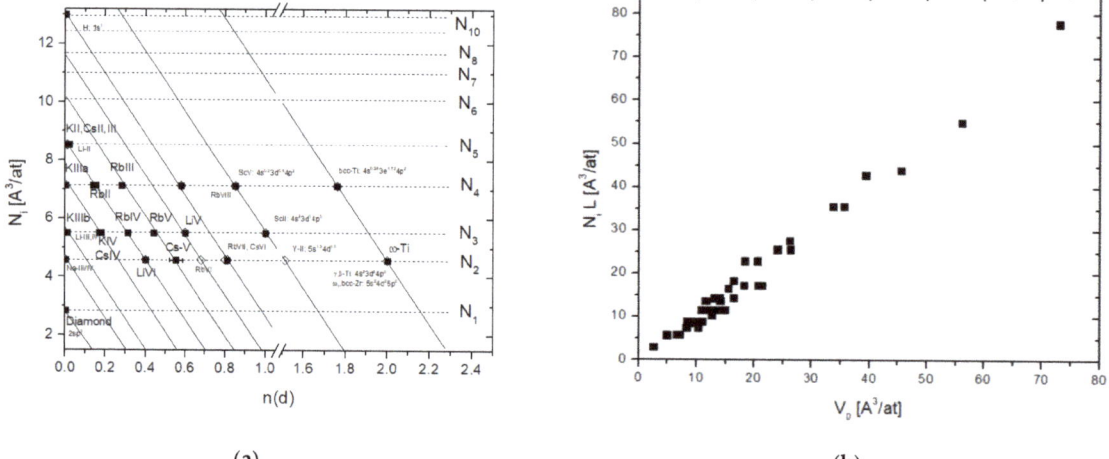

Figure 3. (a) Correlation between the normalized transformation volumes V_{tr} from Equation (1) and d-state occupancy. The volumes N_i from Equation (1) were used to define the V_{tr}/L sets of the alkaline metals [10] and of a selection of other elements as a function of d-state occupancy. Filled symbols: measured d-state occupancies; hollow symbols: estimated occupancies based on Equations (1) and (2). Solid lines: fitted correlation for K-II, Rb-II and -IV, and Cs-II and -III. The same slope has then been applied to all data. Dashed lines give the values N_i and are guides for the eye. (b) Comparison of the atomic volumes of elemental non-molecular solids at ambient pressure and temperature with the calculated volumes $N_i \cdot L$. The correlation is linear with an adjusted $R^2 = 0.9928$ and a slope of 1.015(13), which reflects a general slight contraction of the elemental volumes relative to $N_i \cdot L$.

4.2. Molecular Gap and Forbidden Zone

There are two regions in Figure 1 that contain no values V_{tr}/L: (1) There is no observed V_{tr}/L equal to or smaller than 1.48 Å^3/at at any pressure; 1.48 Å^3/at equals 10 times the cube of the Bohr radius (per atom; this limiting value 10 r_B^3/at is defined here as N_1). Upon compression, elemental solids such as Au and Re reach this limit (see Figure 4) but without further known transformations [24,25]. There is no known elemental solid that undergoes a polymorphic transition between 0 and 300 GPa where the volume at the transition is equal to or smaller than 10 r_B^3.

(2) The area between P ~ 20 GPa for V_{tr}/L ~ 8 Å^3/at and 200–300 GPa for ~3 Å^3/at contains no observations. This empty range has a nearly parabolic shape in Figure 1. It corresponds to volumes and pressures of molecular elemental solids (Figure 4).

For instance, the volumes V_{tr} of the molecular polymorphs β, γ, δ, and ε of oxygen as well as the metallic molecular phase ζ-O$_2$ [26], are not compliant with Equation (1). The isotherm of ε-O$_2$ intersects $V_{tr}/L + N_6$ at 24 GPa without any observed transition ([26]; Figure 4). The isotherm of H$_2$ intersects the $V_{tr}/L + N_6$ at 26 GPa and $V_{tr}/L + N_5$ at 54 and 135 GPa and again $V_{tr}/L + N_6$ around 250 GPa, which is at the pressure of the transition to H$_2$-IV ([27], Figure 4), but neither of these intersections corresponds to the transition to an atomic metal [27]. However, the transition from molecular N$_2$ to the polymeric, semiconducting cg-phase of nitrogen [28] falls onto the N$_4$ trajectory (Figure 1, Table 1). Cl$_2$ intersects $V_{tr}/L + N_6$ at 260 GPa, where it has been reported to become an atomic metal with an In-type structure [29] and thus follows the behavior predicted by Equation (1). Hence, the gap between 50–180 GPa for $V_{tr}/L + N_5$ and N_6 indirectly reflects the strength of the molecular bonds of these materials under compression and should be called a 'molecular gap'. The intersection of the 300 K isotherm of the chlorine phase III (mc8; [29]) with $V_{tr}/L + N_5$ at 180 GPa may be coincidental because this phase is probably still molecular and the V_{tr} of the intermediate phase IV (i-oF3, [29]) falls in between $V_{tr}/L + N_5$ and $V_{tr}/L + N_6$. The isotherms of N, S, P, Br, and I all intersect the trajectories $V_{tr}/L + N_i$ at the pressures of the observed transitions to non-molecular phases [8,27,30,31] and fall mostly below the 'molecular gap' because their transitions to non-molecular solids occur at lower pressure.

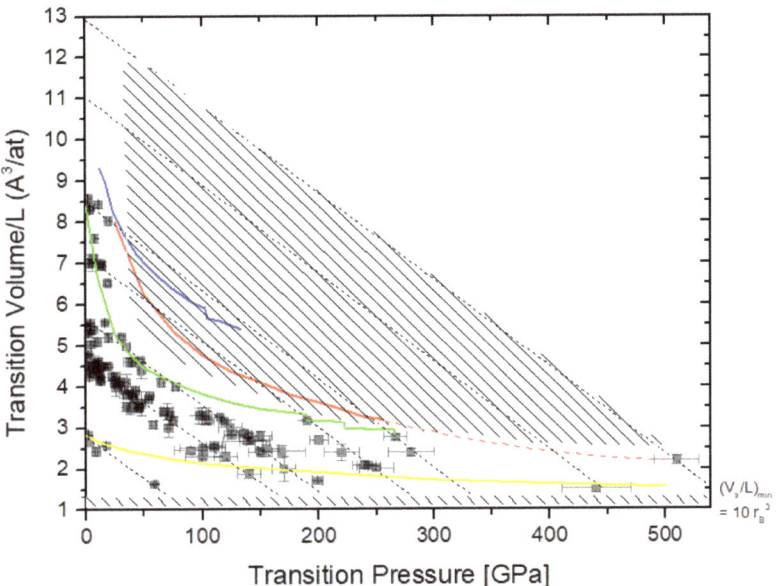

Figure 4. Data from Figure 1 with the equations of state for gold [24], chlorine [29], hydrogen [27], and oxygen [26] shown as yellow, green, red, and blue lines, respectively. Dashed lines indicate the distinct sets that defined V_{tr}/L versus P as defined by Equation (1). The additional data points for hydrogen [32] and the second elastic anomaly of osmium [22] have been added. Hachured areas mark the molecular gap and the forbidden zone below $V_{tr}/L = 10 \cdot r_B^3$.

Interestingly, the extrapolated pressure of the direct gap closure of H_2 that has been assessed through inelastic X-ray spectroscopy of H_2 [32] matches the intersection of the extrapolated isotherm of H_2 [27] and a trajectory $V_{tr}/L + N_9$ (of 13.0 to 13.3 Å^3/at = 90 r_B^3 by using the intersection of the trajectory N_9 in Figures 3a and 4) at around 0.5 TPa, which suggests that the transition from molecular to atomic hydrogen occurs at that pressure at 300 K.

Finally, it is noticed that Equation (1) could be used to assess the atomic volumes of polymorphs with extant structure analyses such as Ca-VII, Sc-III and -IV, Ga-IV, and metastable Ir-II to 8.0(5), 12.08(8), 10.3(11), 9.2(4), and 9.78(12) Å^3/at at 170, 26, 35, 120, and 59 GPa, respectively (Table 1). The volume of hydrogen at the predicted transition at 490–530 GPa is estimated to 2.3(2) Å^3/at. This and above values (marked with an asterisk) and the entire list of volumes V_{tr} and pressures P_{tr} are given in Table 1.

Table 1. Columns from left to right: Name of polymorph, pressure of phase transition at 300 K, volume of phase transition divided by the principal quantum number L, volume of phase transition at 300 K, and reference(s).

Phase	P_{tr} [GPa]	V_{tr}/L [Å^3/at]	V_{tr} [Å^3/at]	Reference(s)
Li-fcc	7.5(1)	7.6(1)	15.2(2)	[33–35]
Li-hR1	39.7(1.0)	4.6(1)	9.2(2)	[34,36]
Li-c16	48(2)	4.4	8.8(4)	[34,35]

Table 1. *Cont.*

Phase	P$_{tr}$ [GPa]	V$_{tr}$/L [Å^3/at]	V$_{tr}$ [Å^3/at]	Reference(s)
Li-oC40	67(2)	3.4(1)	6.8(2)	[35]
Li-oC24	97(3)	3.3(2)	6.6(4)	[35]
Na-fcc	67(1)	4.10(5)	12.3(30)	[36]
Na-cI16	115(2)	3.26(1)	9.77(3)	[3]
Na-oP8	119(2)	3.15(1)	9.45(6)	[3]
Na-h4p	200(8)	2.7(1)	8.1(3)	[2]
K-fcc	11.0(5)	8.4(1)	33.74(4)	[37]
K-III	22(43)	5.64(5)	22.57(3)	[5]
K-IIIb	31(2)	5.2(2)	20.8(4)	[38]
K-IV (oP8)	54(2)	3.69(1)	14.78(4)	[39]
Rb-II	7.0(5)	7.13(6)	35.65(30)	[37]
Rb-III	14.3(2)	6.95(1)	34.77(5)	[5]
Rb-IV	16.7(1)	5.56(1)	27.8(15)	[5,40]
Rb-V	19.6(5)	5.2(1)	26.0(5)	[41]
Rb-VI	48.1(5)	3.6(3)	18.0(1.5)	[41]
Rb-VII	>70	3.4(1)	17(1)	[41]
Rb-VIII	220(15)	2.4(1)	12(1)	[41]
Cs-II	2.37(1)	8.58(1)	51.47(6)	[42]
Cs-III	4.20(2)	8.37(1)	50.22(6)	[5]
Cs-IV	4.30(2)	5.83(1)	32.28(6)	[43]
Cs-V	10(1)	5.1(2)	30.6(1.2)	[44]
Cs-VI	72(2)	2.6(2)	15.6(1.2)	[44,45]
Mg-bcc	47(2)	4.63(3)	13.9(2)	[46]
Ca-II	19.5(15)	8.02(2)	32.08(8)	[47,48]
Ca-III	35(3)	5.0(1)	19.9(4)	[49,50]
Ca-IV	124(10)	3.0(2)	12.0(8)	[50]
Ca-V/VI	155(15)	2.5(2)	10.0(8)	[50]
Ca-VII	170 *	2.0 *	8	this work
Sr-II	3.5(2)	8.30(8)	41.5(4)	[48]
Sr-III	26.0(6)	3.98(19)	19.9(9)	[5,49]
Sr-IV	35(1)	4.1(2)	20.5(10)	[5,49,51]
Sr-V	46(3)	3.5(1)	17.5(5)	[5,49,51]
Ba-II	5.5(1)	7.14(4)	42.84(24)	[48]
Ba-IV	12.6(1)	7.0(1)	42.0(6)	[52]
Ba-V	45(2)	3.5(1)	21.0(6)	[53]
Sc-II	20(1)	4.5(1)	18.0(4)	[54]
Sc-III	104(5)	3.02(2) *	12.08(8)	[54]
Sc-IV	140(10)	2.58(27) *	10.3(1.1)	[54]
Sc-V	240(10)	2.09(7)	3.36(28)	[19]
ω-Zr	12.5(10)	4.15(5)	20.75(25)	[55]
Zr-bcc	35(2)	3.5(10)	17.5(50)	[55,56]
Zr-bccII	58–60	3.08(5)	15.4(3)	[55]
ω-Ti	0(5)	4.375(15)	17.5(6)	[20]
γ-Ti	110(10)	2.55(5)	10.2(2)	[20,21]
δ-Ti	150(10)	2.41(20)	9.64(20)	[20,21]
Ti-bcc	243(10)	2.07(3)	8.28(12)	[21]
V$_{rhombohedral}$	250(20)	2.02(5)	8.08(20)	[57]
Fe-hcp	17.6(1)	2.57(1)	10.28(4)	[58]
Os-hcpII	140(10)	1.88(60)	11.28(60)	[22]
Os-hcpIII	440(20)	1.5(1)	9.0(6)	[22]
Sn-bct	10.8(1)	4.60(1)	23.00(5)	[59]
Sn-bco	32.5(10)	3.93(1)	19.65(5)	[59]
Sn-bcc	40.8(15)	3.78(1)	18.9(5)	[59]
Pb-hcp	14(1)	6.95(10)	41.7(6)	[60]
Pb-bcc	142(10)	2.68(2)	16.08(12)	[5,61]
Ga-II	2.0(2)	4.52(8)	18.08(32)	[5]
Ga-III	2.8(1)	4.24(5)	16.96(24)	[62]
Ga-V	10.5(2)	4.358(5)	17.432(24)	[5]
Ga-IV	120(10)	2.3(1)*	9.2(4)	[62,63]
Tl-fcc	3.5(5)	4.3(2)	25.8(1.2)	[64]
Ge-II	10.6	4.28(8)	17.12(32)	[65]
Ge-hp	75(3)	3.18(10)	17.72(40)	[18,66]
Ge-Cmca	140(120–160) *	2.68(7)	10.72(28)	[66]
Ge-hcp	>170	2.4(1)	9.6(4)	[18]
Diamond	2.0(1)	2.837(1)	5.674(2)	[67]
Si-II	12(1)	4.3(1)	12.9(3)	[68]
Si-V	16(2)	4.43(2)	13.29(6)	[69]
Si-VI	38(2)	3.83(1)	11.49(3)	[69]
Si-VII	42(2)	3.71(2)	11.13(6)	[69]

Table 1. Cont.

Phase	P_{tr} [GPa]	V_{tr}/L [Å^3/at]	V_{tr} [Å^3/at]	Reference(s)
N, cg-phase	110(7)	2.58(3)	5.16(6)	[28]
P-A7	4.5(1)	5.03(5)	15.09(15)	[70]
P-sc	10(1)	4.54(2)	13.62(6)	[70]
P-ph	140(8)	2.8(1)	8.4(3)	[70]
P-bcc	280(20)	2.4(1)	7.2(3)	[70]
As-sc	25(1)	4.1(1)	16.4(4)	[18]
As-III	46(3)	3.5(1)	14.0(4)	[18]
As-IV	125(10)	2.84(2)	11.36(8)	[18]
Sb-II	8.6(2)	4.42(20)	22.1(1)	[71]
Sb-V	28(3)	4.05(6)	20.25(30)	[71]
Bi-II	2.55(1)	5.30(2)	31.80(12)	[8]
Bi-III	2.70(1)	4.48(9)	26.88(54)	[8]
Bi-V	7.7(1)	4.57(10)	27.42(60)	[8]
S-II	3.0(2)	7.0(1)	21.0(3)	[72]
S-III	38(2)	4.6(1)	13.8(3)	[72]
S-IV	103(6)	3.3(1)	9.9(3)	[72]
S-V	150(10)	2.8(1)	8.4(3)	[72]
Se-VII	14(1)	4.5(1)	18.0(4)	[72]
Se-III	26(1)	4.25(7)	17.0(3)	[72]
Se-IV	35(1)	3.8(1)	15.2(4)	[72]
Se-V	77(2)	4.0(1)	16.0(4)	[72]
Se-VI	136(7)	2.87(7)	11.48(28)	[18]
Te-II	4.0(1)	5.56(1)	27.80(5)	[5,73]
Te-III	7.4(1)	5.40(1)	27.00(5)	[5,73]
Te-V	29.2(7)	4.25(2)	21.25(10)	[5,73]
Te-VI	102(5)	3.3(2)	16.5(1)	[74]
I-bct	43(3)	3.9(1)	19.6(1)	[75,76]
I-fcc	55(4)	3.78(2)	18.9(1)	[75,76]
Th-bct	100(30)	2.51(1)	17.57(7)	[77]
γ-Ce	0.5(1)	4.70(1)	28.20(6)	[78]
PrtI2	180(10)		23.90(14)	[79]
Sm-dhcp	0.4(1)	5.51(1)	33.06(6)	[80]
Sm-hR9	13(1)	4.2(1)	25.2(6)	[81]
Sm-oF8	90(10)	2.45(8)	14.70(48)	[81]
Nd-oF8	100(10)	2.30(4)	13.80(24)	[82]
Eu_IV	31.5(20)	3.32(2)	19.9(1)	[83]
Gd-dhcp	9(3)	4.3(1)	26	[84]
Gd-fcc	26(2)	3.7(1)	22.2(6)	[85]
Gd-dfcc	33(2)	3.43(10)	20.5(6)	[85]
Gd-VIII	60.5(3)	2.86(10)	17.15(61)	[85]
Dy-hR9	2.5(2)	4.79(1)	28.74(6)	[86]
Cl-IV	266(10)	2.77(6)	8.31(18)	[29]
Ir-II	59	1.63(2) *	9.78(12)	[5,87]

* Parameters calculated using Equation (1).

5. Conclusions

The volume and pressure of phase transformations of non-molecular elemental solids obeyed a universal relation of discrete corresponding states. These states, as defined in Equation (1), were close to integer multiples of the cube of the Bohr radius. Therefore, the volumes and pressures of polymorphic transitions of these solids were controlled by the number of valence electrons as specified in Equation (4). This finding was in accordance with the commonly used effective single-electron-based computational approaches for modeling elemental solids within the range of 0 to 0.5 TPa. However, Equation (1) can be used for constraining elemental metal volumes for first-principal- and empirical-potential-based calculations, thus removing the weakest point in these approaches (the proper assessment of volumes). The volumes of polymorphs at the phase transitions exhibited a general linear dependence on the pressure of the transition (see Equation (1)). Though not an equation of state, this relation establishes a very simple principle that rules over a vast range of simple and complex solid structures and a range of pressure of 0.5 TPa.

Funding: This research received no external funding.

Supplementary Materials: The following supporting information can be downloaded at: https://www.mdpi.com/article/10.3390/cryst12121698/s1, Figure S1: Volume of elemental polymorphs in Å^3/at divided by the principal quantum number L as function of pressure in GPa and mapped onto the set N$_4$; Figure S2: Correlation between the normalized transformation volumes from Equation (1) and p-state occupancy.

Data Availability Statement: All data are provided in the paper.

Acknowledgments: Comments and suggestions by S. Huang, H.-k. Mao, A. Zerr, and by two anonymous reviewers are gratefully acknowledged.

Conflicts of Interest: The author declares no conflict of interest.

References

1. Mao, H.-K.; Chen, X.J.; Ding, Y.; Li, B.; Wang, L. Solids, liquids, and gases under high pressure. *Rev. Mod. Phys.* **2018**, *90*, 015007. [CrossRef]
2. Ma, Y.M.; Eremets, M.; Oganov, A.R.; Xie, Y.; Trojan, I.; Medvedev, S.; Lyakhov, S.; Valle, M.; Prakapenka, V. Transparent dense sodium. *Nature* **2009**, *458*, 182–185. [CrossRef] [PubMed]
3. Gregoryanz, E.; Lundegaard, L.F.; McMahon, M.I.; Guillaume, C.; Nelmes, R.J.; Mezouar, M. Structural diversity of sodium. *Science* **2009**, *320*, 1054–1057. [CrossRef]
4. Marques, M.; McMahon, M.I.; Gregoryanz, E.; Hanfland, M.; Guillaume, C.L.; Pickard, C.J.; Ackland, G.J.; Nelmes, R.J. Crystal Structures of Dense Lithium: A Metal-Semiconductor-Metal Transition. *Phys. Rev. Lett.* **2011**, *106*, 095502. [CrossRef] [PubMed]
5. McMahon, M.I.; Nelmes, R.J. High-pressure structures and phase transformations in elemental metals. *Chem. Soc. Rev.* **2006**, *35*, 943–963. [CrossRef]
6. Schwarz, U. High-pressure crystallography and synthesis. *Z. Krist.* **2004**, *219*, 943–963.
7. Cammi, R.; Rahm, M.; Hoffmann, R.; Ashcroft, N.W. Varying Electronic Configurations in Compressed Atoms: From the Role of the Spatial Extension of Atomic Orbitals to the Change of Electronic Configuration as an Isobaric Transformation. *J. Chem. Theory Comput.* **2020**, *16*, 5047–5056. [CrossRef]
8. Degtyareva, O.; McMahon, M.I.; Nelmes, R.J. High-pressure structural studies of group 1-5 elements. *High Press. Res.* **2004**, *24*, 319–356. [CrossRef]
9. Winzenick, M.; Vijayakumar, V.; Holzapfel, W.B. High-pressure X-ray-diffraction on potassium and rubidium up to 50 GPa. *Phys. Rev. B* **1994**, *50*, 12381–12385. [CrossRef] [PubMed]
10. Fabbris, G.; Lim, J.; Veiga, L.S.I.; Haskel, D.; Schilling, J.S. Electronic and structural ground state of heavy alkali metals at high pressure. *Phys. Rev. B* **2015**, *91*, 085111. [CrossRef]
11. Duthie, J.C.; Pettifor, D.G. Correlation between d-band occupancy and crystal-structure in rare-earths. *Phys. Rev. Lett.* **1977**, *38*, 564–567. [CrossRef]
12. Holzapfel, W.B. Structural systematics of 4f and 5f elements under pressure. *J. Alloys Comp.* **1995**, *224*, 319–356. [CrossRef]
13. Eriksson, O.; Brooks, M.S.S.; Johansson, B. Orbital polarization in narrow-band systems—Application to volume collapses in light lanthanides. *Phys. Rev. B* **1990**, *41*, 7311–7314. [CrossRef]
14. Neaton, J.B.; Ashcroft, N.W. On the constitution of sodium at higher densities. *Phys. Rev. Lett.* **2001**, *86*, 2830–2833. [CrossRef] [PubMed]
15. Bradley, J.A.; Moore, K.T.; Lipp, M.J.; Mattern, B.A.; Pacold, J.I.; Seidler, G.T.; Chow, P.; Rod, E.; Xiao, Y.M.; Evans, W.J. 4f electron delocalization and volume collapse in praseodymium metal. *Phys. Rev. B* **2021**, *85*, 100102. [CrossRef]
16. Iota, V.; Klepeis, J.H.P.; Yoo, C.S.; Lang, J.; Haskel, D.; Srajer, G. Electronic structure and magnetism in compressed transition metals. *Appl. Phys. Lett.* **2007**, *90*, 042505. [CrossRef]
17. Dewaele, A.; Stutzmann, V.; Bouchet, J.; Bottin, F.; Occelli, F.; Mezouar, M. The $\alpha \to \omega$ phase transformation in zirconium followed with ms-scale time-resolved X-ray absorption spectroscopy. *High Pres. Res.* **2016**, *36*, 237–249.
18. Akahama, Y.; Kamiue, K.; Okawa, N.; Kawaguchi, S.; Hirao, N.; Ohishi, Y. Volume compression of period 4 elements: Zn, Ge, As, and Se above 200GPa: Ordering of atomic volume by atomic number. *J. Appl. Phys.* **2021**, *129*, 02590. [CrossRef]
19. Akahama, Y.; Fujihisa, H.; Kawamura, H. New helical chain structure for scandium at 240 GPa. *Phys. Rev. Lett.* **2005**, *94*, 195503. [CrossRef] [PubMed]
20. Dewaele, A.; Stutzmann, V.; Bouchet, J.; Bottin, F.; Occelli, F.; Mezouar, M. High pressure-temperature phase diagram and equation of state of titanium. *Phys. Rev. B* **2015**, *91*, 134108. [CrossRef]
21. Akahama, Y.; Kawaguchi, S.; Hirao, N.; Ohishi, Y. Observation of high-pressure bcc phase of titanium at 243GPa. *J. Appl. Phys.* **2020**, *128*, 035901. [CrossRef]
22. Dubrovinsky, L.; Dubrovinskaia, N.; Bykova, E.; Bykov, M.; Prakapenka, V.; Prescher, C.; Glazyrin, K.; Liermann, H.-P.; Hanfland, M.; Ekholm, M.; et al. The most incompressible metal osmium at static pressures above 750 gigapascals. *Nature* **2015**, *525*, 226–228. [CrossRef] [PubMed]

23. Tschauner, O. Observations about the pressure-dependence of ionic radii. *Geoscience* **2022**, *12*, 246. [CrossRef]
24. Anzellini, S.; Dewaele, A.; Occelli, F.; Loubeyre, P.; Mezouar, M. Equation of state of rhenium and application for ultra high pressure calibration. *J. Appl. Phys.* **2014**, *115*, 043511. [CrossRef]
25. Fratanduono, D.E.; Millot, M.; Braun, D.G.; Ali, S.J.; Fernandez-Pañella, A.; Seagle, C.T.; Davis, J.-P.; Brown, J.L.; Akahama, Y.; Kraus, R.G.; et al. Establishing gold and platinum standards to 1 Terapascal using shockless compression. *Science* **2021**, *372*, 1063–1068. [CrossRef]
26. Weck, G.; Desgreniers, S.; Loubeyre, P.; Mezouar, M. Single-Crystal Structural Characterization of the Metallic Phase of Oxygen. *Phys. Rev. Lett.* **2009**, *102*, 255503. [CrossRef]
27. Cheng, J.; Li, B.; Liu, W.; Smith, J.S.; Majumdar, A.; Luo, W.; Ahuja, R.; Shu, J.; Wang, J.; Sinogeikin, S.; et al. Ultrahigh-pressure isostructural electronic transitions in hydrogen. *Nature* **2019**, *573*, 558–562.
28. Eremets, M.I.; Gavriliuk, A.G.; Trojan, I.A.; Dzivenko, D.A.; Boehler, R. Single-bonded cubic form of nitrogen. *Nat. Mater.* **2004**, *3*, 558–563. [CrossRef] [PubMed]
29. Dalladay-Simpson, P.; Binns, J.; Pena-Alvarez, M.; Donnelly, M.E.; Greenberg, E.; Prakapenka, V.; Chen, X.J.; Gregoryanz, E.; Howie, R.T. Band gap closure, incommensurability and molecular dissociation of dense chlorine. *Nat. Commun.* **2019**, *10*, 1134. [CrossRef] [PubMed]
30. Fujihisa, H.; Fujii, Y.; Takemura, K.; Shimomura, O. Structural aspects of dense solid halogens under high-pressure studied by X-ray diffraction—Molecular dissociation and metallization. *J. Phys. Chem. Solids* **1995**, *56*, 1439–1444. [CrossRef]
31. Akahama, Y.; Miyakawa, M.; Taniguchi, T.; Sano-Furukawa, A.; Machida, S.; Hattori, T. Structure refinement of black phosphorus under high pressure. *J. Chem. Phys.* **2020**, *153*, 104704. [CrossRef] [PubMed]
32. Li, B.; Ding, Y.; Kim, D.Y.; Wang, L.; Weng, T.-C.; Yang, W.; Yu, Z.; Ji, C.; Wang, J.; Shu, J.; et al. Probing the Electronic Band Gap of Solid Hydrogen by Inelastic X-ray Scattering up to 90 GPa. *Phys. Rev. Lett.* **2021**, *126*, 036402. [CrossRef]
33. Hanfland, M.; Loa, I.; Syassen, K.; Schwarz, U.; Takemura, K. Equation of state of lithium to 21 GPa. *Solid State Commun.* **1999**, *112*, 123–127. [CrossRef]
34. Hanfland, M.; Syassen, K.; Christensen, N.E.; Novikov, D.L. New high-pressure phases of lithium. *Nature* **2000**, *408*, 175–178. [CrossRef] [PubMed]
35. Guillaume, C.L.; Gregoryanz, E.; Degtyareva, O.; McMahon, M.I.; Hanfland, M.; Evans, S.; Guthrie, M.; Sinogeikin, S.V.; Mao, H.-K. Cold melting and solid structures of dense lithium. *Nat. Phys.* **2011**, *7*, 211–214. [CrossRef]
36. Hanfland, M.; Loa, I.; Syassen, K. Sodium und pressure: Bcc to fcc structural transition and pressure-volume relation to 100 GPa. *Phys. Rev. B* **2002**, *65*, 184109. [CrossRef]
37. Olijnyk, H.; Holzapfel, W.B. Phase transitions in K and Rb under pressure. *Phys. Lett.* **1983**, *99A*, 381–383. [CrossRef]
38. Lundegaard, L.F.; Stinton, G.W.; Zelazny, M.; Guillaume, C.L.; Proctor, J.E.; Loa, I.; Gregoryanz, E.; Nelmes, R.J.; McMahon, M.I. Observation of a reentrant phase transition in incommensurate potassium. *Phys. Rev. B* **2013**, *88*, 054106. [CrossRef]
39. Lundegaard, L.F.; Marques, M.; Stinton, G.; Ackland, G.J.; Nelmes, R.J.; McMahon, M.I. Observation of the oP8 crystal structure in potassium at high pressure. *Phys. Rev. B* **2009**, *80*, 020101. [CrossRef]
40. Schwarz, U.; Grzechnik, A.; Syassen, K.; Loa, I.; Hanfland, M. Rubidium-IV: A high pressure phase with complex crystal structure. *Phys. Rev. Lett.* **1999**, *83*, 4085–4088. [CrossRef]
41. Storm, C.V.; McHardy, J.D.; Finnegan, S.E.; Pace, E.J.; Stevenson, M.G.; Duff, M.J.; MacLeod, S.G.; McMahon, M.I. Behavior of rubidium at over eightfold static compression. *Phys. Rev. B* **2021**, *103*, 224103. [CrossRef]
42. Hall, H.T.; Merrill, L.; Barnett, J.D. A high pressure phase of cesium. *Science* **1964**, *146*, 1297–1299. [CrossRef] [PubMed]
43. Takemura, K.; Minomura, S.; Shimomura, O. X-ray Diffraction Study of Electronic Transitions in Cesium under High Pressure. *Phys. Rev. Lett.* **1982**, *49*, 1772–1775. [CrossRef]
44. Schwarz, U.; Takemura, K.; Hanfland, M.; Syassen, K. Crystal Structure of Cesium-V. *Phys. Rev. Lett.* **1998**, *81*, 2711–2713. [CrossRef]
45. Takemura, K.; Shimomura, O.; Fujihisa, H. Cs(VI): A new high-pressure polymorph of cesium above 72 GPa. *Phys. Rev. Lett.* **1991**, *66*, 2014–2017. [CrossRef] [PubMed]
46. Stinton, G.W.; MacLeod, S.G.; Cynn, H.; Errandonea, D.; Evans, W.J.; Proctor, J.E.; Meng, Y.; McMahon, M.I. Equation of state and high-pressure/high-temperature phase diagram of magnesium. *Phys. Rev. B* **2014**, *90*, 134105. [CrossRef]
47. Anzellini, A.; Errandonea, D.; MacLeod, S.G.; Botella, P.; Daisenberger, D.; De'Ath, J.M.; Gonzalez-Platas, J.; Ibáñez, J.; McMahon, M.I.; Munro, K.A.; et al. Phase diagram of calcium at high pressure and high temperature. *Phys. Rev. Mater.* **2022**, *2*, 083608. [CrossRef]
48. Anderson, M.S.; Swenson, C.A.; Peterson, D.T. Experimental equations of state for calcium, strontium, and barium metals to 20 kbar and from 4 to 295 K. *Phys. Rev. B* **1990**, *41*, 3329–3338. [CrossRef] [PubMed]
49. Olijnyk, H.; Holzapfel, W.B. Phase transitions in alkaline earth metals under pressure. *Phys. Lett. A* **1984**, *100*, 191–194. [CrossRef]
50. Fujihisa, H.; Nakamoto, Y.; Sakata, M.; Shimizu, K.; Matsuoka, T.; Ohishi, Y.; Yamawaki, H.; Takeya, S.; Gotoh, Y. Ca-VII: A Chain Ordered Host-Guest Structure of Calcium above 210 GPa. *Phys. Rev. Lett.* **2013**, *110*, 235501. [CrossRef]
51. Bovornratanaraks, T.; Allan, D.R.; Belmonte, S.A.; McMahon, M.I.; Nelmes, R.J. Complex monoclinic superstructure in Sr-IV. *Phys. Rev. B* **2006**, *73*, 144112. [CrossRef]
52. Nelmes, R.J.; Allan, D.R.; McMahon, M.I.; Belmonte, S.A. Self-Hosting Incommensurate Structure of Barium IV. *Phys. Rev. Lett.* **1999**, *83*, 4081–4084. [CrossRef]

53. Kenichi, T. High-pressure structural study of barium to 90 GPa. *Phys. Rev. B* **1994**, *50*, 16238–16246. [CrossRef] [PubMed]
54. Fujihisa, H.; Akahama, Y.; Kawamura, H.; Gotoh, Y.; Yamawaki, H.; Sakashita, M.; Takeya, S.; Honda, K. Incommensurate composite crystal structure of scandium-II. *Phys. Rev. B* **2005**, *72*, 132103. [CrossRef]
55. Stavrou, E.; Yang, L.H.; Söderlind, P.; Aberg, D.; Radousky, H.B.; Armstrong, M.R.; Belof, J.L.; Kunz, M.; Greenberg, E.; Prakapenka, V.B.; et al. Anharmonicity-induced first-order isostructural phase transition of zirconium under pressure. *Phys. Rev. B* **2018**, *98*, 220101. [CrossRef]
56. Anzellini, S.; Bottin, F.; Bouchet, J.; Dewaele, A. Phase transitions and equation of state of zirconium under high pressure. *Phys. Rev. B* **2020**, *102*, 184105. [CrossRef]
57. Akahama, Y.; Kawaguchi, S.; Hirao, N.; Ohishi, Y. High-pressure stability of bcc-vanadium and phase transition to a rhombohedral structure at 200 GPa. *J. Appl. Phys.* **2021**, *129*, 135902. [CrossRef]
58. Dewaele, A.; Torrent, M.; Loubeyre, P.; Mezouar, M. Compression curves of transition metals in the Mbar range: Experiments and projector augmented-wave calculations. *Phys. Rev. B* **2008**, *78*, 104102. [CrossRef]
59. Salamat, A.; Briggs, R.; Bouvier, P.; Petitgirard, S.; Dewaele, A.; Cutler, M.E.; Cora, F.; Daisenberger, D.; Garbarino, G.; McMillan, P.F. High-pressure structural transformations of Sn up to 138 GPa: Angle-dispersive synchrotron X-ray diffraction study. *Phys. Rev. B* **2013**, *88*, 104104. [CrossRef]
60. Kuznetsov, A.; Dmitriev, V.; Dubrovinsky, L.; Prakapenka, V.B.; Weber, H.-P. FCC–HCP phase boundary in lead. *Solid State Commun.* **2018**, *122*, 125–127. [CrossRef]
61. Mao, H.K.; Wu, Y.; Shu, J.F.; Hu, J.Z.; Hemley, R.J.; Cox, D.E. High pressure phase transition and equation of state of lead to 238 GPa. *Solid State Commun.* **1990**, *74*, 1027–1029. [CrossRef]
62. Degtyareva, O.; McMahon, M.I.; Allan, D.R.; Nelmes, R.J. Structural Complexity in Gallium under High Pressure: Relation to Alkali Elements. *Phys. Rev. Lett.* **2004**, *93*, 205502. [CrossRef] [PubMed]
63. Kenichi, T.; Kobayashi, K.; Arai, M. High-pressure bct-fcc phase transition in Ga. *Phys. Rev. B* **1998**, *58*, 2482–2486. [CrossRef]
64. Cazorla, C.; MacLeod, S.G.; Errandonea, D.; Munro, K.A.; McMahon, M.I.; Popescu, C. Thallium under extreme compression. *J. Phys. Condens. Matter* **2016**, *28*, 445401. [CrossRef] [PubMed]
65. Menoni, C.S.; Hu, J.Z.; Spain, I.L. Germanium at high pressures. *Phys. Rev. B* **1986**, *34*, 362–368. [CrossRef] [PubMed]
66. Takemura, K.; Schwarz, U.; Syassen, K.; Hanfland, M.; Christensen, N.E.; Novikov, D.L.; Loa, I. High-pressure Cmca and hcp phases of germanium. *Phys. Rev. B* **2000**, *62*, R10603–R10606. [CrossRef]
67. Kennedy, C.S.; Kennedy, G.C. Equilibrium boundary between graphite and diamond. *J. Geophys. Res.* **1976**, *81*, 2467–2470. [CrossRef]
68. McMahon, M.I.; Nelmes, R.J. New high-pressure phase of Si. *Phys. Rev. B* **1993**, *47*, 8337–8340. [CrossRef]
69. Hanfland, M.; Schwarz, U.; Syassen, K.; Takemura, K. Crystal Structure of the High-Pressure Phase Silicon VI. *Phys. Rev. Lett.* **1999**, *82*, 1197–1199. [CrossRef]
70. Akahama, Y.; Kawamura, H.; Carlson, S.; Le Bihan, T.; Hausermann, D. Structural stability and equation of state of simple-hexagonal phosphorus to 280 GPa: Phase transition at 262 GPa. *Phys. Rev. B* **2000**, *61*, 3139–3142. [CrossRef]
71. Degtyareva, O.; McMahon, M.I.; Nelmes, R.J. Pressure-induced incommensurate-to-incommensurate phase transition in antimony. *Phys. Rev. B* **2004**, *70*, 184119. [CrossRef]
72. Degtyareva, O.; Gregoryanz, E.; Mao, H.K.; Hemley, R.J. Crystal structure of sulfur and selenium at pressures up to 160 GPa. *High Press. Res.* **2005**, *25*, 17–33. [CrossRef]
73. Hejny, C.; Falconi, S.; Lundegaard, L.F.; McMahon, M.I. Phase transitions in tellurium at high pressure and temperature. *Phys. Rev. B* **2006**, *74*, 174119. [CrossRef]
74. Akahama, Y.; Okawa, N.; Sugimoto, T.; Fujihisa, H.; Hirao, N.; Ohishi, Y. Coexistence of a metastable double hcp phase in bcc–fcc structure transition of Te under high pressure. *Jpn. J. Appl. Phys.* **2018**, *57*, 02560. [CrossRef]
75. Reichlin, R.; McMahan, A.K.; Ross, M.; Martin, S. Optical, X-ray, and band-structure studies of iodine at pressures of several megabars. *Phys. Rev. B* **1994**, *49*, 3725–3733. [CrossRef] [PubMed]
76. Fujihisa, H.; Takemura, K.; Onoda, M.; Gotoh, Y. Two intermediate incommensurate phases in the molecular dissociation process of solid iodine under high pressure. *Phys. Rev. Res.* **2021**, *3*, 033174. [CrossRef]
77. Vohra, Y.K.; Akella, J. 5f bonding in thorium metal at extreme compressions—Phase transitions to 300 GPa. *Phys. Rev. B* **1991**, *67*, 3563–3566.
78. Decremps, F.; Belhadi, L.; Farber, D.L.; Moore, K.T.; Occelli, F.; Gauthier, M.; Polian, A.; Antonangeli, D.; Aracne-Ruddle, C.M.; Amadon, B. Diffusionless α-γ Phase Transition in Polycrystalline and Single-Crystal Cerium. *Phys. Rev. Lett.* **2011**, *106*, 065701. [CrossRef]
79. O'Bannon, E.F.; Pardo, O.S.; Soderlind, P.; Sneed, D.; Lipp, M.J.; Park, C.; Jenei, Z. Systematic structural study in praseodymium compressed in a neon pressure medium up to 185 GPa. *Phys. Rev. B* **2022**, *105*, 144107. [CrossRef]
80. Jayaraman, A.; Sherwood, R.C. Phase transformation in samarium induced by high pressure and its effect on the antiferromagnetic ordering. *Phys. Rev.* **1964**, *134*, A691–A692. [CrossRef]
81. Finnegan, S.E.; Pace, E.J.; Storm, C.V.; McMahon, M.I.; MacLeod, S.G.; Liermann, H.P.; Glazyrin, K. High-pressure structural systematics in samarium up to 222 GPa. *Phys. Rev. B* **2020**, *101*, 174109. [CrossRef]
82. Finnegan, S.E.; Storm, C.V.; Pace, E.J.; McMahon, M.I.; MacLeod, S.G.; Plekhanov, E.; Bonini, N.; Weber, C. High-Pressure Structural Systematics in Neodymium to 302 GPa. *Phys. Rev. B* **2021**, *103*, 134117. [CrossRef]

83. Husband, R.J.; Loa, I.; Munro, K.A.; McBride, E.E.; Evans, S.R.; Liermann, H.-P.; McMahon, M.I. Phase transitions in europium at high pressures. *High Press. Res.* **2013**, *33*, 158–164. [CrossRef]
84. Golosova, N.O.; Kozlenko, D.P.; Lukin, E.V.; Kichanov, S.E.; Savenko, B.N. High pressure effects on the crystal and magnetic structure of ^{160}Gd metal. *J. Magn. Magnet. Mater.* **2021**, *540*, 16848515. [CrossRef]
85. Errandonea, D.; Boehler, R.; Schwager, B.; Mezouar, M. Structural studies of gadolinium at high pressure and temperature. *Phys. Rev. B* **2007**, *75*, 014103. [CrossRef]
86. Tschauner, O.; Grubor-Urosevic, O.; Dera, P.; Mulcahy, S. Anomalous elastic behaviour in hcp-Sm-type Dysprosium. *J. Phys. Chem. C* **2012**, *116*, 2090–2095. [CrossRef]
87. Cerenius, Y.; Dubrovinsky, L. Compressibility measurements on iridium. *J. Alloys Comp.* **2000**, *306*, 26–29. [CrossRef]

Article

Stability of FeVO$_4$-II under Pressure: A First-Principles Study

Pricila Betbirai Romero-Vázquez [1], Sinhué López-Moreno [2,3,*] and Daniel Errandonea [4]

1. División de Materiales Avanzados, IPICYT, Camino a la presa de San José 2055 Col. Lomas 4a sección, San Luis Potosí 78126, Mexico
2. CONACYT—División de Materiales Avanzados, IPICYT, Camino a la presa de San José 2055 Col. Lomas 4a sección, San Luis Potosí 78126, Mexico
3. Centro Nacional de Supercómputo, IPICYT, Camino a la presa de San José 2055 Col. Lomas 4a sección, San Luis Potosí 78126, Mexico
4. Departamento de Física Aplicada, Instituto de Ciencias de Materiales, MALTA Consolider Team, Universidad de Valencia, 46100 Valencia, Spain
* Correspondence: sinhue.lopez@ipicyt.edu.mx

Abstract: In this work, we report first-principles calculations to study FeVO$_4$ in the CrVO$_4$-type (phase **II**) structure under pressure. Total-energy calculations were performed in order to analyze the structural parameters, the electronic, elastic, mechanical, and vibrational properties of FeVO$_4$-**II** up to 9.6 GPa for the first time. We found a good agreement in the structural parameters with the experimental results available in the literature. The electronic structure analysis was complemented with results obtained from the Laplacian of the charge density at the bond critical points within the Quantum Theory of Atoms in Molecules methodology. Our findings from the elastic, mechanic, and vibrational properties were correlated to determine the elastic and dynamic stability of FeVO$_4$-**II** under pressure. Calculations suggest that beyond the maximum pressure covered by our study, this phase could undergo a phase transition to a wolframite-type structure, such as in CrVO$_4$ and InVO$_4$.

Keywords: FeVO$_4$ under pressure; CrVO$_4$-type structure; first-principles; mechanical properties; vibrational properties; electronic properties

Citation: Romero-Vázquez, P. B.; López-Moreno, S.; Errandonea, D. Stability of FeVO$_4$-II under Pressure: A First-Principles Study. *Crystals* **2022**, *12*, 1835. https://doi.org/ 10.3390/cryst12121835

Academic Editor: Artem R. Oganov

Received: 18 November 2022
Accepted: 13 December 2022
Published: 15 December 2022

Publisher's Note: MDPI stays neutral with regard to jurisdictional claims in published maps and institutional affiliations.

Copyright: © 2022 by the authors. Licensee MDPI, Basel, Switzerland. This article is an open access article distributed under the terms and conditions of the Creative Commons Attribution (CC BY) license (https:// creativecommons.org/licenses/by/ 4.0/).

1. Introduction

High-pressure studies on ABO_4 compounds have increased in the last two decades due to the evolution of the diamond-anvil cell (DAC) and its impact on other relevant areas, such as materials sciences [1,2]. Among ABO_4 compounds, those with a CrVO$_4$-type structure (Space Group, SG: *Cmcm*, No. 63, Z = 4, with [6 − 4] coordination for the [*A-B* cations]) have received less attention. CrVO$_4$-type compounds are a family formed mainly by phosphates, vanadates, chromates, and sulfates [3]. Given their position in the Bastide diagram [4], it has been reported that they could have a structural phase transition driven by pressure to the wolframite [5,6] or zircon [7,8] structure. This implies a change in the coordination to [6 − 4 + 2] and [8 − 4] in the high-pressure phases, respectively.

According to the literature, AVO_4 vanadates with CrVO$_4$-type structures have been used in applications as photocatalyst materials [1,9,10], cathodoluminescent materials [11,12], and scintillators [13,14], among others. The properties of these compounds are reported to be highly dependent on the occupation of the *d* valence electrons of the transition metal cations [9–14]. Until now, only four vanadates with a CrVO$_4$-type structure have been reported: CrVO$_4$ [6], FeVO$_4$ [15], InVO$_4$ [16], and the less studied TlVO$_4$ [3]. Unlike the other vanadates, FeVO$_4$ presents a great polymorphism, crystallizing at ambient conditions in the triclinic AlVO$_4$-type [17] structure (SG: $P\bar{1}$, No. 2, Z = 6) [15], while the CrVO$_4$-type phase was obtained from high-pressure-high-temperature studies [18] and synthesized at ambient conditions by hydrothermal method [19]. Further, the antiferromagnetic order in FeVO$_4$-**II** was observed after the characterization of its magnetic properties [19]. However, until now, no more experimental or theoretical studies have been reported to characterize this phase. It is worth mentioning that most of the studies carried out at high pressure on this compound have focused on characterizing the crystal structure of the different polymorphs [15,18,20–22].

In this work, we performed a first-principles characterization of FeVO$_4$-II up to 9.6 GPa. The main goal is to determine the elastic and dynamic stability of the studied polymorph at ambient conditions and under pressure. In this sense, a complete report of the elastic constants, mechanical properties, elastic anisotropy, and vibrational properties is presented. Furthermore, given the relevance of the electronic structure of this compound in materials science, we computed the density of states and the band structure together with an analysis of the charge density with the QTAIM methodology [23,24].

2. Computational Details

Total energy calculations were performed within the framework of the density functional theory (DFT) [25] and the projector-augmented wave (PAW) [26,27] method as implemented in the Vienna *Ab initio* Simulation Package (VASP) [28–31]. We used a plane-wave energy cutoff of 520 eV to ensure high precision in our calculations. The exchange-correlation energy was described within the generalized gradient approximation (GGA) in the Perdew–Burke–Ernzerhof for solids (PBEsol) formulation [32]. The GGA+U was used to account for the strong correlation between the electrons in the d shell, on the basis of Dudarev's method [33]. In this method, the Coulomb interaction U and the onsite exchange interaction J_H are treated together as $U_{eff} = U - J_H$. For our GGA+U calculations, we choose $U = 6$ eV and $J_H = 0.95$ eV. Similar values were previously used with relative success in the study of other iron and vanadate compounds [15,20,34–37].

The Monkhorst-Pack scheme was employed for the Brillouin-zone (BZ) integrations [38] with a mesh $4 \times 3 \times 3$, which corresponds to a set of 8 special k-points in the irreducible BZ. In the relaxed equilibrium configuration, the forces are less than 0.3 meV/Å per atom in each Cartesian direction. The highly converged results on forces are required for the calculations of the dynamical matrix using the direct force constant approach [39], which allows us to identify the irreducible representation and the character of the phonon modes at the zone center (Γ point). The phonon density of states (PDOS) were obtained from the calculation of the phonons in the whole BZ with a supercell $2 \times 2 \times 2$ times the conventional unit cell by using the PHONON software. The calculations of elastic constants were made following the methodology validated in Ref. [40] with a k-points mesh of $8 \times 6 \times 6$, a plane-wave energy cutoff of 570 eV, and a POTIM parameter of 0.016. The elastic tensor is determined by performing six finite distortions of the lattice and deriving the elastic constants from the strain–stress relationship [41]. The electron density was computed with a refined grid to be analyzed within the fundaments of Quantum Theory of Atoms in Molecules (QTAIM) [42] with CRITIC [43,44], and AIM [45] codes.

3. Results and Discussion

3.1. Crystal Structure

According to the literature, FeVO$_4$ has been successfully synthesized as a metastable polymorph at ambient conditions in the orthorhombic CrVO$_4$-type structure (SG: *Cmcm*, No. 63, $Z = 4$) [19], also called the FeVO$_4$-II phase, see Figure 1a. In this structure, there are four non-equivalent Wyckoff positions (WP): Fe occupies the 4a (0, 0, 0), V the 4c (0, y, 1/4), and O the 8f (0, y, z), and 8g (x, y, 1/4). Internally, the structure consists of edge-sharing FeO$_6$ distorted octahedra along the c direction. The –FeO$_6$–FeO$_6$– chains are linked together by means of the VO$_4$ distorted tetrahedra, while the VO$_4$ tetrahedra are not linked to each other, as can be seen in Figure 1b. As shown in Figure 1c, there are two non-equivalent interatomic distances in the FeO$_6$ octahedra, the apical Fe-O distance (d_{Fe-O1}) and the equatorial Fe-O distance (d_{Fe-O2}). In the VO$_4$ tetrahedra, there are two V–O interatomic distances (d_{V-O1} and d_{V-O2}). In both polyhedra, the respective interatomic distance d_{X-O2} is larger than d_{X-O1} (X = Fe, V). Inside the polyhedrons, there are important angles to mention that are relevant for the structural changes in the polyhedra under pressure. The α angle is formed between the equatorial plane of the FeO$_6$ octahedra and the polar bond, while the β (γ) angle is between the two d_1 (d_2) interatomic distances of the VO$_4$ tetrahedra.

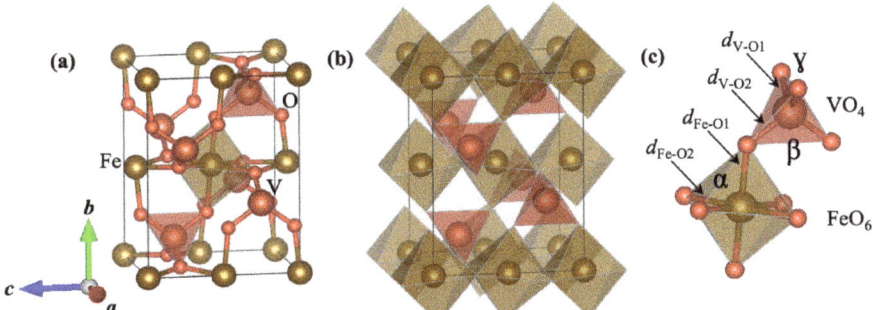

Figure 1. (a) Crystal structure of FeVO$_4$-II polymporph, (b) polyhedral representation of the structure, and (c) irregular polyhedra of FeO$_6$ and VO$_4$ units. The different bond distances mentioned in the text are labeled as well as the bond angles. We used the VESTA software [46] to build the structures.

The equilibrium volume and unit-cell parameters at $P \approx 0$ GPa were computed by minimizing the total crystal energy for different volumes ranging from −4 to 10 GPa while allowing the internal atomic positions and the lattice parameters to relax. We used the third-order Birch–Murnaghan [47] equation-of-state (EOS) to fit the volume–energy curve. Our results showed that the lowest energy configuration is obtained when an antiferromagnetic coupling between Fe ions is obtained, in accordance with the experimental results from Mössbauer effect [19]. The optimized lattice parameters (a, b, c), volume (V), interatomic distances (d_{Fe-O1}, d_{Fe-O2}, d_{V-O1}, d_{V-O2}), internal angles (α, β, γ), bulk modulus (B) and pressure derivative of B (B') appear in Table 1 together with experimental results from the literature [19]. Furthermore, we have included the reported parameters for other CrVO$_4$-type structure vanadates for comparison. According to Table 1, the equilibrium volume at ambient pressure of AVO_4 (A = Cr, Fe, In, Tl) compounds increases with the atomic mass of cation A, while the opposite occurs for bulk modulus. The obtained results for FeVO$_4$–II are in excellent agreement with experiments, with a difference of less than 1% in the lattice parameters. For this volume, the WPs are 4a (0, 0, 0) for Fe, 4c (0, 0.35788, 1/4) for V, 8f (0, 0.76239, 0.96753), and 8g (0.26481, 0.47778, 1/4) for O. Following Table 1, the equilibrium volume at ambient pressure of AVO_4 (A = Cr, Fe, In, Tl) compounds increases with the atomic mass of cation A. In contrast, the opposite occurs for bulk modulus.

Table 1. Calculated structural parameters for FeVO$_4$-II phase. a, b, and c are the lattice parameters, V is the volume, d_{X-On} (X = Fe, V; n = 1, 2) and α, β, and γ, are the interatomic distances and angles inside the crystal structure, respectively (see Figure 1). B is the bulk modulus and B' is the pressure derivative of B. Results from this work are denoted with ⋆. We also include experimental results from the literature of AVO_4 compounds for comparison.

	a (Å)	b (Å)	c (Å)	V (Å^3)	d_{Fe-O1} (Å)	d_{Fe-O2} (Å)	d_{V-O1} (Å)	d_{V-O2} (Å)	α (°)	β (°)	γ (°)	B (GPa)	B'	Ref.
FeVO$_4$	5.6038	8.2624	6.1726	285.80	1.9734	2.0377	1.6705	1.7842	89.271	106.990	112.547	88.8	2.8	⋆
	5.5941	8.3216	6.2252	289.80	1.9784	2.0521	1.6544	1.7762				88.3	3.2	DFT [15]
	5.6284	8.2724	6.1118	284.57	2.0060	2.0290	1.6520	1.7920	89.600	107.200	112.500			Exp. [19]
CrVO$_4$	5.5785	8.2830	6.0576	279.90								93.0	4.0	DFT [6]
	5.5680	8.2080	5.9770	273.16								63.0	4.0	Exp. [6]
InVO$_4$	5.7547	8.6168	6.6751	331.00	2.1877	2.1392	1.6690	1.7824				71.0	4.0	DFT [5]
	5.7380	8.4920	6.5820	320.72								69.0	4.0	Exp. [16]
TlVO$_4$	5.8390	8.6870	6.8000	344.92										Exp. [48]

According to the most recent high-pressure experiments carried out on FeVO$_4$ [15,20], this compound undergoes the following phase transition sequence, $P\bar{1} \rightarrow P\bar{1}' \rightarrow \alpha$-MnMoO$_4$-type, driven by pressure. However, previous high-pressure studies reported the next phase transition sequence $P\bar{1} \rightarrow$ CrVO$_4$-type $\rightarrow$ wolframite in a small range of pressure [49]. Therefore, it is unknown if the CrVO$_4$-type structure of FeVO$_4$ would transition to wolframite when high-pressure experiments are carried out starting from the CrVO$_4$-type structure, as occurs with CrVO$_4$ [6] and InVO$_4$ [5,16], and was predicted through first-principles calculations [15]. Therefore, from now on, we will only deal with studying the behavior of FeVO$_4$-II under pressure without considering whether the CrVO$_4 \rightarrow$ wolframite transition occurs. Figure 2 shows the pressure evolution of the structural parameters for the FeVO$_4$-II phase. According to Figure 2a, a volume reduction of $\Delta V = -9\%$ is observed for a range of pressure of $\approx$10 GPa. Considering the anisotropy of the orthorhombic crystal structure, ΔV is more reflected in the [010] and [001] directions, while the change in [100] is practically negligible, see Figure 2b. Indeed the compressibility of the [001] direction is comparable to the linear compressibility of diamond, making the studied material a near zero linear compressibility material along [001]. The anisotropy can also be understood with the help of the other internal parameters from Figure 2c–f. As seen in Figure 2c, the interatomic distances d_{Fe-On} of FeO$_6$ undergo more changes than the d_{V-On} distances of VO$_4$, which is reflected in the larger compressibility of V_{FeO_4} octahedra against the minimal change in V_{VO_4}, see Figure 2d. Hence, the large compressibility of the y-axis is due primarily to the compressibility of octahedra due to the reduction in d_{Fe-O2}. Whereas, despite the reduction in d_{V-O2}, the β and γ angles adjust in such a way that V_{VO_4} remains almost constant with pressure. On the other hand, the distortion parameter Δ_d presents an important change at 6.2 GPa, which will be discussed in detail in Section 3.3.

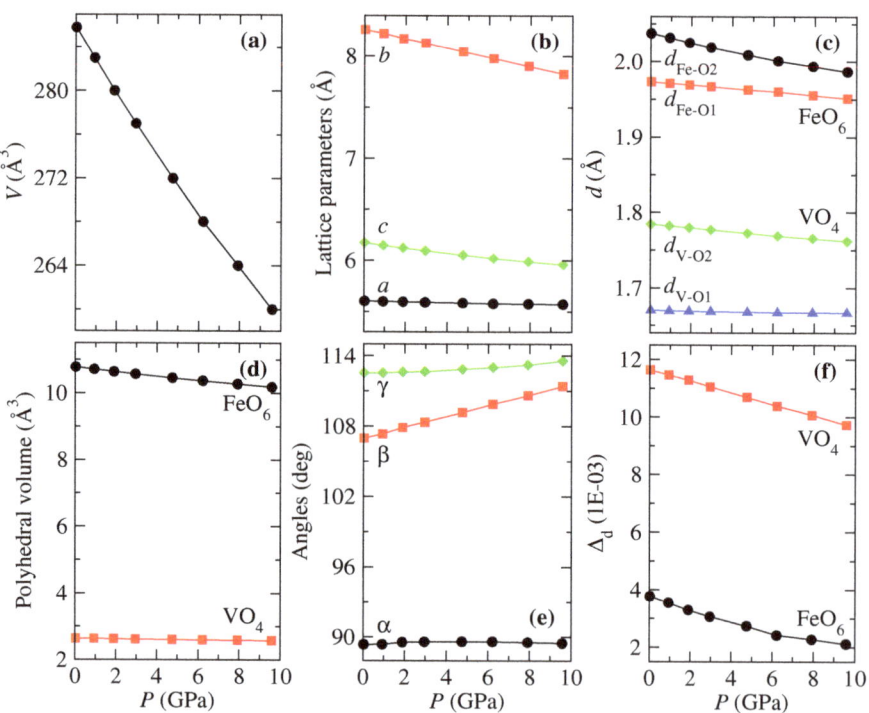

Figure 2. Structural parameters for FeVO$_4$-II as a function of pressure: (**a**) volume, (**b**) lattice parameters, (**c**) interatomic distances, (**d**) polyhedral volumes, (**e**) angles, and (**f**) polyhedral distortion parameter.

3.2. Electronic Structure

Figure 3a,b displays our calculated density of states (DOS) and band structures along the high-symmetry path Z–Γ–S–R–Z–T–Y–Γ of the FeVO$_4$-II phase at ambient pressure and 9.6 GPa, respectively. Red dashed lines at 0 eV indicate the Fermi level. At ambient pressure, the valence band maximum (VBM) is located at an H ($\frac{1}{2}, \frac{1}{2}, -u$) point (between T and Y), mainly occupied by $2p$ (p_x) states from oxygen. In contrast, the conduction band minima (CBM) is located at the Γ point with contributions of the $3d$ states from Fe, mostly from d_{yz} sub-orbital. Therefore, FeVO$_4$-II phase presents an indirect energy band-gap E_g = 2.11 eV. In comparison, the direct Γ–Γ energy band-gap is 2.17 eV. In this case, the maximum at Γ below the Fermi level is occupied by $2p$ (p_x and p_y) states from oxygen, as happens in other ABO$_4$ compounds [40]. It has been reported that CBM of many rare-earth AVO$_4$ vanadates is dominated by vanadium $3d$ states, which makes these compounds present energy band-gaps close to 3.8 eV [50,51]. However, in the case of vanadates InVO$_4$ (phase II), CrVO$_4$ (phase II), and FeVO$_4$ (phase I), the energy band-gap lowers to 3.2, 2.6, and 2.1 eV, respectively, due to contribution from cation A to the valence and conduction band [5,6,50,52]. Hence, our E_g value follows the observed results for those vanadates. According to partial electronic DOS of Figure 3a, the states below the Fermi level mainly belong to the O atoms. In contrast, the states above the conduction bands are due to Fe, followed by V and O.

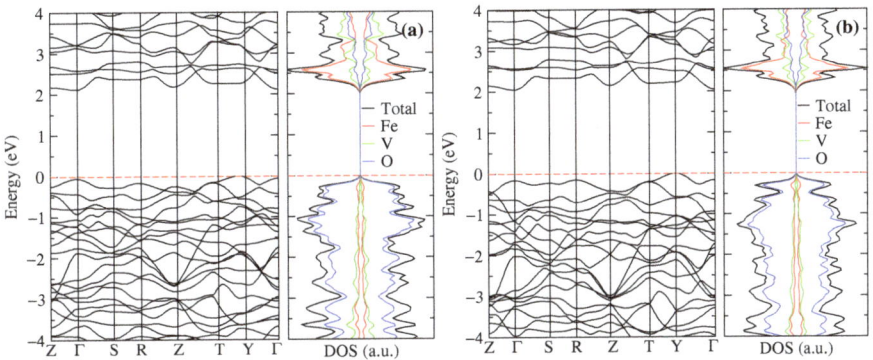

Figure 3. Band structure and electronic density of states at (**a**) ambient pressure and (**b**) 9.6 GPa of FeVO$_4$-II phase.

As pressure increases up to 9.6 GPa, the VBM shifts to the high-symmetry point Y while the CBM is still at the Γ point with the same band occupation observed at ambient pressure, see Figure 3b. Hence, at this pressure, the band-gap is still indirect with E_g = 2.05 eV, which gives a pressure coefficient of dE_g/dp = −7.1 meV/GPa. In contrast, the pressure coefficients for InVO$_4$ and CrVO$_4$ are positive, with values of 13.5 [52] (8.9 [5]) meV for InVO$_4$ and 1.9 (4.1) meV for CrVO$_4$ [6], obtained by optical absorption measurements (DFT calculations). On the other hand, the forbidden direct Γ–Γ band-gap increases with pressure up to 2.21 eV at 9.6 GPa, which gives a value of dE_g/dP = 4.2 meV, similar to the obtained values for the other AVO$_4$ vanadates.

We performed a topological analysis of the charge density (ρ) with the QTAIM methodology to complement the study of the electronic structure of FeVO$_4$-II. Figure 4a,b show the gradient vector field of the electron density ($\nabla \rho$) of FeO$_6$ and VO$_4$ polyhedra, respectively (gold for Fe, purple for V, and red for O). Blue points indicate the atomic nucleus, corresponding to the maximum critical points (MCP). The minimal CP (mCP) are in green, and the bond CPs (BCPs) are in black and are located in the bond path (blue dashed lines) between Fe (V) and O. The distance between Fe (V) and the BCP (d_{BCP}) is plotted as a function of pressure in Figure 4c. As seen, the distances d_{BCP} follow the same trend observed in the interatomic distances of Figure 2c. Therefore, the crystal structure can be interpreted in terms of charge density. In this sense, the Laplacian of the charge density ($\nabla^2 \rho_{\text{BCP}}$) at the BCP gives an idea of the ionicity ($\nabla^2 \rho_{\text{BCP}} > 0$) or covalency ($\nabla^2 \rho_{\text{BCP}} < 0$) of a system [53]. According to Figure 4d, the Fe–O and V–O bonds have an ionic nature. The obtained values of $\nabla^2 \rho_{\text{BCP}}$ (0.214–0.295 a.u.) for Fe–O bonds in the distorted octahedra are larger than those reported for A–O bonds in AWO$_4$ (0.086–0.147 a.u.) [54], AMoO$_4$ (0.085–0.22 a.u.) [53], and

$ATcO_4$ (0.045–0.153 a.u.) [40] compounds. In the case of the VO_4 distorted tetrahedra, there is a significative difference in the values of $\nabla^2\rho_{BCP_1}$ and $\nabla^2\rho_{BCP_2}$, which is related to the interatomic distances d_{V-O_n} (see Table 1) being significantly larger than those obtained for the FeO_6 distorted octahedra. This is the reason for the low compressibility of the VO_4 polyhedra. As the pressure grows, the Laplacian values increase, strengthening the Fe–O and V–O bonds and the resistance of the polyhedrons to compression.

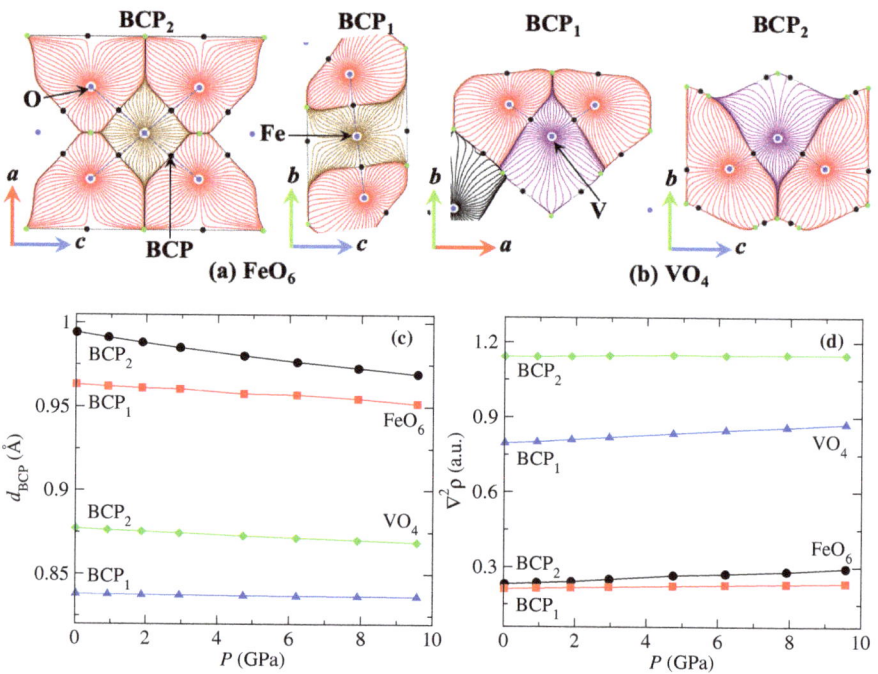

Figure 4. Gradient vector field of the electron density ($\nabla\rho$) for (**a**) FeO_6 and (**b**) VO_4 polyhedra. (**c**) Evolution of the interatomic distances from the cations to the critical bond points and (**d**) Laplacian of the charge density at the critical bond point as a function of pressure for $FeVO_4$.

3.3. Elastic constants and mechanical properties

The elastic constants (c_{ij}) determine the resistance to deformation in the directions and planes of a crystal, making them highly dependent on the system's symmetry. For the orthorhombic crystalline system, there are nine independent elastic constants: c_{11}, c_{22} and c_{33} for longitudinal compression, c_{12}, c_{13} and c_{23} for transverse expansion, c_{44}, c_{55}, and c_{66} for pure shear [55]. Table 2 shows the elastic constants c_{ij} of $FeVO_4$-II at 0 GPa and the respective pressure coefficients dc_{ij}/dP. The necessary and sufficient Born stability criteria for orthorhombic systems are [56]: $c_{11} > 0$, $c_{11}c_{22} > c_{12}^2$, $c_{11}c_{22}c_{33} + 2c_{12}c_{13}c_{33} + 2c_{12}c_{13}c_{23} - c_{11}c_{23}^2 - c_{22}c_{13}^2 - c_{33}c_{12}^2 > 0$, $c_{44} > 0$, $c_{55} > 0$, and $c_{66} > 0$, which are fulfilled for $FeVO_4$-II. Under hydrostatic pressure (P), the stability criteria are [57]: $c_{ii} - P > 0$ ($i = 1$ to 6), $c_{11} + c_{22} + c_{33} + 2c_{12} + 2c_{13} + 2c_{23} + 3P > 0$, $c_{11} + c_{22} - 2c_{12} - 4P > 0$, $c_{11} + c_{33} - 2c_{13} - 4P > 0$, and $c_{22} + c_{33} - 2c_{23} - 4P > 0$, which are also meet. According to Table 2, the elastic constant c_{11} presents the higher value, followed by c_{33} and c_{22}, indicating that longitudinally the crystal is more resistant to deformation in the a-axis; this was also observed in the evolution of the lattice parameters under pressure in Figure 2b. As would be expected, the largest transverse expansion constant is c_{12}, followed by c_{13} and c_{23}, c_{33} being the smallest of the nine elastic constants. At the same time, the shear elastic constants are generally small against those from longitudinal compression and transverse expansion. We found that elastic constants follow a linear trend with pressure, where only c_{22} presents a negative pressure coefficient related to the trends observed in the structural parameters with pressure.

Table 2. Elastic constants c_{ij} (in GPa) at 0 GPa and pressure coefficients.

	c_{11}	c_{22}	c_{33}	c_{44}	c_{55}	c_{66}	c_{12}	c_{13}	c_{23}
c_{ij}	234.2	125.5	144.8	62.98	52.49	47.04	78.46	68.21	39.81
dc_{ij}/dP	2.41	−0.87	3.78	0.74	1.04	0.15	2.39	2.18	2.57

The elastic constants and compliances were used to compute the bulk (B) and shear (G) modulus with the Hill procedure [58], which is the average of the Voigt (V) [59] and Reuss (R) [60] methods. The bulk modulus in the Voigt and Reuss methods is computed as follows: $9B_V = (c_{11} + c_{22} + c_{33} + 2c_{12} + 2c_{13} + 2c_{23})$, and $B_R^{-1} = (s_{11} + s_{22} + s_{33} + 2s_{12} + 2s_{13} + 2s_{23})$, respectively. While the shear modulus is obtained with the following equations: $15G_V = (c_{11} + c_{22} + c_{33} - c_{12} - c_{23} - c_{13} + 3c_{44} + 3c_{55} + 3c_{66})$, and $15G_R^{-1} = [4(s_{11} + s_{22} + s_{33}) - 4(s_{12} + s_{13} + s_{23}) + 3(s_{44} + s_{55} + s_{66})]$, respectively. B and G in the Hill average are obtained by means of $B = (B_V + B_R)/2$ and $G = (G_V + G_R)/2$, respectively. On the other hand, Young's modulus (E) and Poisson's ratio (ν) are given by $E = 9BG/(3B + G)$ and $\nu = (3B - 2G)/(6B + 2G)$, respectively. The Vickers hardness was calculated with the approximation of Tian et al. [61]: $H_v = 0.92\, k^{1.137} G^{0.708}$, $k = G/B$.

The mechanical properties and their evolution with pressure are provided in Figure 5. As seen, B increases almost linearly with pressure, which means that the resistance to deformation increases for this pressure range. Whereas G undergoes a change in the linear trend above 3 GPa, reaching a maximum value of 53.9 at 6.2 GPa and falling down to 53.3 GPa at 9.6 GPa. There is no significant change in G for this pressure range, but the change in trend will substantially affect the other properties that depend on G. Interestingly, in other materials, such change in the slope of the pressure dependence of the shear modulus G has been identified as a precursor effect of a phase transition observed at higher pressures [62]. According to the equations of Voigt and Reuss for G, the behavior of G is due to the negative pressure coefficient for c_{22} and the significant values of dc_{ij}/dP for the transverse expansion elastic constants. On the other hand, B behaves linearly because B_V (B_G) only sums the longitudinal and transverse elastic constants (compliances), while c_{22} (s_{22}) decreases with pressure, so the value of c_{22} affects B by a small constant factor.

The behavior of Young's modulus is similar to G due to its relation with this mechanical property. According to Figure 5c, E increases by 4.3 % in a pressure range of 7.86 GPa; above this pressure, it decreases from 138.1 to 137.6 GPa. This behavior indicates that $FeVO_4$ could undergo a phase transition from a $CrVO_4$-type phase to another one at a relatively small pressure since it has been observed that $FeVO_4$ undergoes several phase transitions below 10 GPa, see Refs. [15,20] and references therein. The Poisson's ratio follows an increasing trend, taking values between 0.26 and 0.29, which are in the range for ceramic semiconductors (0.25–0.42) [63] and are similar for other ABO_4 compounds [40,54,64]. These values indicate that ionic contribution dominates in $FeVO_4$-II. The Pugh relation, Figure 5e, represents the ductility or brittleness of the system. A value below 1.75 [11] reveals a tendency to brittleness, while a value greater than this indicates that the material behaves in a ductile manner. $FeVO_4$-II presents a value of 1.745 at ambient pressure and increases to 2.06 at 9.6 GPa, which means that ductility increases with pressure. As was expected, $FeVO_4$-II, as other ABO_4 compounds, presents a small Vickers hardness due to the relatively small values of B and G. Given the relation between H_V and B/G, it is natural that the Vickers hardness decreases with pressure, Figure 5f.

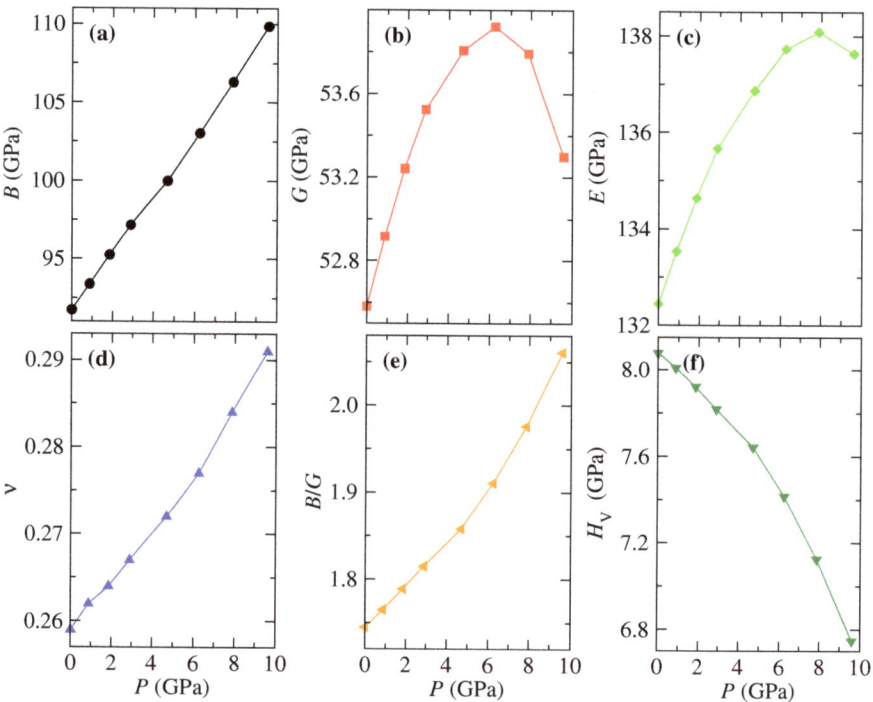

Figure 5. Pressure dependence of mechanical properties obtained from elastic constants with the Hill average: (**a**) bulk modulus, B, (**b**) shear modulus, G, (**c**) Young's modulus, E, (**d**) Poisson's ratio, ν, (**e**) Pugh's ratio, B/G, and (**f**) Vickers hardness, H_V, for FeVO$_4$-**II**.

On the other hand, elastic anisotropy has important implications in engineering science and crystal physics because it is commonly responsible for the formation of microcracks in materials [65]. In this work, we analyzed several anisotropic factors discussed below. The anisotropic shear factors provide a measure of the degree of anisotropy in the bonding between atoms in different crystallographic planes. The shear anisotropic factor for the {100} shear planes between the $\langle 011 \rangle$ and $\langle 010 \rangle$ directions is $A_{\{100\}} = 4c_{44}/(c_{11} + c_{33} - 2c_{13})$. For the {010} shear planes between $\langle 101 \rangle$ and $\langle 001 \rangle$ directions it is $A_{\{010\}} = 4c_{55}/(c_{22} + c_{33} - 2c_{23})$. For the {001} shear planes between $\langle 110 \rangle$ and $\langle 010 \rangle$ directions, it is $A_{\{001\}} = 4c_{66}/(c_{11} + c_{22} - 2c_{12})$. The factors $A_{\{100\}}$, $A_{\{010\}}$, and $A_{\{001\}}$ must be one for isotropic crystals, and any value greater or smaller than one measures the degree of elastic anisotropy. According to Figure 6a, $A_{\{100\}}$ presents the smallest deviation from one, followed by $A_{\{010\}}$ and $A_{\{001\}}$. However, once pressure is applied, $A_{\{010\}}$ is more affected than the others. Factor $A_{\{010\}}$ is related to the equatorial interatomic distances of the distorted octahedra [$d_{\text{Fe-O2}}$, see Figure 2c], which makes quite a bit of sense since it is $d_{\text{Fe-O2}}$ that undergoes the greatest change with pressure. While $A_{\{001\}}$ is more related to $d_{\text{Fe-O1}}$, with this interatomic distance being the second one most affected by pressure.

Other important factors are the universal anisotropy $A_U = 5G_V/G_R + B_V/B_R - 6$, the percentage of elastic anisotropy in bulk [$A_B = (B_V - B_R)/(B_V + B_R)$] and shear [$A_G = (G_V - G_R)/(G_V + G_R)$] modulus, where a value of zero is associated with isotropic elastic constants, which means that $B_V = B_R$ and $G_V = G_R$. For the entire range of pressure, A_B is larger than A_G because the difference between B_V and B_R is bigger than between G_V and G_R. However, the pressure affects more A_G than A_B as happens with G. This difference increases above 6.2 GPa as G_R presents the same behavior of G, while G_V is always increasing. A_U is larger than A_G and A_B since the anisotropies of both modules are added.

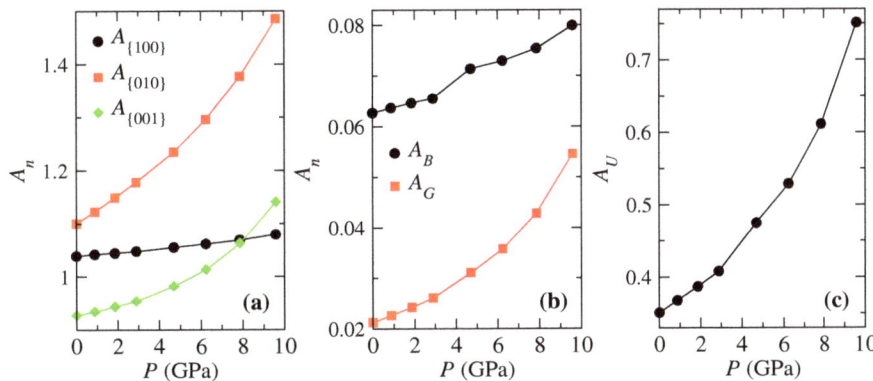

Figure 6. (a) Pressure dependence of anisotropy factors: (a) by crystallographic direction, (b) percentage bulk and shear anisotropy, and (c) universal anisotropy for FeVO$_4$-II.

3.4. Vibrational Properties

According to the group theory, the space group *Cmcm* has the following Raman, infrared (IR), and silent phonon modes at the zone center $\Gamma = 5A_g + 4B_{1g} + 2B_{2g} + 4B_{3g}$, $\Gamma = 6B_{1u} + 7B_{2u} + 5B_{3u}$, and $\Gamma = 3A_u$, respectively. These modes can be further classified into internal (symmetric stretching, ν_1, symmetric bending, ν_2, asymmetric stretching, ν_3, and asymmetric bending, ν_4) and external (translational, T, and rotational, R) modes of VO$_4$ units. The calculated Raman, IR, and silent phonon frequencies, the pressure coefficients ($d\omega/dP$), and the Grüneisen parameters ($\gamma = -\partial \ln(\omega)/\partial \ln V$) at the Γ point appear in Table 3.

Table 3. Calculated Raman, infrared, and silent phonon frequencies (ω in cm^{-1}), pressure coefficients ($d\omega/dp$), and Grüneisen parameters (γ) of FeVO$_4$-II.

	Raman				Infrared				Silent		
Mode	ω	$d\omega/dp$	γ	Mode	ω	$d\omega/dp$	γ	Mode	ω	$d\omega/dp$	γ
$T(B_{3g})$	153.7	1.37	0.57	$T(B_{1u})$	153.7	0.24	0.15	$T(A_u)$	164.4	1.29	0.77
$T(B_{1g})$	177.4	−1.05	−0.82	$T(B_{1u})$	180.8	1.68	0.86	$\nu_2(A_u)$	220.3	0.55	0.23
$R(B_{1g})$	219.6	0.52	0.22	$R(B_{3u})$	201.2	−0.01	−0.03	$\nu_2(A_u)$	404.9	7.45	1.68
$T(A_g)$	221.8	3.23	1.29	$\nu_2(B_{2u})$	274.9	1.12	0.37				
$R(B_{2g})$	261.8	4.25	1.47	$\nu_4(B_{3u})$	298.8	1.51	0.46				
$\nu_2(A_g)$	355.1	0.30	0.07	$T(B_{3u})$	315.1	3.82	0.25				
$\nu_4(B_{3g})$	361.7	1.36	0.33	$\nu_4(B_{2u})$	327.9	−0.88	−1.06				
$\nu_2(B_{2g})$	366.4	3.78	0.94	$R(B_{1u})$	363.1	2.41	0.48				
$\nu_4(B_{1g})$	374.0	3.58	0.88	$\nu_4(B_{2u})$	381.5	6.61	1.61				
$\nu_4(A_g)$	394.2	1.55	0.36	$\nu_4(B_{1u})$	398.3	4.93	1.13				
$\nu_4(B_{3g})$	431.3	5.90	1.25	$\nu_2(B_{2u})$	420.6	4.38	0.92				
$\nu_3(B_{1g})$	652.8	5.71	0.78	$\nu_3(B_{3u})$	654.2	5.64	0.77				
$\nu_3(A_g)$	751.0	5.74	0.72	$\nu_3(B_{2u})$	735.8	5.47	0.67				
$\nu_1(A_g)$	918.9	1.33	0.12	$\nu_3(B_{1u})$	866.8	2.58	0.26				
$\nu_3(B_{3g})$	922.0	2.44	0.24	$\nu_3(B_{2u})$	924.3	1.38	0.13				

According to Table 3, one Raman, $R(B_{1g})$, and two IR modes, $R(B_{3u})$ and $\nu_4(B_{2u})$, have negative pressure coefficients and Grüneisen parameters (soften with pressure), which are related to the instability of the CrVO$_4$-type structure under pressure [2]. Such behavior was already observed in InVO$_4$ [5] and CrVO$_4$ [20]. The soft translational Raman $T(B_{1g})$ mode involves a translation of the VO$_4$ polyhedra accompanied by a rotation of the FeO$_6$ units around the z-axis as rigid units, see Figure 7, whose freedom of movement is affected by the decrease in the lattice parameter *b*, which is the most compressible under pressure. In contrast, the remaining Raman modes harden with pressure. The IR soft $R(B_{3u})$ mode is quite similar to the $R(B_{1g})$, but in this case, the FeO$_6$ vibrates asymmetrically. While the other soft B_{2u} mode implies the asymmetric bending in the VO$_4$ polyhedra with a translation of the Fe cation in the *c* direction, which produces a stretching of the apical d_{Fe-O1} interatomic distances. Therefore, the softening of R and IR modes is related to the compression of the y-axis. Interestingly, phosphates with CrVO$_4$-type structure, such as InPO$_4$ and TiPO$_4$, only present some IR soft modes. The main difference being that Ti(In)PO$_4$ is more compressible in the $c(b)$-axis [7].

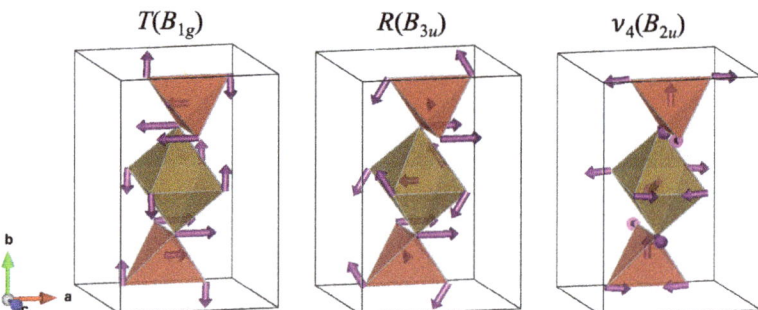

Figure 7. Eigenvectors of the phonon frequencies with negative pressure coefficients of FeVO$_4$-II. We used the VESTA software [46] to build the structures.

The phonon dispersion relation and the phonon density of states (DOS) are displayed in Figure 8 at (a) ambient pressure and (b) 9.6 GPa. As can be seen in Figure 8a, the CrVO$_4$-type structure of FeVO$_4$ complains with the dynamic stability criteria at $P = 0$, which implies that ω^2 (**q**, s) > 0 for all wave vectors **q** and polarization s (longitudinal and transverse modes) [66]. Therefore, FeVO$_4$-II complains with the elastic and dynamic stability criteria to be experimentally synthesized [19], such as InVO$_4$ [5] and CrVO$_4$ [20]. We can see no significant differences in the phonon spectrum at 9.6 GPa, which ensures that the FeVO$_4$-II phase would be stable in a pressure range where the isostructural vanadates InVO$_4$ and CrVO$_4$ undergo a phase transition to the wolframite-type structure [5,20].

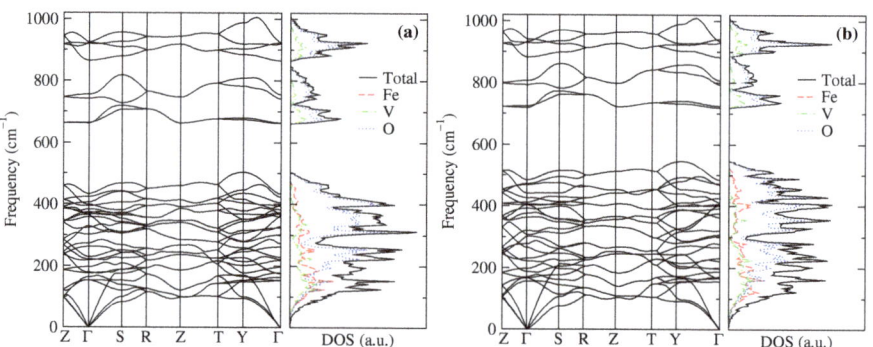

Figure 8. Phonon spectrum and phonon DOS at (**a**) 0 GPa, and (**b**) 9.6 GPa of FeVO$_4$-II.

The phonon DOS of FeVO$_4$-II is more similar to CrVO$_4$ than InVO$_4$ because In is more than double the mass of Cr and Fe. The partial phonon DOS of the first lower frequency zone (up to ≈ 505 cm^{-1}) is due to external (R and T) and internal (ν_2 and ν_4) phonon modes, with Fe and V similar contributions. In contrast, the two high-frequency zones are due to pure internal vibrations ν_1 and ν_3, which indicates that phonon DOS is only due to O and V.

4. Conclusions

First-principles calculations were performed to study FeVO$_4$ in the CrVO$_4$-type structure up to 9.6 GPa. Our results indicate that this phase is elastically and dynamically stable for this pressure range. FeVO$_4$ presents an indirect band-gap $E_g = 2.11$ eV, similar to the value reported for the triclinic phase of FeVO$_4$ at ambient pressure. The Laplacian charge density at BCPs in both polyhedra is larger than those reported for AMoO$_4$ and

AWO_4 compounds and increases with pressure. It was observed that the y-axis is the most compressible, related to the reduction in the interatomic distances of the FeO_6 polyhedra. These changes in the structural parameters have strong effects on the phonon modes with negative pressure coefficients. We found that the shear modulus G softens with pressure, which suggests that the $FeVO_4$-II phase could undergo a phase transition below 10 GPa as was predicted in a previous study and happen in $CrVO_4$ and $InVO_4$, where these compounds have a phase transition from $CrVO_4$-type to the wolframite structure. We hope this work encourages other research groups to perform high-pressure characterization studies with the diamond-anvil cell device in order to increase the knowledge of less studied orthovanadates with $CrVO_4$-type structure, such as $FeVO_4$ and $TlVO_4$.

Author Contributions: Methodology, P.B.R.-V., S.L.-M.; calculations, P.B.R.-V., S.L.-M.; writing—original draft preparation, P.B.R.-V., S.L.-M.; supervision, S.L.-M.; discussion, S.L.-M., D.E.; writing—review and editing, P.B.R.-V., S.L.-M., D.E. All authors have read and agreed to the published version of the manuscript.

Funding: S.L.-M. thanks CONACYT of Mexico for financial support through the program "Programa de Investigadoras e Investigadores por México". D.E. thanks the financial support from the Spanish Ministerio de Ciencia e Investigación (10.13039/501100011033) under Project PID2019-106383GB-41, as well as through the MALTA Consolider Team research network (RED2018-102612-T) and from Generalitat Valenciana under Grants PROMETEO CIPROM/2021/075-GREENMAT and MFA/2022/007.

Institutional Review Board Statement: Not applicable.

Informed Consent Statement: Not applicable.

Data Availability Statement: All relevant data that support the findings of this study are available from the corresponding authors upon request.

Acknowledgments: The authors gratefully acknowledge the computing time granted by LANCAD and CONACYT on the supercomputer Miztli at LSVP DGTIC UNAM. Furthermore, the IPICYT Supercomputing National Center for Education and Research, grant TKII-R2022-SLM1/PBRV1.

Conflicts of Interest: The authors declare no conflict of interest.

References

1. Errandonea, D.; Manjon, F. Pressure effects on the structural and electronic properties of ABX_4 scintillating crystals. *Prog. Mater. Sci.* **2008**, *53*, 711–773. [CrossRef]
2. Errandonea, D.; Garg, A.B. Recent progress on the characterization of the high-pressure behaviour of AVO_4 orthovanadates. *Prog. Mater. Sci.* **2018**, *97*, 123–169. [CrossRef]
3. Baran, E. Materials belonging to the $CrVO_4$ structure type: Preparation, crystal chemistry and physicochemical properties. *J. Mater. Sci.* **1998**, *33*, 2479–2497. [CrossRef]
4. de Jesus Pereira, A.L.; Santamaría-Pérez, D.; Vilaplana, R.; Errandonea, D.; Popescu, C.; da Silva, E.L.; Sans, J.A.; Rodríguez-Carvajal, J.; Muñoz, A.; Rodríguez-Hernández, P.; et al. Experimental and Theoretical Study of $SbPO_4$ under Compression. *Inorg. Chem.* **2020**, *59*, 287–307. [CrossRef] [PubMed]
5. López-Moreno, S.; Rodríguez-Hernández, P.; Munoz, A.; Errandonea, D. First-principles study of $InVO_4$ under pressure: phase transitions from $CrVO_4$-to $AgMnO_4$-type structure. *Inorg. Chem.* **2017**, *56*, 2697–2711. [CrossRef]
6. Botella, P.; López-Moreno, S.; Errandonea, D.; Manjón, F.J.; Sans, J.; Vie, D.; Vomiero, A. High-pressure characterization of multifunctional $CrVO_4$. *J. Phys. Condens. Matter.* **2020**, *32*, 385403. [CrossRef]
7. López-Moreno, S.; Errandonea, D. Ab initio prediction of pressure-induced structural phase transitions of $CrVO_4$-type orthophosphates. *Phys. Rev. B* **2012**, *86*, 104112. [CrossRef]
8. Dwivedi, A.; Kaiwart, R.; Varma, M.; Velaga, S.; Poswal, H. High-pressure structural investigations on $InPO_4$. *J. Solid State Chem.* **2020**, *282*, 121065. [CrossRef]
9. Butcher J., D.P.; Gewirth, A.A. Photoelectrochemical response of $TlVO_4$ and $InVO_4$: $TlVO_4$ composite. *Chem. Mater.* **2010**, *22*, 2555–2562. [CrossRef]
10. Zhao, C.; Tan, G.; Yang, W.; Xu, .; Liu, T.; Su, Y.; Ren, H.; Xia, A. Fast interfacial charge transfer in α-Fe_2O_3-$\delta C \delta$/$FeVO_{4-x}$+ δC x-δ bulk heterojunctions with controllable phase content. *Sci. Rep.* **2016**, *6*, 38603. [CrossRef]
11. Balamurugan, M.; Yun, G.; Ahn, K.S.; Kang, S.H. Revealing the beneficial effects of $FeVO_4$ nanoshell layer on the $BiVO_4$ inverse opal core layer for photoelectrochemical water oxidation. *J. Phys. Chem. C* **2017**, *121*, 7625–7634. [CrossRef]

12. Marberger, A.; Elsener, M.; Ferri, D.; Sagar, A.; Schermanz, K.; Krocher, O. Generation of NH_3 selective catalytic reduction active catalysts from decomposition of supported $FeVO_4$. *ACS Catal.* **2015**, *5*, 4180–4188. [CrossRef]
13. Yi, X.; Li, J.; Chen, Z.; Tok, A. Single-crystalline $InVO_4$ nanotubes by self-template-directed fabrication. *J. Am. Ceram. Soc.* **2010**, *93*, 596–600. [CrossRef]
14. Yu, Y.; Ju, P.; Zhang, D.; Han, X.; Yin, X.; Zheng, L.; Sun, C. Peroxidase-like activity of $FeVO_4$ nanobelts and its analytical application for optical detection of hydrogen peroxide. *Sens. Actuators B* **2016**, *233*, 162–172. [CrossRef]
15. López-Moreno, S.; Errandonea, D.; Pellicer-Porres, J.; Martínez-García, D.; Patwe, S.J.; Achary, S.N.; Tyagi, A.K.; Rodríguez-Hernández, P.; Muñoz, A.; Popescu, C. Stability of $FeVO_4$ under Pressure: An X-ray Diffraction and First-Principles Study. *Inorg. Chem.* **2018**, *57*, 7860–7876. [CrossRef]
16. Errandonea, D.; Gomis, O.; García-Domene, B.; Pellicer-Porres, J.; Katari, V.; Achary, S.N.; Tyagi, A.K.; Popescu, C. New polymorph of $InVO_4$: A high-pressure structure with six-coordinated vanadium. *Inorg. Chem.* **2013**, *52*, 12790–12798. [CrossRef] [PubMed]
17. Arisi, E.; Sánchez, S.; Leccabue, F.; Watts, B.; Bocelli, G.; Calderón, F.; Calestani, G.; Righi, L. Preparation and characterization of $AlVO_4$ compound. *J. Mater. Sci.* **2004**, *39*, 2107–2111. [CrossRef]
18. Muller, J.; Joubert, J. Synthese sous haute pression d'oxygene d'une forme dense ordonne´e de $FeVO_4$ et mise en evidence d'une varié té´ allotropique de structure $CrVO_4$. *J. Solid State Chem.* **1975**, *14*, 8–13. [CrossRef]
19. Oka, Y.; Yao, T.; Yamamoto, N.; Ueda, Y.; Kawasaki, S.; Azuma, M.; Takano, M. Hydrothermal synthesis, crystal structure, and magnetic properties of $FeVO_4$-II. *J. Solid State Chem.* **1996**, *123*, 54–59. [CrossRef]
20. Gonzalez-Platas, J.; López-Moreno, S.; Bandiello, E.; Bettinelli, M.; Errandonea, D. Precise characterization of the rich structural landscape induced by pressure in multifunctional $FeVO_4$. *Inorg. Chem.* **2020**, *59*, 6623–6630. [CrossRef]
21. Young, A.; Schwartz, C. High pressure forms of $CrVO_4$ and $FeVO_4$. *Acta Crystallogr. Sect. A Found. Crystallogr.* **1962**, *15*, 1305–1305. [CrossRef]
22. Laves, F.; Young, A.P.; Schwartz, C.M. On the high-pressure form of $FeVO_4$. *Acta Cryst.* **1964**, *17*, 1476–1477. [CrossRef]
23. Ouahrani, T.; Garg, A.B.; Rao, R.; Rodríguez-Hernández, P.; Muñoz, A.; Badawi, M.; Errandonea, D. High-Pressure Properties of Wolframite-Type $ScNbO_4$. *J. Phys. Chem. C* **2022**, *126*, 4664–4676. [CrossRef]
24. Bader, R.F. Atoms in molecules. *Accounts Chem. Res.* **1985**, *18*, 9–15. [CrossRef]
25. Jones, R.O. Density functional theory: Its origins, rise to prominence, and future. *Rev. Mod. Phys.* **2015**, *87*, 897–923. [CrossRef]
26. Blöchl, P.E. Projector augmented-wave method. *Phys. Rev. B* **1994**, *50*, 17953–17979. [CrossRef] [PubMed]
27. Kresse, G.; Joubert, D. From ultrasoft pseudopotentials to the projector augmented-wave method. *Phys. Rev. B* **1999**, *59*, 1758–1775. [CrossRef]
28. Kresse, G.; Hafner, J. Ab initio molecular dynamics for liquid metals. *Phys. Rev. B* **1993**, *47*, 558–561. [CrossRef]
29. Kresse, G.; Hafner, J. Ab initio molecular-dynamics simulation of the liquid-metal–amorphous-semiconductor transition in germanium. *Phys. Rev. B* **1994**, *49*, 14251–14269. [CrossRef]
30. Kresse, G.; Furthmüller, J. Efficient iterative schemes for ab initio total-energy calculations using a plane-wave basis set. *Phys. Rev. B* **1996**, *54*, 11169–11186. [CrossRef]
31. Kresse, G.; Furthmüller, J. Efficiency of ab-initio total energy calculations for metals and semiconductors using a plane-wave basis set. *Comput. Mater. Sci.* **1996**, *6*, 15–50. [CrossRef]
32. Csonka, G.I.; Perdew, J.P.; Ruzsinszky, A.; Philipsen, P.H.T.; Lebègue, S.; Paier, J.; Vydrov, O.A.; Ángyán, J.G. Assessing the performance of recent density functionals for bulk solids. *Phys. Rev. B* **2009**, *79*, 155107. [CrossRef]
33. Dudarev, S.L.; Botton, G.A.; Savrasov, S.Y.; Humphreys, C.J.; Sutton, A.P. Electron-energy-loss spectra and the structural stability of nickel oxide: An LSDA+U study. *Phys. Rev. B* **1998**, *57*, 1505–1509. [CrossRef]
34. Ruiz-Fuertes, J.; Errandonea, D.; López-Moreno, S.; González, J.; Gomis, O.; Vilaplana, R.; Manjón, F.J.; Muñoz, A.; Rodríguez-Hernández, P.; Friedrich, A.; et al. High-pressure Raman spectroscopy and lattice-dynamics calculations on scintillating $MgWO_4$: Comparison with isomorphic compounds. *Phys. Rev. B* **2011**, *83*, 214112. [CrossRef]
35. López, S.; Romero, A.H.; Mejía-López, J.; Mazo-Zuluaga, J.; Restrepo, J. Structure and electronic properties of iron oxide clusters: A first-principles study. *Phys. Rev. B* **2009**, *80*, 085107. [CrossRef]
36. López-Moreno, S.; Romero, A.H.; Mejía-López, J.; Muñoz, A.; Roshchin, I.V. First-principles study of electronic, vibrational, elastic, and magnetic properties of FeF_2 as a function of pressure. *Phys. Rev. B* **2012**, *85*, 134110. https://doi.org/10.1103/PhysRevB.85.134110.
37. Mejía-López, J.; Mazo-Zuluaga, J.; López-Moreno, S.; Muñoz, F.; Duque, L.F.; Romero, A.H. Physical properties of quasi-one-dimensional MgO and Fe_3O_4-based nanostructures. *Phys. Rev. B* **2014**, *90*, 035411. [CrossRef]
38. Monkhorst, H.J.; Pack, J.D. Special points for Brillouin-zone integrations. *Phys. Rev. B* **1976**, *13*, 5188–5192. [CrossRef]
39. Parlinski, K. Computer Code PHONON. **2008**. Available online: http://wolf.ifj.edu.pl/phonon (accessed on 5 December 2022).
40. Romero-Vázquez, P.B.; López-Moreno, S.; Errandonea, D. First-principles study of $ATcO_4$ pertechnetates. *J. Phys. Chem. Solids* **2022**, *171*, 110979. [CrossRef]
41. Le Page, Y.; Saxe, P. Symmetry-general least-squares extraction of elastic data for strained materials from ab initio calculations of stress. *Phys. Rev. B* **2002**, *65*, 104104. [CrossRef]
42. Popelier, P.L.A. The QTAIM Perspective of Chemical Bonding. In *The Chemical Bond: Fundamental Aspects of Chemical Bonding*; John Wiley & Sons, Ltd: Hoboken, NJ, USA, 2014; chapter 8, pp. 271-308. [CrossRef]

43. Otero-de-la Roza, A.; Blanco, M.; Pendás, A.M.; Luaña, V. Critic: A new program for the topological analysis of solid-state electron densities. *Comput. Phys. Commun.* **2009**, *180*, 157–166. [CrossRef]
44. Otero-de-la Roza, A.; Johnson, E.R.; Luaña, V. Critic2: A program for real-space analysis of quantum chemical interactions in solids. *Comput. Phys. Commun.* **2014**, *185*, 1007–1018. [CrossRef]
45. Vega, D.; Almeida, D. AIM-UC: An application for QTAIM analysis. *J. Comput. Methods Sci. Eng.* **2014**, *14*, 131–136. [CrossRef]
46. Momma, K.; Izumi, F. VESTA3 for three-dimensional visualization of crystal, volumetric and morphology data. *J. Appl. Crystallog.* **2011**, *44*, 1272–1276. [CrossRef]
47. Birch, F. Finite elastic strain of cubic crystals. *Phys. Rev.* **1947**, *71*, 809–824. [CrossRef]
48. Touboul, M.; Ingrain, D. Syntheses et propriétés thermiques de InVO$_4$ et TlVO$_4$. *J. Less Common Met.* **1980**, *71*, 55–62. [CrossRef]
49. Hotta, Y.; Ueda, Y.; Nakayama, N.; Kosuge, K.; Kachi, S.; Shimada, M.; Koizumi, M. Pressure-products diagram of FeVO$_4$ system ($0 \leq x \leq 0.5$). *J. Solid State Chem.* **1984**, *55*, 314–319. [CrossRef]
50. Errandonea, D. High pressure crystal structures of orthovanadates and their properties. *J. Appl. Phys.* **2020**, *128*, 040903. [CrossRef]
51. Díaz-Anichtchenko, D.; Errandonea, D. Comparative Study of the Compressibility of M$_3$V$_2$O$_8$ (M = Cd, Zn, Mg, Ni) Orthovanadates. *Crystals* **2022**, *12*. [CrossRef]
52. Botella, P.; Errandonea, D.; Garg, A.; Rodriguez-Hernandez, P.; Muñoz, A.; Achary, S.; Vomiero, A. High-pressure characterization of the optical and electronic properties of InVO$_4$, InNbO$_4$, and InTaO$_4$. *SN Appl. Scien.* **2019**, *1*, 389. [CrossRef]
53. Monteseguro, V.; Ruiz-Fuertes, J.; Contreras-García, J.; Rodríguez-Hernández, P.; Muñoz, A.; Errandonea, D. High pressure theoretical and experimental analysis of the bandgap of BaMoO$_4$, PbMoO$_4$, and CdMoO$_4$. *Appl. Phys. Lett.* **2019**, *115*, 012102. [CrossRef]
54. Romero-Vázquez, P.B.; López-Moreno, S. Ab initio study of RaWO$_4$: Comparison with isoelectronic tungstates. *J. Solid State Chem.* **2023**, *317*, 123709. [CrossRef]
55. Nye, J.F.; et al. *Physical properties of crystals: their representation by tensors and matrices*; Oxford university press, 1985.
56. Mouhat, F.; Coudert, F.X. Necessary and sufficient elastic stability conditions in various crystal systems. *Phys. Rev. B* **2014**, *90*, 224104. [CrossRef]
57. ping Feng, L.; tang Liu, Z.; jun Liu, Q. Structural, elastic and mechanical properties of orthorhombic SrHfO3 under pressure from first-principles calculations. *Phys. B Condens. Matter* **2012**, *407*, 2009–2013. [CrossRef]
58. Najafvandzadeh, N.; López-Moreno, S.; Errandonea, D.; Pavone, P.; Draxl, C. First-principles study of elastic and thermal properties of scheelite-type molybdates and tungstates. *Mater. Today Commun.* **2020**, *24*, 101089. [CrossRef]
59. Voigt, W. *Lehrbuch Kristallphysik*; Springer: Berlin/Heidelberg, Germany, 1928. [CrossRef]
60. Reuss, A. Berechnung der Fliebgrenze von Mischkristallen auf Grund der Plastizitatsbedigung für Einkristalle. *Angew Appl. Math. Mech.* **1929**, *9*, 49–58. [CrossRef]
61. Tian, Y.; Xu, B.; Zhao, Z. Microscopic theory of hardness and design of novel superhard crystals. *Int. J. Refract. Met. Hard. Mater.* **2012**, *33*, 93–106. [CrossRef]
62. Singh, J.; Sahoo, S.S.; Venkatakrishnan, K.; Vaitheeswaran, G.; Errandonea, D. High-pressure study of the aurophilic topological Dirac material AuI. *J. Alloys Compd.* **2022**, *928*, 167178. [CrossRef]
63. Prawoto, Y. Seeing auxetic materials from the mechanics point of view: A structural review on the negative Poisson's ratio. *Comput. Mater. Sci.* **2012**, *58*, 140–153. [CrossRef]
64. Liu, Y.; Jia, D.; Zhou, Y.; Zhou, Y.; Zhao, J.; Li, Q.; Liu, B. Discovery of ABO4 scheelites with the extra low thermal conductivity through high-throughput calculations. *J. Mater.* **2020**, *6*, 702–711. [CrossRef]
65. Tvergaard, V.; Hutchinson, J.W. Microcracking in Ceramics Induced by Thermal Expansion or Elastic Anisotropy. *J. Am. Ceram. Soc.* **1988**, *71*, 157–166. [CrossRef]
66. Grimvall, G.; Magyari-Köpe, B.; Ozoliņš, V.; Persson, K.A. Lattice instabilities in metallic elements. *Rev. Mod. Phys.* **2012**, *84*, 945. [CrossRef]

Article

Phase Relations of Ni$_2$In-Type and CaC$_2$-Type Structures Relative to Fe$_2$P-Type Structure of Titania at High Pressure: A Comparative Study

Khaldoun Tarawneh * and Yahya Al-Khatatbeh

Department of Basic Sciences, Princess Sumaya University for Technology, Amman 11941, Jordan
* Correspondence: khaldoun@psut.edu.jo

Abstract: Density functional theory (DFT) based on *first-principles* calculations was used to study the high-pressure phase stability of various phases of titanium dioxide (TiO$_2$) at extreme pressures. We explored the phase relations among the following phases: the experimentally identified nine-fold hexagonal Fe$_2$P-type phase, the previously predicted ten-fold tetragonal CaC$_2$-type phase of TiO$_2$, and the recently proposed eleven-fold hexagonal Ni$_2$In-type phase of the similar dioxides zirconia (ZrO$_2$) and hafnia (HfO$_2$). Our calculations, using the generalized gradient approximation (GGA), predicted the Fe$_2$P $\to$ Ni$_2$In transition to occur at 564 GPa and Fe$_2$P $\to$ CaC$_2$ at 664 GPa. These transitions were deeply investigated with reference to the volume reduction, coordination number decrease, and band gap narrowing to better determine the favorable post-Fe$_2$P phase. Furthermore, it was found that both transitions are mostly driven by the volume reduction across transitions in comparison with the small contribution of the electronic energy gain. Additionally, our computed Birch–Murnaghan equation of state for the three phases reveals that CaC$_2$ is the densest phase, while Ni$_2$In is the most compressible phase.

Keywords: phase transitions; enthalpy difference components; equation of state; first principles; phase relations; band gap

Citation: Tarawneh, K.; Al-Khatatbeh, Y. Phase Relations of Ni$_2$In-Type and CaC$_2$-Type Structures Relative to Fe$_2$P-Type Structure of Titania at High Pressure: A Comparative Study. *Crystals* **2023**, *13*, 9. https://doi.org/10.3390/cryst13010009

Academic Editors: Daniel Errandonea and Enrico Bandiello

Received: 27 November 2022
Revised: 17 December 2022
Accepted: 18 December 2022
Published: 21 December 2022

Copyright: © 2022 by the authors. Licensee MDPI, Basel, Switzerland. This article is an open access article distributed under the terms and conditions of the Creative Commons Attribution (CC BY) license (https://creativecommons.org/licenses/by/4.0/).

1. Introduction

The nature of bonding in titania (TiO$_2$) has attracted great interest over the last few decades due to its interesting industrial applications such as photocatalysts, energy generation and storage, environmental protection, and many more [1,2]. One of the important and promising research directions in studying this dioxide is investigating the high-pressure behavior of TiO$_2$ polymorphs, both experimentally and theoretically, due to their interesting properties [3–39]. However, much attention, using measurements and calculations, has been given to exploring the high-pressure phase stability of TiO$_2$ phases and the transition pressures between different phases and their equation of state parameters as well as to searching for new possible phases (e.g., [3,8,11,15,18,19,21,22,29,31–33,37]). Over the past decades, density functional theory has become the workhorse theory for the identification of the pressure-driven phase transitions [40,41].

As mentioned above, titania can be found in many structural forms with increasing pressure. In this regard, the well-known transition sequence that is experimentally observed and theoretically predicted is as follows: Orthorhombic OI $\to$ orthorhombic cotunnite OII $\to$ Fe$_2$P-type [3,15,29,31]. Thus, the hexagonal Fe$_2$P-type structure (Figure 1a) is the highest-pressure phase experimentally observed for TiO$_2$ which has been found to be stable at (210 GPa, 4000 K) using diamond-anvil experiments [31]. Recently, the tetragonal CaC$_2$-type structure (Figure 1c) has been theoretically predicted as a post-Fe$_2$P phase of titanium dioxide at pressures beyond 647–689 GPa [33,36]. Additionally, recent density functional theory (DFT) calculations have predicted the hexagonal Ni$_2$In-type structure

(Figure 1b) to be the post-Fe$_2$P phase in the similar dioxides ZrO$_2$ and HfO$_2$ at pressures that exceed 300 GPa [42–44].

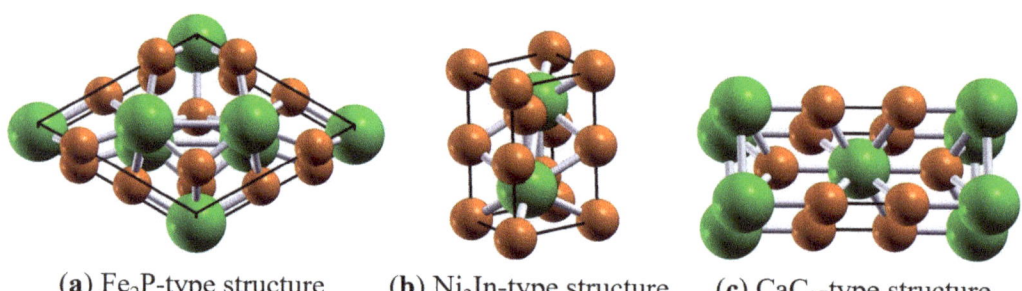

(a) Fe$_2$P-type structure **(b)** Ni$_2$In-type structure **(c)** CaC$_2$-type structure

Figure 1. Crystal structures of TiO$_2$ phases (generated using XCrySDen software [45]). The green spheres represent the titanium atom, while the alloy orange spheres represent the oxygen atom. (**a**) Fe$_2$P-type: crystal structure: hexagonal, space group: $P\bar{6}2m$, coordination number: 9, lattice parameters: $a = b = 5.3274$ Å and $c = 3.1234$ Å, atomic coordinates: Ti1 (1/3, 2/3, 1/2), Ti2 (0, 0, 0), O1 (0.262, 0, 1/2), O2 (0.601, 0, 0). (**b**) Ni$_2$In-type: crystal structure: hexagonal, space group: $P6_3/mmc$, coordination number: 11, lattice parameters: $a = b = 3.2632$ Å and $c = 6.0335$ Å, atomic coordinates: Ti (1/3, 2/3, 1/4), O1 (0, 0, 0), O2 (1/3, 2/3, 3/4). (**c**) CaC$_2$-type: crystal structure: tetragonal, space group: $I4/mmm$, coordination number: 10, lattice parameters: $a = b = 2.7407$ Å and $c = 6.7179$ Å, atomic coordinates: Ti (1/2, −1/2, 0), O (1, −1, 0.152).

However, it is important to note that the three dioxides (TiO$_2$, ZrO$_2$, and HfO$_2$) share obvious similarities, especially in their high-pressure behavior (e.g., see Ref. [26] and references therein). In this regard, the overlapping high-pressure phase transition sequence, experimentally observed and theoretically confirmed in the three dioxides is as follows: Monoclinic baddeleyite MI → OI → OII → Fe$_2$P [3,15,26,29,31,46–50].

The obvious and close similarities of the high-pressure behavior of titania, zirconia, and hafnia have motivated us to test the stability of Fe$_2$P-TiO$_2$ with respect to CaC$_2$ and Ni$_2$In phases. In detail, until recently, the theoretical work performed on TiO$_2$ predicted the Fe$_2$P → CaC$_2$ transition [33,36], while the Fe$_2$P → Ni$_2$In transition was predicted in ZrO$_2$ and HfO$_2$ [42–44]. Therefore, to better understand the upper part of the phase transition sequence in TiO$_2$, we have performed DFT calculations to investigate the phase relations among Fe$_2$P, CaC$_2$, and Ni$_2$In phases at megabar pressures. Consequently, the long-range target of this study is to draw the similarities and differences between TiO$_2$ and other similar transition-metal dioxides in terms of predicting the ultrahigh-pressure phase transition sequence, which will hopefully lead to a better understanding of the high-pressure behavior of such dioxides.

2. Computational Details

In order to investigate the phase stability and the equations of state (EOSs) of all tested phases of TiO$_2$, we used static *first-principles* computations performed within the framework of density functional theory (DFT) [51]. The projector-augmented wave (PAW) formalism [52,53] was used to treat the interactions between the titanium (Ti) and oxygen (O) atoms with the valence configuration of $3s^23p^63d^24s^2$ for Ti and $2s^22p^4$ for O. Following previous theoretical high-pressure studies carried out on TiO$_2$ and similar dioxides [29,43,44,46–48], the electronic exchange and correlation effects were treated within the GGA [54]. We performed our calculations using the Quantum ESPRESSO package [55] with an energy cutoff of 80 Ry and Γ-centered k-point meshes [56]. Our calculations yielded sufficient convergence to better than 10^{-5} Ry in the total energies for both phases, and pressures were converged to better than 0.1 GPa. The Brillouin zone integration was performed using the following k-point meshes for the ZrO$_2$ phases: 8 × 8 × 12 for Fe$_2$P,

12 × 12 × 8 for CaC$_2$, and 20 × 20 × 16 for Ni$_2$In. For a fixed volume, all internal degrees of freedom and unit-cell parameters of the structure were optimized simultaneously during the geometry optimizations. The ground-state energy for each phase was determined for 13–16 volumes, and the EOS parameters for each phase were obtained by fitting the total energy as a function of volume to a second-order Birch–Murnaghan equation of state (BM-EOS) [57] (Table 1). Phonon dispersion calculations have not been performed and will be the scope of future studies.

Table 1. Calculated equations of state for the Fe$_2$P, Ni$_2$In, and CaC$_2$ phases of TiO$_2$. Our EOS is determined from GGA calculations using the second-order BM-EOS [57]. For comparison, we list other calculated EOSs for TiO$_2$ and the similar dioxides ZrO$_2$ and HfO$_2$. 1σ uncertainties are given in parentheses.

Phase	Equation of State			Reference
	V_0 (Å^3)	K_0 (GPa)	K_0'	
Fe$_2$P-TiO$_2$	25.7	272.1	4	[31]
	25.53	287	4.1	[32]
	25.59 (0.03)	284 (2)	4 (fixed)	This work
Fe$_2$P-ZrO$_2$	30.17	248	3.76	[42]
	30.94 (0.03)	272 (2)	4 (fixed)	[44]
	30.34	272	4 (fixed)	[48]
Fe$_2$P-HfO$_2$	29.73 (0.02)	282 (2)	4 (fixed)	[43]
	29.69 (0.03)	288 (2)	4 (fixed)	[48]
	29.8	284	4.2	[50]
Ni$_2$In-TiO$_2$	27.82 (0.19)	173 (6)	4 (fixed)	This work
Ni$_2$In-ZrO$_2$	29.21	239	3.86	[42]
	31.81 (0.13)	200 (5)	4 (fixed)	[44]
Ni$_2$In-HfO$_2$	30.49 (0.14)	213 (6)	4 (fixed)	[43]
CaC$_2$-TiO$_2$	25.23 (0.04)	264 (3)	4 (fixed)	This work

3. Results and Discussion

3.1. Phase Stability and Equation of State

We computed our EOS parameters for each phase by fitting the energy-volume (E-V) data to the following second-order BM-EOS [57] for which the first pressure derivative of the bulk modulus at zero pressure (K_0') was fixed to 4 and the zero-pressure volume (V_0) and the zero-pressure bulk modulus (K_0) were used as fitting parameters. The determination of the EOS for the three phases allowed us to investigate their compressibilities and phase stability at high pressure as well as the volume change across different transitions.

The EOS parameters for all phases are summarized in Table 1 along with results from previous calculations [31,32]. We note that our calculated EOS for the Fe$_2$P phase agrees well with previous studies [31,32]. On the other hand, the EOS of either Ni$_2$In or CaC$_2$ has not been previously obtained. In this regard, we should emphasize that the Ni$_2$In-type structure has not been tested for TiO$_2$, while CaC$_2$ has been proposed for TiO$_2$ [33,36]. Therefore, to our knowledge, we provide the EOS of the Ni$_2$In and CaC$_2$ phases of TiO$_2$ for the first time (Table 1). When comparing the EOS for the three phases, we notice that Fe$_2$P has the highest bulk modulus and thus is the most incompressible phase, followed by CaC$_2$ and finally Ni$_2$In that exhibits the most compressibility among all tested phases. In detail, a small bulk modus decrease (~7%) is predicted across Fe$_2$P → CaC$_2$, while we note an obviously large decrease (~39%) across Fe$_2$P → Ni$_2$In (Table 1). Furthermore, the density of CaC$_2$ at zero pressure is higher than that of Fe$_2$P, as expected, which is not the case for Ni$_2$In where the density is less than that of Fe$_2$P by ~9% (Figure 2). We should note that such predicted large changes in K_0 and V_0 across Fe$_2$P → Ni$_2$In are likely to be unexpected across transitions to higher-pressure phases, whereas the corresponding changes across Fe$_2$P → CaC$_2$ are much more reasonable.

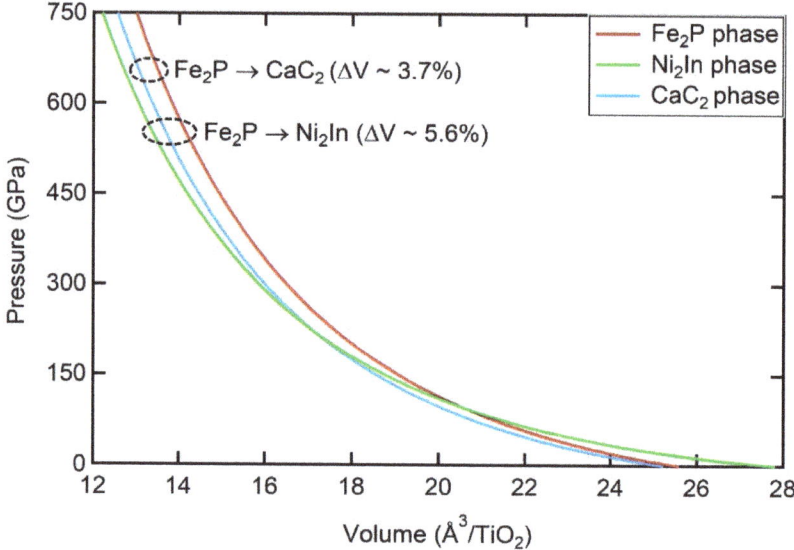

Figure 2. Pressure versus volume of TiO_2 phases as determined by GGA calculations using BM-EOS [57]. The dashed circles show the large volume reduction across the $Fe_2P \rightarrow Ni_2In$ and $Fe_2P \rightarrow CaC_2$ transitions.

To obtain the transition pressures, we calculated the enthalpy change relative to the Fe_2P phase (Figure 3). First, we note that the transition from Ni_2In to CaC_2 or vice versa is not possible at ultrahigh pressures as their enthalpy curves are unlikely to intersect. Furthermore, although the enthalpy curves of the two phases seem to cross at lower pressures, this transition is not expected to occur within the stability field of the Fe_2P phase. Our enthalpy calculations indicate that the transition pressure across the $Fe_2P \rightarrow Ni_2In$ transition is 564 GPa. On the other hand, our calculated transition pressure across the $Fe_2P \rightarrow CaC_2$ transition is 664 GPa, in agreement with previous findings (Table 2) [33,36]. We should note that although the Ni_2In phase has a lower enthalpy than the CaC_2 phase, this does not necessarily mean that $Fe_2P \rightarrow Ni_2In$ is the favorable transition when compared to the $Fe_2P \rightarrow CaC_2$ transition. Consequently, there are two possible scenarios for the transition from the Fe_2P phase, and thus one must be eliminated from the transition sequence of TiO_2. Therefore, in the next sections, we further explore the two transitions to predict the favored post-Fe_2P phase of TiO_2.

3.2. Enthalpy Difference and Volume Collapse across Phase Transitions

In this section, we discuss the enthalpy difference across the two transitions and the contribution of the volume change in this difference as well as the correlation between the volume decrease and the coordination number increase. We note that the $Fe_2P \rightarrow Ni_2In$ and $Fe_2P \rightarrow CaC_2$ transitions are associated with a large enthalpy difference (Figure 3, Table 3) across the phase transition. Our calculations reveal that such noticeable enthalpy difference is mainly due to the obvious volume reduction across both transitions (Figure 2, Table 3), while the contribution of electronic energy gain has a minimal effect.

Although both transitions are driven by a large enthalpy change as discussed above, it has been found that the $Fe_2P \rightarrow Ni_2In$ transition requires a larger enthalpy change compared to the $Fe_2P \rightarrow CaC_2$ transition. This result is not unexpected and can be explained in view of the coordination number change across transitions; in this regard, we notice that the $Fe_2P \rightarrow Ni_2In$ is related to a 2-coordination-number increase (from 9 to 11) whereas $Fe_2P \rightarrow CaC_2$ is driven by a 1-coordination-number increase (from 9 to 10).

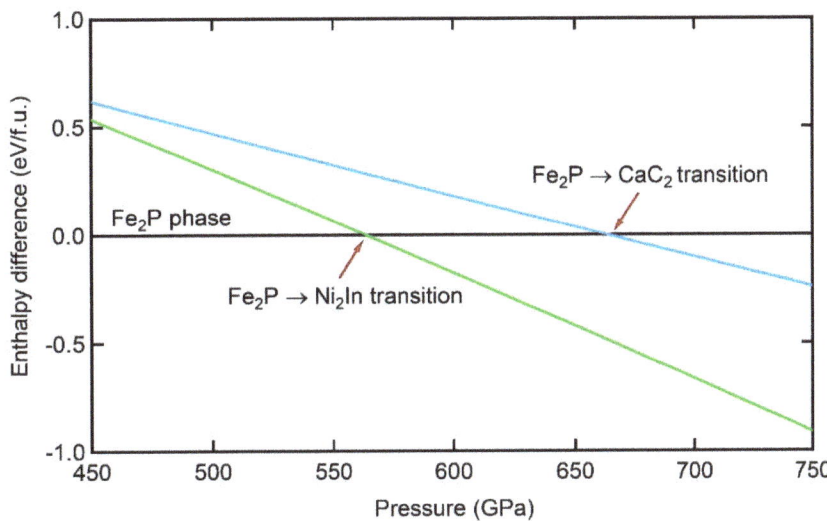

Figure 3. Change in enthalpy with respect to Fe_2P phase versus pressure of one formula unit as determined by GGA calculations for TiO_2. The transition pressures across transitions from Fe_2P to Ni_2In and CaC_2 phases are shown.

Table 2. Calculated transition pressures across the $Fe_2P \rightarrow Ni_2In$ and $Fe_2P \rightarrow CaC_2$ phase transitions for TiO_2. For comparison, we list other calculated results.

Phase Transition	Transition Pressure (GPa)	Reference
$Fe_2P \rightarrow Ni_2In$	564 GPa	This work
$Fe_2P \rightarrow CaC_2$	647 GPa	[36]
	689 GPa	[33]
	664 GPa	This work

Table 3. Enthalpy difference, band gap difference, volume change, and coordination number change across the $Fe_2P \rightarrow Ni_2In$ and $Fe_2P \rightarrow CaC_2$ phase transitions for TiO_2.

Phase Transition	$\Delta H/\Delta P$		Band Gap Difference (eV)	Volume Change (%)	Coordination Number Change
	$eV \cdot GPa^{-1}$ ($\times 10^{-4}$)	$kJ \cdot mol^{-1} \cdot GPa^{-1}$			
$Fe_2P \rightarrow Ni_2In$	−48.102	−0.46411	0.04	5.6	2
$Fe_2P \rightarrow CaC_2$	−28.135	−0.27416	0	3.7	1

However, regardless of the coordination number increase (either 1 or 2), the calculated volume collapse across both transitions does not show a large difference (3.7% for CaC_2 vs. 5.6% for Ni_2In), where $Fe_2P \rightarrow Ni_2In$ is expected to show a much larger volume reduction due to the large coordination increase across this transition. In detail, it has been previously found that for TiO_2, the 2-coordination increase (from 7 to 9 in OI → OII transition) results in an ~ 8% volume reduction [29] compared to a 5.6% reduction for the same coordination increase in $Fe_2P \rightarrow Ni_2In$. On the other hand, the 3.7% volume decrease in $Fe_2P \rightarrow CaC_2$ transition looks more appropriate for a 1-coordination increase and agrees well with previous studies that reported values of 3.3–3.4% [33,36]. Therefore, based on the interplay between the volume reduction and coordination number change, the $Fe_2P \rightarrow CaC_2$ transition is likely favorable in TiO_2 when compared to the $Fe_2P \rightarrow Ni_2In$ transition.

3.3. Band Gap Calculations at the Transition Pressures

To further investigate the proposed ultrahigh-pressure phases of titania, we explored the pressure dependence of the band gap by analyzing the band structure of each phase at different pressures (Figures 4 and 5). Often, DFT calculations systematically underestimate band gaps; however, they accurately describe the pressure dependence [58,59]. We speculate that such analysis might be helpful in gaining a deeper insight into the $Fe_2P \to Ni_2In$ and $Fe_2P \to CaC_2$ transitions to better determine the favorable phase transition. Our band gap calculations show that the band gap of the Fe_2P-type structure at zero pressure is ~0.94 eV and decreases as pressure increases, where the drop in the band gap becomes more obvious at megabar pressures (Figure 4). However, regardless of such band-gap collapse, we should note that the Fe_2P phase remains a semiconductor up to multi-megabar pressures before its metallization begins at pressures greater than 650 GPa (Figure 4), in good agreement with previous predictions obtained for Fe_2P-TiO_2 [33,36].

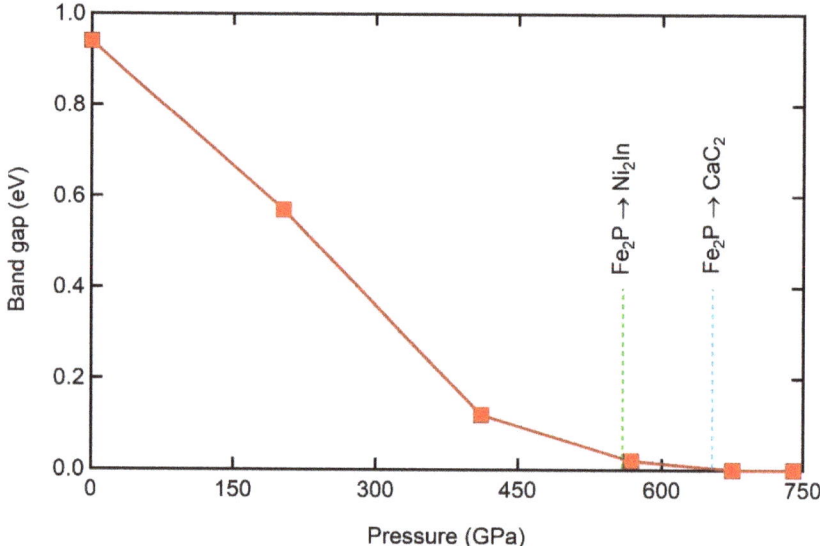

Figure 4. Calculated band gap of the Fe_2P phase as a function of pressure. The vertical dashed lines indicate the transition pressures of the $Fe_2P \to Ni_2In$ and $Fe_2P \to CaC_2$ transitions.

Additionally, our analysis of the band structures of the Ni_2In and CaC_2 phases reveals that both structures have metallic characteristics (band gap = 0 eV) at their predicted transition pressures from the Fe_2P phase (Figures 4 and 5, Table 2). In this regard, we note that the band gap difference (Table 2, Figure 4) across $Fe_2P \to CaC_2$ transition is zero (0 eV → 0 eV) compared to a ~0.04 eV difference across $Fe_2P \to Ni_2In$ transition (0.04 eV → 0 eV). Consequently, our band gap calculations likely provide further evidence that supports the $Fe_2P \to CaC_2$ transition being the favorable transition when compared to the $Fe_2P \to CaC_2$ transition. Thus, we confirm previous theoretical findings that predict CaC_2 to be the post-Fe_2P phase for TiO_2 [33,36]. Finally, it should be noted that such confirmation is important since the Ni_2In phase was a highly possible scenario for TiO_2 as it has been predicted to be the post-Fe_2P phase in the similar dioxides ZrO_2 and HfO_2 [42–44]. In this regard, we should emphasize that the three dioxides undergo the same transition sequence at high pressures (MI → OI → OII → Fe_2P), but the post-Fe_2P phase in TiO_2 (i.e., CaC_2-type) is predicted to be different from that of ZrO_2 and HfO_2 (i.e., Ni_2In-type).

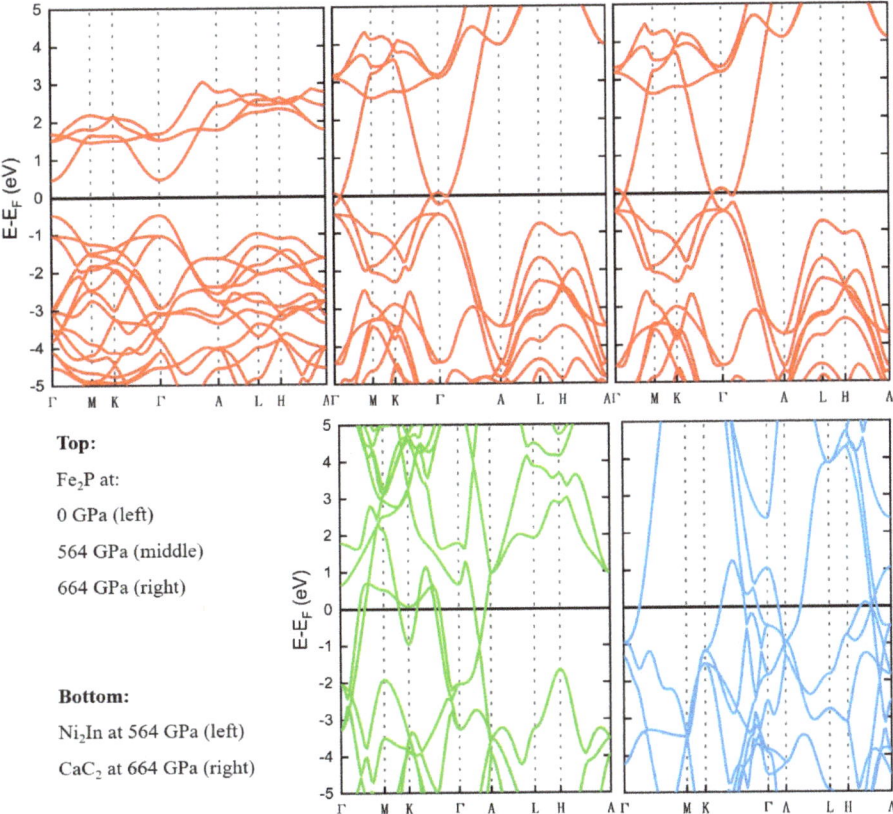

Figure 5. Electronic band structure for: (**Top**) the Fe_2P phase at 0 GPa, 564 GPa, and 664 GPa; (**Bottom**) the Ni_2In and CaC_2 phases at 564 GPa and 664 GPa, respectively.

4. Conclusions

In conclusion, we employed DFT computations to investigate the ultrahigh-pressure phase stability of the Ni_2In and CaC_2 structures with respect to the Fe_2P structure of titania. We explored the $Fe_2P \rightarrow Ni_2In$ and $Fe_2P \rightarrow CaC_2$ transitions to better predict the favored post-Fe_2P of TiO_2. These transitions were thoroughly studied in terms of the volume decrease, the coordination number increase, and the band gap reduction. Our analysis favored the $Fe_2P \rightarrow CaC_2$ transition, and thus we predict that the CaC_2-type structure is likely the post-Fe_2P phase of TiO_2, and therefore the most stable phase of titania at pressures that exceed 664 GPa. This result is also evidenced by a similar conclusion we recently obtained for ZrO_2, where the $Fe_2P \rightarrow CaC_2$ transition was predicted to be more favorable than the $Fe_2P \rightarrow Ni_2In$ transition (unpublished), thus emphasizing the close similarities of the high-pressure behavior of titania and zirconia. Finally, our equation of state determination shows that CaC_2 is the densest phase with a high bulk modulus that is comparable to that of the experimentally observed Fe_2P phase, while the Ni_2In phase reveals a low density and high compressibility.

Author Contributions: Conceptualization, Y.A.-K. and K.T.; methodology, Y.A.-K.; software, K.T.; formal analysis, Y.A.-K. and K.T.; writing—original draft preparation, K.T. and Y.A.-K.; writing—review and editing, Y.A.-K. and K.T. All authors have read and agreed to the published version of the manuscript.

Funding: This research received no external funding.

Data Availability Statement: The data presented in this study are available in the article.

Conflicts of Interest: The authors declare no conflict of interest.

References

1. Wu, X. Applications of Titanium Dioxide Materials. In *Titanium Dioxide—Advances and Applications*; IntechOpen: London, UK, 2022.
2. Basheer, C. Application of Titanium Dioxide-Graphene Composite Material for Photocatalytic Degradation of Alkylphenols. *J. Chem.* **2013**, *2013*, 456586. [CrossRef]
3. Nishio-Hamane, D.; Shimizu, A.; Nakahira, R.; Niwa, K.; Sano-Furukawa, A.; Okada, T.; Yagi, T.; Kikegawa, T. The Stability and Equation of State for the Cotunnite Phase of TiO_2 up to 70 GPa. *Phys. Chem. Miner.* **2009**, *37*, 129–136. [CrossRef]
4. Swamy, V.; Kuznetsov, A.Y.; Dubrovinsky, L.S.; Kurnosov, A.; Prakapenka, V.B. Unusual Compression Behavior of Anatase TiO_2 Nanocrystals. *Phys. Rev. Lett.* **2009**, *103*, 75505. [CrossRef] [PubMed]
5. Shul'ga, Y.M.; Matyushenko, D.V.; Golyshev, A.A.; Shakhrai, D.V.; Molodets, A.M.; Kabachkov, E.N.; Kurkin, E.N.; Domashnev, I.A. Phase Transformations in Nanostructural Anatase TiO_2 under Shock Compression Conditions Studied by Raman Spectroscopy. *Tech. Phys. Lett.* **2010**, *36*, 841–843. [CrossRef]
6. Ding, Y.; Chen, M.; Wu, W. Mechanical Properties, Hardness and Electronic Structures of New Post-Cotunnite Phase (Fe_2P-Type) of TiO_2. *Phys. B Condens. Matter* **2014**, *433*, 48–54. [CrossRef]
7. Wang, Z.; Saxena, S.K. Raman Spectroscopic Study on Pressure-Induced Amorphization in Nanocrystalline Anatase (TiO_2). *Solid State Commun.* **2001**, *118*, 75–78. [CrossRef]
8. Gerward, L.; Olsen, J.S. Post-Rutile High-Pressure Phases in TiO_2. *J. Appl. Crystallogr.* **1997**, *30*, 259–264. [CrossRef]
9. Wang, Z.; Saxena, S.K.; Pischedda, V.; Liermann, H.P.; Zha, C.S. X-Ray Diffraction Study on Pressure-Induced Phase Transformations in Nanocrystalline Anatase/Rutile (TiO_2). *J. Phys. Condens. Matter* **2001**, *13*, 8317–8323. [CrossRef]
10. Swamy, V.; Dubrovinsky, L.S.; Dubrovinskaia, N.A.; Langenhorst, F.; Simionovici, A.S.; Drakopoulos, M.; Dmitriev, V.; Weber, H.-P. Size Effects on the Structure and Phase Transition Behavior of Baddeleyite TiO_2. *Solid State Commun.* **2005**, *134*, 541–546. [CrossRef]
11. Dubrovinskaia, N.A.; Dubrovinsky, L.S.; Ahuja, R.; Prokopenko, V.B.; Dmitriev, V.; Weber, H.-P.; Osorio-Guillen, J.M.; Johansson, B. Experimental and Theoretical Identification of a New High-Pressure TiO_2 Polymorph. *Phys. Rev. Lett.* **2001**, *87*, 275501. [CrossRef]
12. Caravaca, M.A.; Mino, J.C.; Perez, V.J.; Casali, R.A.; Ponce, C.A. Ab Initio Study of the Elastic Properties of Single and Polycrystal TiO_2, ZrO_2 and HfO_2 in the Cotunnite Structure. *J. Phys. Condens. Matter* **2009**, *21*, 15501. [CrossRef] [PubMed]
13. Hearne, G.R.; Zhao, J.; Dawe, A.M.; Pischedda, V.; Maaza, M.; Nieuwoudt, M.K.; Kibasomba, P.; Nemraoui, O.; Comins, J.D. Effect of Grain Size on Structural Transitions in Anatase TiO_2: A Raman Spectroscopy Study at High Pressure. *Phys. Rev. B* **2004**, *70*, 134102. [CrossRef]
14. Swamy, V.; Kuznetsov, A.; Dubrovinsky, L.S.; McMillan, P.F.; Prakapenka, V.B.; Shen, G.; Muddle, B.C. Size-Dependent Pressure-Induced Amorphization in Nanoscale TiO_2. *Phys. Rev. Lett.* **2006**, *96*, 135702. [CrossRef] [PubMed]
15. Ahuja, R.; Dubrovinsky, L.S. High-Pressure Structural Phase Transitions in TiO_2 and Synthesis of the Hardest Known Oxide. *J. Phys. Condens. Matter* **2002**, *14*, 10995–10999. [CrossRef]
16. Wang, Y.; Zhao, Y.; Zhang, J.; Xu, H.; Wang, L.; Luo, S.; Daemen, L.L. In Situ Phase Transition Study of Nano- and Coarse-Grained TiO_2 under High Pressure/Temperature Conditions. *J. Phys. Condens. Matter* **2008**, *20*, 125224. [CrossRef]
17. Flank, A.-M.; Lagarde, P.; Itie, J.-P.; Hearne, G.R. Pressure-Induced Amorphization and a Possible Polyamorphism Transition in Nanosized TiO_2: An x-Ray Absorption Spectroscopy Study. *Phys. Rev. B* **2008**, *77*, 224112. [CrossRef]
18. Arlt, T.; Bermejo, M.; Blanco, M.A.; Gerward, L.; Jiang, J.Z.; Olsen, J.S.; Recio, J.M. High-Pressure Polymorphs of Anatase TiO_2. *Phys. Rev. B* **2000**, *61*, 14414. [CrossRef]
19. Mattesini, M.; de Almeida, J.S.; Dubrovinsky, L.; Dubrovinskaia, N.; Johansson, B.; Ahuja, R. High-Pressure and High-Temperature Synthesis of the Cubic TiO_2 Polymorph. *Phys. Rev. B* **2004**, *70*, 212101. [CrossRef]
20. Swamy, V.; Muddle, B.C. Ultrastiff Cubic TiO_2 Identified via First-Principles Calculations. *Phys. Rev. Lett.* **2007**, *98*, 35502. [CrossRef]
21. Swamy, V.; Dubrovinskaia, N.A.; Dubrovinsky, L.S. Compressibility of Baddeleyite-Type TiO_2 from Static Compression to 40 GPa. *J. Alloys Compd.* **2002**, *340*, 46–48. [CrossRef]
22. Muscat, J.; Swamy, V.; Harrison, N.M. First-Principles Calculations of the Phase Stability of TiO_2. *Phys. Rev. B* **2002**, *65*, 224112. [CrossRef]
23. Lyakhov, A.O.; Oganov, A.R. Evolutionary Search for Superhard Materials: Methodology and Applications to Forms of Carbon and TiO_2. *Phys. Rev. B* **2011**, *84*, 92103. [CrossRef]

24. Al-Khatatbeh, Y.; Lee, K.K.M.; Kiefer, B. Compressibility of Nanocrystalline TiO$_2$ Anatase. *Phys. Chem. C* **2012**, *116*, 21635–21639. [CrossRef]
25. Swamy, V.; Dubrovinsky, L.S.; Dubrovinskaia, N.A.; Simionovici, A.S.; Drakopoulos, M.; Dmitriev, V.; Weber, H.-P. Compression Behavior of Nanocrystalline Anatase TiO$_2$. *Solid State Commun.* **2003**, *125*, 111–115. [CrossRef]
26. Al-Khatatbeh, Y.; Lee, K.K.M. From Superhard to Hard: A Review of Transition Metal Dioxides TiO$_2$, ZrO$_2$, and HfO$_2$ Hardness. *J. Superhard Mater.* **2014**, *36*, 231–245. [CrossRef]
27. Van Gestel, T.; Sebold, D.; Kruidhof, H.; Bouwmeester, H.J.M. ZrO$_2$ and TiO$_2$ Membranes for Nanofiltration and Pervaporation. Part 2. Development of ZrO$_2$ and TiO$_2$ Toplayers for Pervaporation. *J. Memb. Sci.* **2008**, *318*, 413–421. [CrossRef]
28. Swamy, V.; Muddle, B.C. Pressure-Induced Polyamorphic Transition in Nanoscale TiO$_2$. *J. Aust. Cer. Soc.* **2008**, *44*, 1–5.
29. Al-Khatatbeh, Y.; Lee, K.K.M.; Kiefer, B. High-Pressure Behavior of TiO$_2$ as Determined by Experiment and Theory. *Phys. Rev. B* **2009**, *79*, 134114. [CrossRef]
30. Meng, X.; Wang, L.; Liu, D.; Wen, X.; Zhu, Q.; Goddard, W.A.; An, Q. Discovery of Fe$_2$P-Type Ti(Zr/Hf)2O$_6$ Photocatalysts toward Water Splitting. *Chem. Mater.* **2016**, *28*, 1335–1342. [CrossRef]
31. Dekura, H.; Tsuchiya, T.; Kuwayama, Y.; Tsuchiya, J. Theoretical and Experimental Evidence for a New Post-Cotunnite Phase of Titanium Dioxide with Significant Optical Absorption. *Phys. Rev. Lett.* **2011**, *107*, 45701. [CrossRef]
32. Fu, Z.; Liang, Y.; Wang, S.; Zhong, Z. Structural Phase Transition and Mechanical Properties of TiO$_2$ under High Pressure. *Phys. Status Solidi Basic Res.* **2013**, *250*, 2206–2214. [CrossRef]
33. Zhong, X.; Wang, J.; Zhang, S.; Yang, G.; Wang, Y. Ten-Fold Coordinated Polymorph and Metallization of TiO$_2$ under High Pressure. *RSC Adv.* **2015**, *5*, 54253–54257. [CrossRef]
34. Liang, Y.; Zhang, B.; Zhao, J. Mechanical Properties and Structural Identifications of Cubic TiO$_2$. *Phys. Rev. B* **2008**, *77*, 94126. [CrossRef]
35. Swamy, V.; Kuznetsov, A.; Dubrovinsky, L.S.; Caruso, R.A.; Shchukin, D.G.; Muddle, B.C. Finite-Size and Pressure Effects on the Raman Spectrum of Nanocrystalline Anatase TiO$_2$. *Phys. Rev. B* **2005**, *71*, 184302. [CrossRef]
36. Lyle, M.J.; Pickard, C.J.; Needs, R.J. Prediction of 10-Fold Coordinated TiO$_2$ and SiO$_2$ Structures at Multimegabar Pressures. *Proc. Natl. Acad. Sci. USA* **2015**, *112*, 6898–6901. [CrossRef]
37. Luo, W.; Yang, S.F.; Wang, Z.C.; Ahuja, R.; Johansson, B.; Liu, J.; Zou, G.T. Structural Phase Transitions in Brookite-Type TiO$_2$ under High Pressure. *Solid State Commun.* **2005**, *133*, 49–53. [CrossRef]
38. Machon, D.; Daniel, M.; Pischedda, V.; Daniele, S.; Bouvier, P.; LeFloch, S. Pressure-Induced Polyamorphism in TiO$_2$ Nanoparticles. *Phys. Rev. B* **2010**, *82*, 140102. [CrossRef]
39. Wang, Y.; Zhang, J.; Zhao, Y. Strength Weakening by Nanocrystals in Ceramic Materials. *Nano Lett.* **2007**, *7*, 3196–3199. [CrossRef]
40. Baty, S.R.; Burakovsky, L.; Errandonea, D. Ab Initio Phase Diagram of Copper. *Crystals* **2021**, *11*, 537. [CrossRef]
41. Diaz-Anichtchenko, D.; Errandonea, D. Comparative Study of the Compressibility of M$_3$V$_2$O$_8$ (M = Cd, Zn, Mg, Ni) Orthovanadates. *Crystals* **2022**, *12*, 1544. [CrossRef]
42. Durandurdu, M. Novel High-Pressure Phase of ZrO$_2$: An Ab Initio Prediction. *J. Solid State Chem.* **2015**, *230*, 233–236. [CrossRef]
43. Al-Khatatbeh, Y.; Tarawneh, K.; Hamad, B. The Prediction of a New High-Pressure Phase of Hafnia Using First-Principles Computations. *IOP Conf. Ser. Mater. Sci. Eng.* **2018**, *305*, 012006. [CrossRef]
44. Al-Taani, H.; Tarawneh, K.; Al-Khatatbeh, Y.; Hamad, B. The High-Pressure Stability of Ni$_2$In-Type Structure of ZrO$_2$ with Respect to OII and Fe$_2$P-Type Phases: A First-Principles Study. *IOP Conf. Ser. Mater. Sci. Eng.* **2018**, *305*, 012016. [CrossRef]
45. Kokalj, A. Computer Graphics and Graphical User Interfaces as Tools in Simulations of Matter at the Atomic Scale. *Comput. Mater. Sci.* **2003**, *28*, 155–168. [CrossRef]
46. Al-Khatatbeh, Y.; Lee, K.K.M.; Kiefer, B. Phase Diagram up to 105 GPa and Mechanical Strength of HfO$_2$. *Phys. Rev. B* **2010**, *82*, 144106. [CrossRef]
47. Al-Khatatbeh, Y.; Lee, K.K.M.; Kiefer, B. Phase Relations and Hardness Trends of ZrO$_2$ Phases at High Pressure. *Phys. Rev. B* **2010**, *81*, 214102. [CrossRef]
48. Al-Khatatbeh, Y.; Tarawneh, K.; Al-Taani, H.; Lee, K.K.M. Theoretical and Experimental Evidence for a Post-Cotunnite Phase Transition in Hafnia at High Pressures. *J. Superhard Mater.* **2018**, *40*, 374–383. [CrossRef]
49. Nishio-Hamane, D.; Dekura, H.; Seto, Y.; Yagi, T. Theoretical and Experimental Evidence for the Post-Cotunnite Phase Transition in Zirconia at High Pressure. *Phys. Chem. Miner.* **2015**, *42*, 385–392. [CrossRef]
50. Dutta, R.; Kiefer, B.; Greenberg, E.; Prakapenka, V.B.; Duffy, T.S. Ultrahigh-Pressure Behavior of AO$_2$ (A = Sn, Pb, Hf) Compounds. *J. Phys. Chem. C* **2019**, *123*, 27735–27741. [CrossRef]
51. Hohenberg, P.; Kohn, W. Inhomogeneous Electron Gas. *Phys. Rev.* **1964**, *136*, B864–B871. [CrossRef]
52. Kresse, G.; Joubert, D. From Ultrasoft Pseudopotentials to the Projector Augmented-Wave Method. *Phys. Rev. B* **1999**, *59*, 1758–1775. [CrossRef]
53. Blochl, P.E. Projector Augmented-Wave Method. *Phys. Rev. B* **1994**, *50*, 17953. [CrossRef] [PubMed]
54. Perdew, J.P.; Burke, K.; Ernzerhof, M. Generalized Gradient Approximation Made Simple. *Phys. Rev. Lett.* **1996**, *77*, 3865–3868. [CrossRef]
55. Giannozzi, P.; Baroni, S.; Bonini, N.; Calandra, M.; Car, R.; Cavazzoni, C.; Ceresoli, D.; Chiarotti, G.L.; Cococcioni, M.; Dabo, I.; et al. QUANTUM ESPRESSO: A Modular and Open-Source Software Project for Quantum Simulations of Materials. *J. Phys. Condens. Matter* **2009**, *21*, 395502. [CrossRef] [PubMed]

56. Monkhorst, H.J.; Pack, J.D. Special Points for Brillouin-Zone Integrations. *Phys. Rev. B* **1976**, *13*, 5188–5192. [CrossRef]
57. Birch, F. Elasticity and Constitution of the Earth's Interior. *J. Geophys. Res.* **1952**, *57*, 227–234. [CrossRef]
58. Bandiello, E.; Errandonea, D.; Martinez-Garcia, D.; Santamaria-Perez, D.; Manjón, F.J. Effects of High Pressure on the Structural, Vibrational, and Electronic Properties of Monazite-Type PbCrO4. *Phys. Rev. B* **2012**, *85*, 024108. [CrossRef]
59. Borlido, P.; Schmidt, J.; Huran, A.W.; Tran, F.; Marques, M.A.L.; Botti, S. Exchange-Correlation Functionals for Band Gaps of Solids: Benchmark, Reparametrization and Machine Learning. *npj Comput. Mater.* **2020**, *6*, 96. [CrossRef]

Disclaimer/Publisher's Note: The statements, opinions and data contained in all publications are solely those of the individual author(s) and contributor(s) and not of MDPI and/or the editor(s). MDPI and/or the editor(s) disclaim responsibility for any injury to people or property resulting from any ideas, methods, instructions or products referred to in the content.

Article

First-Principle Study of Ca₃Y₂Ge₃O₁₂ Garnet: Dynamical, Elastic Properties and Stability under Pressure

Alfonso Muñoz * and Plácida Rodríguez-Hernández

Departamento de Física, Instituto de Materiales y Nanotecnología, Malta Consolider Team, Universidad de La Laguna, 38200 La Laguna, Tenerife, Spain
* Correspondence: amunoz@ull.edu.es; Tel.: +34-922-318-275

Abstract: We present here an ab initio study under the framework of the Density Functional Theory of the $Ca_3Y_2Ge_3O_{12}$ garnet. Our study focuses on the analysis of the structural, electronic, dynamic, and elastic properties of this material under hydrostatic pressure. We report information regarding the equation of state, the compressibility, and the structural evolution of this compound. The dynamical properties and the evolution under pressure of infrared, silent, and Raman frequencies and their pressure coefficients are also presented. The dependence on the pressure of the elastic constants and the mechanical and elastic properties are analyzed. From our results, we conclude that this garnet becomes mechanically unstable at 45.7 GPa; moreover, we also find evidence of soft phonons at 34.4 GPa, showing the dynamical instability of this compound above this pressure.

Keywords: ab initio; garnet; equation of state; elasticity; phonons; Raman

1. Introduction

The study of garnet minerals under high pressure is an essential topic in Earth Sciences; garnets are among the major components of Earth's deep interior. The physical properties of these materials at extreme conditions have an important effect on the dynamics of the mantle transition zone [1].

In technological applications, garnets are potential candidates for many devices. Their excellent thermal conductivity, hardness, mechanical and chemical stability, and high optical transparency suggest that these materials can be used as laser materials. Doping garnets with Rare Earth (RE^{3+}) elements allows the use of their great luminescence properties. For example, YAG, a $Y_3Al_5O_3$ garnet, used as a solid-state laser doped with Nd, is well-known in medical and commercial applications [2]. Under pressure, these garnets can be employed as sensors for high-pressure experiments [3,4], and can also be used with adequate doping as scintillators or light-emitting diodes [5]. Recently, a $Ca_3Y_2Ge_3O_{12}$ garnet doped with Cr^{+3} was studied as a near-infrared LED combined with a 455 nm LED chip [6]. In addition, the near-infrared phosphor $Ca_3Y_2Ge_3O_{12}$: Cr^{+3} garnet has been synthesized, showing a broadband NIR emission of 700–1100 nm, peaking around 800 nm [7].

The oxide garnets have the general formula $A_3B_2C_3O_{12}$, where A denotes the dodecahedral, B the octahedral, and C the tetrahedral sites with different coordination. They usually crystallize in the body-centered cubic (bcc) structure ($Ia3d$ space group), with a unit cell of 160 atoms and a primitive cell of 80 atoms. In our case, Ca atoms (A) occupy the 24 c positions with coordination 8, Y atoms (B) occupy the 16 a positions with coordination 6 and octahedral point symmetry (C_{3i}), and Ge atoms (C) occupy the 24 d positions with coordination 4 and tetrahedral point symmetry (S_4). Finally, the O atoms occupy the 96 h positions (Figure 1).

Citation: Muñoz, A.; Rodríguez-Hernández, P. First-Principle Study of Ca₃Y₂Ge₃O₁₂ Garnet: Dynamical, Elastic Properties and Stability under Pressure. *Crystals* **2023**, *13*, 29. https://doi.org/10.3390/cryst13010029

Academic Editor: Claudio Cazorla

Received: 5 December 2022
Revised: 19 December 2022
Accepted: 21 December 2022
Published: 24 December 2022

Copyright: © 2022 by the authors. Licensee MDPI, Basel, Switzerland. This article is an open access article distributed under the terms and conditions of the Creative Commons Attribution (CC BY) license (https://creativecommons.org/licenses/by/4.0/).

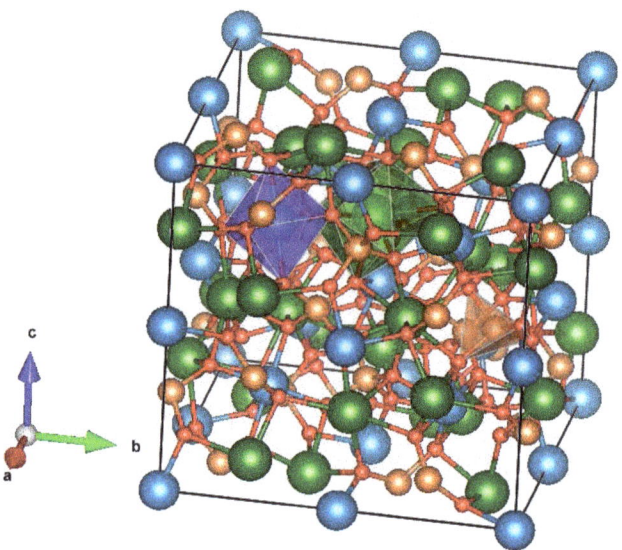

Figure 1. The conventional unit cell with polyhedral units CaO_8 (green), YO_6 (blue), and GeO_4 (orange).

Due to the large number of atoms in the unit cell and hence the complexity of the simulation, few ab initio studies of the structural, electronic, dynamical, or mechanical properties both at room and high pressures are available for garnets [8,9].

In the last decades, the use of Density Functional Theory (DFT) simulations of material under high pressure has become a very well-established technique. The quality and precision of these studies make possible the prediction of many properties of materials under extreme conditions, providing information on the stability of these compounds.

The application of pressure allows modifications of the bonds and interatomic distances of material and, therefore, changes in their elastic and vibrational properties. On the other hand, the optical properties of Re3+ doped oxides garnets are affected by changes in the host structure due to the reduction in interatomic distances and possible coordination changes. Thus, it is worth studying garnets' structural and electronic properties under pressure to help understand their luminescence properties and improve their practical applications. A study of this garnet under pressure is required to explore potential applications of the $Ca_3Y_2Ge_3O_{12}$ garnet, for example, as a pressure optical sensor. The knowledge of its mechanical properties under pressure is also essential in geological studies due to the abundance of this mineral phase in the upper mantle and Earth's transition zone [10].

Since it is important to study how the properties of garnets evolve under high pressure, in this work, we report an extensive first-principle study of the electronic, structural, dynamic, and elastic properties of the $Ca_3Y_2Ge_3O_{12}$ garnet under pressure. To the best of our knowledge, there are no reported ab initio studies of this compound under high pressure. We hope the information provided in this work will stimulate further experimental studies on this garnet.

2. Simulation Details

Ab initio total energy simulations at zero temperature were performed in the framework of the Density Functional Theory, DFT, with the plane-wave method and the pseudopotential theory implemented in the Vienna ab initio simulations package [11,12]. The Projector Augmented-Wave (PAW) [13] pseudopotentials provided in this package were employed to take into account the full nodal character of the all-electron charge density in the core region. For the Ca atoms, 8 valence electrons ($3p^64s^2$) were included, for Y atoms

11 valence electrons ($4s^2 4p^6 5s^2 4d^1$), and for Ge atoms 14 valence electrons ($3d^{10} 4s^2 4p^2$), whereas for O atoms 6 valence electrons ($2s^2 2p^4$) were used. The basis set included plane waves up to a cut-off energy of 520 eV in order to achieve highly converged results. The exchange-correlation energy was described with the generalized gradient approximation, GGA, using the PBEsol functional [14]. The integrations over the Brillouin zone, BZ, were performed with a $4 \times 4 \times 4$ Monkorts-Pack grid [15] k-points sampling to obtain accurate converged energies and forces. At a set of selected volumes, the structure was fully optimized to obtain its equilibrium configuration through the calculations of forces and stress tensors. The convergence criteria were to achieve forces smaller than 0.004 eV/Å and deviations of stress tensor diagonal lower than 0.1 GPa. The data set of volume (V) and energy (E) obtained as result of our simulations was fitted with a standard equation of state, EOS, to determine the bulk modulus and its pressure derivative.

Lattice-dynamic calculations of the phonon modes were carried out at the zone center (Γ point) of the BZ with the direct force-constant approach [16]. These calculations provide the frequency of the normal modes, their symmetry, and their polarization vectors. This allowed to identify the irreducible representations and the character of the phonon modes at the Γ-point. The study of the dynamical stability was performed through the phonon density of state and the phonon dispersion calculated using a $2 \times 2 \times 2$ supercell.

The mechanical and elastic properties were studied by obtaining the elastic constants with the strain–stress method implemented in the VASP code following Le Page method [17].

3. Results and Discussion

3.1. Structural Properties under Pressure

From our ab initio simulations, the obtained unit cell volume at zero pressure is 2095.5 Å^3 with a lattice constant of a_0 = 12.7966 Å, and the O 96 h positions are defined by x = 0.9646, y = 0.0573, and z = 0.1610 (experiment a = 12.8059 Å, O 96h positions: 0.9637, 0.0567, and 0.1609) [18]. Our results are in excellent agreement with the available experimental data, and they differ by less than 0.3%. The Ca, Y, and Ge atoms occupy fixed Wyckoff positions already described in a previous section. In Table 1, we report the calculated structural parameter, the bulk modulus, and its pressure derivative. The bulk modulus, B_0 = 111.24 GPa, with a pressure derivative B_0' of 4.75, was obtained from the theoretical energy and volume data using a Birch–Murnaghan EOS [19]. This bulk modulus is smaller than those of most silicate garnets (between 150 and 190 GPa) [10,20], due to the smaller ionic radius of Ca.

Table 1. The lattice constant, a_0, the volume, V_0, at cero pressure, and the bulk modulus, B_0, and its pressure derivative B_0'.

	a_0 (Å)	V_0 (Å^3)	B_0 (GPa)	B_0'
This work	12.7966	2095.5	111.24	4.75
Experiment [1]	12.8059	2100.05	-	-

[1] Reference [18].

The volume versus pressure evolution of the Ca$_3$Y$_2$Ge$_3$O$_{12}$ garnet is presented in Figure 2, and the inset shows the lattice constant evolution. The linear axial compressibility, $k_a = 0.255 \times 10^{-3}$ GPa^{-1}, was obtained from these data.

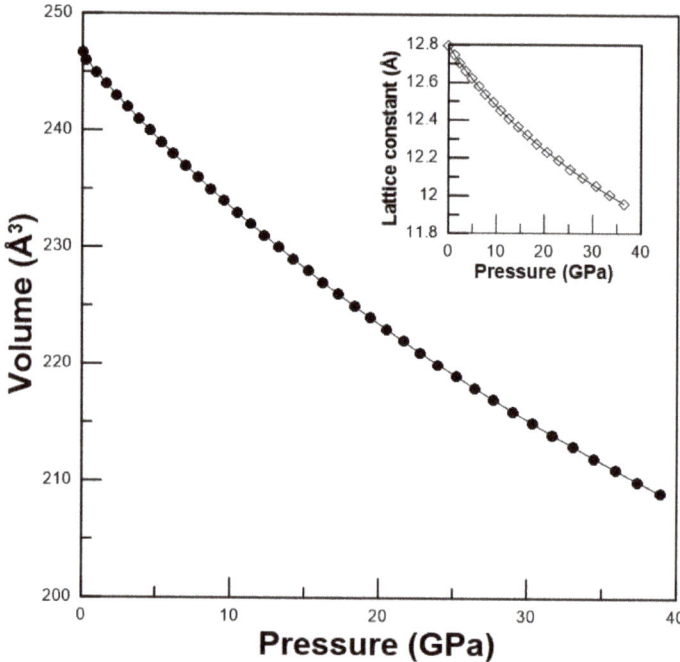

Figure 2. Pressure dependence of the volume. The inset shows the pressure dependence of the lattice constant.

The structural changes of $Ca_3Y_2Ge_3O_{12}$ under pressure are illustrated in Figures 3 and 4. The pressure dependence of the interatomic distances is shown in Figure 3. It can be observed that even at zero pressure, CaO_8 dodecahedra are slightly distorted with two different Ca_{dod}-O distances (2.4587 and 2.5489 Å), while Y_{oct}-O and Ge_{tet}-O distances are 2.2327 and 1.7739 Å, respectively. When compared to distances in $Sr_3Y_2Ge_3O_{12}$, a garnet of the same family, the main difference is that Ca_{dod}-O distances are slightly lower than Sr_{dod}-O ones (5-3 %), while the Y_{oct}-O and Ge_{tet}-O distances are approximately the same in both compounds [21]. All the interatomic distances decrease with increasing pressure. However, whereas Y_{oct}–O and Ge_{tet}-O distances decrease at similar rates, -3.8×10^{-3} Å/GPa and -2.1×10^{-3} Å/GPa, respectively, one of the distances between Ca and O atoms varies more significantly than the other. The smaller distance $(Ca_{dod}-O)2$ changes at a rate of -4.95×10^{-3} Å/GPa, while the larger $(Ca_{dod}-O)1$ distance changes faster, at about -11.7×10^{-3} Å/GPa. Both become equal at around 15.9 GPa. This effect is common among garnets, e.g., the Lu-O distances become equal in $Lu_3Ga_5O_{12}$ [22] or the Y-O distances in $Y_3Al_5O_{12}$ [9].

The evolution with pressure of the polyhedral units of $Ca_3Y_2Ge_3O_{12}$ are plotted in Figure 4. The CaO_8 dodecahedra are the polyhedra with the major reduction in volume: 23.4% in the pressure range under study. In the case of the GeO_4 tetrahedra, the volume is almost constant along the pressure range with a small change of -0.09%. Finally, the YO_6 octahedra volume decreases by 13.9%.

The volume changes of the CaO_8 dodecahedra seem to be responsible for most of the volume reduction induced by pressure in $Ca_3Y_2Ge_3O_{12}$. In fact, the GeO_4 tetrahedra can be treated as rigid units within the structure.

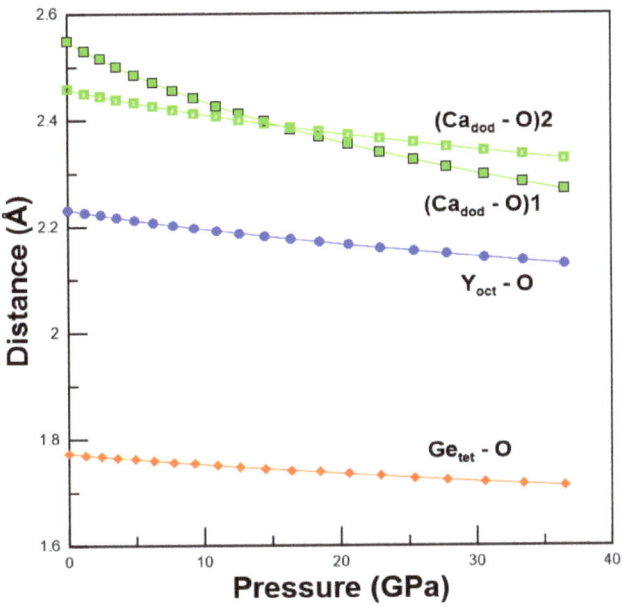

Figure 3. Evolution of interatomic distances as a function of pressure. Ca_{dod} stands for Ca atoms with dodecahedral coordination, Y_{oct} for Y atoms with octahedral coordination, and Ge_{tet} for Ge atoms with tetrahedral coordination.

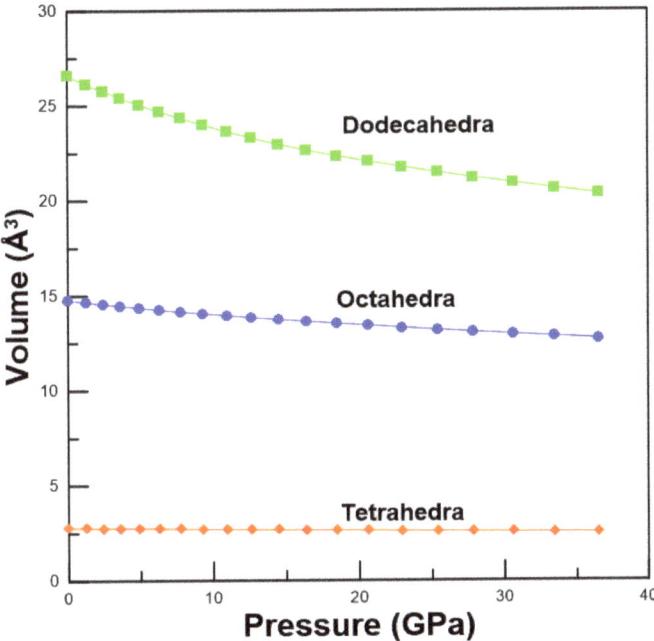

Figure 4. Pressure evolution of the polyhedral units: CaO_8 (green), YO_6 (blue), and GeO_4 (orange).

3.2. Electronic Structure

The electronic structure of $Ca_3Y_2Ge_3O_{12}$ has been scarcely investigated. Even though Baklanova et al. [23] have studied the electronic structure of this garnet at ambient pressure, the electronic properties under pressure are still unknown. Therefore, we performed band structure calculations at zero and different pressures of the $Ca_3Y_2Ge_3O_{12}$ garnet.

Figure 5 shows the electronic band structure at zero pressure. The valence band maxima and the conduction band minima are located at the Γ point, with a direct band gap of Eg = 3.33 eV. This value is in good agreement with the DFT result of 3.32 eV given in reference [23]. In the same reference, the experimental value obtained for the band gap is 5.56 eV, larger than the calculated value. As it is well known, the energy gap is underestimated by DFT compared with the experiments, although the order of magnitude of the pressure coefficients is in agreement.

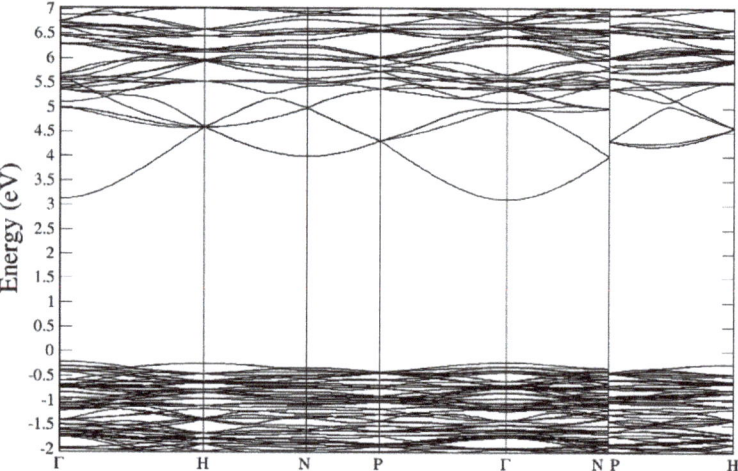

Figure 5. The band structure of $Ca_3Y_2Ge_3O_{12}$ along high symmetry directions in the Brillouin zone, at zero pressure.

As observed, the dispersion of the upper valence bands, VB, is small, similar to other garnets [9,22]. On the other hand, at the Γ point, the band at the bottom of the conduction bands, CB, is well separated with respect to the other conduction bands. To understand the nature of the electronic structure of $Ca_3Y_2Ge_3O_12$, we analyzed the total density of states (DOS) and the projected density of state (PDOS). The PDOS corresponding to the top of the VBs and bottom of CBs, from −2 to 7eV (Figure 6), agrees with previous results [23]. O-2p states dominate the top of the valence bands. The bottom of the CBs is dominated by Y-4d states, with a contribution of O-2p states and a minor contribution of O-2s states near 5.5 eV.

As stated above, although the DFT calculations underestimate the band gap energy, they provide a good description of the pressure dependence of the band gap. To study the effect of pressure on electronic properties, we obtained the band structure of this compound under pressure. According to our calculations, the band gap remains direct in all the stability ranges and increases with a pressure derivative of 29.5 meV/GPa (Figure 7).

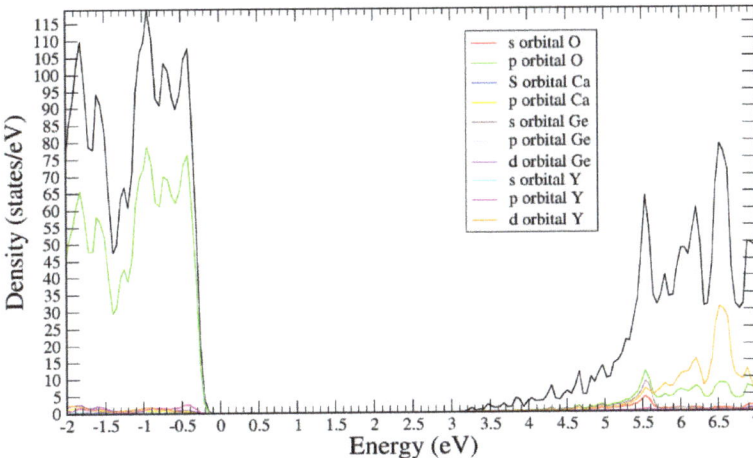

Figure 6. The total (DOS) and partial (PDOS) density of states of $Ca_3Y_2Ge_3O_{12}$, at zero pressure. The black curve represents the DOS.

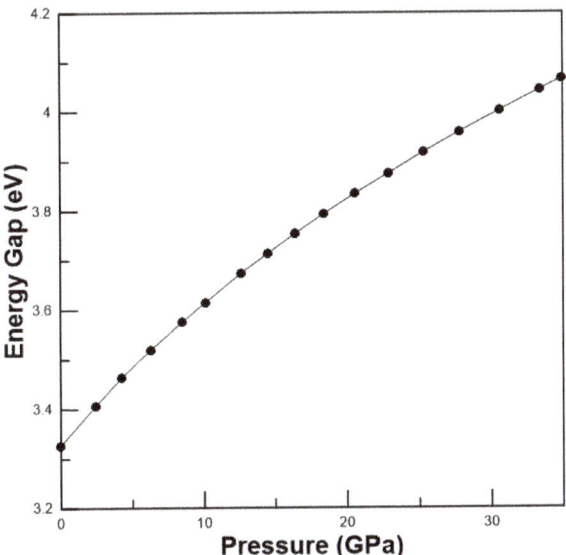

Figure 7. The pressure dependence of the band gap energy of $Ca_3Y_2Ge_3O_{12}$.

3.3. Elastic Properties

Since $Ca_3Y_2Ge_3O_{12}$ has a cubic structure, it has, therefore, only three independent elastic constants: C_{11}, C_{12}, and C_{44}. The elastic constants allow us to obtain the elastic moduli, bulk modulus (B), shear modulus (G), Young's modulus, and Poisson's ratio (ν) that describe the significant elastic properties of a compound along with the B/G ratio.

Cubic crystals the bulk modulus, B, are defined as the following:

$$B = \frac{C_{11} + 2C_{12}}{3}$$

The shear modulus in the scheme of Hill [24], G_H, can be expressed as the average between the upper bond, in the scheme of Voigt, G_V, and the lower bond, in scheme of Reuss, G_R, [25,26]. In the present case [8,27]:

$$G_H = \frac{1}{2}\left[\frac{C_{11} - C_{12} + 3C_{44}}{5} + \frac{5C_{44}(C_{11} - C_{12})}{4C_{44} + 3(C_{11} - C_{12})}\right]$$

The Young's modulus and the Poisson's ratio can be defined as the following:

$$E_x = \frac{9B_x G_x}{3B_x + G_x}$$

and

$$\nu_x = \frac{E_x - 2G_x}{2G_x}$$

where the subscript x refers to the symbols V, R, and H.

The computed elastic constants, C_{ij}, and the elastic moduli at 0 GPa are presented in Tables 2 and 3. The bulk modulus, the inverse of the compressibility, is related to the material resistance to hydrostatic pressure. The bulk modulus computed through the elastic constants, 112.430 GPa, is in excellent agreement with the value obtained with the EOS using total energy calculations, 111.24 GPa. This is an indication of the quality and consistency of our simulations. The Young modulus, 129.94 GPa, provides a measure of the material stiffness. Young's modulus, Poisson's ratio, and the B/G ratio are essential for engineering applications. B/G is 2.26; therefore, according to the Pugh criterion (B/G > 1.75), $Ca_3Y_2Ge_3O_{12}$ behaves like a ductile material at 0 GPa [28]. The Poisson's ratio is $\nu = 0.31$; hence, the ionic bonding is predominant against the covalent bonding, and the interatomic forces can still be considered predominantly central [29,30].

Table 2. Elastic constants in GPa of $Ca_3Y_2Ge_3O_{12}$ at 0 Gpa.

C_{11}	C_{12}	C_{44}
199.43	68.93	41.38

Table 3. Elastic moduli B, E, and G, the B/G and the Poisson's ratio, ν, in the Voigt, Reuss, and Hill approximations, at 0 GPa.

	Voigt	Reuss	Hill
Bulk modulus (GPa)	112.4	112.43	112.43
Shear modulus (GPa)	50.93	48.47	49.70
Young modulus (GPa)	132.74	127.14	129.94
Poisson's ratio	0.30	0.31	0.31
Bulk/shear ratio	2.21	2.32	2.26

This compound is stable at zero pressure; hence, the elastic constants fulfill the Born stability criteria [31]:

$$C_{11} - C_{12} > 0,\ C_{11} + 2C_{12} > 0,\ \text{and}\ C_{44} > 0$$

However, when hydrostatic pressure is applied to the crystal, the above criteria to describe the stability limits of a cubic crystal are not adequate. Moreover, the relevant magnitudes that describe the elastic properties are the elastic stiffness coefficients, B_{ij}, related to the elastic constants through the following relationships [32,33]:

$$B_{11} = C_{11} - P,\ B_{12} = C_{12} + P,\ \text{and}\ B_{44} = C_{44} - P$$

The evolution of the B_{ij} with pressure is plotted in Figure 8. The diagonal coefficient B_{11} and the B_{12} coefficient increase with pressure in the studied pressure range. However, the diagonal coefficient B_{44} decreases rapidly, becoming negative at about 45 GPa. B_{11} is higher than B_{44}; this indicates that the resistance to a shear deformation is weaker than the resistance to compression.

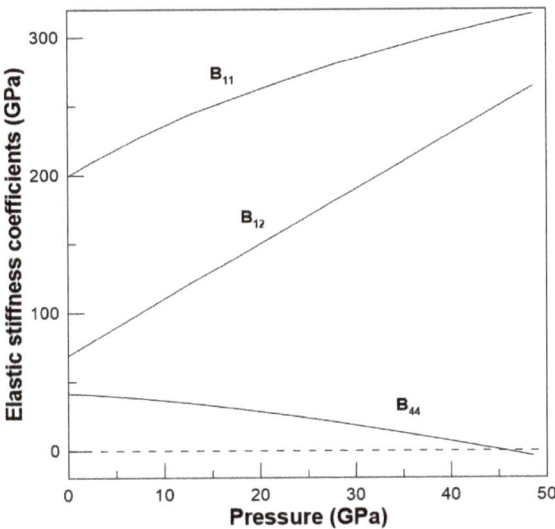

Figure 8. Pressure evolution of the elastic stiffness coefficients, Bij.

The elastic stiffness coefficients enable us to study the evolution of the elastic properties under pressure. Replacing the C_{ij} elastic constants with the B_{ij} elastic stiffness coefficients in the above expressions, we obtain the analytical expressions for the elastic moduli, valid under pressure [31,34]. The pressure dependence of the elastic moduli with pressure up to 50 GPa is displayed in Figure 9. The bulk modulus increases almost linearly with increasing pressure. Both the shear modulus and the Young modulus decrease, and their values converge at high pressure. The Poisson's ratio becomes larger, and this can be interpreted as an increase in metallization. The B/G ratio indicates that the material is ductile until approximately 45 GPa; above this pressure, there is a dramatic change in the slope (Figure 10a). This could indicate a possible mechanical instability at higher pressures.

To study the mechanical stability of a crystal under pressure, the generalized Born stability criteria must be employed [31,35]. In the present case, as stated above, these criteria can be written as the following:

$$M_1 = B_{11} - B_{12} > 0, \quad M_2 = B_{11} + 2B_{12} > 0, \quad M_3 = B_{44} > 0 \tag{1}$$

These generalized stability criteria are plotted as a function of pressure in Figure 10b. The M_3 stability criterion is violated at 45.7 GPa. This result indicates that the $Ca_3Y_2Ge_3O_{12}$ garnet is mechanically unstable above this pressure. Experimental studies on garnets report that an amorphous phase appears at high pressure [8,22,36]. However, in the present case, as we will see in the next section, the $Ca_3Y_2Ge_3O_{12}$ garnet becomes dynamically unstable at pressure over 34.4 GPa.

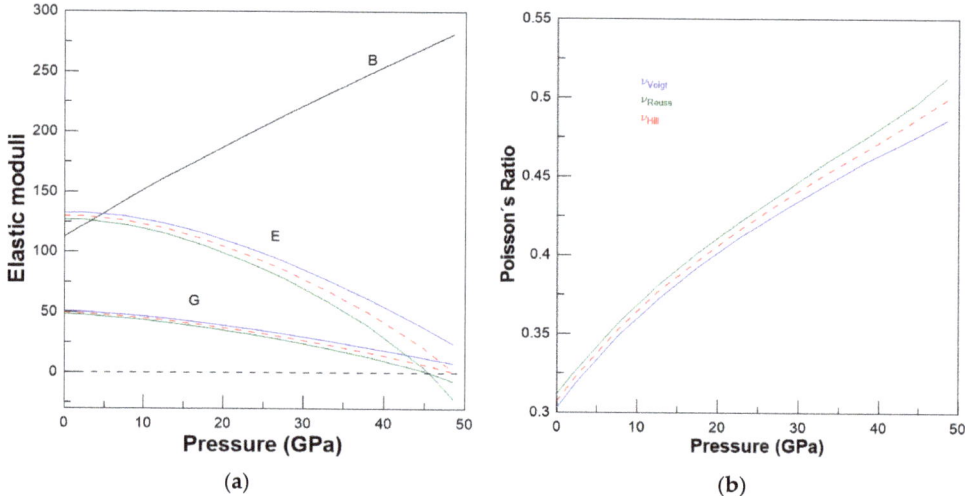

Figure 9. (a) Pressure evolution of the elastic moduli: bulk modulus, B, Young Modulus, E, and shear modulus, G. The blue line represents the moduli in the Voigt approximation, the green line in the Reuss approximation, and the red one in the Hill approximation. (b) Pressure evolution of the Poisson's ratio in the three approximations: Voigt, Reuss, and Hill.

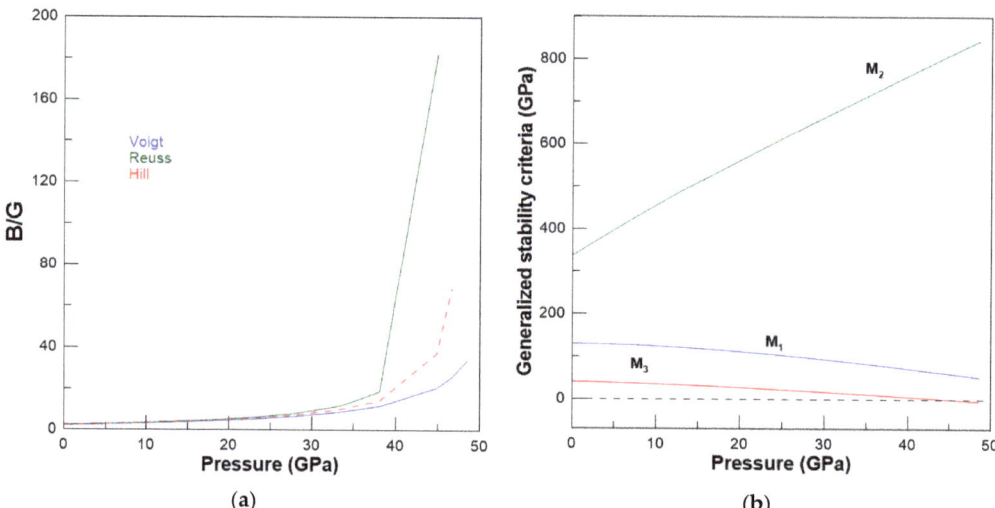

Figure 10. (a) Pressure evolution of the B/G ratio in the approximations of Voigt, Reuss, and Hill. (b) Generalized stability criteria as a function of pressure.

3.4. Vibrational Properties

The $Ca_3Y_2Ge_3O_{12}$ garnet belongs to the *Ia-3d* space group and, according to group theory, has 98 vibrational modes. At the zone center, they can be classified as 25 Raman active modes ($3A_{1g} + 8E_g + 14T_{2g}$), 17 infrared active modes ($17T_{1u}$), and 55 silent modes ($16T_{2u} + 14T_{1g} + 5A_{2u} + 5A_{2g} + 10E_u + 5A_{1u}$), which are optically inactive.

Figure 11 shows the phonon dispersion. It can be observed that the phonon spectrum could be divided into three regions. The low-frequency region ranges up to approximately 400 cm^{-1} and is separated by a small gap from the medium-frequency region, which is very narrow and reaches up to 510 cm^{-1}. The high-frequency region extends from 625 to 780 cm^{-1}.

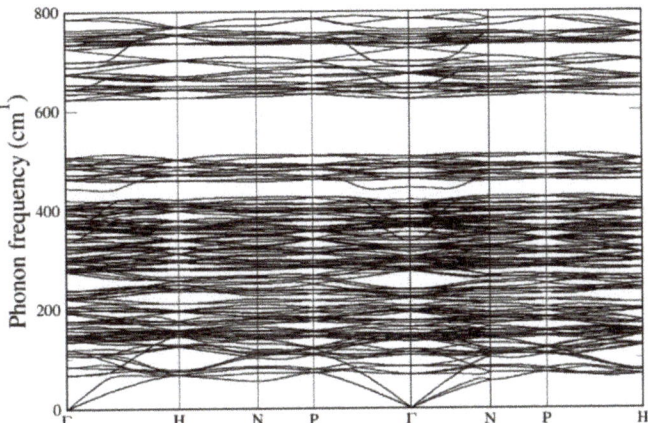

Figure 11. Phonon dispersion of Ca$_3$Y$_2$Ge$_3$O$_{12}$ garnet at 0 GPa along the high symmetry directions.

The Raman modes are connected with the vibration of the different polyhedral units strongly coupled to each other; therefore, the attribution of each mode to a single unit is not straightforward. To analyze the dynamical contribution of each atom, the total phonon density of states (DOS) and the partial density of states (PDOS) were calculated (see Figure 12). As shown, in the case of Ca$_3$Y$_2$Ge$_3$O$_{12}$, the PDOS is very compact at low and medium frequencies, between 100 and 510 cm^{-1}, and also in the high-frequency region between 625 and 780 cm^{-1}. The contributions of Y atoms are located at the low- and medium-frequency regions, mainly up to 200 cm^{-1}. Ca atoms also contribute in these regions, although the contribution is small in the region of very low frequencies. Ge atoms have a small contribution in the region from 200 to 500 cm^{-1} and a larger one in the high-frequency region. Finally, the O atoms have more significant contributions, spread over all frequencies, mainly in the high-frequency region.

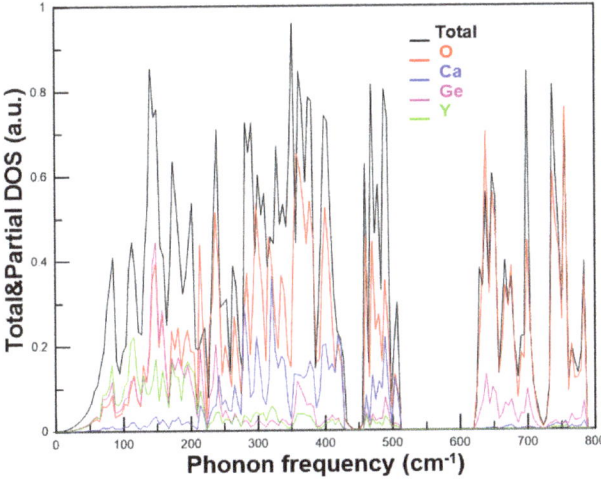

Figure 12. Total and partial phonon density of states at zero pressure. The black line represents the total DOS, and the contributions of O, Ca, Ge, and Y atoms are represented by the red, blue, magenta, and green lines, respectively.

The pressure dependence of the calculated frequencies has been fitted with a second-order polynomial. All theoretical Raman active modes are reported at zero pressure in Table 4, along with the pressure derivative coefficients, $d\omega/dP$, and the Grüneisen parameters, $\gamma = (d\omega/dP)/(B_0/\omega_0)$, where B_0 is the bulk modulus and ω_0 the frequency of the corresponding mode at zero pressure.

Table 4. Wavenumbers of Raman active vibrational frequencies at zero pressure, ω_0, their pressure coefficients, $d\omega/dP$, and Grüneisen parameters, γ, of $Ca_3Y_2Ge_3O_{12}$ garnet.

Mode Symmetry	ω_0 (cm^{-1})	$d\omega/dP$ (cm^{-1}/GPa)	γ
E_g	141.47	0.958	0.75
T_{2g}	142.84	1.539	1.20
T_{2g}	152.62	2.190	1.59
T_{2g}	194.70	1.430	0.82
T_{2g}	234.25	2.320	1.10
T_{2g}	295.08	3.626	1.37
E_g	301.61	2.169	0.80
T_{2g}	306.04	5.405	1.96
E_g	304.89	3.666	1.34
A_{1g}	325.85	3.768	1.28
T_{2g}	354.64	3.495	1.10
T_{2g}	350.29	6.711	2.13
E_g	380.55	3.140	0.92
E_g	400.63	5.960	1.65
T_{2g}	408.54	3.975	1.08
A_{1g}	444.83	4.085	1.02
T_{2g}	466.97	4.222	1.00
E_g	485.32	3.118	0.71
T_{2g}	483.25	3.839	0.88
T_{2g}	655.21	4.998	0.85
T_{2g}	676.95	4.594	0.75
E_g	701.16	4.755	0.76
E_g	736.78	4.354	0.66
A_{1g}	759.40	4.295	0.63
T_{2g}	786.11	4.062	0.57

The evolution of the Raman modes of this garnet with pressure is presented in Figure 13. Most modes harden under compression; however, the frequencies of the lowest Raman modes change slightly. The modes on the low- and medium-frequency regions are related to movements of Ca, Ge, Y, and O atoms and exhibit very different pressure coefficients, since they involve several types of bonds with diverse compressibility. Due to the significant differences between the pressure dependence of these modes, crossing and anti-crossing phenomena are observed in the medium-frequency region. The highest modes, however, increase with pressure with similar coefficients and exhibit similar and low Grüneisen parameters.

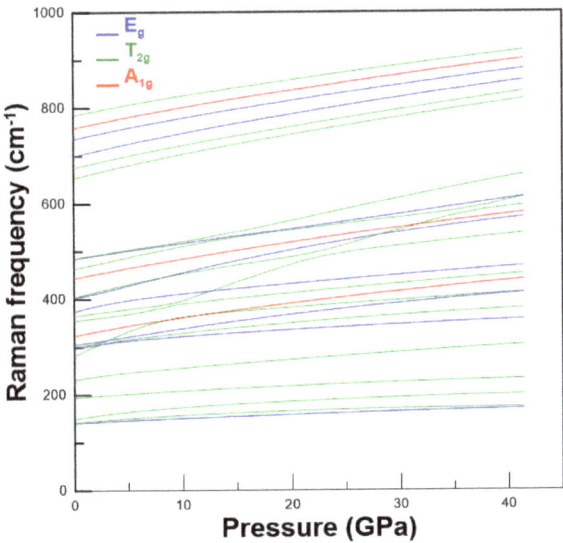

Figure 13. Theoretical evolution with pressure of Raman modes of the $Ca_3Y_2Ge_3O_{12}$ garnet.

Additionally, the pressure evolution of the infrared modes is plotted in Figure 14. As in the case of Raman modes, there is clearly a large gap between the medium- and high-frequency region. The phonons in the medium- and high-frequency regions harden under pressure, and there are several anti-crossing phonons in the medium region. Except for the lowest mode ($\omega_0 = 67.96$ cm^{-1}), all the phonons in the low-frequency region are very slightly affected by pressure. The frequency values, pressure derivative, and Grüneisen parameters for the infrared modes are reported in Table 5.

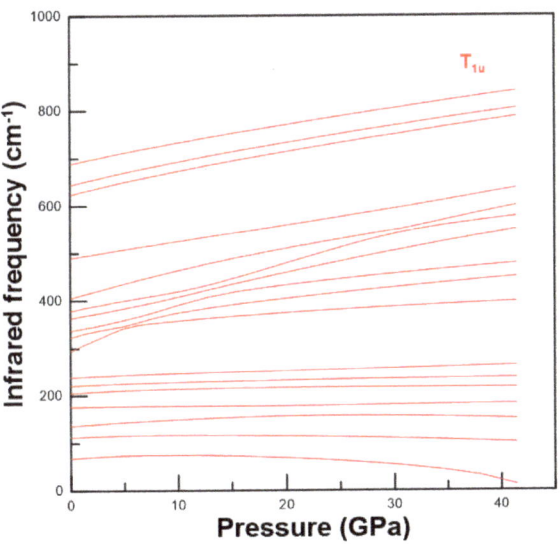

Figure 14. Theoretical evolution with the pressure of infrared modes of the $Ca_3Y_2Ge_3O_{12}$ garnet.

Table 5. Infrared active vibrational modes, ω_0, pressure coefficients, $d\omega/dP$, and Grüneisen parameters, γ, of $Ca_3Y_2Ge_3O_{12}$ garnet.

Mode Symmetry	ω_0 (cm^{-1})	$d\omega/dP$ (cm^{-1}/GPa)	γ
T_{1u}	67.96	1.418	2.32
T_{1u}	113.97	0.397	0.39
T_{1u}	136.53	1.566	1.27
T_{1u}	176.95	−0.017	−0.01
T_{1u}	207.04	0.706	0.38
T_{1u}	221.62	0.661	0.33
T_{1u}	239.60	0.656	0.30
T_{1u}	311.77	4.571	1.63
T_{1u}	326.13	4.881	1.66
T_{1u}	333.97	6.179	2.06
T_{1u}	359.84	5.185	1.60
T_{1u}	372.30	5.558	1.66
T_{1u}	409.26	5.375	1.46
T_{1u}	490.98	3.269	0.74
T_{1u}	625.86	4.871	0.865
T_{1u}	645.88	3.118	0.71
T_{1u}	690.02	4.383	0.70

Finally, the pressure evolution of the 55 silent modes is presented in Figure 15. The most crucial point is the softening of the lower silent mode (T_{2u}).

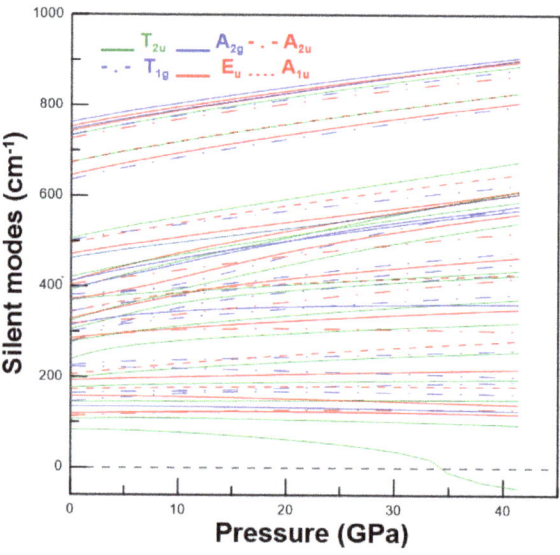

Figure 15. Theoretical evolution with the pressure of silent modes of $Ca_3Y_2Ge_3O_{12}$ garnet.

At 34.4 Gpa, this T_{2u} mode becomes imaginary, as shown in Figure 16. It is important to note that this indicates a dynamical instability at 34.4 GPa. Therefore, our results evidence that, under hydrostatic pressure, this compound becomes dynamically unstable.

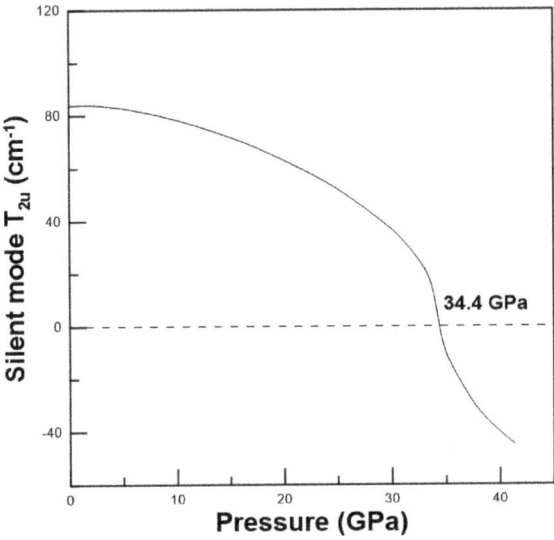

Figure 16. Theoretical evolution with pressure of silent mode T_{2u} of the $Ca_3Y_2Ge_3O_{12}$ garnet.

We analyzed this soft phonon and obtained a triclinic *P-1* structure with 80 atoms in the cell at positions *1a* (x y z). By studying the stability of this new structure, we obtained that it is also dynamically unstable, so we discard it as a possible high-pressure phase of this garnet.

4. Conclusions

An ab initio study of the structural, electronic, elastic, and vibrational properties of the $Ca_3Y_2Ge_3O_{12}$ garnet was performed both at ambient and under hydrostatic pressure.

The lattice parameter compares nicely with the experimental value at 0 GPa. The evolution of the lattice parameter and interatomic distances under compression, the equation of state, and the compressibility are reported. The dodecahedral units (CaO_8 in the present case) that account for most of the volume changes are distorted and behave under pressure in a common way among garnets.

The electronic band structure agrees with previous theoretical data and shows a direct band gap at Γ that increases with increasing pressure.

The elastic constants and elastic stiffness coefficients were accurately determined and the elastic behavior of the $Ca_3Y_2Ge_3O_{12}$ garnet was analyzed. The pressure evolution of B, G, and E elastic moduli and the Poisson's ratio were reported. The B/G relation indicates that the material is ductile in the whole pressure stability range and more sensitive to volume compression than to shear deformation (B>G). The study of the stability criteria indicates that the garnet is mechanically unstable above 45.7 GPa.

The total and partial phonon density of states were analyzed to determine each atom's contribution to the different vibrational modes. The pressure evolutions of the Raman, infrared, and silent modes were presented. At 34.4 GPa, the lower silent mode (T_{2u}) softens and becomes imaginary; this means that the garnet structure is unstable at a pressure above 34.4 GPa.

Author Contributions: Both authors contributed equally to this work: simulations, analysis and interpretation of results, and manuscript writing, A.M., P.R.-H. All authors have read and agreed to the published version of the manuscript.

Funding: This research was funded by the Spanish Research Agency (AEI) and the Spanish Ministry of Science and Investigation (MCIN) under grant PID2019-106383GB-43 (DOI: 10.13039/501100011033) and RED2018-102612-T (MALTA Consolider-Team Network).

Data Availability Statement: All relevant data that support the findings of this study are available from the corresponding authors upon request.

Acknowledgments: The authors thank the financial support from the Spanish Research Agency (AEI) and Spanish Ministry of Science and Investigation (MCIN) under grant PID2019-106383GB-43 (DOI: 10.13039/501100011033) and RED2018-102612-T (MALTA Consolider-Team Network).

Conflicts of Interest: The authors declare no conflict of interest.

References

1. Karato, S.-I.; Wang, Z.; Liu, B.; Fujino, K. Plastic deformation of garnets: Systematics and implications for the rheology of the mantle transition zone. *Earth Planet. Sci. Lett.* **1995**, *130*, 13–30. [CrossRef]
2. Speghini, A.; Piccinelli, F.; Bettinelli, M. Synthesis, characterization and luminescence spectroscopy of oxide nanopowders activated with trivalent lanthanide ions: The garnet family. *Opt. Mater.* **2011**, *33*, 247–257. [CrossRef]
3. Rodriguez-Mendoza, U.R.; León-Luis, S.F.; Muñoz-Santiuste, J.E.; Jaque, D.; Lavín, V. Nd^{3+}-doped $Ca_3Ga_2Ge_3O_{12}$ garnet: A new optical pressure sensor. *J. Appl. Phys.* **2013**, *113*, 213517. [CrossRef]
4. León-Luis, S.F.; Muñoz-Santiuste, J.E.; Lavín, V.; Rodriguez-Mendoza, U. Optical pressure and temperature sensor based on the luminescence properties of Nd^{3+} ion in a gadolinium scandium gallium garnet crystal. *Opt. Express* **2012**, *20*, 10393–10398. [CrossRef]
5. Nishiura, S.; Tanabe, A.; Fujioka, F.Y. Properties of transparent Ce:YAG ceramic phosphors for white LED. *Opt. Mater.* **2011**, *33*, 688–691. [CrossRef]
6. Cui, J.; Zheng, Y.; Wang, Z.; Cao, L.; Wang, X.; Yao, Y.; Zhang, M.; Zheng, M.; Yang, Z.; Li, P. Improving the luminescence thermal stability of $Ca_3Y_2Ge_3O_{12}$: Cr^{+3} based on cation substitution and its application in NIR LEDs. *Mater. Adv.* **2022**, *3*, 2772–2778. [CrossRef]
7. Mao, N.; Liu, S.; Song, Z.; Yu, Y.; Liu, Q. A broadband near-infrared phosphor $Ca_3Y_2Ge_3O_{12}:Cr^{3+}$ with garnet structure. *J. Alloys Compd.* **2021**, *863*, 158699. [CrossRef]
8. Monteseguro, V.; Rodríguez-Hernández, P.; Lavín, V.; Manjón, F.J.; Muñoz, A. Electronic and elastic properties of Yttrium gallium garnet under pressure from ab initio studies. *J. Appl. Phys.* **2013**, *113*, 183505. [CrossRef]
9. Monteseguro, V.; Rodríguez-Hernández, P.; Muñoz, A. Yttrium aluminun garnet under pressure: Structural, elastic, and vibrational properties from ab initio studies. *J. Appl. Phys.* **2015**, *118*, 245902. [CrossRef]
10. Mookherjee, M. High-pressure elasticity of sodium majorite garnet, $Na_2MgSi_5O_{12}$. *Am. Mineralogist.* **2014**, *99*, 2416–2423. [CrossRef]
11. Kresse, G.; Furthmuller, J. Efficient iterative schemes for ab initio total-energy calculations using a plane-wave basis set. *Phys. Rev. B* **1996**, *54*, 11169–11186. [CrossRef] [PubMed]
12. Kresse, G.; Furthmüller, J. Efficiency of ab-initio total energy calculations for metals and semiconductors using a plane-wave basis set. *Comput. Mater. Sci.* **1996**, *6*, 15–50. [CrossRef]
13. Blöchl, P.E. Projector augmented-wave method. *Phys. Rev. B* **1994**, *50*, 17953–17979. [CrossRef] [PubMed]
14. Perdew, J.P.; Ruzsinszky, A.; Csonka, G.I.; Vydrov, O.A.; Scuseria, G.E.; Constantin, L.A.; Zhou, X.; Burke, K. Restoring the Density-Gradient Expansion for Exchange in Solids and Surfaces. *Phys. Rev. Lett.* **2008**, *100*, 136406. [CrossRef]
15. Monkhorst, H.J.; Pack, J.D. Special points for Brillouin-zone integrations. *Phys. Rev. B* **1976**, *13*, 5188–5192. [CrossRef]
16. Togo, A.; Tanaka, I. First principles phonon calculations in materials science. *Scr. Mater.* **2015**, *108*, 1–5. [CrossRef]
17. Le Page, Y.; Saxe, P. Symmetry-General Least-Squares Extraction of Elastic Data for Strained Materials from Ab Initio Calculations of Stress. *Phys. Rev. B* **2002**, *65*, 104104. [CrossRef]
18. Levy, D.; Barbier, J. Normal and inverse garnets: $Ca_3Fe_2Ge_3O_{12}$, $Ca_3Y_2Ge_3O_{12}$ and $Mg_3Y_2Ge_3O_{12}$. *Acta Cryst.* **1999**, *55*, 1611–1614. [CrossRef]
19. Birch, F. Finite Elastic Strain of Cubic Crystals. *Phys. Rev.* **1947**, *71*, 809–824. [CrossRef]
20. Milman, V.; Nobes, R.H.; Akhmatskaya, E.V.; Winkler, B.; Pickard, C.J.; White, J.A. Ab initio study of the structure and compressibility of garnets. In *Properties of Complex Inorganic Solids*; Meike, A., Gonis, A., Turchi, P.E.A., Rajan, K., Eds.; Kluwer Academic/Plenum: New York, NY, USA, 2000; Volume 3, pp. 417–427.
21. Marin, S.J.; O´Keeffe, M.; Young, V.G., Jr.; Von Dreele, R.B. The crystal structure of $Sr_3Y_2Ge_3O_{12}$. *J. Solid State Chem.* **1991**, *91*, 173–175. [CrossRef]
22. Monteseguro, V.; Rodríguez-Hernández, P.; Ortiz, H.M.; Venkatramu, V.; Manjón, F.J.; Jayasankar, C.K.; Lavín, V.; Muñoz, A. Structural, elastic and vibrational properties of nanocrystalline lutetium gallium garnet under high pressure. *Phys. Chem. Chem. Phys.* **2015**, *17*, 9454–9464. [CrossRef] [PubMed]
23. Baklanova, Y.V.; Enyashin, A.N.; Maksimova, L.G.; Tyutyunnik, A.P.; Chufarov, A.Y.; Gorbatov, E.V.; Baklanova, I.V.; Zubkov, V.G. Sensitized IR luminescence in $Ca_3Y_2Ge_3O_{12}$: Nd^{3+}, Ho^{3+} under 808 nm laser excitation. *Ceram. Int.* **2018**, *44*, 6959–6967. [CrossRef]
24. Hill, R. The Elastic Behaviour of a Crystalline Aggregate. *Proc. Phys. Soc. Sect. A* **1952**, *65*, 349. [CrossRef]
25. Voigt, W. *Lehrbuch der Kristallphysik (Mit Ausschluss der Kristalloptik)*; B.G. Teubner: Leipzig/Berlin, Germany, 1928.

26. Reuss, A. Berechnung der Fließgrenze von Mischkristallen auf Grund der Plastizitätsbedingung für Einkristalle. *J. Appl. Math. Mech.* **1929**, *9*, 49–58. [CrossRef]
27. Erba, A.; Mahmoud, A.; Orlando, R.; Dovesi, R. Elastic properties of six silicate garnet end members from accurate ab initio simulations. *Phys. Chem. Miner.* **2014**, *41*, 151–160. [CrossRef]
28. Pugh, S.F. XCII. Relations between the elastic moduli and the plastic properties of polycrystalline pure metals. *Lond. Edinb. Dublin Philos. Mag. J. Sci.* **1954**, *45*, 823–843. [CrossRef]
29. Brazhkin, V.V.; Lyapin, A.G.; Hemley, R.J. Harder than diamond: Dreams and reality. *Philos. Mag. A* **2002**, *82*, 231–253. [CrossRef]
30. Greaves, G.N.; Greer, A.L.; Lakes, R.S.; Rouxel, T. Poisson's ratio and modern materials. *Nat. Mater.* **2011**, *10*, 823–837. [CrossRef]
31. Born, M.; Huang, K. *Dynamical Theory of Crystal Lattices*; Clarendon Press: London, UK, 1954.
32. Wallace, D.C. *Thermodynamics of Crystals*; Dover Publications: New York, NY, USA, 1998.
33. Grimvall, G.; Magyari-Köpe, B.; Ozolins, V.; Persson, K.A. Lattice instabilities in metallic elements. *Rev. Mod. Phys.* **2012**, *84*, 945–986. [CrossRef]
34. Wallace, D.C. Thermoelasticity of Stressed Materials and Comparison of Various Elastic Constants. *Phys. Rev.* **1967**, *162*, 776–789. [CrossRef]
35. Wang, J.; Yip, S.; Phillpot, S.R.; Wolf, D. Crystal instabilities at finite strain. *Phys. Rev. Lett.* **1993**, *71*, 4182–4185. [CrossRef] [PubMed]
36. Hua, H.; Mirov, S.; Vohra, Y.K. High-pressure and high-temperature studies on oxide garnets. *Phys. Rev. B* **1996**, *54*, 6200–6209. [CrossRef] [PubMed]

Disclaimer/Publisher's Note: The statements, opinions and data contained in all publications are solely those of the individual author(s) and contributor(s) and not of MDPI and/or the editor(s). MDPI and/or the editor(s) disclaim responsibility for any injury to people or property resulting from any ideas, methods, instructions or products referred to in the content.

Article

Structural and Luminescence Properties of Cu(I)X-Quinoxaline under High Pressure (X = Br, I)

Javier Gonzalez-Platas [1,*], Ulises R. Rodriguez-Mendoza [2], Amagoia Aguirrechu-Comeron [3], Rita R. Hernandez-Molina [4], Robin Turnbull [5], Placida Rodriguez-Hernandez [2] and Alfonso Muñoz [2]

1. Instituto Universitario de Estudios Avanzados en Física Atómica, Molecular y Fotónica (IUDEA), MALTA Consolider Team, Departamento de Física, Universidad de La Laguna, E38204 La Laguna, Tenerife, Spain
2. Instituto de Materiales y Nanotecnología (IMN), MALTA Consolider Team, Departamento de Física, Universidad de La Laguna, E38204 La Laguna, Tenerife, Spain
3. Departamento de Física, Universidad de La Laguna, E38204 La Laguna, Tenerife, Spain
4. Instituto Universitario de Bioorgánica Antonio González, Departamento de Química, Unidad Departamental de Química Inorgánica, Universidad de La Laguna, E38204 La Laguna, Tenerife, Spain
5. Departamento de Física Aplicada, Instituto Ciencia de Materiales, MALTA Consolider Team, Universidad de Valencia, E46100 Burjassot, Valencia, Spain
* Correspondence: jplatas@ull.edu.es; Tel.: +34-922-318-251

Abstract: A study of high-pressure single-crystal X-ray diffraction and luminescence experiments together with ab initio simulations based on the density functional theory has been performed for two isomorphous copper(I) halide compounds with the empirical formula $[C_8H_6Cu_2X_2N_2]$ (X = Br, I) up to 4.62(4) and 7.00(4) GPa for X-ray diffraction and 6.3(4) and 11.6(4) GPa for luminescence, respectively. An exhaustive study of compressibility has been completed by means of determination of the isothermal equations of state and structural changes with pressure at room temperature, giving bulk moduli of K_0 = 14.4(5) GPa and K'_0 = 7.7(6) for the bromide compound and K_0 = 13.0(2) GPa and K'_0 = 7.4(2) for the iodide compound. Both cases exhibited a phase transition of second order around 3.3 GPa that was also detected in luminescence experiments under the same high-pressure conditions, wherein redshifts of the emission bands with increasing pressure were observed due to shortening of the Cu–Cu distances. Additionally, ab initio studies were carried out which confirmed the results obtained experimentally, although unfortunately, the phase transition was not predicted.

Keywords: equation of state; luminescence; high pressure; phase transition; copper halides

1. Introduction

Coordination complexes based on copper(I) halides with N-donor ligands have been some of the most attractive and widely studied complexes over the last three decades [1–8]. This has been due to the variety of physical and chemical properties exhibited by the copper(I) halide complexes, such as photo- and electroluminescence, nonlinear optics, and electrical conductivity. In particular, the electrical conductivity properties lead to many potential technological applications [9–17], such as light-emitting diodes (LEDs) [18], organic light-emitting diodes (OLEDs) [19–22], simpler light-emitting electrochemical cells (LECs) [23–27], biosensors [28], or solar energy conversion [29–31].

Systems based on these copper(I) halide complexes are very interesting because copper (Cu) is more abundant and less expensive than noble or rare earth metals. Copper halides also exhibit remarkable structural diversity [2,4,32]. Specifically, the d^{10} electron configuration of Cu^I does not lead to a pre-defined spatial configuration of the ligands; rather, the coordination sphere is determined by molecular mechanics and electrostatic factors. The halide ions exhibit four pairs of electrons in the outer shell and can therefore coordinate with four Cu^I ions. When a nitrogen-centered Lewis base is added to a Cu^I halide, the coordination number of the halide ions is reduced, resulting in a cluster of Cu^I salt $(CuX)_n$. This can give rise to interesting geometries such as mononuclear copper(I) complexes [33], square dimers [34], cubane tetramers [35], or polymers in one or two dimensions [36–39].

Regarding the photo-physical properties of metal complexes, they are intimately related to the electronic configuration of the metal center and the ligands around it. Cu^I with a d^{10} electronic configuration has a completely filled d-subshell, which excludes the possibility of d–d electronic transitions. Therefore, the luminescence of these d^{10} copper(I) complexes arises from other types of electronic transitions, including metal-to-ligand charge transfer (MLCT), ligand-to-metal charge transfer (LMCT), and ligand-to-ligand charge transfer (LLCT). The presence of other types of interactions is also possible. For example, transitions involving the CuX cluster core, whereby a combination of halide-to-metal charge transfer (XMCT) and metal-cluster-centered $d^{10} \rightarrow d^9s^1$ (MCC) charge transfer occurs due to the weak attractive d^{10}–d^{10} interactions between the Cu^I ions in the cluster. In this latter case, known as cuprophilic interaction, the intensity of the interaction is directly connected with the Cu–Cu distances, which play an important role. Cuprophilic interaction has been observed experimentally in systems which exhibit luminescence for Cu–Cu distances shorter than twice the Cu van der Waals radius (2.8 Å). In addition, the d^{10} configuration favors tetrahedral four-coordinate environments of the ligands around the metal center Cu^I [40]. Pressure is a valuable thermodynamic variable which can be used as a tool to induce, in a continuous way, changes in the crystal structure and the immediate coordination complex environment, thereby provoking changes in the emission spectra [41,42]. Therefore, high-pressure techniques allow us to relate structural changes with optical properties.

Despite the large number of studies into copper(I) halides, there has historically been a deficit of research into these systems under extreme conditions of temperature or pressure, with still fewer publications utilizing ab initio studies for comparison with experimental results. Pressure and temperature are external stimuli which both affect intermolecular interactions, molecular packing, and structural parameters, thereby affecting emergent physical and chemical properties. Knowledge of how the properties of copper(I) halide complexes are affected by these different external stimuli is essential for technological applications [15,43]. In recent years, this situation seems to be changing with the publication of studies into how temperature [12,44,45] and pressure [31,36–39,46,47] affect the luminescent properties of Cu^I iodide complexes.

In this work, we present the pressure-induced changes in the crystal structure and luminescence properties of the 2D isomorphous coordination polymers $[(CuX)_2(Quin)]$ (X = Br, I and Quin = quinoxaline) at room temperature.

2. Materials and Methods

2.1. Synthesis of $[(CuX)_2(Quin)]$ with X = Br, I

The compound $[(CuBr)_2(Quin)]$ was prepared by mixing an equimolar amount of CuBr (0.1 g, 6.9 mmol) dissolved in 20 mL of acetonitrile with quinoxaline (0.09 g, 6.9 mmol) dissolved in 10 mL of acetonitrile. A red-orange precipitate immediately formed. The precipitate was filtered off. This powder was recrystallized from acetonitrile, and orange single crystals suitable for X-ray diffraction were obtained by slow evaporation of the solvent over 1 week. The compound $[(CuI)_2(Quin)]$ was prepared from CuI (0.2g, 1.05mmol) and quinoxaline (0.13g, 1mmol) following the same procedure as described above for the equivalent bromine compound $[(CuBr)_2(Quin)]$.

Elemental analysis. $[(CuBr)_2(Quin)]$ Found: C, 22.9; H, 1.4; N, 6.6%. Calcd: C, 23.0; H, 1.5; N, 6.7%. $[(CuI)_2(Quin)]$ Found: C, 18.5; H, 1.2; N, 5.4%. Calcd: C, 18.8; H, 1.2; N, 5.5%.

2.2. X-ray Diffraction Measurements

At ambient conditions, we used an Agilent SuperNOVA diffractometer equipped with an EOS detector (CCD) and a Mo radiation micro-source. Data were collected and processed using CrysAlisPro software [48]. The structures for both compounds were solved and refined using the SHELXT and SHELXL programs [49,50]. The PLATON program [51] was used for geometric calculations.

For high-pressure (HP) X-ray measurements, we used the Agilent SuperNOVA diffractometer for the bromine compound [(CuBr)$_2$(Quin)]. The equivalent iodine compound [(CuI)$_2$(Quin)] was measured at the MSPD beamline at ALBA synchrotron with a focused beam of 15 × 15 μm^2 (FWHM) and a Rayonix SX165 CCD detector (Figure S1 in Supporting Information). The synchrotron X-ray energy used was 38.9 keV (λ = 0.3185 Å), selected from the La absorption K-edge. The sample-to-detector distance (240 mm), beam center, and detector tilt were calibrated from LaB$_6$ (NIST) diffraction data measured under the same experimental conditions as the HP experiments. For each pressure, data collection was performed by rotating the DAC around the omega axis in small steps (0.2°) from −30 to 30°. The data were processed using CrysAlisPro software through Esperanto conversion.

For both samples, a single crystal of the compound was placed into a Diacell Bragg-Mini diamond anvil cell (DAC) from Almax-EasyLab (Figure S2 in Supporting Information), with an opening angle of 90° and anvil culets of 500 μm in diameter. The DAC was fitted with a stainless gasket of 60 μm thickness containing a hole of 200 μm in diameter. A small ruby sphere was placed next to the sample on one of the diamond anvils (diffraction side) as a pressure sensor and measured using the ruby R_1 fluorescence line [52]. A methanol–ethanol–water mixture (16:3:1) was used as the pressure-transmitting medium (PTM). This PTM remains hydrostatic in the range of pressures used in our experiments [53,54], thereby minimizing deviatoric stresses which can cause incorrect values for the bulk modulus to be determined [55].

Structures were refined using results from the previous pressure as a starting point, and they were refined on F^2 by full-matrix least-squares refinement using the SHELXL program. Due to limitations of the opening angle of our DAC, it was only possible to observe about 30% of the reflections which were observed outside of the DAC at ambient conditions. In general, structure refinements were performed with isotropic displacement parameters for all atoms except for the heavy atoms (Cu, Br, or I) that were refined with anisotropic displacement parameters whenever they did not become non-positive. In the case of the [(CuBr)$_2$(Quin)] compound, constraints were introduced to maintain the planarity of the organic ligand as well as to adjust certain thermal factors influenced by the quality of the intensities measured during the HP experiment, which significantly decreased after the phase transition. Hydrogen atoms were included in the final procedure in the same way as for ambient conditions.

2.3. High-Pressure Optical Measurements

Two different setups were used to obtain the high-pressure emission spectra. The first setup consisted of a 375-nanometer continuum diode laser as the excitation source and a 0.75-m single grating monochromator (Spex 750 M) equipped with a cooled photomultiplier (PMT) (Hamamatsu 928b) for detection. The second setup consisted of a 532-nanometer diode laser as the excitation course and a commercial scanning confocal Raman instrument (Renishaw InVia) with a cooled CCD for detection, where a 20× SLWD objective was used to achieve a laser-spot diameter of less than 5 μm. The luminescence of the [(CuBr)$_2$(Quin)] compound was obtained with the second setup, while for the [(CuI)$_2$(Quin)] compound, both setups were used. All the spectra have been corrected for the instrument response. A miniature DAC (Figure S2 in Supplementary Information) designed at the University of Paderborn (Germany) was used for high-pressure experiments using the same hydrostatic pressure-transmitting medium and pressure sensor as for the high-pressure X-ray diffraction experiments. Lifetime measurements were carried out by exciting with an Optical Parametric Oscillator (OPO) (EKSPLA/NT342/3UVE). The emission was focused onto a 0.32-m monochromator (Jobin Yvon Triax 320) coupled with a cooled PMT (Hamamatsu 928P) and recorded and averaged using a digital storage oscilloscope (LeCroy WS424).

2.4. Computer Simulations

Combining ab initio simulations with experimental studies of materials under high pressure has proven to be a very powerful technique [56]. In this work, we have performed first principles simulations in the framework of density functional theory (DFT) using the Vienna ab initio simulation package (VASP) [57–59]. A generalized gradient approximation (GGA) with the Perdew–Burke–Ernzerhof (PBE) functional [60] was used to describe the exchange-correlation energy. We have also included the van der Waals dispersion energy correction employing the Grimme D3 method [61] to take into account the weak interactions. Interactions among the core and the valence electrons were treated with pseudopotentials through the projector-augmented wave scheme (PAW) [62] to solve the Schrödinger equation. A plane wave basis with an energy cut-off of 520 eV was used, which assures high accuracy in the results. The integration over the Brillouin zone (BZ) was performed with a Monkhorst–Pack scheme [63] grid of $8 \times 2 \times 2$, which ensures high accuracy in the results since we are working with a primitive cell of 80 atoms in general positions. The structural parameters of the crystalline structures and the atomic positions have been relaxed at selected volumes. During the process of relaxation and optimization of the structures, it was required that the forces on the atoms were less than 0.003 eV/Å, and the stress tensor was diagonal with differences below 0.1 GPa, to ensure hydrostaticity.

3. Results and Discussion

3.1. Structural Analysis

The main single-crystal X-ray diffraction data and structure refinement parameters at ambient conditions for [(CuBr)$_2$(Quin)] and [(CuI)$_2$(Quin)] are reported in Table 1. Only the structure of [(CuI)$_2$(Quin)] has been previously reported [2,32]. Both compounds are mutually isomorphous and can therefore be described using the iodide compound as the reference. The asymmetric unit is shown in Figure 1. When expanded, the asymmetric unit results in polymeric 2D layers with a staircase motif as copper bromide/iodide is situated parallel to the *a*-axis (Figure 2).

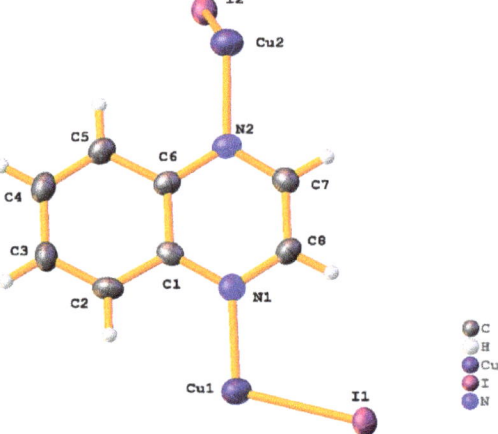

Figure 1. Thermal ellipsoids plot (50%) and labeling scheme for compound [(CuI)$_2$(Quin)] at ambient conditions.

Table 1. Summary of single-crystal X-ray diffraction crystal data and structure refinement parameters for [(CuBr)$_2$(Quin)] and [(CuI)$_2$(Quin)] at ambient conditions.

	[(CuBr)$_2$(Quin)]	[(CuI)$_2$(Quin)]
Formula	C$_8$H$_6$N$_2$Cu$_2$Br$_2$	C$_8$H$_6$Cu$_2$I$_2$N$_2$
D_{calc}/g cm^{-3}	2.806	3.164
μ/mm^{-1}	12.355	9.678
Formula weight	417.05	511.03
Color	Dark orange	Dark yellow
Size/mm^3	0.14 × 0.05 × 0.03	0.07 × 0.03 × 0.02
T/K	293(2)	293(2)
Crystal system	Monoclinic	Monoclinic
Space group	$P2_1/n$	$P2_1/n$
a/Å	4.1080(2)	4.3722(2)
b/Å	17.6594(7)	17.7218(7)
c/Å	13.6139(9)	13.8630(5)
α/°	90	90
β/°	91.496(6)	92.886(4)
γ/°	90	90
V/Å^3	987.28(9)	1072.79(8)
Z	4	4
Wavelength/Å	0.71073	0.71073
Radiation type	Mo K$_\alpha$	Mo K$_\alpha$
Θ_{min}/°	1.889	1.867
Θ_{max}/°	29.163	26.370
Measured refl's.	4422	4010
Independent refl's.	1974	2199
Refl's I ≥ 2 σ(I)	1629	1835
R_{int}	0.0236	0.0292
Param./Restr.	127/0	127/0
Largest peak	0.658	0.684
Deepest hole	−0.798	−0.795
GooF	1.138	1.057
wR_2 (all data)	0.0699	0.0638
wR_2	0.0656	0.0598
R_1 (all data)	0.0529	0.0517
R_1	0.0386	0.0376

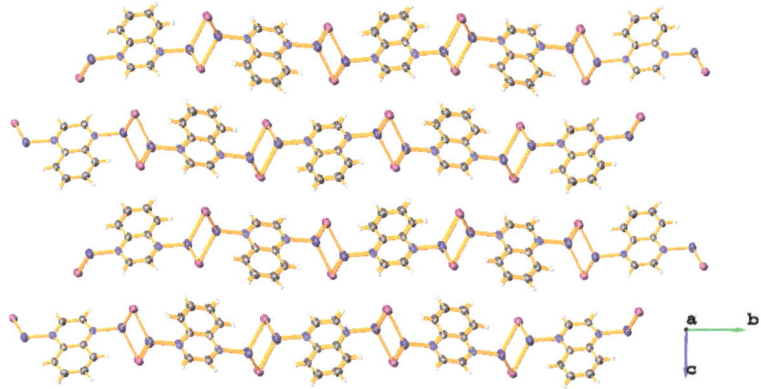

Figure 2. *Cont.*

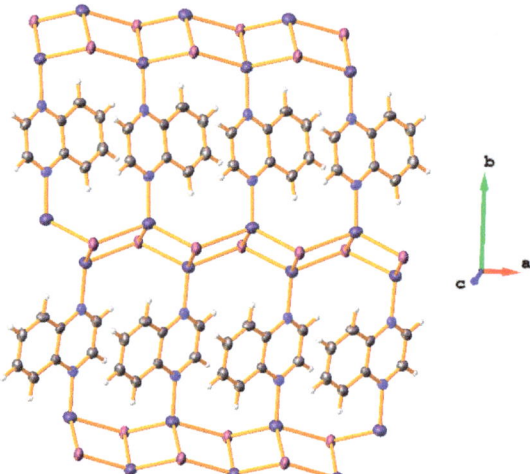

Figure 2. View of the stacking 2D layers along the *c*-axis (**upper**) and view of the staircase motif parallel to the *a*-axis (**lower**) for the complex [(CuI)$_2$(Quin)].

Each Cu^I atom is coordinated to three adjacent halogen ions over a range of 2.4584(8) Å to 2.5942(11) Å for [(CuBr)$_2$(Quin)] and 2.6275(10) Å to 2.7030(10) Å for [(CuI)$_2$(Quin)]. In both cases, the Cu^I atom is also coordinated by one of the nitrogen donors from the quinoxaline ligand, which bridges the adjacent Cu-X staircase, resulting in the formation of 2D sheets. Typically, a single geometric index [64] called τ4 is used to describe the shape of this four-fold coordination environment, which takes a range of values from 1.0 for perfect tetrahedral configuration to 0.0 in the case of perfect square planar configuration. In this study, the τ4 values were 0.88 (for Cu1) and 0.93 (for Cu2) in [(CuBr)$_2$(Quin)] and 0.87 (for Cu1) and 0.92 (for Cu2) in [(CuI)$_2$(Quin)], indicating a distorted trigonal pyramidal configuration. The quinoxaline adopts an arrangement in order to maximize the π–π interactions. Neighboring 2D sheets are related by inversion symmetry and are separated by half the *c*-axis.

The geometric parameters for the [(CuX)$_2$(Quin)] at ambient conditions agree with the values obtained in a search in the CSD (v5. 43) [65] for similar structures. The average values found in the CSD for Cu-Br, Cu-I, and Cu-N were 2.50(8) Å, 2.63(5) Å, and 2.06(7) Å, respectively. Similarly, the Cu–Cu distance was 2.9(3)Å when bromine atoms were involved and 2.8(3) Å when iodide atoms were involved. According to the structural analysis of the present work, all values fall within the expected ranges: 2.53 Å (Cu-Br), 2.02 Å (Cu-N), and 2.83 Å (Cu–Cu) for [(CuBr)$_2$(Quin)], and 2.68 Å (Cu-I), 2.05 Å (Cu-N), and 2.79 Å (Cu–Cu) for [(CuI)$_2$(Quin)].

The bond angles determined in this work are also consistent with those reported in the literature. The average value for the I-Cu-N angle calculated from the CSD was 110(7)°. In this work, it was found to be 111.3(2)° and 112.2(2)° for [(CuBr)$_2$(Quin)] and [(CuI)$_2$(Quin)], respectively. In the case of X-Cu-X (with X = Br, I), the range found in the CSD is between 93 and 120°, and all values for both compounds fall within this range.

With respect to the intermolecular interactions, the perpendicular stacking distance between quinoxaline layers was around 3.33 Å (4.1080(3) Å for [(CuBr)$_2$(Quin)] and 4.3722(2) Å for [(CuI)$_2$(Quin)] Cg–Cg distance), indicating weak π–π interactions.

High-pressure experiments were conducted at room temperature. The dependence of the unit cell parameters on pressure is shown in Figure 3. Although the behavior of the volume does not show any discontinuity, there was a significant change in the behavior of the angles in the unit cell parameters, where the α and γ angles differed from their original values of 90° at around 3.5 GPa for [(CuBr)$_2$(Quin)] and 3.2 GPa for [(CuI)$_2$(Quin)], with the β angle increasing smoothly with pressure. Therefore, it can be concluded that there was a phase transition of second order, since there is no discontinuity in the dependence of volume with pressure, where the sample transitioned from a monoclinic phase ($P2_1/n$) to a triclinic phase ($P\text{-}1$).

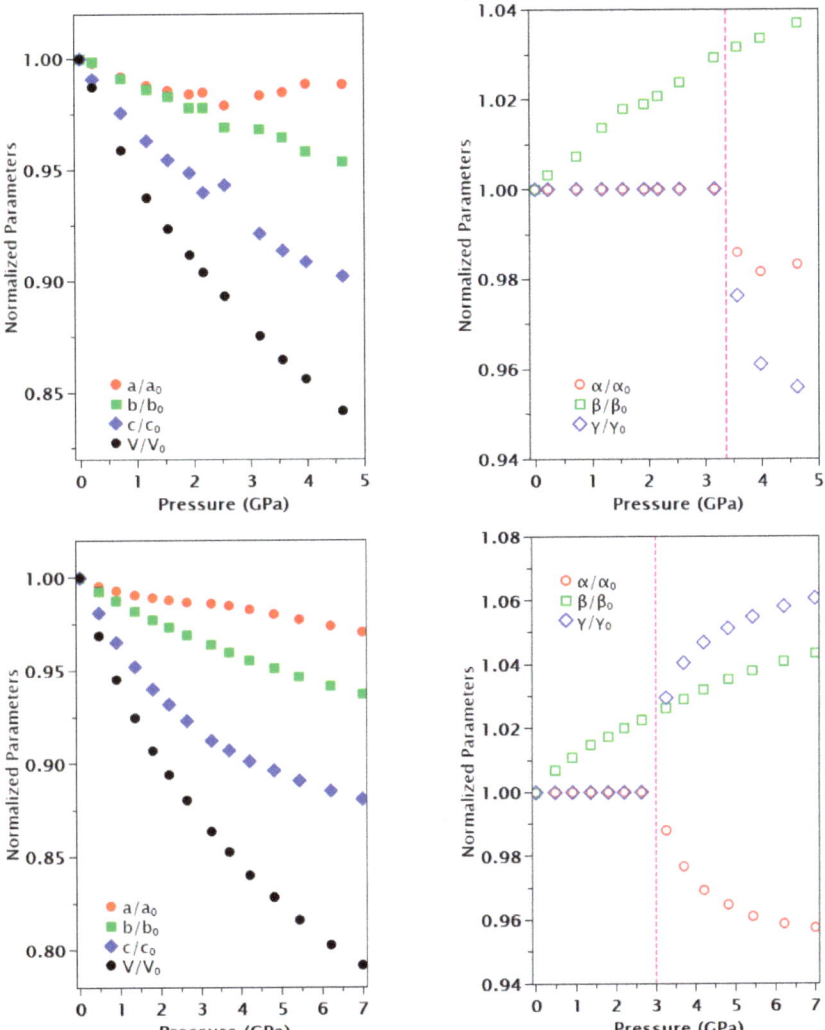

Figure 3. Pressure dependence of the cell parameters and volume for [(CuBr)$_2$(Quin)] (**top**) and [(CuI)$_2$(Quin)] (**bottom**). The dashed magenta line marks the phase transition zone. Error bars are smaller than their respective symbols.

Cu–Cu interactions are shown in Figure 4. In the low-pressure monoclinic phase ($P2_1/n$), we observed three different Cu–Cu interactions imposed by the symmetry (red, black, and dashed blue lines). In the high-pressure triclinic phase ($P\text{-}1$), the number of different Cu–Cu interactions increased up to five (red, black, green, magenta, and dashed blue lines). A similar situation occurred with the other bond distances (Cu-I and Cu-N, Figures S3 and S4 in Supplementary Information). Regarding the bond angles, the N-Cu-X (X = Br, I) angles presented the most significant changes after the phase transition (Figure 5). The other parameters produced only a slight distortion (gliding movement) of the Cu-X staircases which is associated with a slight decrease in the space between the quinoxaline ligands, which maintained their planarity and orientation, with an additional small displacement of the ligands relative to one another (Figure S5 in Supplementary Information).

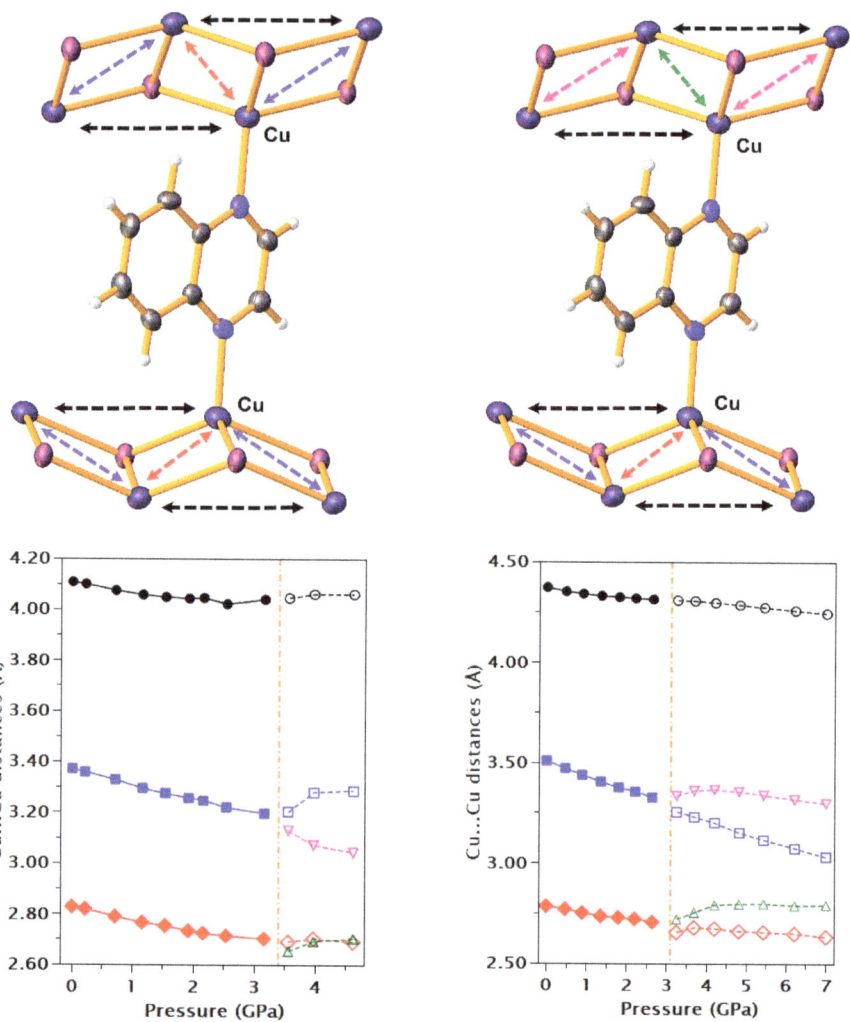

Figure 4. Variation in Cu–Cu distances with pressure for [(CuX)$_2$(Quin)] (X = Br, I). Upper plots correspond with the scheme of distance variation before and after the phase transition: monoclinic (**left**) and triclinic (**right**). Different colored arrows correspond to different Cu–Cu interactions. The bottom part of the figure shows the Cu–Cu distances for [(CuBr)$_2$(Quin)] (**left**) and for [(CuI)$_2$(Quin)] (**right**) as a function of pressure. Error bars are smaller than respective symbols. Colored lines correspond with the different Cu–Cu interactions shown in the upper part of the figure.

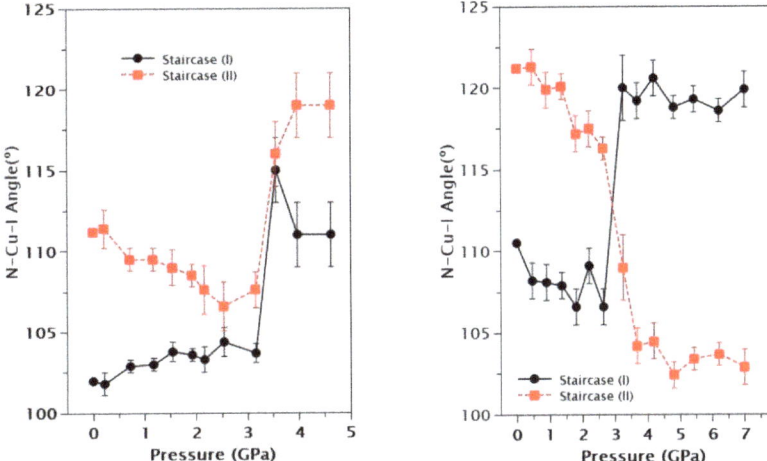

Figure 5. Variation in N-Cu-X (X = Br, I) angles with pressure for [(CuBr$_2$(Quin)] (**left**) and [(CuI$_2$(Quin)] (**right**). In the right-hand figure, the label 'Staircase (I)' corresponds to the Cu1 atoms, while 'Staircase (II)' corresponds to the Cu2 atoms.

3.2. The Equation of State Analysis

The determination of the variation in volume with hydrostatic pressure at a fixed temperature is known as isothermal equation of state (EoS) analysis, whereby the volume variation is characterized by the bulk modulus (K) and its pressure derivatives (K', K'', ...). We calculated such parameters for both compounds using a Birch–Murnaghan (BM) EoS model using EoSFit7-GUI software [66,67]. In order to assign the correct order to be used in the fitting, we have to represent the normalized pressure (F) against finite strain (f) (i.e., f-F plot). In both cases, we can observe a linear behavior with a positive slope, indicating that volume can be fitted with a third-order BM EoS, as can be seen in Figure 6.

The bulk modulus (K_0) and its first derivative (K'_0) were 14.4(5) GPa and 7.7(6) for [(CuBr)$_2$(Quin)] and 13.0(2) GPa and 7.4(2) for [(CuI)$_2$(Quin)], respectively.

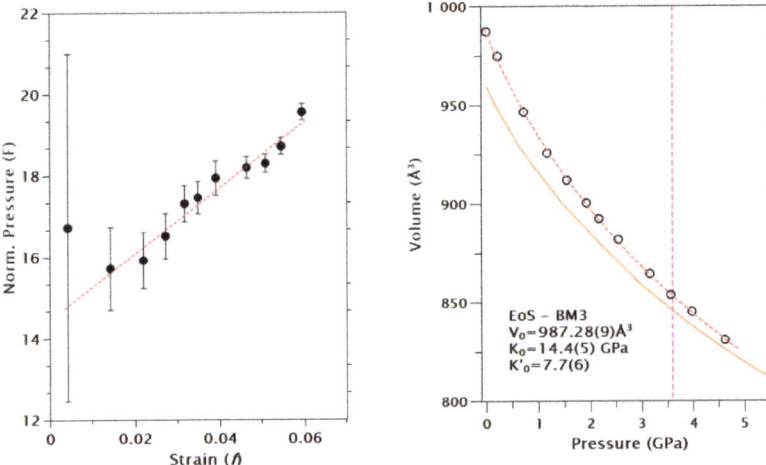

Figure 6. *Cont.*

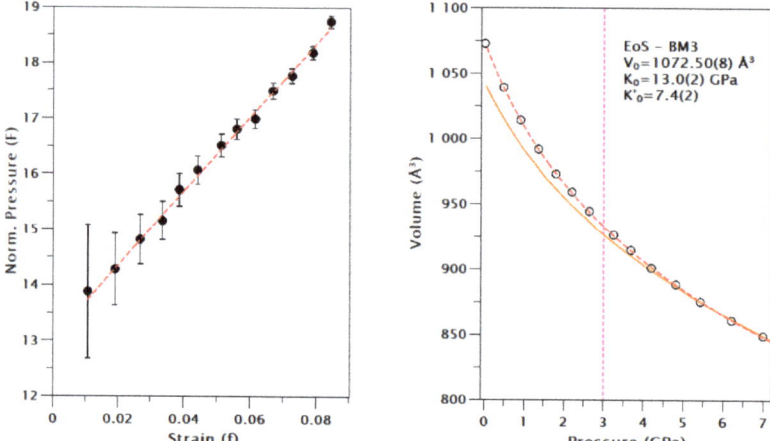

Figure 6. *f-F* plots and EoS BM3 fit for volume variation for [(CuBr)$_2$(Quin)] (**upper**) and [(CuI)$_2$(Quin)] (**lower**). Dashed red curves represent the EoS fitting results. Orange curves are the DFT simulations using the PBE + D3 model.

In general, this type of compound has a very anisotropic compression, especially when the compounds crystallize in monoclinic or triclinic systems. From Figure 3, we can observe this phenomenon, whereby the *c*-axis is substantially softer than the *b*- and *a*-axes. The bulk modulus obtained for both compounds falls in the range of 10–20 GPa, which is typical for organometallic compounds [68]. A quite similar case is the 2D coordination polymer with the formula Cu$_2$I$_2$(2-aminopyrazine) [36], where the bulk modulus (K_0) and its first derivative (K'_0) were 14.1(3) GPa and 7.4(2), respectively, using the same EoS model (BM3) in the fitting calculations. For similar staircase polymers of Cu^I iodides (1D), the bulk moduli were slightly smaller at around 10 GPa [37,46], and the exceptionally small bulk modulus of 7.5(4) GPa for CuI(3,5-dichloropyridine) [38] was probably due to the existence of a phase transition of the first order at 6 GPa. For 0D cases, such as a Cu$_4$I$_4$ cluster with different organic ligands [47], the trend is that the bulk modulus is even smaller (K_0 around 9 GPa or less). Therefore, we can conclude that the coordination dimensionality in this type of compound influences the compressibility values. In cases where the compressibility is high, this can be attributed to the deformability of the intermolecular interactions.

Unfortunately, the simulations carried out in this investigation (not only with the PBE + D3 functional but also with other commonly used functionals such as PBESol) have not reproduced the phase transition that was experimentally observed in both compounds. The simulation study shows us that in both compounds, the most favorable energy configuration is the monoclinic one rather than the triclinic one, as can be seen in Figure 7.

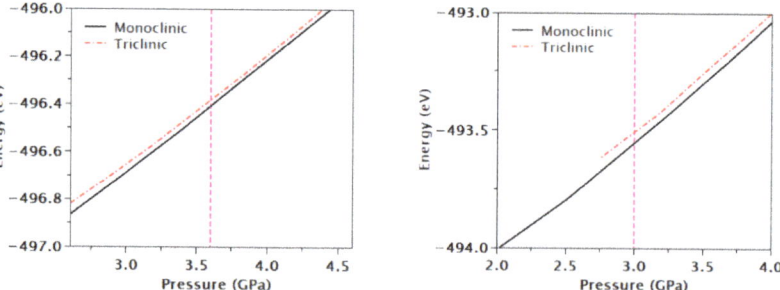

Figure 7. Energy curves of DFT simulations for [(CuBr)$_2$(Quin)] (**left**) and [(CuI)$_2$(Quin)] (**right**) considering monoclinic or triclinic phases.

3.3. Luminescence Properties under Pressure

In both samples, asymmetric emission bands were observed, exhibiting their multi-component character. To follow the pressure-induced evolution of the peak maxima of the emission bands, the center of gravity (centroid) (N_1) was used, defined as follows:

$$N_1 = \frac{\int_{-\infty}^{+\infty} \lambda I(\lambda) d\lambda}{\int_{-\infty}^{+\infty} I(\lambda) d\lambda}. \quad (1)$$

The [(CuBr)$_2$(Quin)] emission spectra with increasing pressure were recorded up to 6.3 GPa, with an excitation source wavelength of 532 nm (Figure 8 left). The spectra consist of a single asymmetric band that can be deconvoluted into three Gaussians (see Figure S5 in Supplementary Information).

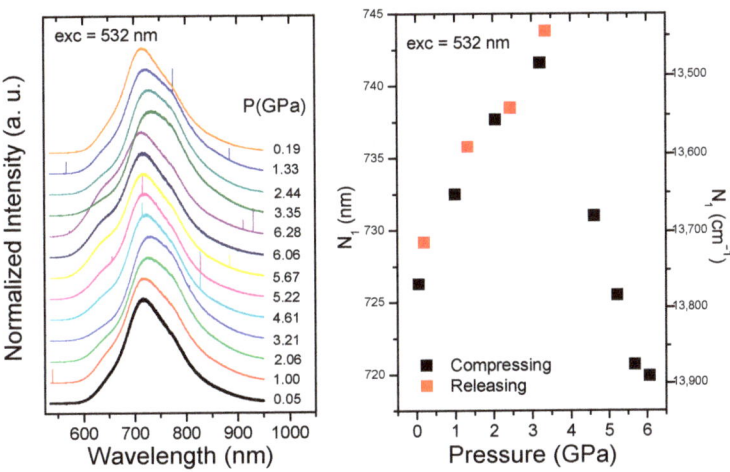

Figure 8. Pressure dependence of luminescence spectra of [(CuBr)$_2$(Quin)] compound obtained at RT under 532 nm laser excitation (**left**), and pressure evolution of the centroid (**right**).

At ambient conditions, the centroid of the band was around 720 nm, and when the pressure was increased, two different behaviors were observed (Figure 8 right) in the pressure interval from ambient to 6.3 GPa. In the first interval (0–3.2 GPa), a linear redshift was obtained with a slope of around −89 cm^{-1}/GPa. This tendency drastically changed in the interval of 3.2–6.3 GPa with an opposite behavior, giving rise to a linear blue shift around 149.2 cm^{-1}/GPa. Upon releasing pressure, the original peak positions were almost recovered.

A similar behavior was observed in the pressure evolution of the emission intensity, with a pressure-induced decrease in intensity observed at pressures higher than 3.2 GPa up to 6.0 GPa. The initial intensity was almost recovered after releasing pressure (see Figure 9).

In the [(CuI)$_2$(Quin)] compound, two different excitation wavelengths were used to obtain the emission spectra, resulting in similar observed behaviors.

Under 375 nm excitation, the spectra consisted of two bands: one asymmetric band centered at 750 nm, denoted as the low-energy band (LE), and another more symmetric band centered at 510 nm, denoted as the high-energy band (HE). The LE band could be deconvoluted into two Gaussian profiles (Figure S5). It is important to point out that in this setup, the limit imposed by the response of the photomultiplier was around 850 nm, which precluded measurements at higher wavelengths.

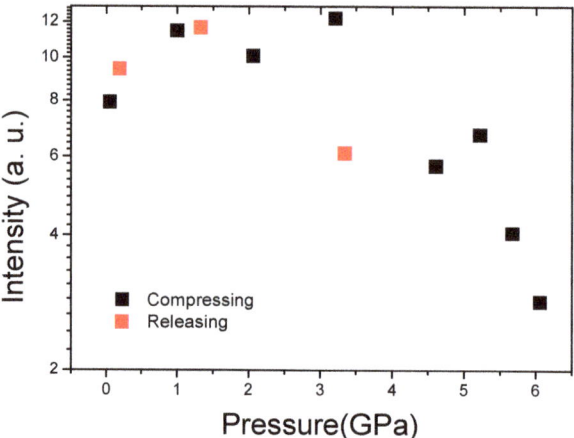

Figure 9. Integrated intensity pressure dependence of the [(CuBr)$_2$(Quin)] compound.

Upon compressing the copper(I) halide coordination complexes, two different regimes could be distinguished. The first regime is in the pressure interval between ambient and 2.9 GPa, where both the integrated intensity and the centroid of the band only exhibit slight changes of 20% and −49 cm^{-1}/GPa, respectively. The second regime is in the pressure interval between 3.0 and 5.4 GPa, where a progressive decrease in the integrated intensity of the band occurs until it is no longer observable beyond 6 GPa along with an enhancement of the redshift rate up to −209 cm^{-1}/GPa. These pressure-induced behaviors are depicted in Figure 10, where clear changes can be seen around 3.0 GPa in the slopes of the peak positions and in the integrated intensities (Figure 11). Upon decompression, the emission behavior was almost fully recovered.

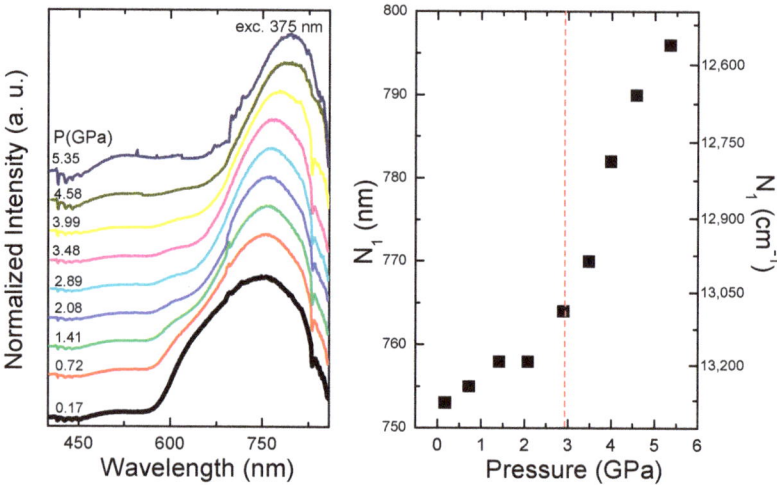

Figure 10. Pressure dependence of luminescence spectra of [(CuI)$_2$(Quin)] compound obtained at ambient temperature under 375 nm laser excitation.

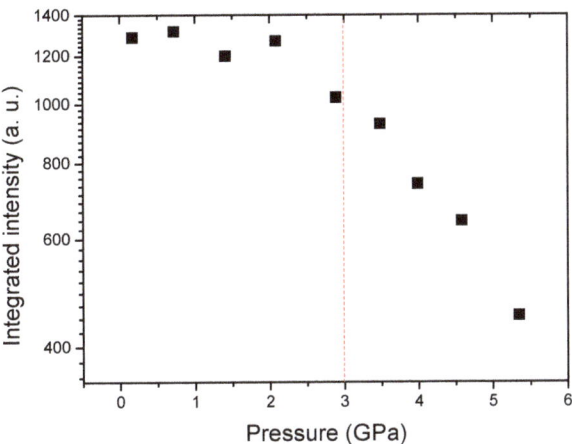

Figure 11. Integrated intensity pressure dependence of the [(CuI)$_2$(Quin)] compound.

Concerning the luminescence measurements obtained using an excitation wavelength of 532 nm, similar pressure behavior was found, with bands that can be deconvoluted into two Gaussian profiles (see Figure S5 in Supplementary Information). In this case, a change in tendency was observed around 4.0 GPa. Initially, a linear redshift of around −66 cm^{-1}/GPa was found up to 4.0 GPa; then, up to 11.6 GPa, an increase in the redshift to −112 cm^{-1}/GPa was observed (Figure 12). Due to the 532 nm excitation, it was impossible to observe the HE band appearing with 375 nm excitation.

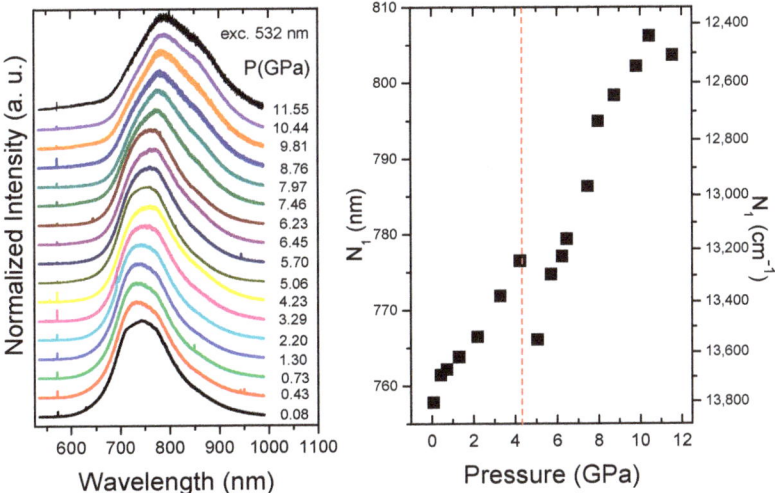

Figure 12. Pressure dependence of luminescence spectra of [(CuI)$_2$(Quin)] compound obtained at ambient temperature under 532 nm laser excitation.

Additionally, the decay curves of [(CuBr)$_2$(Quin)] and [(CuI)$_2$(Quin)] at ambient conditions with an excitation wavelength of 375 nm were also obtained, showing non-exponential behaviors, as expected (Figure 13), that can be related to the multicomponent character of the emission bands. For comparison purposes, an average lifetime (<τ>) was considered according to Equation (2), and values of 0.7 µs and 0.3 µs for [(CuBr)$_2$(Quin)] and [(CuI)$_2$(Quin)] were found, respectively:

$$\langle \tau \rangle = \frac{\int_{-\infty}^{+\infty} t I(t) dt}{\int_{-\infty}^{+\infty} I(t) dt}. \qquad (2)$$

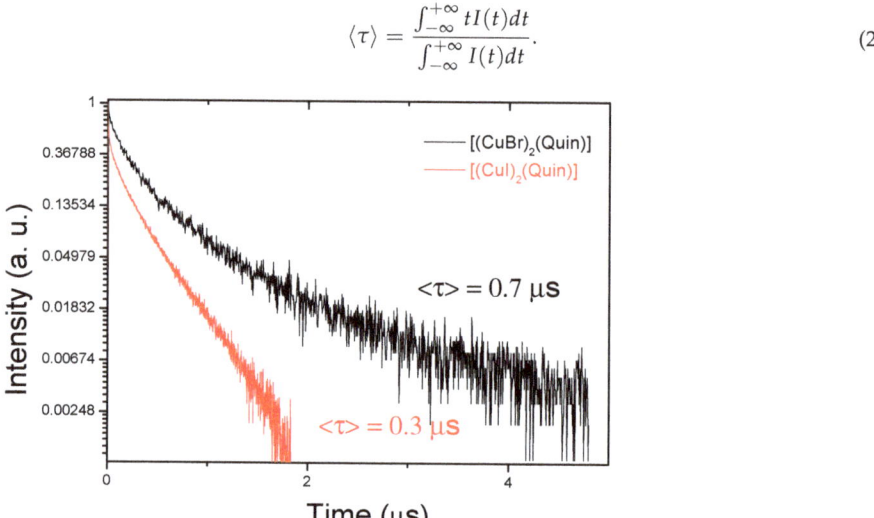

Figure 13. Decay curves of [(CuBr)$_2$(Quin)] (red) and [(CuI)$_2$(Quin)] (black) obtained at ambient conditions.

The average lifetime obtained for the emission bands, around the order of microseconds, allows us to consider the luminescence process as phosphorescence, which reflects the triplet character of the emitter state [69].

In previous studies, the correlation between the structure of Cu^I halide complexes and their luminescent properties has been demonstrated [31,36–39,41–43,46,70].

As commented on in the Introduction section, the electronic configuration for Cu^I corresponds to d^{10}. Therefore, since the d–d electronic transitions are excluded, charge transfer transitions can be the only mechanisms responsible for the electronic transitions. On the basis of DFT calculations, the HE band observed in the Cu^I-based complexes is usually ascribed to a mixed triplet metal-to-ligand charge transfer and triplet iodide-to-ligand charge transfer transition ^{3}(M + X)LCT. Moreover, the LE band is connected with transitions in the CuX cluster core, exhibiting a combination of halide-to-metal ^{3}XMCT and metal-cluster-centered (^{3}MCC) $d^{10} \rightarrow d^9s^1$, which is independent of the nature of the ligands.

The presence of structure in the emission bands reveals their multicomponent character and clearly indicates the contribution of different excited states involved in this mechanism. From the high-pressure X-ray experiments, some important information on luminescent properties can be inferred. Firstly, at around 3.5 GPa and 3.0 GPa, a second-order phase transition from monoclinic ($P2_1/n$) to triclinic phase (P-1) took place for the [(CuBr)$_2$(Quin)] and [(CuI)$_2$(Quin)], respectively. In the low-pressure $P2_1/n$ phase, three different Cu–Cu distances were observed, of which only one was equal to or shorter than the minimum distance required for so-called cuprophilic interactions, established as 2.8 Å for the Cu(I) ions [70]. This is depicted by solid red diamond symbols in Figure 4, with distances at ambient conditions of 2.8293 Å and 2.7869 Å for [(CuBr)$_2$(Quin)] and [(CuI)$_2$(Quin)], respectively. The pressure evolution of the Cu–Cu distances consisted of shortening at different rates depending on the sample and the pressure range. For [(CuBr)$_2$(Quin)], the change in Cu–Cu distance was −0.0396 Å/GPa for the low-pressure $P2_1/n$ phase (0–3.5 GPa), which then stabilized or showed a slight increase in the high-pressure P-1 phase (Figure 4, left). Unfortunately, the X-ray diffraction pressure measurements were limited to 4.5 GPa. For the [(CuI)$_2$(Quin)] sample, the change in Cu–Cu distance in the

$P2_1/n$ phase (0–3.0 GPa) was around −0.0289 Å/GPa and was around −0.0134 Å/GPa in the P-1 phase (3.2–7.0 GPa), thereby revealing the same tendency over the whole pressure range studied.

The redshift of the emission bands with the shortening of the Cu–Cu distances is one of the main features of the cuprophilic interaction, since from molecular orbital theory, it is known that the Cu–Cu distances in the lowest unoccupied molecular orbital (LUMO) have a bonding character [71] which increases with the shortening of the Cu–Cu distances. In this context, some correlation with the observed experimental results can be drawn. For example, the [(CuBr)$_2$(Quin)] complex in the low-pressure $P2_1/n$ phase showed a redshift of the emission centroid concurrently with shortening of the Cu–Cu distances of around 5% in the 0–3.5 GPa pressure interval. In the high-pressure P-1 phase, the opposite tendency was observed—that is, a blueshift occurred while the Cu–Cu distances stabilized or even increased. The latter behavior can be connected with the rigidochromism effect, which is observed when the environment becomes more rigid and is connected to metal-to-ligand charge transfer (MLCT) excited states [72] and consists of a blue shift of the emission band. It seems that in the [(CuBr)$_2$(Quin)] compound, at least two independent excited states are competing: the metal-cluster-centered (^{3}MCC) $d^{10} \rightarrow d^9 s^1$ transition and the ^{3}MLCT. Considering the fact that the emission data have been interpreted using the centroid, which is itself an average value, it can be interpreted that in the $P2_1/n$ phase, the ^{3}MCC state predominates over the ^{3}MLCT state, and that in the P-1 phase, the roles are reversed.

For the other complex, [(CuI)$_2$(Quin)], the same tendency was observed over the whole pressure interval (0 GPa to 11.5 GPa). In the $P2_1/n$ phase, as can be seen in Figure 4, the variation in the Cu–Cu distances was less than 5% in the 0.0–3.0 GPa interval. Accordingly, small differences in the shifts of the centroids of the emission bands were observed (−49 cm^{-1}/GPa and −70 cm^{-1}/GPa) in this interval for excitations at 375 nm and 532 nm, respectively, that were more pronounced in the triclinic (P-1) phase (−209 cm^{-1}/GPa and −100 cm^{-1}/GPa) in the 4.0–7.0 GPa interval. Unfortunately, beyond 7.0 GPa, there is no information about the Cu–Cu distances since X-ray diffraction data were not acquired at these pressures, although the evolution of the emission spectra characterized by redshifts suggests a progressive shortening of the Cu–Cu distances. These distances correspond to the blue closed and open square symbols in Figure 4, which suggest that the Cu–Cu distances decrease below 2.8 Å for pressures over 7.0 GPa since decreases of around 14%, from 3.5105 Å (AC) to 3.0306 Å (7 GPa), were already observed. Therefore, in this complex, the metal-cluster-centered (^{3}MCC) $d^{10} \rightarrow d^9 s^1$ transition is expected to be the dominant charge transfer mechanism.

4. Conclusions

We have performed a structural and luminescence study of a 2D polymeric staircase copper(I) halogen (halogen = bromine or iodine) with quinoxaline as an organic ligand. The high-pressure X-ray diffraction showed a phase transition of second order at around 3.3 GPa for both compounds whereby the cell parameters changed from the monoclinic to the triclinic crystal system. The compressibility study determined similar values in both compounds for the bulk modulus and its first derivate. The geometric changes revealed that the main structural variations occurred in the distortion of the ladder, with predominantly gliding versus folding movements affecting the Cu–Cu interactions. In general, the luminescence spectra showed asymmetric bands that reflect their multicomponent origin relating to the excited states involved in the transitions. The non-exponential decay curves confirmed this hypothesis, with averaged decay time data on the order of microseconds indicating the triplet character of the emitter states. Different behaviors of the pressure evolutions were observed for both compounds. With regard to [(CuBr)$_2$(Quin)], a change in the tendency of the emission centroid from redshift to blueshift was observed around 3.5 GPa, which agrees with the phase transition pressure from the monoclinic $P2_1/n$ phase to the triclinic P-1 phase. From this, a combination of metal-cluster-centered (^{3}MCC) $d^{10} \rightarrow d^9 s^1$ (cuprophilic) and ^{3}MLCT excited states were considered competing emitter states. In the 0–3.5 GPa interval, the former dominated, but after the phase transition, the rigidochromism effect became more important. In the [(CuI)$_2$(Quin)] compound, the same tendency (redshift) was observed over the whole pressure range studied (0–11.5 GPa), although different rates were observed around the phase transition. In this case, metal-cluster-centered (^{3}MCC) $d^{10} \rightarrow d^9s^1$ transition seemed to be the dominant mechanism.

Supplementary Materials: The following supporting information can be downloaded at: https://www.mdpi.com/article/10.3390/cryst13010100/s1, Figure S1: Instrumental setup used for HP experiments. On the left, a SuperNOVA diffractometer with EoS CCD detector. On the right, HP Station in MSPD-BL04 (ALBA Synchrotron) with Rayonix SX165 CCD Detector; Figure S2: Bragg-Mini DAC (Almax EasyLab company) with 500 μm diameter in the culets of diamonds (top). Paderborn Mini DAC with 400 μm diameter in the culets of diamonds (bottom); Figure S3: Normalized distances for Cu-N for (CuX)2-Quin (X = Br, I) compounds; Figure S4: Normalized distances for Cu-X for (CuX)2-Quin (X = Br, I) compounds; Figure S5: Room temperature emission spectrum fitted to three Gaussian profiles for (CuBr)2-Quin (top) and (CuI)2-Quin (bottom) compounds; Table S1: Crystal data and structure refinements for (CuBr)$_2$-Quin sample at different pressures (0.00–4.67GPa) at room temperature; Table S2: Crystal data and structure refinements for (CuI)$_2$-Quin sample at different pressures (0.00–7.00GPa) at room temperature.

Author Contributions: J.G.-P. worked on the X-ray diffraction and EoS calculations; U.R.R.-M. performed luminescence measurements; R.R.H.-M. and A.A.-C. conducted the synthesis of compounds; P.R.-H. and A.M. performed the simulations calculations; R.T. conducted the discussion and writing. All authors have read and agreed to the published version of the manuscript.

Funding: The authors acknowledge financial support from the Spanish Research Agency (AEI) and the Spanish Ministry of Science and Investigation (MCIN) under project PID2019-106383GB-C43/C44 (DOI:10.13039/5011000011033) and the MALTA Consolider Team network under project RED2018-102612-T. J.G.-P. thanks the Servicios Generales de Apoyo a la Investigación (SEGAI) at La Laguna University. U.R.R.-M. thanks the Gobierno de Canarias and EU-FEDER for grants ProID2020010067 and ProID2021010102.

Informed Consent Statement: Not applicable.

Data Availability Statement: CCDC 2226239-2226264 contains the supplementary crystallographic data for this paper. The data can be obtained free of charge from The Cambridge Crystallographic Data Centre via www.ccdc.cam.ac.uk/structures.

Acknowledgments: Some parts of the HP diffraction experiments were performed with the BL04-MSPD beamline at ALBA Synchrotron with the collaboration of ALBA staff. J.G.-P. thanks Servicio de Difracción de Rayos X (SIDIX) at La Laguna University for complementary support with the HP diffraction experiments. R.T. acknowledges funding from the Generalitat Valenciana through the APOSTD postdoctoral Fellowship No. CIAPOS/2021/20.

Conflicts of Interest: The authors declare no conflict of interest.

References

1. Blake, A.; Brooks, N.; Champness, N.; Hanton, L.; Hubberstey, P.; Schroder, M. Copper(I) halide supramolecular networks linked by N-heterocyclic donor bridging ligands. *Pure Appl. Chem.* **1998**, *70*, 2351–2357. [CrossRef]
2. Blake, A.J.; Brooks, N.R.; Champness, N.R.; Cooke, P.A.; Crew, M.; Deveson, A.M.; Hanton, L.R.; Hubberstey, P.; Fenske, D.; Schröder, M. Copper(I) iodide coordination networks-controlling the placement of (CuI)∞ ladders and chains within two-dimensional sheets. *Cryst. Eng.* **1999**, *2*, 181–195. [CrossRef]
3. Li, D.; Shi, W.-J.; Hou, L. Coordination polymers of copper(I) halides and neutral heterocyclic thiones with new coordination modes. *Inorg. Chem.* **2005**, *44*, 3907–3913. [CrossRef] [PubMed]
4. Peng, R.; Li, M.; Li, D. Copper(I) halides: A versatile family in coordination chemistry and crystal engineering. *Coord. Chem. Rev.* **2010**, *254*, 1–18. [CrossRef]
5. Troyano, J.; Perles, J.; Amo-Ochoa, P.; Zamora, F.; Delgado, S. Strong luminescent copper(i) halide coordination polymers and dinuclear complexes with thioacetamide and N,N'-donor ligands. *CrystEngComm* **2016**, *18*, 1809–1817. [CrossRef]
6. Aguirrechu-Comerón, A.; Hernández-Molina, R.; González-Platas, J. Structure of two new compounds of copper(I) iodide with N-donor and P-donor ligands. *J. Struct. Chem.* **2018**, *59*, 943–948. [CrossRef]

7. Neshat, A.; Aghakhanpour, R.B.; Mastrorilli, P.; Todisco, S.; Molani, F.; Wojtczak, A. Dinuclear and tetranuclear copper(I) iodide complexes with P and P^N donor ligands: Structural and photoluminescence studies. *Polyhedron* **2018**, *154*, 217–228. [CrossRef]
8. Masahara, S.; Yokoyama, H.; Suzaki, Y.; Ide, T. Convenient synthesis of copper(i) halide quasi-one-dimensional coordination polymers: Their structures and solid-state luminescent properties. *Dalton Trans.* **2021**, *50*, 8889–8898. [CrossRef] [PubMed]
9. Cariati, E.; Roberto, D.; Ugo, R.; Ford, P.C.; Galli, S.; Sironi, A. New structural motifs, unusual quenching of the emission, and second harmonic generation of copper(I) iodide polymeric or oligomeric adducts with para-substituted pyridines or trans-stilbazoles. *Inorg. Chem.* **2005**, *44*, 4077–4085. [CrossRef]
10. Liu, J.B.; Li, H.H.; Chen, Z.R.; Li, J.B.; Chen, X.B.; Huang, C.C. A new semi-conductive copper(I) halide coordination polymer: Synthesis, structure, and theoretical study. *J. Clust. Sci.* **2009**, *20*, 515–523. [CrossRef]
11. Perruchas, S.; Le Goff, X.F.; Maron, S.; Maurin, I.; Guillen, F.; Garcia, A.; Gacoin, T.; Boilot, J.P. Mechanochromic and thermochromic luminescence of a copper iodide cluster. *J. Am. Chem. Soc.* **2010**, *132*, 10967–10969. [CrossRef] [PubMed]
12. Perruchas, S.; Tard, C.; Le Goff, X.F.; Fargues, A.; Garcia, A.; Kahlal, S.; Gacoin, T.; Boilot, J.P. Thermochromic luminescence of copper iodide clusters: The case of phosphine ligands. *Inorg. Chem.* **2011**, *50*, 10682–10692. [CrossRef]
13. Gallego, A.; Castillo, O.; Gómez-García, C.; Zamora, F.; Delgado, S. Electrical conductivity and luminescence in coordination polymers based on copper(I)-halides and sulfur-pyrimidine ligands. *Inorg. Chem.* **2012**, *51*, 718–727. [CrossRef]
14. Hassanein, K.; Conesa-Egea, J.; Delgado, S.; Benmansour, S.; Martinez, I.; Abellan, G.; Gomez-Garcia, C.J.; Zamora, F.; Amo-Ochoa, P. Electrical conductivity and strong luminescence in copper iodide double chains with isonicotinato derivatives. *Chem. Eur. J.* **2015**, *21*, 17282–17292. [CrossRef] [PubMed]
15. Khatri, N.M.; Pablico-Lansigan, M.H.; Boncher, W.L.; Mertzman, J.E.; Labatete, A.C.; Grande, L.M.; Wunder, D.; Prushan, M.J.; Zhang, W.; Halasyamani, P.S.; et al. Luminescence and nonlinear optical properties in copper(I) halide extended networks. *Inorg. Chem.* **2016**, *55*, 11408–11417. [CrossRef]
16. Schlachter, A.; Harvey, P.D. Properties and applications of copper halide-chalcogenoether and -chalcogenone networks and functional materials. *J. Mater. Chem. C* **2021**, *9*, 6648–6685. [CrossRef]
17. Mensah, A.; Shao, J.-J.; Ni, J.-L.; Li, G.-J.; Wang, F.-M.; Chen, L.-Z. Recent progress in luminescent Cu(I) halide complexes: A mini-review. *Front. Chem.* **2022**, *9*, 816363. [CrossRef]
18. Ma, Z.; Shi, Z.; Qin, C.; Cui, M.; Yang, D.; Wang, X.; Wang, L.; Ji, X.; Chen, X.; Sun, J.; et al. Stable yellow light-emitting devices based on ternary copper halides with broadband emissive self-trapped excitons. *ACS Nano* **2020**, *14*, 4475–4486. [CrossRef]
19. Hashimoto, M.; Igawa, S.; Yashima, M.; Kawata, I.; Hoshino, M.; Osawa, M. Highly efficient green organic light-emitting diodes containing luminescent three-coordinate copper(I) complexes. *J. Am. Chem. Soc.* **2011**, *133*, 10348–10351. [CrossRef]
20. Xie, M.; Han, C.; Zhang, J.; Xie, G.; Xu, H. White electroluminescent phosphine-chelated copper iodide nanoclusters. *Chem. Mater.* **2017**, *29*, 6606–6610. [CrossRef]
21. Ravaro, L.P.; Zanoni, K.P.S.; de Camargo, A.S.S. Luminescent copper(I) complexes as promising materials for the next generation of energy-saving OLED devices. *Energy Rep.* **2020**, *6* (Suppl. 4), 37–45. [CrossRef]
22. Guo, B.K.; Yang, F.; Wang, Y.Q.; Wei, Q.; Liu, L.; Zhong, X.X.; Wang, L.; Gong, J.K.; Li, F.B.; Wong, W.Y.; et al. Efficient TADF-OLEDs with ultra-soluble copper(I) halide complexes containing non-symmetrically substituted bidentate phosphine and PPh3 ligands. *J. Lumin.* **2020**, *220*, 116963-1–116963-11. [CrossRef]
23. Jia, W.L.; McCormick, T.; Tao, Y.; Lu, J.-P.; Wang, S. New phosphorescent polynuclear Cu(I) compounds based on linear and star-shaped 2-(2′-Pyridyl)benzimidazolyl derivatives: Syntheses, structures, luminescence, and electroluminescence. *Inorg. Chem.* **2005**, *44*, 5706–5712. [CrossRef]
24. Armaroli, N.; Accorsi, G.; Holler, M.; Moudam, O.; Nierengarten, J.-F.; Zhou, Z.; Wegh, R.T.; Welter, R. Highly luminescent Cu(I) complexes for light-emitting electrochemical cells. *Adv. Mater.* **2006**, *18*, 1313–1316. [CrossRef]
25. Chuaysong, R.; Chooto, P.; Pakawatchai, C. Electrochemical properties of copper(I) halides and substituted thiourea complexes. *ScienceAsia* **2008**, *34*, 440–442. [CrossRef]
26. Ezealigo, B.N.; Nwanya, A.C.; Simo, A.; Osuji, R.U.; Bucher, R.; Maaza, M.; Fabian, I.; Ezema, F.I. Optical and electrochemical capacitive properties of copper (I) iodide thin film deposited by SILAR method. *Arab. J. Chem.* **2019**, *12*, 5380–5391. [CrossRef]
27. Yin, J.; Lei, Q.; Han, Y.; Bakr, O.M.; Mohammed, O.F. Luminescent copper(I) halides for Optoelectronic applications. *Phys. Status Solidi RRL* **2021**, *15*, 2100138. [CrossRef]
28. Liu, G.-N.; Xu, R.-D.; Zhao, R.-Y.; Sun, Y.; Bo, Q.-B.; Duan, Z.-Y.; Li, Y.-H.; Wang, Y.; Wu, Q.; Li, C. Hybrid copper iodide cluster-based pellet sensor for highly selective optical detection of o-nitrophenol and tetracycline hydrochloride in aqueous solution. *ACS Sustain. Chem. Eng.* **2019**, *7*, 18863–18873. [CrossRef]
29. Peng, Y.; Yaacobi-Gross, N.; Perumal, A.K.; Faber, H.A.; Vourlias, G.; Patsalas, P.A.; Bradley, D.D.C.; He, Z.; Anthopoulos, T.D. Efficient organic solar cells using copper(I) iodide (CuI) hole transport layers. *Appl. Phys. Lett.* **2015**, *106*, 243302-1–243302-4. [CrossRef]
30. Sepalage, G.A.; Meyer, S.; Pascoe, A.; Scully, A.D.; Huang, F.; Bach, U.; Cheng, Y.-B.; Spiccia, L. Copper(I) iodide as hole-conductor in planar perovskite solar cells: Probing the origin of J–V hysteresis. *Adv. Funct. Mater.* **2015**, *25*, 5650–5661. [CrossRef]
31. Lopez, J.; Gonzalez-Platas, J.; Rodriguez-Mendoza, U.R.; Martinez, J.I.; Delgado, S.; Lifante-Pedrola, G.; Cantelar, E.; Guerrero-Lemus, R.; Hernandez-Rodriguez, C.; Amo-Ochoa, P. Cu(I)–I-2,4-diaminopyrimidine coordination polymers with optoelectronic properties as a proof of concept for solar cells. *Inorg. Chem.* **2021**, *60*, 1208–1219. [CrossRef] [PubMed]

32. Graham, P.M.; Pike, R.D.; Sabat, M.; Bailey, R.D.; Pennington, W.T. Coordination polymers of copper(I) halides. *Inorg. Chem.* **2000**, *39*, 5121–5132. [CrossRef] [PubMed]
33. Molina, R.H.; Aguirretxu, A.; González-Platas, J. Synthesis and structure of [CuI(3-methyl-2-phenylpiridine)2] with intermolecular stacking interactions. *J. Struct. Chem.* **2014**, *55*, 1478–1483. [CrossRef]
34. Bath, E.R.; Golz, C.; Knorr, M.; Strohmann, C. Crystal structure of di-μ-iodido-bis[bis(acetonitrile-κN)copper(I)]. *Acta Cryst. E* **2015**, *71*, m189–m190. [CrossRef]
35. Yang, S.; Li, Y.; Cui, Y.; Pan, J. A new coordination tetramer of copper(I) iodide and benzyldimethylamine: Tetra-μ3-iodido-tetrakis[(benzyldimethylamine-κN)copper(I)]. *Acta Cryst. E* **2009**, *765*, m906. [CrossRef]
36. Conesa-Egea, J.; Gallardo-Martínez, J.; Delgado, S.; Martínez, J.I.; Gonzalez-Platas, J.; Fernández-Moreira, V.; Rodríguez-Mendoza, U.R.; Ocón, P.; Zmora, F.; Amo-Ochoa, P. Multistimuli response micro- and nanolayers of a coordination polymer based on Cu_2I_2 chains linked by 2-aminopyrazine. *Small* **2017**, *13*, 1700965. [CrossRef]
37. Conesa-Egea, J.; Nogal, N.; Martínez, J.I.; Fernández-Moreira, V.; Rodríguez-Mendoza, U.R.; Gonzalez-Platas, J.; Gómez-García, C.J.; Delgado, S.; Zamora, F.; Amo-Ochoa, P. Smart composite films of nanometric thickness based on copper–iodine coordination polymers. Toward sensors. *Chem. Sci.* **2018**, *9*, 8000–8010. [CrossRef]
38. Conesa-Egea, J.; González-Platas, J.; Rodríguez-Mendoza, U.R.; Martínez, J.I.; Ocon, P.; Fernández-Moreira, V.; Costa, R.D.; Fernández-Cestau, J.; Zamora, F.; Amo-Ochoa, P. Cunning defects: Emission control by structural point defects on Cu(i)I double chain coordination polymers. *J. Mater. Chem. C* **2020**, *8*, 1448–1458. [CrossRef]
39. López, J.; Murillo, M.; Lifante-Pedrola, G.; Cantelar, E.; Gonzalez-Platas, J.; Rodríguez-Mendoza, U.R.; Amo-Ochoa, P. Multi-stimulus semiconductor Cu(I)–I-pyrimidinecoordination polymer with thermo- and mechanochromic sensing. *CrystEngComm* **2022**, *24*, 341–349. [CrossRef]
40. Armaroli, N.; Accorsi, G.; Cardinali, F.; Listorti, A. Photochemistry and photophysics of coordination compounds: Copper. *Top. Curr. Chem.* **2007**, *280*, 69–115. [CrossRef]
41. Benito, Q.; Maurin, I.; Cheisson, T.; Nocton, G.; Fargues, A.; Garcia, A.; Martineau, C.; Gacoin, T.; Boilot, J.-P.; Perruchas, S. Mechanochromic luminescence of copper iodide clusters. *Chem. Eur. J.* **2015**, *21*, 5892–5897. [CrossRef]
42. Benito, Q.; Le Goff, X.F.; Nocton, G.; Fargues, A.; Garcia, A.; Berhault, A.; Kahlal, S.; Saillard, J.Y.; Martineau, C.; Trébosc, J.; et al. Geometry flexibility of copper iodide clusters: Variability in luminescence thermochromism. *Inorg. Chem.* **2015**, *54*, 4483–4494. [CrossRef]
43. Benito, Q.; Baptiste, B.; Polian, A.; Delbes, L.; Martinelli, L.; Gacoin, T.; Boilot, J.P.; Perruchas, S. Pressure control of cuprophilic interactions in a luminescent mechanochromic copper cluster. *Inorg. Chem.* **2015**, *54*, 9821–9825. [CrossRef]
44. Fu, Z.; Lin, J.; Wang, L.; Li, C.; Yan, W.; Wu, T. Cuprous iodide pseudopolymorphs based on imidazole ligand and their luminescence thermochromism. *Cryst. Growth Des.* **2016**, *16*, 2322–2327. [CrossRef]
45. Troyano, J.; Perles, J.; Amo-Ochoa, P.; Martines, J.I.; Concepcion-Gimeno, M.; Fernandez-Moreira, V.; Zamora, F.; Delgado, S. Luminescent thermochromism of 2D coordination polymers based on copper(I) halides with 4-Hydroxythiophenol. *Chem. Eur. J.* **2016**, *22*, 18027–18035. [CrossRef]
46. Aguirrechu-Comerón, A.; Hernández-Molina, R.; Rodríguez-Hernández, P.; Muñoz, A.; Rodríguez-Mendoza, U.R.; Lavín, V.; Angel, R.J.; Gonzalez-Platas, J. Experimental and ab initio study of catena(bis(μ2-iodo)-6-methylquinoline-copper(I)) under pressure: Synthesis, crystal structure, electronic, and luminescence properties. *Inorg. Chem.* **2016**, *55*, 7476–7484. [CrossRef]
47. Aguirrechu-Comerón, A.; Rodríguez-Hernández, P.; Rodríguez-Mendoza, U.R.; Vallcorba, O.; Muñoz, A.; Perruchas, S.; Gonzalez-Platas, J. Equation of state and structural characterization of $Cu_4I_4\{PPh_2(CH_2CH=CH_2)\}_4$ under pressure. *High Press. Res.* **2019**, *39*, 69–80. [CrossRef]
48. Rigaku Oxford Diffraction. *CrysAlisPro Software System, Version 1.171.42.71*; Rigaku Corporation: Oxford, UK, 2022.
49. Sheldrick, G.M. SHELXT—Integrated space-group and crystal-structure determination. *Acta Cryst. A* **2015**, *71*, 3–8. [CrossRef]
50. Sheldrick, G.M. Crystal structure refinement with SHELXL. *Acta Cryst. C* **2015**, *71*, 3–8. [CrossRef]
51. Spek, A.L. Structure validation in chemical crystallography. *Acta Cryst. D* **2009**, *65*, 148–155. [CrossRef]
52. Shen, G.; Wang, Y.; Dewaele, A.; Wu, C.; Fratanduono, D.E.; Eggert, J.; Klotz, S.; Dziubek, K.F.; Loubeyre, P.; Fatyanov, O.V.; et al. Toward an international practical pressure scale: A proposal for an IPPS ruby gauge. *High Pres. Res.* **2020**, *40*, 299–314. [CrossRef]
53. Angel, R.J.; Bujak, M.; Zhao, J.; Gatta, G.D.; Jacobsen, S.D. Effective hydrostatic limits of pressure media for high-pressure crystallographic studies. *J. Appl. Cryst.* **2007**, *40*, 26–32. [CrossRef]
54. Klotz, S.; Chervin, J.; Munsch, P.; Le Marchand, G. Hydrostatic limits of 11 pressure transmitting media. *J. Phys. D Appl. Phys.* **2009**, *42*, 075413. [CrossRef]
55. Errandonea, D.; Muñoz, A.; Gonzalez-Platas, J. Comment on High-pressure x-ray diffraction study of YBO_3/Eu^{3+}, $GdBO_3$, and $EuBO_3$: Pressure-induced amorphization in $GdBO_3$. *J. Appl. Phys.* **2014**, *115*, 043507. [CrossRef]
56. Mujica, A.; Rubio, A.; Muñoz, A.; Needs, R.J. High-pressure phases of group-IV, III-V, and II-VI compounds. *Rev. Mod. Phys.* **2003**, *75*, 863–912. [CrossRef]
57. Kresse, G.; Hafner, J. Ab initio molecular dynamics for liquid metals. *Phys. Rev. B* **1993**, *47*, 558–561. [CrossRef]
58. Kresse, G.; Furthmüller, J. Efficiency of ab-initio total energy calculations for metals and semiconductors using a plane-wave basis set. *Comput. Mater. Sci.* **1996**, *6*, 15–50. [CrossRef]
59. Kresse, G.; Furthmüller, J. Efficient iterative schemes for ab initio total-energy calculations using a plane-wave basis set. *Phys. Rev. B* **1996**, *54*, 11169–11186. [CrossRef]

60. Perdew, J.P.; Burke, K.; Ernzerhof, M. Generalized gradient approximation made simple. *Phys. Rev. Lett.* **1997**, *77*, 3865–3868. [CrossRef]
61. Grimme, S.; Ehrlich, S.; Goerigk, L. Effect of the damping function in dispersion corrected density functional theory. *J. Comput. Chem.* **2011**, *32*, 1456–1465. [CrossRef]
62. Blöchl, P.E. Projector augmented-wave method. *Phys. Rev. B* **1994**, *50*, 17953–17979. [CrossRef]
63. Monkhorst, H.J.; Pack, J.D. Special points for brillouin-zone integration. *Phys. Rev. B* **1976**, *13*, 5188–5192. [CrossRef]
64. Yang, L.; Powell, D.R.; Houser, R.P. Structural variation in copper(I) complexes with pyridylmethylamide ligands: Structural analysis with a new four-coordinate geometry index, τ4. *Dalton Trans.* **2007**, *9*, 955–964. [CrossRef]
65. Groom, C.R.; Bruno, I.J.; Lightfoot, M.P.; Ward, S.C. The cambridge structural database. *Acta Cryst. B* **2016**, *72*, 171–179. [CrossRef]
66. Gonzalez-Platas, J.; Alvaro, M.; Nestola, F.; Angel, R.J. EoSFit7-GUI: A new graphical user interface for equation of state calculations, analyses and teaching. *J. Appl. Cryst.* **2016**, *49*, 1377–1382. [CrossRef]
67. Angel, R.J.; Gonzalez-Platas, J.; Alvaro, M. EosFit7c and a fortran module (library) for equation of state calculations. *Z. Kristallogr.* **2014**, *229*, 405–419. [CrossRef]
68. Moggach, S.A.; Parsons, S. High pressure crystallography of inorganic and organometallic complexes. *Spectrosc. Prop. Inorg. Organomet. Compd.* **2009**, *40*, 324–354. [CrossRef]
69. Zhang, Q.; Komino, T.; Matsunami, S.; Goushi, K.; Adachi, C.; Huang, S. Triplet exciton confinement in green organic light-emitting diodes containing luminescent charge-transfer Cu(I) complexes. *Adv. Funct. Mater.* **2012**, *22*, 2327–2336. [CrossRef]
70. Ford, P.C.; Cariati, E. Bourassa, photoluminescence properties of multinuclear copper(I) compounds. *J. Chem. Rev.* **1999**, *99*, 3625–3648. [CrossRef]
71. Kim, T.H.; Shin, Y.W.; Jung, J.H.; Kim, J.S.; Kim, J. Crystal-to-crystal transformation between three cui coordination polymers and structural evidence for luminescence thermochromism. *J. Angew. Chem.* **2008**, *120*, 697–700. [CrossRef]
72. Lees, A.J. The luminescence rigidochromic effect exhibited by organometallic complexes: Rationale and applications. *Comments Inorg. Chem.* **1995**, *17*, 319–346. [CrossRef]

Disclaimer/Publisher's Note: The statements, opinions and data contained in all publications are solely those of the individual author(s) and contributor(s) and not of MDPI and/or the editor(s). MDPI and/or the editor(s) disclaim responsibility for any injury to people or property resulting from any ideas, methods, instructions or products referred to in the content.

Article

Ab Initio Theoretical Study of DyScO$_3$ at High Pressure

Enrique Zanardi, Silvana Radescu, Andrés Mujica, Plácida Rodríguez-Hernández and Alfonso Muñoz *

Departamento de Física and Instituto de Materiales y Nanotecnología, MALTA Consolider Team, Universidad de La Laguna, E38200 San Cristóbal de La Laguna, Tenerife, Spain
* Correspondence: amunoz@ull.edu.es

Abstract: DyScO$_3$ is a member of a family of compounds (the rare-earth scandates) with exceptional properties and prospective applications in key technological areas. In this paper, we study theoretically the behavior of DyScO$_3$ perovskite under pressures up to about 65 GPa, including its structural and vibrational properties (with an analysis of the Raman and infrared activity), elastic response, and stability. We have worked within the ab initio framework of the density functional theory, using projector-augmented wave potentials and a generalized gradient approximation form to the exchange-correlation functional, including dispersive corrections. We compare our results with existing theoretical and experimental published data and extend the range of previous studies. We also propose a candidate high-pressure phase for this material.

Keywords: DyScO$_3$; high pressure; density functional theory; stability

Citation: Zanardi, E.; Radescu, S.; Mujica, A.; Rodríguez-Hernández, P.; Muñoz, A. Ab Initio Theoretical Study of DyScO$_3$ at High Pressure. *Crystals* **2023**, *13*, 165. https://doi.org/10.3390/cryst13020165

Academic Editor: Artem R. Oganov

Received: 21 December 2022
Revised: 12 January 2023
Accepted: 14 January 2023
Published: 17 January 2023

Copyright: © 2023 by the authors. Licensee MDPI, Basel, Switzerland. This article is an open access article distributed under the terms and conditions of the Creative Commons Attribution (CC BY) license (https://creativecommons.org/licenses/by/4.0/).

1. Introduction

On account of their dielectric and optical properties, as well as their optimal chemical stability, the rare-earth (RE) scandates with formula REScO$_3$, and among them DyScO$_3$, have been researched for the past two decades with a view mainly to replace SiO$_2$ in next-generation MOSFETs [1–3]. Other usages of RE-scandates include their employment as optimal substrates for the epitaxial growth and fine-tuning of high-quality perovskite-type thin films, as well as their application in the terahertz range when embedded in perovskite heterostructures [4–10]. The enticing properties and promising scope of applications of DyScO$_3$ suggest extending the fundamental research on this material to include as broad a range of pressures as possible, and very recently Bura and co-workers have experimentally undertaken this goal in the range from ambient conditions up to 40 GPa [11]. In the present work we aim to provide an ab initio theoretical study, performed within the framework of the density functional theory (DFT), of the so far only observed phase of DyScO$_3$, in the range up to 65 GPa, with a focus on its structural, elastic, and vibrational response, to compare with the experimental results of Bura et al. [11] as well as with other previously existing works and to provide a bound to its local stability. Our results indicate the existence of local instabilities, both elastic and dynamic, above 60–65 GPa. We also provide a possible candidate high-pressure phase, a well-known post-perovskite structure. We hope that these results will stimulate future experiments which, combined with theoretical studies, may lead to a better understanding of the properties of this technologically relevant, emerging material under high pressure.

2. Theoretical Method and Details of the Calculations

All the calculations were performed within the ab initio framework of the density functional theory (DFT) [12,13], using a projector augmented wave scheme (PAW) [14,15] and the VASP computational package [16–19]. The outermost, valence electrons of each species were explicitly considered in the calculations whereas their innermost, closed-shell core electrons were treated at the PAW level. For dysprosium we have used the recommended VASP Dy_3 PAW potential, which represents Dy with partially frozen 4f electrons and formal valency 3, a standard way to cope with localized f electrons in DFT calculations. The cutoff in the kinetic energy of the basis's plane waves was 540 eV and the required Brillouin-zone integrations were performed with a 4 × 4 × 4 grid [20]. Several levels of approximations for the exchange-correlation functional were considered in our preliminary calculations, with optimal results obtained for the PBE generalized gradient approximation form [21] including dispersive corrections within the so-called Grimme's D3 scheme [22], which was accordingly the level of approximation adopted for the production of all the results shown in this paper. The material was structurally relaxed at different volumes, sampling the region between negative pressures just below 0 GPa to about 60 GPa. Hydrostatic conditions were assumed in all cases and the criteria for stopping the structural relaxations was set to stress anisotropies less than 0.1 GPa and residual forces on atoms less than 5 meV/Å.

Further to the total energy calculations just described, which provide information on the pressure dependence of the structural properties of $DyScO_3$, we also performed a general study of its dynamical and elastic properties and their variation with pressure, and draw conclusions on the stability of the system based on them, as described in Sections 3.2 and 3.3. For the study of the lattice vibrations, we used the small-displacement method as implemented in the phonopy package [23], employing a 2 × 2 × 2 supercells for sampling the phonons across the whole Brillouin zone. At the zone center, the symmetry of the calculated modes was analyzed to determine their possible spectroscopical activity, Raman or infra-red (IR). For the calculation of the elastic constants, we used the Le Page stress-strain method as implemented in the VASP code [24].

3. Results and Discussion

3.1. Crystallographic Description of $DyScO_3$ Perovskite and Evolution of Its Structural Properties with Pressure

Like many other ABO_3 oxide compounds, $DyScO_3$ crystallizes in the perovskite-type structure. The ideal cubic perovskite structure, space group (SG) *Pm-3m* No 221, is characterized by corner-sharing BO_6 octahedra and 12-fold coordinated A sites. For non-ideal or distorted perovskites, octahedron tilting lowers the space group symmetry of the structure [25]. For $DyScO_3$ that means an orthorhombic phase with SG *Pbnm* No 62 and four formula units on the conventional unit cell (see Figure 1).

Our calculated lattice parameters *a*, *b*, *c*, and unit cell volume *V* at 0 GPa are in good agreement with the experimental results, as shown in Table 1. The calculated variation with the pressure of the lattice parameters (Figure 2) also reproduces the experimental results within a wide range of pressures up to approximately 10–12 GPa, although for larger pressures the calculated values of *a* and *c* are somewhat lower than the experimental values while *b* still follows the experimental trend. These deviations may be related to the loss of hydrostaticity above 10–12 GPa [26] in the methanol/ethanol pressure-transmitting medium used in the experiments [11] but more experimental and theoretical research would be needed to shed light on this point. The calculated linear axial compressibilities are 1.469×10^{-3} GPa^{-1} for *a*, 0.731×10^{-3} GPa^{-1} for *b*, and 1.411×10^{-3} GPa^{-1} for *c*, and the value for the bulk modulus B_0 of 164.7 GPa obtained by fitting the calculated energy-volume curve to a third-order Birch-Murnaghan equation of state [27] is a bit below the experimental one in Ref. [11] (see Table 2) whereas it is in good agreement with the calculated value of 165 GPa retrieved from the Materials Project database [28,29].

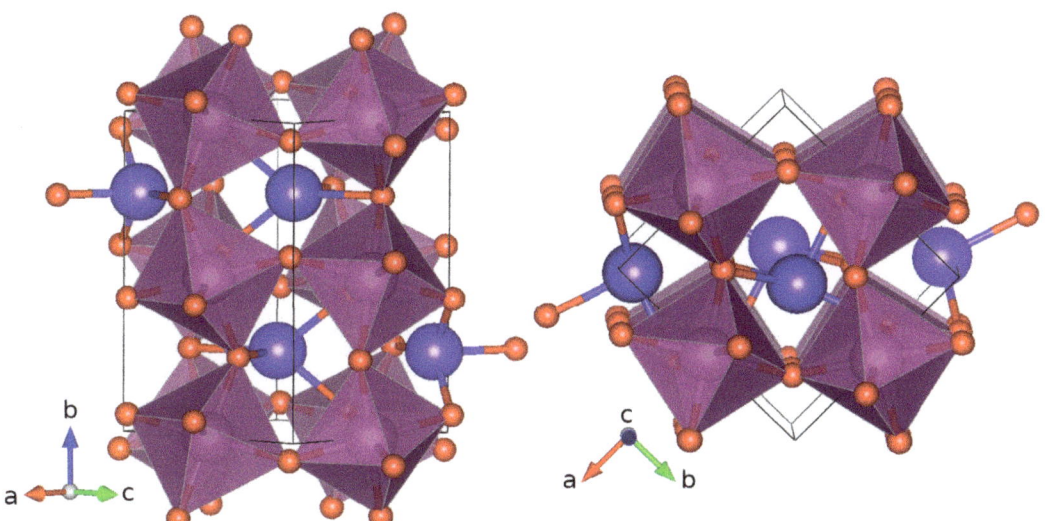

Figure 1. The structure of DyScO$_3$ is of the orthorhombic perovskite type. In this plot the ScO$_6$ octahedra of the structure are shown in magenta, with oxygen sites (in red) at their corners and Sc sites at the centers. The Dy cation sites are shown in blue.

Table 1. Calculated and experimental lattice parameters and unit cell volume at 0 GPa. Also provided are the calculated values at 60 GPa, for reference.

	This Work 0 GPa	Experiment Ref. [25]	Experiment Ref. [30]	This Work 60 GPa
a (Å)	5.421	5.443 (2)	5.442417 (54)	4.944
b (Å)	5.735	5.717 (2)	5.719357 (52)	5.484
c (Å)	7.935	7.901 (2)	7.904326 (98)	7.266
V (Å^3)	246.68	245.9 (1)	246.04	197.00

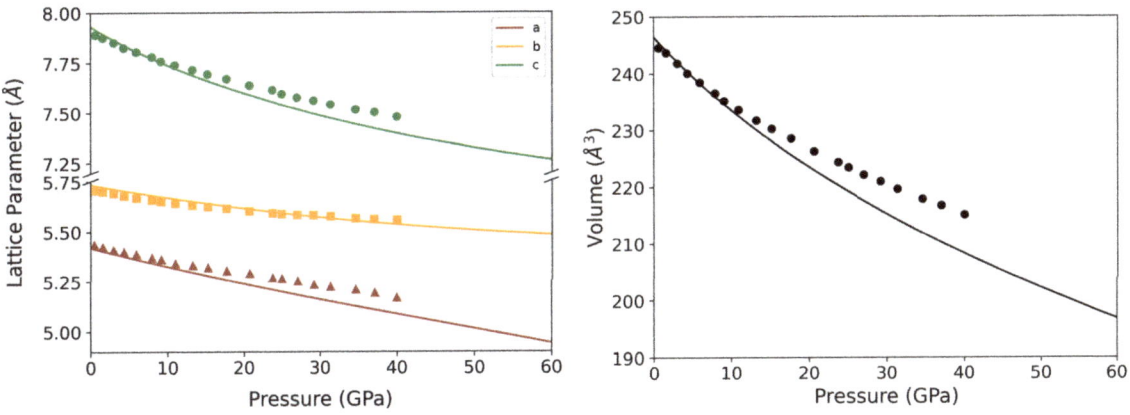

Figure 2. (**left**) Variation of the lattice parameters a, b, c of DyScO$_3$ perovskite with pressure. (**right**) Variation of the unit cell volume with pressure. In both panels, the curves represent our calculated values and the symbols represent the experimental values from Ref. [11].

Table 2. Comparison of the calculated and experimental unit cell volume, bulk modulus, and pressure derivative of the bulk modulus at zero pressure.

	$V_0(\text{Å}^3)$	$B_0(\text{GPa})$	B'_0
This work	246.7	164.7	4.42
Experiment Ref. [11]	245.3	189.4	7.68

The physical properties of the ABO_3 perovskites and the instabilities associated with soft modes found near the phase transitions are related to the distortions from the ideal cubic structure. Those distortions can be broken up into three components: two independent tilts of the BO_6 octahedra (with tilt angles respectively θ and ϕ), the distortion of those octahedra, and the displacement of the A cations [31–34]. In Table 3 we summarize the calculated values of the atomic positions in $DyScO_3$, showing, in particular, the displacement of the Dy cation, the tilt angles, and the octahedra distortion index, compared to the values of the ideal cubic perovskite structure, and the experimental values published in Ref. [25]. (The tilt angles were calculated using Equations (A1) and (A2) in Ref. [33].)

Table 3. Summary of the atomic sites of $DyScO_3$ perovskite, showing the values of the internal parameters and tilt angles. The values for the ideal cubic perovskite structure are also provided, for reference.

		Ideal Cubic Perovskite	This Work 0 GPa	This Work 60 GPa	Exp. Ref. [25]
Wyckoff positions					
Dy: 4c	u_{Dy}	½	0.4788	0.4733	0.48262(5)
(u_{Dy}, v_{Dy}, ¼)	v_{Dy}	½	0.4350	0.4165	0.43844(5)
Sc: 4b					
(½, 0, 0)					
O1: 4c	u_{O1}	½	0.6313	0.6312	0.6261(1)
(u_{O1}, v_{O1}, ¼)	v_{O1}	0	0.0562	0.0464	0.0560(2)
O2: 8d	u_{O2}	¼	0.1906	0.1717	0.1885(5)
(u_{O2}, v_{O2}, w_{O2})	v_{O2}	¼	0.1936	0.1863	0.1937(6)
	w_{O2}	0	0.0684	0.0622	0.0658(6)
Tilt angles					
θ in [110]		0	21.26	19.10	21.00
ϕ in [001]		0	13.39	17.00	13.23

The variation of those values with pressure is shown in Figure 3. We observe that as the applied pressure increases the Dy cations move further away from the cubic perovskite positions, as the octahedra rotate further away around the [001] axes.

In Figure 4 we show the calculated pressure-evolution of near-neighbor distances in $DyScO_3$. Our calculated values for the Sc-O interatomic distances, which are related to the ScO_6 octahedra, are typical for octahedrally coordinated scandium, and the octahedra show little distortion over the studied pressure range. The Dy-O distances show a larger variation and there is a gap between the first eight and the last four such distances, which makes it possible to distinguish between a first and second shell for the Dy sites and has led some authors to consider Dy as being 8-fold coordinated [25,35].

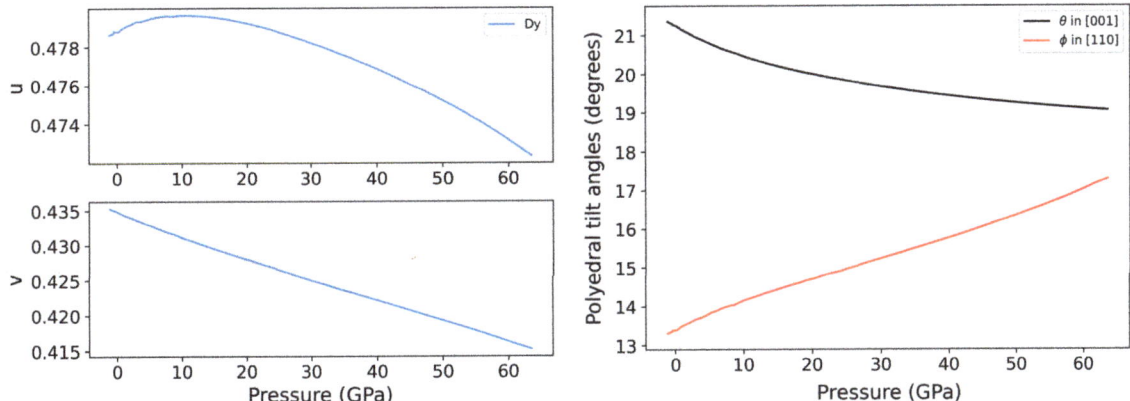

Figure 3. (**left**) Evolution of the internal parameters of the 4c Wickoff positions of Dy with pressure. (The values for the ideal cubic perovskite are ½, ½) (**right**) Tilt angles change with pressure.

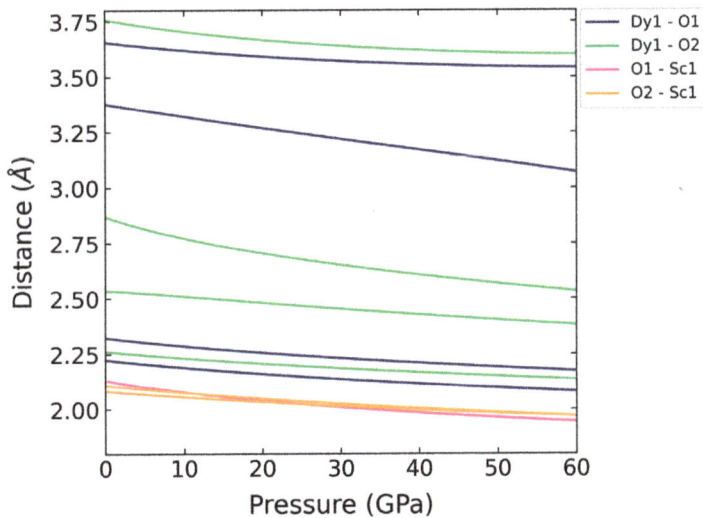

Figure 4. Evolution of interatomic distances with pressure in $DyScO_3$ perovskite. The O1-Sc1, O2-Sc1, and Dy1-O2 distances have all a multiplicity of two (each represented curve corresponds to two symmetrically equivalent but different bonds).

Related to the observed behavior for the cell volumes and the lattice parameters, we find that the calculated values of the polyhedra volumes stray somewhat away from the experimental ones at large pressures, but nonetheless the volume ratio of DyO_{12} to ScO_6 stays above the value of four as suggested in Ref. [34] for a stable material. The ScO_6 octahedra are less compressible than the DyO_{12} polyhedra, so the latter are the main contributors to the volume change of the material, see Figure 5.

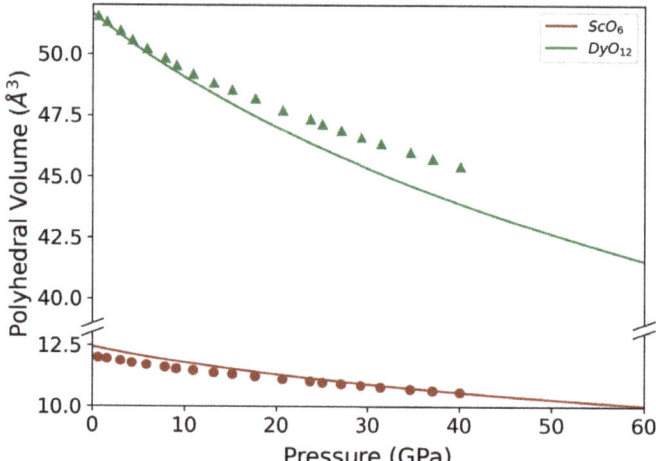

Figure 5. Variation of the polyhedral volumes of DyScO$_3$ perovskite with pressure. Solid curves represent the calculated values and symbols represent the experimental values from Ref. [11].

3.2. Elastic Properties

As described in Section 3.1, DyScO$_3$ has an orthorhombic structure with SG *Pbnm* No 62 and therefore there are nine independent elastic constants: C_{11}, C_{22}, C_{33}, C_{44}, C_{55}, C_{66}, C_{12} C_{13}, and C_{23} [36]. The study of elastic constants is useful to understand both the mechanical properties and the structural stability of the material. Our calculated values at zero pressure of these constants for DyScO$_3$ are compared with other ab initio and experimental results in Table 4.

Table 4. Elastic constants (in GPa) of DyScO$_3$ perovskite at 0 GPa. (The *Pbnm* setting is used.)

	This Work	Experiment Ref. [37]	Theory Ref. [37]
C_{11}	281	302.4	310
C_{22}	332	345.3	339
C_{33}	213	254.6	236
C_{44}	100	103.9	96
C_{55}	76	86.0	73
C_{66}	81	84.4	77
C_{12}	121	124.4	96
C_{13}	105	130.0	132
C_{23}	119	132.5	106

The C_{11}, C_{22}, and C_{33} elastic constants, which are related to unidirectional compressions along the principal crystallographic directions, are much larger than the C_{44}, C_{55}, and C_{66} constants that are related to resistance against shear deformations. Consequently, there is a larger resistance to unidirectional compression than to shear deformation. The structure is stable at 0 GPa as it fulfills the Born stability criteria [38] given by the following expressions [39]:

$$C_{11} > 0 \; ; \; C_{11}C_{22} - C_{12}^2 > 0$$

$$C_{11}C_{22}C_{33} + 2C_{12}C_{13}C_{23} - C_{11}C_{23}^2 - C_{22}C_{13}^2 - C_{33}C_{12}^2 > 0$$

$$C_{44} > 0 \; ; \; C_{55} > 0 \; ; \; C_{66} > 0$$

When stress is applied to the crystal, the elastic properties of the material are no longer described by the elastic constants C_{ij}, and the elastic stiffness coefficients B_{ij} must be employed instead. If the external stress corresponds to a hydrostatic pressure p, the elastic stiffness coefficients for a crystal are expressed as [40,41]:

$$B_{ii} = C_{ii} - p, \ i = 1, \ldots, 6; \ B_{ij} = C_{ij} + p, \ i \neq j, \ i, j = 1, 2, 3$$

where the C_{ij} are the elastic constants evaluated at the current stressed state. (As can be seen, the B_{ij}, coefficients reduce to the C_{ij} when no pressure is applied.) The evolution of the elastic stiffness coefficients B_{ij} under pressure is depicted in Figure 6. The coefficients B_{22}, B_{33}, B_{12}, B_{23}, and B_{13} increase with pressure in the whole pressure range studied. On the other hand, B_{44}, B_{55}, and B_{66} decrease slowly. The B_{11} coefficient increases slightly up to 30 GPa and then decreases smoothly.

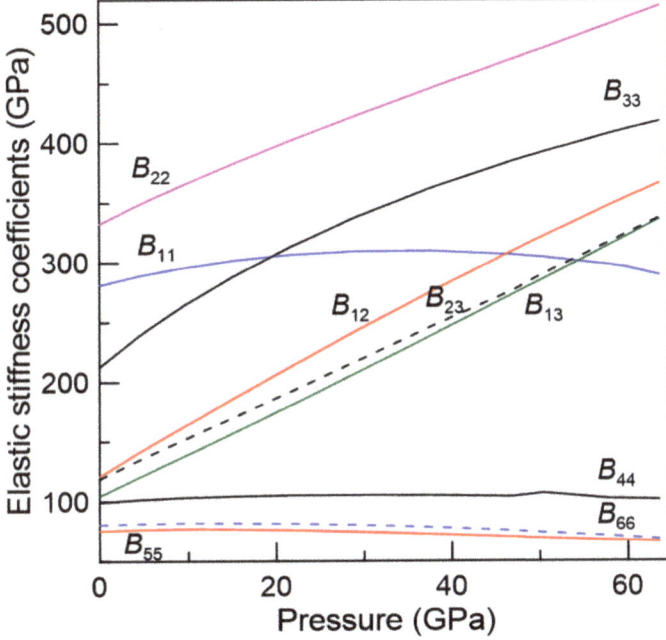

Figure 6. Calculated pressure evolution of the elastic stiffness coefficients B_{ij} of DyScO$_3$ perovskite.

The stiffness coefficients allow us to obtain the elastic moduli that describe the major elastic properties of the material at any hydrostatic pressure. Employing standard relations for orthorhombic crystals [42], we have obtained the bulk modulus B, the shear modulus G, the young modulus E, the Poison's ratio ν, and the B/G ratio. The values of the above-mentioned elastic moduli at zero pressure as calculated within the standard schemes of Voigt, Reuss, and Hill (with the latter one defined as the arithmetic average of the other two models [43–45]) are displayed in Table 5.

Table 5. Elastic moduli B_0, E_0 and G_0; B_0/G_0 ratio; and Poisson's ratio ν_0, calculated using the Voigt, Reuss, and Hill approximations from the calculated elastic coefficients at 0 GPa of DyScO$_3$ perovskite.

	Voigt	Reuss	Hill
Bulk modulus B_0 (GPa)	168.4	161.5	164.9
Shear modulus G_0 (GPa)	83.3	80.9	82.1
Young modulus E_0 (GPa)	214.5	207.9	211.2
Poisson's ratio ν_0	0.288	0.285	0.286
Bulk/Shear ratio B_0/G_0	2.02	2.00	2.01

It is worth noting that the bulk modulus, an important parameter related to the resistance of a material to compression, yields a similar value in the Hill formalism (164.9 GPa) than the one obtained from the EOS fitting (164.7 GPa), which reflects the consistency of the calculations. As the shear modulus G is associated with the resistance to plastic deformation, it is also interesting to analyze the B/G ratio. According to Pugh's criterion, if B/G > 1.75 the material is ductile, otherwise the material is brittle [46]. In the present case, our results indicate that DyScO$_3$ is a ductile material at zero pressure. The Poisson's coefficient ν is related to volume changes during uniaxial deformation. Lower coefficients indicate large volume changes during deformation; if ν = 0.5 no volume changes occur during elastic deformation [40]. This parameter also provides information about the characteristics of the bonding forces. A value of ν = 0.25 is a lower limit for central forces while ν = 0.5 is an upper limit [40]. For DyScO$_3$, the calculated value of ν suggests that the forces in this material are predominantly central.

The calculated pressure evolution of the elastic moduli of DyScO$_3$ is presented in Figure 7. The bulk modulus B as calculated using Voigt's prescription increases in the whole pressure range under study. However, using Reuss' scheme the bulk modulus decreases dramatically around 50 GPa. This could be related to mechanical instability, as we will show in the next paragraph. As for the shear and Young moduli (G and E), E reaches a shallow maximum at about 14 GPa, then both moduli decrease with pressure and tend to get closer in values.

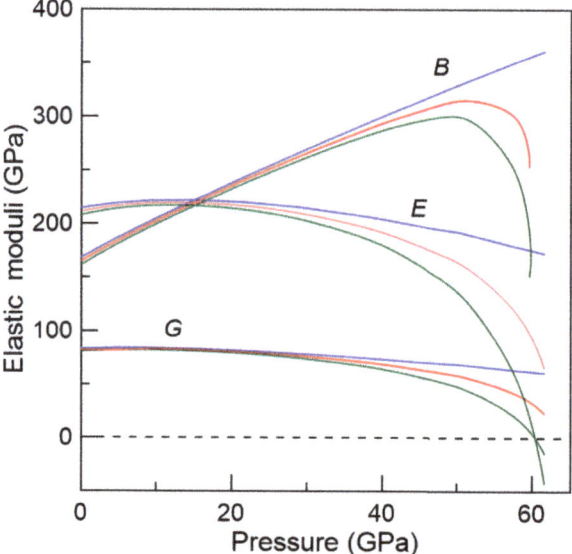

Figure 7. Pressure evolution of the elastic moduli: bulk modulus B, Young modulus E, and Shear modulus G of DyScO$_3$ perovskite. The blue line represents the moduli in the Voigt approximation, the green line in the Reuss approximation, and the red one in the Hill approximation.

The B/G ratio becomes larger under pressure and so the material becomes more ductile under compression. The Poisson's ratio increases rapidly with pressure, taking a value of 0.45 in the Hill expression at 60 GPa, see Figure 8.

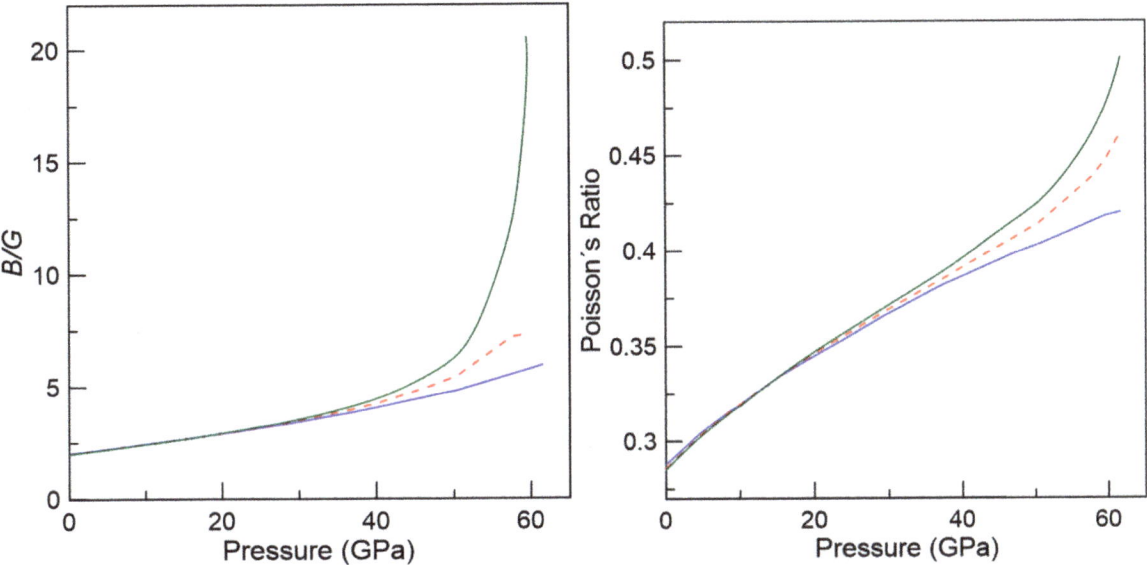

Figure 8. (**left**) Pressure evolution of the B/G ratio of $DyScO_3$ perovskite. (**right**) Pressure evolution of the Poisson's ratio. The blue line represents the moduli in the Voigt approximation, the green line in the Reuss approximation, and the red one in the Hill approximation.

To conclude, we analyze the mechanical stability of $DyScO_3$ perovskite under pressure. As stated before, when hydrostatic pressure is applied, the elastic stiffness coefficients must be employed instead of the elastic constants [38]. Therefore, the stability conditions are modified, and one needs to use the following generalized stability conditions for an orthorhombic crystal [38,47]:

$$M_1 = B_{11} > 0 \,;\, M_2 = B_{11}B_{22} - B_{12}^2 > 0$$

$$M_3 = B_{11}B_{22}B_{33} + 2B_{12}B_{13}B_{23} - B_{11}B_{23}^2 - B_{22}B_{13}^2 - B_{33}B_{12}^2 > 0$$

$$M_4 = B_{44} > 0 \,;\, M_5 = B_{55} > 0 \,;\, M_6 = B_{66} > 0$$

The crystalline structure is mechanically stable when all the above criteria are fulfilled. The evolution of these generalized Born stability criteria M_i ($i = 1$ to 6) as functions of pressure is depicted in Figure 9 where it can be observed that the M_3 criterion is violated above 60 GPa, indicating that $DyScO_3$ would become mechanically unstable at such pressure. Above this pressure an amorphization or a pressure-driven phase transition can occur, a result that may entice further experimental and theoretical research along this line.

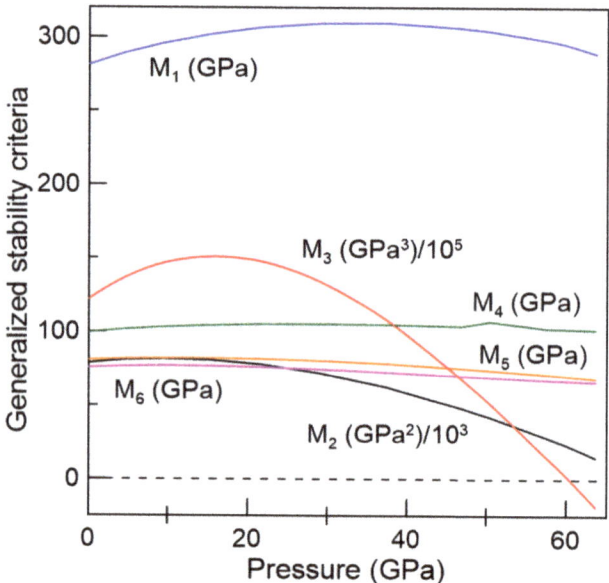

Figure 9. Generalized stability criteria as a function of pressure for DyScO$_3$ perovskite.

3.3. Vibrational Properties

We have also explored the vibrational properties of DyScO$_3$ perovskite at high pressure using the small-displacements method. The mechanical representation of DyScO$_3$ yields the following multiplicities for the irreducible representations of the modes at the zone center: M = 7A$_g$ (R) + 8A$_u$ (S) + 5B$_{1g}$ (R) + 10B$_{1u}$ (IR) + 7B$_{2g}$ (R) + 8B$_{2u}$ (IR) + 5B$_{3g}$ (R) + 10B$_{3u}$ (IR); of which the A$_g$, B$_{1g}$, B$_{2g}$, and B$_{3g}$ modes are active in Raman (R); the B$_{1u}$, B$_{2u}$, and B$_{3u}$ modes are infrared-active (IR); and the A$_u$ modes are silent (S). In Figure 10 we display the behavior of the phonon modes at the Γ point (Raman, infrared and silent modes) under applied hydrostatic pressure.

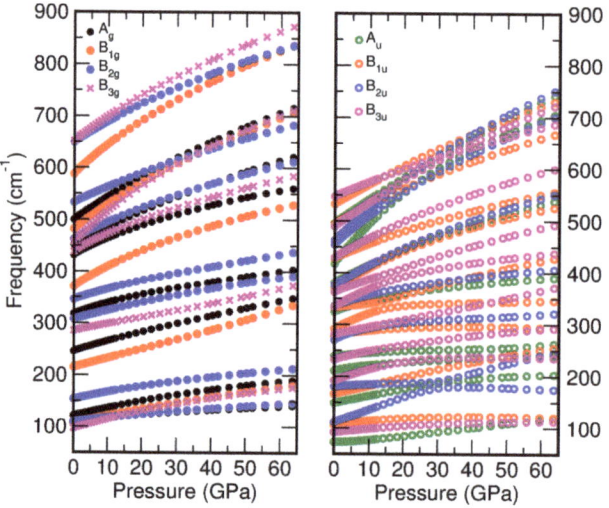

Figure 10. (**left**) Pressure dependence of the calculated Raman-active modes of DyScO$_3$ perovskite; (**right**) idem for the infrared and silent modes.

The present calculated values of the frequencies are in overall agreement with previous theoretical values obtained by Delugas et al. [48]. The pressure behavior in the very large region investigated in our work displays some curvature as evidenced in Figure 10 but up to about 15 GPa it can be approximated fairly well by a linear relation. The corresponding fit to the theoretical data yields the values of the pressure coefficients and calculated mode Grüneisen parameters shown in Table 6. As can be seen, they are in overall agreement with the experimental data obtained by Bura et al. [11].

The calculated phonon-dispersion curves along selected paths within the Brillouin zone with endpoints at the high symmetry points for SG 62 are plotted in Figure 11. The phonon dispersion curves at ambient pressure in these figures correspond to a dynamically stable behavior where all the frequencies are positive whereas above around 63 GPa the structure becomes dynamically unstable, displaying phonon branches with negative (imaginary) frequencies around the zone-boundary X point. This instability becomes more accentuated as the pressure is further increased, a pressure regime that we have not considered in this paper.

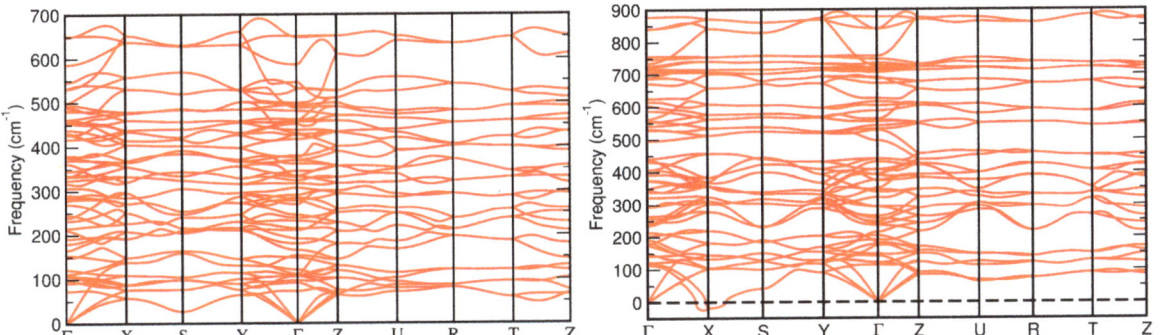

Figure 11. (**left**) Phonon dispersion curves of DyScO$_3$ perovskite at ambient pressure and (**right**) at 66 GPa.

To provide more detail about the nature of the lattice vibrations, the spectral distribution of the phonon density of states (phDOS) was obtained along with the partial phonon density of states, to extract the contribution of each atomic species in the local lattice configuration. Figure 12 shows phDOS plots at ambient pressure. The shape of the phDOS curve above 150 cm^{-1} is dominated by the oxygen atoms, which are present in larger numbers. At low frequencies (up to 150 cm^{-1}) the phonon peaks correspond mainly to the heavy Dy atoms, whereas the contribution of these atoms is very small at higher frequencies. In the intermediate frequency region (from 150 cm^{-1} to 450 cm^{-1}) there are significant contributions from both the Sc and O atoms.

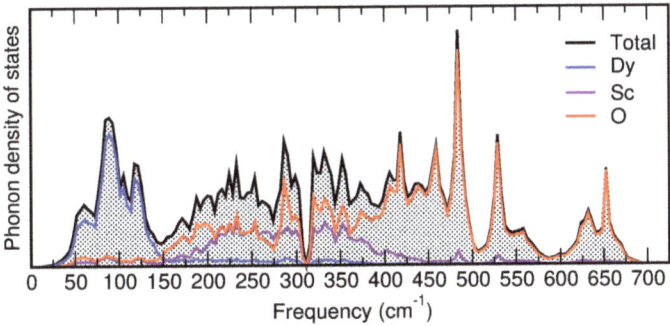

Figure 12. Total phonon density of states and contribution per species, as calculated at ambient pressure for DyScO$_3$ perovskite.

Table 6. Raman (R), infrared (IR), and silent (S) mode frequencies as calculated in this work for DyScO$_3$ perovskite, and their respective pressure coefficients obtained by fitting in the range up to 15 GPa with a linear equation $\omega(p) = \omega_o + \alpha p$. Mode Grüneisen parameters (γ) are also provided using the calculated bulk modulus B_o (164.7 GPa). Experimental values from Ref. [11], where available, are given in parenthesis.

Mode (R)	ω_o (cm^{-1})	α (cm^{-1} GPa^{-1})	γ
A_g	107.6	0.91	1.39
	122.3	1.46	1.96
	246.6 (253.2)	1.92 (1.48)	1.28 (1.11)
	319.2 (327.7)	2.02 (1.59)	1.04 (0.92)
	433.6	3.05	1.16
	452.9 (458.3)	3.28 (2.74)	1.19 (1.13)
	501.3	4.32	1.42
B_{1g}	108.7	1.34	2.03
	215.4 (226.4)	1.75 (1.62)	1.34 (1.36)
	373.3 (380.1)	3.65 (2.90)	1.61 (1.45)
	484.2 (476)	4.61 (2.50)	1.57 (1.00)
	590.8	5.42	1.51
B_{2g}	110.7	0.75	1.12
	154.7 (157.4)	1.30 (1.15)	1.38 (1.38)
	308.3 (308.7)	1.65 (1.80)	0.88 (1.10)
	347.2	1.89	0.90
	464.1	3.11	1.10
	534.4	2.92	0.90
	651.2	3.83	0.97
B_{3g}	97.8	1.74	2.93
	286.0	1.29	0.74
	441.5 (458.3)	3.15 (2.74)	1.17 (1.13)
	459.8	5.66	2.03
	654.0	4.57	1.15
Mode (IR)	ω_o (cm^{-1})	α (cm^{-1} GPa^{-1})	γ
B_{1u}	107.2	0.73	1.12
	166.2	1.58	1.56
	280.6	0.76	0.45
	294.0	2.68	1.50
	332.1	0.89	0.44
	372.2	3.71	1.64
	429.2	2.69	1.03
	495.7	3.49	1.16
	533.1	3.73	1.15
B_{2u}	112.2	2.41	3.53
	181.6	0.33	0.30
	274.7	1.81	1.08
	337.7	1.89	0.92
	377.6	3.69	1.61
	457.9	5.19	1.87
	464.0	6.03	2.14
B_{3u}	92.3	0.66	1.18
	193.4	2.02	1.72
	234.0	1.49	1.05
	281.7	1.37	0.80
	336.0	3.27	1.60
	363.5	2.39	1.08
	430.7	3.64	1.39
	490.2	4.58	1.54
	547.7	2.76	0.83

Table 6. *Cont.*

Mode (S)	ω_o (cm^{-1})	α (cm^{-1} GPa^{-1})	γ
A_u	72.6	0.48	1.09
	149.3	1.34	1.48
	212.5	0.91	0.70
	239.7	0.58	0.40
	326.5	1.90	0.96
	381.7	3.37	1.45
	421.7	7.31	2.85
	496.8	4.51	1.49

3.4. Prospective High-Pressure Post-Perovskite Phase

The recent experimental work of Bura et al. [11] shows that the DyScO$_3$ perovskite has a large stability range, up to at least 40 GPa, which are the highest pressures for which experimental results on this compound exist. For this reason, up to now we have been concerned with the local stability of the perovskite phase of DyScO$_3$, the only one so far observed. However, other perovskites, like MgSiO$_3$, undergo a pressure-induced transition to a post-perovskite phase with a *Cmcm*-type structure [49–51]. We have performed simulations on this post-perovskite structure for DyScO$_3$ (see Table 7) and our results show that this new phase has lower enthalpy above 14.4 GPa, see Figure 13. The study of the phonon dispersion curves at 15 GPa (Figure 14) shows that this phase is also dynamically stable. The existence of large kinetic barriers (and the enhanced local stability of the perovskite phase) might explain why this high-pressure phase was not observed in the experimental study of Ref. [11], but more experimental work, involving heating at high pressure to overcome barriers, would be needed to test this hypothesis. In the Supplementary Material we provide some extra results for this proposed high-pressure phase of DyScO$_3$, but its full study will not be pursued here.

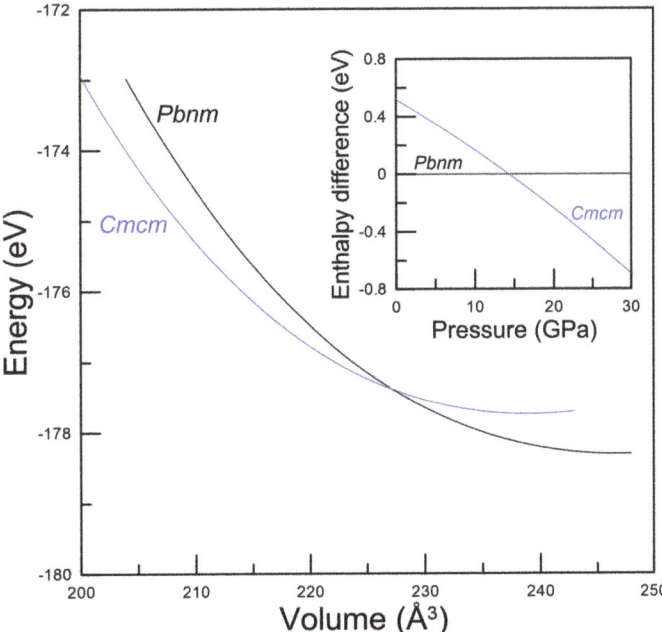

Figure 13. Energy-volume curves and (inset) enthalpy-pressure curves for the *Pbnm* (perovskite) and *Cmcm* (post-perovskite) phases of DyScO$_3$ (black and blue curves, respectively). The enthalpies are given with respect to that of the perovskite phase.

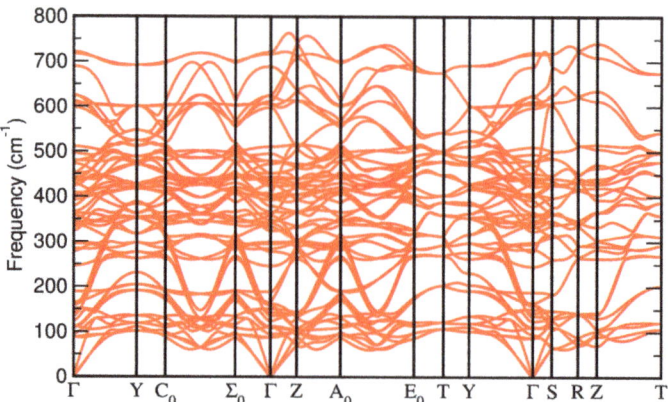

Figure 14. Calculated phonon dispersion curves of the *Cmcm* phase at 15 GPa.

Table 7. Crystallographic description of *Cmcm* at 15 GPa.

Space Group and Lattice Parameters	Wyckoff Positions
No 63 - C 2/m 2/c 2_1/m (*Cmcm*)	Dy 4c (0.000, 0.756, 0.250)
$a = 3.081$ Å; $b = 9.901$ Å; $c = 7.278$ Å	Sc 4b (0.000, 0.500, 0.000)
$\alpha = \beta = \gamma = 90°$	O1 4c (0.000, 0.414, 0.250)
$V = 222.00$ Å^3	O2 8f (0.000, 0.139, 0.063)

3.5. Summary and Conclusions

We have performed an ab initio study of DyScO$_3$ perovskite both at ambient and under hydrostatic pressure up to just above 60 GPa. We have analyzed the evolution of its structural, dynamic, and elastic properties. The calculated lattice parameters compare well with experimental values up to 10–15 GPa, above which pressures the experiments are probably affected by the loss of hydrostatic conditions of the pressure transmitting medium. The equation of state, the evolution of interatomic distances with pressure, and the compressibility were calculated. The DyO$_{12}$ polyhedra account for most of the volume reduction under compression. The phDOS and partial phDOS were analyzed to determine the contribution of each atomic species to the vibrational spectrum. The evolution of all the Raman, infrared and silent phonon modes was obtained along with their corresponding Grüneisen parameters. At about 63 GPa, imaginary frequencies appear around the X point of the Brillouin zone and therefore the structure becomes dynamically unstable. The elastic constants and the elastic stiffness coefficients were also determined. This allowed us to study the major elastic moduli (B, G, E, ν). The value of the B/G ratio points out that this material is ductile and that its ductility increases under pressure. Above 60 GPa the phase becomes mechanically unstable. We have also considered the possibility of a high-pressure post-perovskite phase with *Cmcm* symmetry. Our calculations suggest that it is favored over the perovskite phase above 14.4 GPa, which is a pressure within the range of observation of the perovskite phase and well below its predicted local (i.e., dynamical and elastic) instabilities. This prediction should thus entice further experimental and theoretical studies on DyScO$_3$.

Supplementary Materials: The following supporting information can be downloaded at: https://www.mdpi.com/article/10.3390/cryst13020165/s1, Figure S1. Variation of the lattice parameters a, b, c with pressure for the proposed *Cmcm* phase. Table S1. Elastic moduli B, E and G; B/G ratio; and Poisson's ratio ν for the proposed *Cmcm* phase., calculated using the Voigt, Reuss, and Hill approximations from the calculated elastic coefficients at 15 GPa. Figure S2. Calculated pressure evolution of the elastic stiffness coefficients for the proposed *Cmcm* phase. Figure S3. M$_3$ generalized stability criteria as a function of pressure for the proposed *Cmcm* phase. Figure S4. Pressure dependence of the calculated Raman-active modes of the *Cmcm* phase. Figure S5. Pressure dependence of the calculated infrared modes of the *Cmcm* phase. Figure S6. Pressure dependence of the calculated silent modes of the *Cmcm* phase. Table S2. Calculated Raman (R), infrared (IR), and silent (S) mode frequencies for the proposed *Cmcm* phase as calculated in this work, and their respective pressure coefficients obtained by fitting in the range up to 60 GPa with a cuadratic equation $\omega(p) = \omega_o + \alpha p + \beta p^2$. Figure S7. Phonon dispersion curves at 65.7 GPa for the proposed *Cmcm* phase.

Author Contributions: All authors contributed to this work doing simulations, analysis, interpretation of results, and/or manuscript writing. All authors have read and agreed to the published version of the manuscript.

Funding: This research was funded by the Spanish Research Agency (AEI) and the Spanish Ministry of Science and Investigation (MCIN) under grant PID2019-106383GB-43 (DOI: 10.13039/501100011033) and RED2018-102612-T (MALTA Consolider-Team Network).

Data Availability Statement: All relevant data that support the findings of this study are available from the corresponding author upon request.

Conflicts of Interest: The authors declare no conflict of interest. The funders had no role in the design of the study; in the collection, analyses, or interpretation of data; in the writing of the manuscript; or in the decision to publish the results.

References

1. Coh, S.; Heeg, T.; Haeni, J.H.; Biegalski, M.D.; Lettieri, J.; Edge, L.F.; O'Brien, K.E.; Bernhagen, M.; Reiche, P.; Uecker, R.; et al. Si-compatible candidates for high-κ dielectrics with the Pbnm perovskite structure. *Phys. Rev. B* **2010**, *82*, 064101. [CrossRef]
2. Schlom, D.G.; Haeni, J.H. A Thermodynamic Approach to Selecting Alternative Gate Dielectrics. *MRS Bull.* **2002**, *27*, 198–204. [CrossRef]
3. Christen, H.M.; Jellison, G.E.; Ohkubo, I.; Huang, S.; Reeves, M.E.; Cicerrella, E.; Freeouf, J.L.; Jia, Y.; Schlom, D.G. Dielectric and optical properties of epitaxial rare-earth scandate films and their crystallization behavior. *Appl. Phys. Lett.* **2006**, *88*, 262906. [CrossRef]
4. Haeni, J.H.; Irvin, P.; Chang, W.; Uecker, R.; Reiche, P.; Li, Y.L.; Choudhury, S.; Tian, W.; Hawley, M.E.; Craigo, B.; et al. Room-temperature ferroelectricity in strained $SrTiO_3$. *Nature* **2004**, *430*, 4. [CrossRef]
5. Choi, K.J.; Biegalski, M.; Li, Y.L.; Sharan, A.; Schubert, J.; Uecker, R.; Reiche, P.; Chen, Y.B.; Pan, X.Q.; Gopalan, V.; et al. Enhancement of Ferroelectricity in Strained $BaTiO_3$ Thin Films. *Science* **2004**, *306*, 1005–1009. [CrossRef] [PubMed]
6. Biegalski, M.D.; Haeni, J.H.; Trolier-McKinstry, S.; Schlom, D.G.; Brandle, C.D.; Graitis, A.J.V. Thermal expansion of the new perovskite substrates $DyScO_3$ and $GdScO_3$. *J. Mater. Res.* **2005**, *20*, 952–958. [CrossRef]
7. Vasudevarao, A.; Kumar, A.; Tian, L.; Haeni, J.H.; Li, Y.L.; Eklund, C.J.; Jia, Q.X.; Uecker, R.; Reiche, P.; Rabe, K.M.; et al. Multiferroic Domain Dynamics in Strained Strontium Titanate. *Phys. Rev. Lett.* **2006**, *97*, 257602. [CrossRef]
8. Catalan, G.; Janssens, A.; Rispens, G.; Csiszar, S.; Seeck, O.; Rijnders, G.; Blank, D.H.A.; Noheda, B. Polar Domains in Lead Titanate Films under Tensile Strain. *Phys. Rev. Lett.* **2006**, *96*, 127602. [CrossRef]
9. Wördenweber, R.; Hollmann, E.; Kutzner, R.; Schubert, J. Induced ferroelectricity in strained epitaxial $SrTiO_3$ films on various substrates. *J. Appl. Phys.* **2007**, *102*, 044119. [CrossRef]
10. Kužel, P.; Kadlec, F.; Petzelt, J.; Schubert, J.; Panaitov, G. Highly tunable $SrTiO_3/DyScO_3$ heterostructures for applications in the terahertz range. *Appl. Phys. Lett.* **2007**, *91*, 232911. [CrossRef]
11. Bura, N.; Srihari, V.; Bhoriya, A.; Yadav, D.; Singh, J.; Poswal, H.K.; Dilawar Sharma, N. Structural stability of orthorhombic $DyScO_3$ under extreme conditions of pressure and temperature. *Phys. Rev. B* **2022**, *106*, 024113. [CrossRef]
12. Hohenberg, P.; Kohn, W. Inhomogeneous Electron Gas. *Phys. Rev.* **1964**, *136*, B864–B871. [CrossRef]
13. Kohn, W.; Sham, L.J. Self-Consistent Equations Including Exchange and Correlation Effects. *Phys. Rev.* **1965**, *140*, A1133–A1138. [CrossRef]
14. Blöchl, P.E. Projector augmented-wave method. *Phys. Rev. B* **1994**, *50*, 17953–17979. [CrossRef] [PubMed]
15. Kresse, G.; Joubert, D. From ultrasoft pseudopotentials to the projector augmented-wave method. *Phys. Rev. B* **1999**, *59*, 1758–1775. [CrossRef]
16. Kresse, G.; Hafner, J. Ab Initio molecular dynamics for liquid metals. *Phys. Rev. B* **1993**, *47*, 558–561. [CrossRef] [PubMed]
17. Kresse, G.; Hafner, J. Ab Initio molecular-dynamics simulation of the liquid-metal–amorphous-semiconductor transition in germanium. *Phys. Rev. B* **1994**, *49*, 14251–14269. [CrossRef] [PubMed]
18. Kresse, G.; Furthmüller, J. Efficiency of ab-initio total energy calculations for metals and semiconductors using a plane-wave basis set. *Comput. Mater. Sci.* **1996**, *6*, 15–50. [CrossRef]
19. Kresse, G.; Furthmüller, J. Efficient iterative schemes for Ab Initio Total-Energy Calc. Using A Plane-Wave Basis Set. *Phys. Rev. B* **1996**, *54*, 11169–11186. [CrossRef] [PubMed]
20. Monkhorst, H.J.; Pack, J.D. Special points for Brillouin-zone integrations. *Phys. Rev. B* **1976**, *13*, 5188–5192. [CrossRef]

21. Perdew, J.P.; Burke, K.; Ernzerhof, M. Generalized Gradient Approximation Made Simple. *Phys. Rev. Lett.* **1996**, *77*, 3865–3868. Erratum in *Phys. Rev. Lett.* **1997**, *78*, 1396–1396. [CrossRef]
22. Grimme, S.; Antony, J.; Ehrlich, S.; Krieg, H. A consistent and accurate Ab Initio Parametr. Density Funct. Dispers. Correct. (DFT-D) 94 Elem. H-Pu. *J. Chem. Phys.* **2010**, *132*, 154104. [CrossRef] [PubMed]
23. Togo, A.; Tanaka, I. First principles phonon calculations in materials science. *Scr. Mater.* **2015**, *108*, 1–5. [CrossRef]
24. Le Page, Y.; Saxe, P. Symmetry-general least-squares extraction of elastic data for strained materials from Ab Initio Calc. Stress. *Phys. Rev. B* **2002**, *65*, 104104. [CrossRef]
25. Veličkov, B.; Kahlenberg, V.; Bertram, R.; Bernhagen, M. Crystal chemistry of $GdScO_3$, $DyScO_3$, $SmScO_3$ and $NdScO_3$. *Z. Für Krist.* **2007**, *222*, 466–473. [CrossRef]
26. Celeste, A.; Borondics, F.; Capitani, F. Hydrostaticity of pressure-transmitting media for high pressure infrared spectroscopy. *High Press. Res.* **2019**, *39*, 608–618. [CrossRef]
27. Birch, F. Finite Elastic Strain of Cubic Crystals. *Phys. Rev.* **1947**, *71*, 809–824. [CrossRef]
28. Jain, A.; Ong, S.P.; Hautier, G.; Chen, W.; Richards, W.D.; Dacek, S.; Cholia, S.; Gunter, D.; Skinner, D.; Ceder, G.; et al. Commentary: The Materials Project: A materials genome approach to accelerating materials innovation. *APL Mater.* **2013**, *1*, 011002. [CrossRef]
29. Materials Data on $DyScO_3$ by Materials Project. Type: Dataset. 2020. Available online: https://materialsproject.org/materials/mp-31120 (accessed on 15 December 2022).
30. Schmidbauer, M.; Kwasniewski, A.; Schwarzkopf, J. High-precision absolute lattice parameter determination of $SrTiO_3$, $DyScO_3$ and $NdGaO_3$ single crystals. *Acta Crystallogr. B Struct. Sci.* **2012**, *68*, 8–14. [CrossRef]
31. Glazer, A.M. Simple ways of determining perovskite structures. *Acta Cryst. A* **1975**, *31*, 756–762. [CrossRef]
32. Zhao, Y.; Weidner, D.J.; Parise, J.B.; Cox, D.E. Thermal expansion and structural distortion of perovskite—Data for $NaMgF_3$ perovskite. Part I. *Phys. Earth Planet. Inter.* **1993**, *76*, 1–16. [CrossRef]
33. Zhao, Y.; Weidner, D.J.; Parise, J.B.; Cox, D.E. Critical phenomena and phase transition of perovskite—Data for $NaMgF_3$ perovskite. Part II. *Phys. Earth Planet. Inter.* **1993**, *76*, 17–34. [CrossRef]
34. Martin, C.D.; Parise, J.B. Structure constraints and instability leading to the post-perovskite phase transition of $MgSiO_3$. *Earth Planet. Sci. Lett.* **2008**, *265*, 630–640. [CrossRef]
35. Liferovich, R.P.; Mitchell, R.H. A structural study of ternary lanthanide orthoscandate perovskites. *J. Solid State Chem.* **2004**, *177*, 2188–2197. [CrossRef]
36. Nye, J.F. *Physical Properties of Crystals: Their Representation by Tensors and Matrices*; Clarendon Press: Oxford, NY, USA; Oxford University Press: Oxford, NY, USA, 1984.
37. Janovská, M.; Sedlák, P.; Seiner, H.; Landa, M.; Marton, P.; Ondrejkovič, P.; Hlinka, J. Anisotropic elasticity of $DyScO_3$ substrates. *J. Phys. Condens. Matter* **2012**, *24*, 385404. [CrossRef] [PubMed]
38. Born, M.; Huang, K. *Dynamical Theory of Crystal Lattices*; Oxford Classic Texts in the Physical Sciences; Clarendon Press: Oxford, NY, USA; Oxford University Press: Oxford, NY, USA, 1954.
39. Mouhat, F.; Coudert, F.X. Necessary and sufficient elastic stability conditions in various crystal systems. *Phys. Rev. B* **2014**, *90*, 224104. [CrossRef]
40. Wallace, D.C. Pressure-Volume Variables, Stress-Strain Variables & Wave Propagation. In *Thermodynamics of Crystals*; Dover Publications: Mineola, NY, USA, 1998.
41. Grimvall, G.; Magyari-Köpe, B.; Ozoliņš, V.; Persson, K.A. Lattice instabilities in metallic elements. *Rev. Mod. Phys.* **2012**, *84*, 945–986. [CrossRef]
42. Ravindran, P.; Fast, L.; Korzhavyi, P.A.; Johansson, B.; Wills, J.; Eriksson, O. Density functional theory for calculation of elastic properties of orthorhombic crystals: Application to $TiSi_2$. *J. Appl. Phys.* **1998**, *84*, 4891–4904. [CrossRef]
43. Voigt, W. *Lehrbuch der Kristallphysik (mit Ausschluss der Kristalloptik)*; B.G. Teubner: Leipzig/Berlin, Germany, 1928.
44. Reuss, A. Berechnung der Fließgrenze von Mischkristallen auf Grund der Plastizitätsbedingung für Einkristalle. *Z. Angew. Math. Mech.* **1929**, *9*, 49–58. [CrossRef]
45. Hill, R. The Elastic Behaviour of a Crystalline Aggregate. *Proc. Phys. Soc. A* **1952**, *65*, 349–354. [CrossRef]
46. Pugh, S. XCII. Relations between the elastic moduli and the plastic properties of polycrystalline pure metals. *Lond. Edinb. Dublin Philos. Mag. J. Sci.* **1954**, *45*, 823–843. [CrossRef]
47. Wallace, D.C. Thermoelasticity of Stressed Materials and Comparison of Various Elastic Constants. *Phys. Rev.* **1967**, *162*, 776–789. [CrossRef]
48. Delugas, P.; Fiorentini, V.; Filippetti, A.; Pourtois, G. Cation charge anomalies and high-κ dielectric behavior in $DyScO_3$: Ab Initio Density-Funct. Self-Interact. Calc. *Phys. Rev. B* **2007**, *75*, 115126. [CrossRef]
49. Murakami, M.; Hirose, K.; Kawamura, K.; Sata, N.; Ohishi, Y. Post-Perovskite Phase Transition in $MgSiO_3$. *Science* **2004**, *304*, 855–858. [CrossRef] [PubMed]

50. Oganov, A.R.; Ono, S. Theoretical and experimental evidence for a post-perovskite phase of MgSiO$_3$ in Earth's D″layer. *Nature* **2004**, *430*, 445–448. [CrossRef]
51. Tsuchiya, T.; Tsuchiya, J.; Umemoto, K.; Wentzcovitch, R.M. Phase transition in MgSiO$_3$ perovskite in the earth's lower mantle. *Earth Planet. Sci. Lett.* **2004**, *224*, 241–248. [CrossRef]

Disclaimer/Publisher's Note: The statements, opinions and data contained in all publications are solely those of the individual author(s) and contributor(s) and not of MDPI and/or the editor(s). MDPI and/or the editor(s) disclaim responsibility for any injury to people or property resulting from any ideas, methods, instructions or products referred to in the content.

Article

Pressure-Induced Structural Phase Transitions in the Chromium Spinel LiInCr$_4$O$_8$ with Breathing Pyrochlore Lattice

Meera Varma [1], Markus Krottenmüller [1], H. K. Poswal [2,*] and C. A. Kuntscher [1,*]

[1] Experimentalphysik II, Augsburg University, 86159 Augsburg, Germany
[2] High Pressure & Synchrotron Radiation Physics Division, Bhabha Atomic Research Centre, Trombay, Mumbai 400085, India
* Correspondence: himanshu@barc.gov.in (H.K.P.); christine.kuntscher@physik.uni-augsburg.de (C.A.K.)

Abstract: This study reports high-pressure structural and spectroscopic studies on polycrystalline cubic chromium spinel compound LiInCr$_4$O$_8$. According to pressure-dependent X-ray diffraction measurements, three structural phase transitions occur at ∼14 GPa, ∼19 GPa, and ∼36 GPa. The first high-pressure phase is indexed to the low-temperature tetragonal phase of the system which coexists with the ambient phase before transforming to the second high-pressure phase at ∼19 GPa. The pressure-dependent Raman and infrared spectroscopic measurements show a blue-shift of the phonon modes and the crystal field excitations and an increase in the bandgap under compression. During pressure release, the sample reverts to its ambient cubic phase, even after undergoing multiple structural transitions at high pressures. The experimental findings are compared to the results of first principles based structural and phonon calculations.

Keywords: high-pressure studies; Raman spectroscopy; infrared spectroscopy; X-ray diffraction

1. Introduction

The chromium spinels ACr$_2$O$_4$ belong to the widely studied geometrically frustrated systems, owing to their varied magnetic couplings, magnetostructural transitions, and exotic ground states. With the A-site occupied by a non-magnetic ion, the magnetic chromium ions (Cr^{3+}) at the B-site form a network of corner-linked Cr$_4$ tetrahedra, i.e., the pyrochlore lattice. The dominant antiferromagnetic (AFM) interactions between the Cr^{3+} ions in the pyrochlore network leads to strong magnetic frustrations, which result in Jahn–Teller-driven structural distortion and antiferromagnetic ordering at low temperatures. Due to their varied response to external parameters, the chromium-based spinels could be used in magnetic sensing devices, data storage or spintronic devices [1]. There have also been studies reporting their use as a potential electrode support material [2]. A study also suggested cobalt containing chromium spinels as possible candidates for catalytic combustion [3].

An archetype derived by substituting two different ions at the A-site of ACr$_2$O$_4$ was first reported by Joubert and Durif in 1966 [4]. The difference in ionic radii at the A-site led to an alternate arrangement of small and large Cr$_4$ tetrahedra at the B-sites, known as the breathing pyrochlore lattice (see Figure 1) [5–8]. This new family of Cr spinels, namely LiMCr$_4$O$_8$ (with M = In, Ga, Fe), are being actively investigated due to the geometrical frustration and the Cr–Cr bond alternations in these materials. The substitution of different ions at the A-site leads to the loss of inversion symmetry found in the conventional spinels, and the crystal symmetry is reduced to $F\bar{4}3m$. The alternating smaller and larger Cr–Cr bonds between Cr$_4$ tetrahedra also cause a difference in the nearest-neighbour magnetic interactions, without relieving the frustration in the system. The magnitude of AFM interactions between neighbouring Cr ions is denoted as J and J' for small and large tetrahedra, respectively. The ratio of J and J' is defined as the breathing factor B$_f$, i.e., B$_f$ = J'/J, and determines the degree of frustration in the system [5].

The physical properties of this family of frustrated breathing pyrochlores $LiMCr_4O_8$ (with M = In, Ga, Fe) are highly dependent on the cationic radii at the A-site, which define the breathing factor and in turn have an overwhelming influence on the response of the material to different thermodynamic conditions. Recent studies have been conducted to improving the understanding of the structural and magnetic response of these materials at low temperatures. A temperature-dependent neutron diffraction experiment on $LiInCr_4O_8$ indicated the opening of a spin gap below 65 K, which is followed by a long-range magnetic ordering at ~15.9 K [5]. An NMR study on $LiInCr_4O_8$ suggested a singlet ground state with a gap at 18 K and a structural transition at 16 K followed by a second-order AFM transition at 13 K [6]. However, another study reported a structural transition at 18 K followed by magnetic ordering at 12 K [7]. The related compound $LiGaCr_4O_8$ has been reported to show short-range AFM ordering close to ~50 K, which is followed by a first-order magnetostructural transition around ~15 K [5,6]. However, another study reported it as two consecutive events of magnetic and structural transitions at 14.1 K and 14.5 K, respectively, [8]. As opposed to $LiInCr_4O_8$ and $LiGaCr_4O_8$, the compound $LiFeCr_4O_8$ has been reported to undergo a ferrimagnetic transition at 94 K. Furthermore, the opening of a spin gap at ~60 K and a magnetostructural transition at ~23 K were found [9]. It is to be noted that all three compounds, having different degrees of frustration, show a structural instability at low temperature.

Although low-temperature-induced structural and magnetic responses have been explored and reported for this class of compounds, high-pressure studies have not been reported to the best of our knowledge. In this work, we study the structural phase transitions in $LiInCr_4O_8$ induced by high pressure using synchrotron-based X-ray diffraction (XRD) and Raman and infrared (IR) spectroscopic measurements supplemented by density functional theory (DFT)-based simulations.

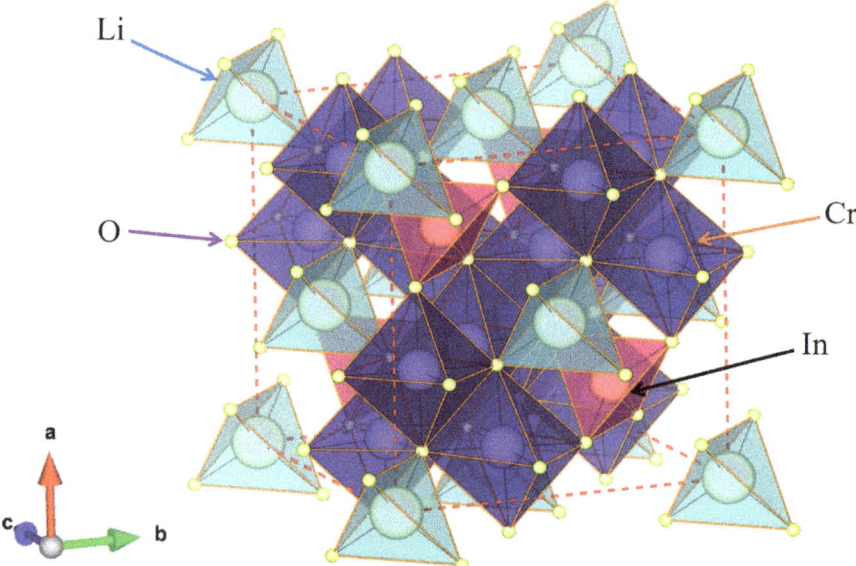

Figure 1. Sketch of the breathing chromium spinel $LiInCr_4O_8$ crystal structure with LiO_4 and InO_4 tetrahedra and CrO_6 octahedra as structural units.

2. Materials and Methods

Polycrystalline $LiInCr_4O_8$ was synthesized using the solid-state reaction method reported previously [5,8], where the stoichiometric amounts of In_2O_3 and Cr_2O_3 and 10% excess of Li_2CO_3 were thoroughly ground, pelleted, and heated in a furnace at 1100 °C for 48 h in an alumina crucible with intermittent grinding. As lithium is volatile, a slow rate of cooling was maintained. The phase purity of the synthesized compound LiInCr$_4$O$_8$ was characterized using XRD measurements performed in the angle-dispersive mode at the ECXRD beamline (BL-11), Indus-2, RRCAT. X-rays with energies of 19.7 keV (λ = 0.6280 Å) were incident on the synthesized polycrystalline sample. NIST standard LaB$_6$ was used to calibrate the distance from sample to detector.

High-pressure XRD (HPXRD) measurements on LiInCr$_4$O$_8$ were performed at the XPRESS beamline of the Elettra synchrotron radiation source, Italy. Powdered sample was loaded into a Mao–Bell type diamond anvil (DAC) along with gold (Au) as a marker to determine the pressure with an accuracy of $\sim$0.1 GPa. [10,11]. The diamonds had a culet diameter of $\sim$400 µm. A tungsten gasket of thickness $\sim$180 µm was pre-indented to a thickness of $\sim$50 µm before drilling a hole of 150 µm at the centre of the gasket. A methanol–ethanol mixture in a 4:1 ratio served as a quasi-hydrostatic pressure transmitting medium (PTM) [12]. Monochromatic X-rays of energy 25 keV (λ = 0.4957 Å) were incident on the sample. A MAR345 detector was used to record the diffraction patterns. Standard LaB$_6$ loaded in the DAC was used to calibrate the experimental setup. The diffraction images were reduced into 2θ-intensity patterns using the program Fit2D [13], and the refinement of the XRD patterns obtained were performed using GSAS [14].

Raman spectroscopy measurements were performed with a confocal micro-Raman setup (Jobin–Yvon T64000 spectrograph, single stage mode, 1800 groves/mm grating, and a resolution of 2 cm^{-1}) using a 50$\times$ objective in the back scattering geometry. A 488 nm argon ion laser was used as excitation source. The pressure-dependent Raman scattering measurements were carried out using a gas membrane type DAC, with diamonds with $\sim$500 µm culet diameter. The pressure inside the DAC was monitored using the well-known ruby fluorescence shift with an accuracy of 0.1 GPa [15]. A methanol–ethanol mixture in a 4:1 ratio was used as PTM.

The infrared spectroscopic measurements were performed using a Bruker Vertex FTIR spectrometer coupled to an IR microscope (Bruker Hyperion). The ambient pressure infrared reflectivity measurements were carried out in the spectral range 100–25,000 cm^{-1} on a polycrystalline sample pressed into a pellet of thickness 72 µm. The resolution for the various frequency ranges, viz., FIR, MIR, NIR-VIS, amount to 2 cm^{-1}, 4 cm^{-1}, and 8 cm^{-1}, respectively. Reflection from an aluminium mirror was used for the reference measurement, for normalizing the sample spectrum. The optical conductivity σ_1 was obtained by Kramer–Kronig (KK) transformation. For the KK transformation, the reflectivity spectrum was fitted with the Lorentz model and extrapolated to zero frequency based on the fitting model; in the high-energy range, a constant extrapolation up to 10^5 cm^{-1} was used, and beyond this, an extrapolation following a $1/w^4$ dependency was chosen.

For the pressure-dependent infrared transmittance measurements in the spectral range 500–20,500 cm^{-1} with a resolution of 4 cm^{-1}, the powder sample was diluted with CsI in the ratio 1:20 and pressed into a pellet of thickness 64 µm. A small piece of this diluted pellet was loaded into a membrane type DAC (500 µm culets) along with ruby spheres and well-ground CsI powder serving as reference for normalizing the sample spectrum while at the same time acting as a quasihydrostatic PTM [16]. The absorption spectrum was calculated using the formula A = $-\log_{10}$(T), where A is the absorbance and T is the measured transmittance spectrum.

Ab initio-based simulations for structural relaxations were performed using DFT within the framework of the projected augmented wave (PAW) method [17], as implemented in the Quantum Espresso 6.4.1 package [18]. Calculations were performed using Perdew, Burke, and Ernzerhof (PBE) [19] generalized gradient approximations (GGA) for exchange and correlation functional. Structural relaxations were performed on a primitive cell of LiInCr$_4$O$_8$ with fourteen atoms of four different types in a non-magnetic configuration. Simulations were performed by considering one valence electron of Li ($2s^1$), thirteen valence electrons of In ($5s^2\ 5p^1\ 4d^{10}$), six valence electrons of Cr ($4s^2\ 3d^4$), and six valence electrons of O ($2s^2\ 2p^4$). A plane wave cut-off energy of 160 Ry was used for expanding the basis set. The Brillouin zone integration was performed at the zone centre on a 6 $\times$ 6 $\times$ 6 Monkhorst–

Pack k-point mesh [20]. The Hellman–Feynman forces were converged until the largest force component was less than 1×10^{-5}. For the purpose of phonon-mode assignments, density functional perturbation theory (DFPT) [21] calculations were performed using Martins–Trouilier [22] pseudopotentials with local density approximation (LDA) on the primitive lattice of $LiInCr_4O_8$ at ambient volume.

3. Results

3.1. Ambient Pressure Results

$LiInCr_4O_8$ crystallizes in the cubic space group $F\bar{4}3m$ with four formula units per unit cell [5]. The lattice parameter, shape profile parameters, and asymmetry corrections were refined during the Rietveld refinement of the ambient pressure XRD pattern, shown in Figure 2. The background was fitted using a Chebyschev polynomial. The synthesized sample contains less than 2% unreacted Cr_2O_3 as observed from Figure 2. The refined lattice parameter is $a = 8.4038(1)$ Å and the unit cell volume amounts to $V = 593.52(2)$ Å^3, with $R_p = 0.075$ and $R_{wp} = 0.109$. These values are in good agreement with the reported values of $a = 8.4205$ Å and $V = 597.05$ Å^3 [4,5]. The fractional coordinates, occupations, and Wyckoff sites as reported by an earlier study [5] and the refined thermal parameter obtained from the Rietveld analysis are given in Table 1.

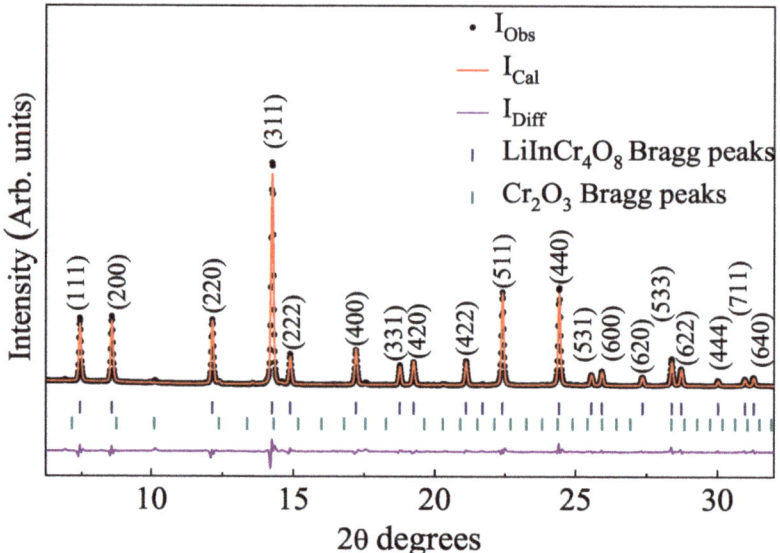

Figure 2. Rietveld refinement of the X-ray diffraction pattern of $LiInCr_4O_8$ at ambient pressure.

Table 1. The Wyckoff sites, fractional coordinates, and occupations as reported in an earlier study [5] and thermal parameter U as obtained from Rietveld refinement of $LiInCr_4O_8$ at ambient conditions. Space group $F\bar{4}3m$, $Z = 4$, $a = 8.4038(1)$ Å, $V = 593.52(2)$ Å^3.

Atoms	Wyckoff Sites	x	y	z	Occupancy	U
Li1	4a	0.0000	0.0000	0.0000	0.9920	1.090
In1	4a	0.0000	0.0000	0.0000	0.0080	0.350
Li2	4d	0.7500	0.7500	0.7500	0.0080	1.090
In2	4d	0.7500	0.7500	0.7500	0.9920	0.350
Cr	16e	0.3719	0.3719	0.3719	1.0000	0.140
O1	16e	0.1377	0.1377	0.1377	1.0000	0.380
O2	16e	0.6107	0.6107	0.6107	1.0000	0.180

DFT-based simulations were performed using LDA and GGA pseudopotentials and the results for ambient volume calculations are compared in Table 2. The A-site cations Li^{1+} and In^{3+} at 4a and 4d crystallographic sites form LiO_4 and InO_4 tetrahedra, respectively, and share corners with CrO_6 at the B-site (see Figure 1). The CrO_6 octahedral units at the general 16e Wyckoff positions have shared edges between them, and the Cr_4 tetrahedra form the breathing pyrochlore lattice. The Cr–Cr distances are usually short enough to facilitate electron hopping in some of the conventional spinels (Mott insulators). The quality of synthesized sample was verified by determining the degree of distortion (d'/d) due to the two different Cr–Cr bond lengths of Cr_4 tetrahedra from the Rietveld refinement, where the ratio d'/d is found to be 1.051, which matches with the reported value [4,5,8]. This ratio of dissimilar Cr–Cr distances can be directly associated with the breathing factor B_f in $LiInCr_4O_8$.

Table 2. Comparison of experimental and theoretical primitive cell volume together with the distortion parameter and the reported values from the literature.

	Reported from [5]	Present Study	Simulations	
			LDA	GGA
Volume of Primitive cell (Å³)	149.24	148.38	149.9925	149.0278
distortion parameter (d'/d)	1.051	1.05	1.32	1.33

For a primitive cell containing one formula unit (14 atoms), the factor group analysis gives the following irreducible representations:

$$\Gamma_{total} = 3A_1 + 3E + 3T_1 + 8T_2.$$

This can further be classified as:

$$\Gamma_{acoustic} = T_2$$

$$\Gamma_{IR} = 7T_2$$

$$\Gamma_{Raman} = 3A_1 + 3E + 7T_2$$

From the expected 13 Raman-active modes at ambient conditions, we experimentally observe nine modes, which are denoted as $M_1, M_2, \ldots M_9$ in Figure 3. Due to the loss of inversion symmetry in comparison with conventional spinels, $LiInCr_4O_8$ has T_2 modes which are both IR- and Raman-active. The weak mode at ~530 cm^{-1} has been assigned to Cr_2O_3 [23,24]. The peak observed at ~220 cm^{-1} is a plasma line from the excitation source used, which served as an internal calibrant. The ambient pressure Raman spectrum of $LiInCr_4O_8$ in Figure 3 matches well with a recently reported study [25], except for the low-energy Raman mode M_1, which is not observed in the recorded spectral range of Ref. [25]. Because there has been no detailed report on the vibrational properties of $LiInCr_4O_8$, DFPT calculations were carried out for explicit assignment of Raman and IR modes. These calculations were performed on a primitive cell consisting of 14 atoms using the Martins–Trouilier pseudopotentials with LDA approximations. The phonon modes were assigned with the help of Molden, a visualization software [26]. Simulations were also performed using PAW potentials with GGA approximations at ambient volume. The theoretical results are compared with experimental findings, summarized in Table 3.

Predominantly, the observed Raman modes can be classified as the internal vibrations of polyhedral units (viz., LiO_4, InO_4 and CrO_6). Due to the covalent nature of In–O and Cr–O bonds, they are expected to have stronger contributions in the Raman spectrum. The symmetry of a free CrO_6 ion is reduced inside the crystal site. Table 4 shows the changes in the internal modes of a free CrO_6 ion at the crystal site (C3v) of the ambient pressure cubic phase (T_d). Of these CrO_6 modes, only the A_1 and T_2 modes are Raman active in $LiInCr_4O_8$.

From the DFPT calculations, the Raman-active T_2 mode observed at 164 cm^{-1} is assigned to the translation motion of In. The modes M_2 and M_3 at 310 cm^{-1} and 443 cm^{-1}, respectively, are both assigned to the translation motion of Li ions. The Raman modes at 460 cm^{-1} and 491 cm^{-1} (M_4 & M_5) are associated with asymmetric and symmetric O–Li–O bending vibrations, respectively. The strongest Raman mode observed at $\sim$590 cm^{-1} is related to the O–Cr–O symmetric bending and O–In symmetric stretching vibrations, whereas the adjacent T_2 mode at $\sim$582 cm^{-1} is related to the O–Cr–O asymmetric bending and In-O asymmetric stretching vibrations. In a recent study, the DFT-based calculations for $LiGaCr_4S_8$ show large differences in the calculated Cr–Cr distances in magnetic and non-magnetic configurations of the system [27]. This could explain the large difference in the calculated and observed M_6 and M_7 modes, as all the calculations are performed in a non-magnetic configuration in the present study. The Raman modes at 718 cm^{-1} and 739 cm^{-1} (M_8 and M_9) are assigned to asymmetric and symmetric stretching vibrations of Li–O. A study on $LiFeCr_4O_8$ has assigned the strongest observed mode to Cr–O stretching vibration [9]. However, as the CrO_6 octahedra have shared edges, they restrict the Cr–O stretching motion in this structure, and hence only the bending modes are prominent.

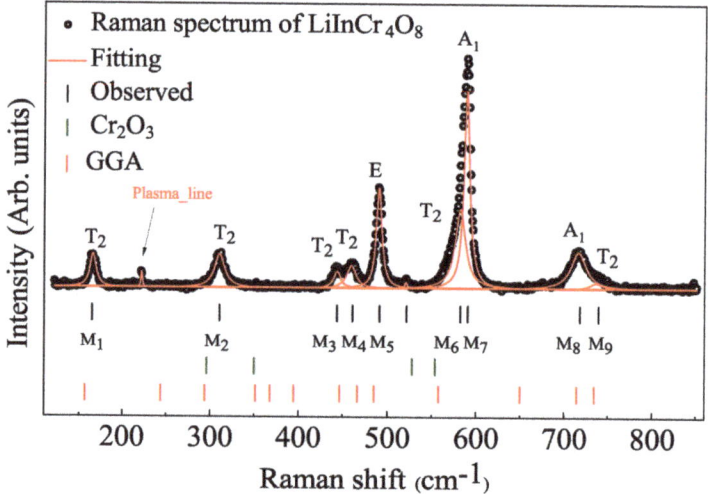

Figure 3. Ambient pressure Raman spectrum of $LiInCr_4O_8$, together with the fitting curve. Shown also are the theoretical mode frequencies (GGA) and the Raman mode frequencies for Cr_2O_3 [23,24].

Table 3. Mode assignment and comparison of experimental and theoretical mode frequencies (in cm^{-1}). (w), (m), and (s) denote the strength of active mode, viz., weak, medium and strong, respectively.

Modes Assigned	Raman			IR		
	Obs	LDA	GGA	Obs	LDA	GGA
T_2	165 (m)	160	157	-	160	157
E	-	245	244			
T_2	310 (m)	300	294	-	300	294
A_1	-	349	352			

Table 3. Cont.

Modes Assigned	Raman			IR		
	Obs	LDA	GGA	Obs	LDA	GGA
T_2	-	359	368	406	359	368
E	-	398	395			
T_2	444 (m)	471	447	473	471	447
T_2	461 (m)	475	467	529	475	467
E	491 (s)	503	485			
T_2	582 (m)	561	558	581	561	558
A_1	591 (s)	642	650			
A_1	718 (m)	731	715			
T_2	739 (w)	756	735	651	756	735

Table 4. Internal modes of CrO_6 octahedra.

Modes		Free CrO_6 Ion (O_h)		Site Symmetry (C_{3v})		Crystal Symmetry (T_d)
ν_1	→	A_{1g}	→	A_1	→	A_1
ν_2	→	E_g	→	E	→	E
ν_3	→	F_{1u}	→	A_1+E	→	T_2
ν_4	→	F_{1u}	→	A_1+E	→	T_2
ν_5	→	F_{2g}	→	A_1+E	→	T_2
ν_6	→	F_{2u}	→	A_2+E	→	T_1
ν_{rot}	→	F_{1g}	→	A_1+E	→	T_2
ν_{tran}	→	F_{1u}	→	A_1+E	→	T_2

Figure 4a depicts the ambient pressure reflectivity spectrum of $LiInCr_4O_8$ over a broad frequency range, together with the Lorentz fitting. The inset shows the low-frequency range up to 800 cm^{-1}, where the phonon modes are located. The corresponding optical conductivity σ_1 obtained from the KK transformation is depicted in Figure 4b. The σ_1 spectrum shows strong phonon contributions in the far-infrared range, which is followed by the onset of electronic excitations around ~0.1 eV, which gradually increases to two prominent absorption bands centred at ~1.65 and 2.4 eV. Based on previous studies, the absorption bands can be assigned to intra-atomic d-d excitations, i.e., crystal field (CF) excitations, of the Cr^{3+} ions in an octahedral environment. Electronic excitations from the $^4A_{2g}$ ground state to the $^4T_{2g}$ and $^4A_{1g}$ excited states are expected in the spectral ranges of ~13,000 to 17,000 cm^{-1} and ~13,000 to 17,000 cm^{-1}, respectively, and the spin-forbidden transitions from $^4A_{2g}$ to ^2E_g and $^2E_{2g}$ are expected in the spectral ranges of 13,000 to 14,400 cm^{-1} and 18,000 to 19,200 cm^{-1}, respectively, [28–36].

The inset in Figure 4b shows the observed ten phonon modes fitted with Lorentzian oscillators. However, the group theoretical analysis predicts only seven T_2 modes for LICO. Therefore, modes were assigned to the strongest modes after comparison with DFPT calculations to the closest calculated values (see Table 3). An earlier infrared spectroscopic study on $LiFeCr_4O_8$ has assigned the observed five phonon modes to the internal vibrations of the polyhedral units, viz., Li–O stretching modes in the range 400–500 cm^{-1} and CrO_6 and FeO_4 vibrations around 500 cm^{-1} and ~640 cm^{-1}, respectively [37]. In the present study, based on the DFPT calculations, the phonon modes above 600 cm^{-1} are assigned to the Li–O stretching vibrations. The low-energy modes are assigned to the In–O vibrations and those observed around ~500 cm^{-1} belong to the CrO_6 internal vibrations.

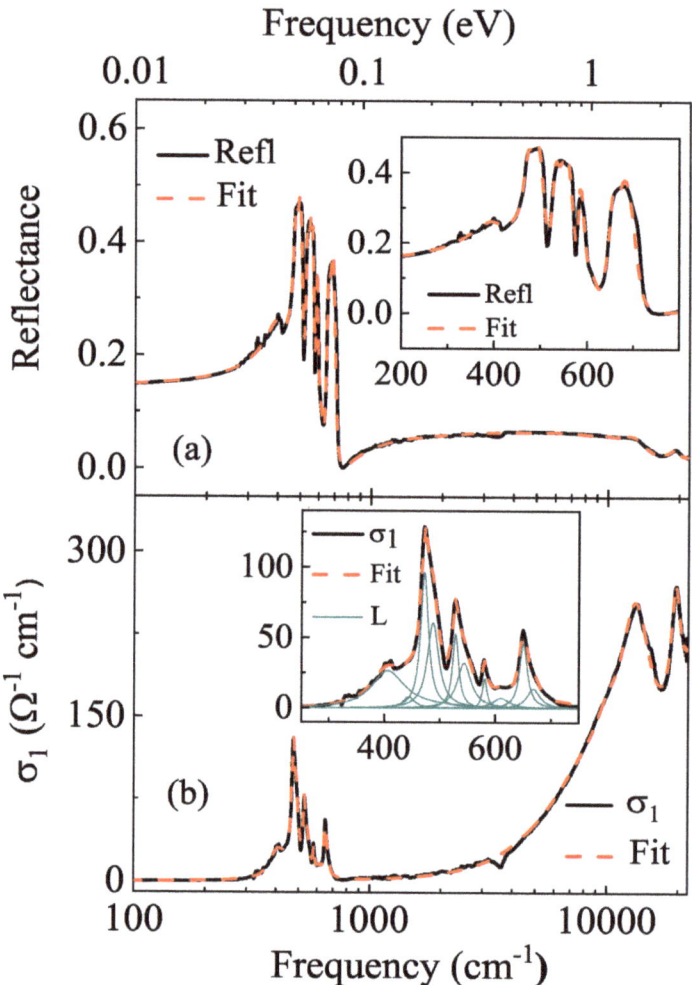

Figure 4. Ambient pressure (**a**) reflectance and (**b**) optical conductivity spectrum of LiInCr$_4$O$_8$ in the range 100–22,000 cm^{-1}, together with the Lorentz fitting. The insets of (**a**,**b**) depict the corresponding low-frequency range (200–800 cm^{-1}) of the reflectance and optical conductivity spectrum, respectively, where the phonon modes are located, together with the Lorentz fitting.

3.2. High-Pressure Results

3.2.1. X-ray Diffraction Measurements

A few XRD patterns at selected pressures are presented in Figure 5. According to these results, the sample remains in the ambient phase up to ∼14 GPa. Above 10 GPa, i.e., beyond the hydrostatic pressure limits of the PTM used, the peaks at higher values of 2θ become weaker and undergo a pressure-induced broadening, thereby making it difficult to trace their behaviour precisely with pressure. However, the peaks at lower angles provide clear signatures of phase transitions. Above 10 GPa, the (200) and (220) reflections show a broadening. Beyond ∼14 GPa, they undergo a splitting, which is an indication of a first-order structural phase transition to a high-pressure phase (HP-1). The clear splitting of these off-diagonal planes into two peaks while the diagonal planes (111) remain intact is an indication of a cubic-to-tetragonal structural phase transition. Previous temperature-dependent XRD studies on LiMCr$_4$O$_8$ (M = In, Ga, Fe) compounds have reported a cubic-to-tetragonal phase transition at very low temperatures. Based on earlier reported studies, the HP-1 phase was indexed to a tetragonal structure (space group I$\bar{4}$m2) with two formula units per unit cell (Z = 2). The Rietveld refinement of our XRD data collected at ~14.3 GPa indicated the coexistence of the ambient and HP-1 phases (see Figure 6). The refined lattice parameters of the HP-1 phase at ~14 GPa are a = 5.9030(26) Å, c = 8.0681(60) Å, and V = 281.13(22) Å^3, with R_p = 0.038 and R_{wp} = 0.058. The structural information from the Rietveld analysis for the coexisting ambient and the HP-1 phases are given in Table 5 and Table 6, respectively. This information is also shown in Figure 8. For both the ambient and HP-1 phases, the atomic positions and their occupations were not refined and are used as reported in earlier studies [5,8]. The structural refinement of the diffraction patterns above 14 GPa could not be performed; hence, only one data point of the HP-1 phase is shown in Figure 8. Detailed information regarding the coordination of polyhedral units cannot be extracted from this powder diffraction as the oxygen positions remain unrefined to maintain the reliability of the Rietveld refinement. The volume per formula unit in the ambient pressure phase is 148.38 Å^3 and that of the HP-1 phase is 140.57 Å^3, which gives a compression of 5.3% per formula unit.

Table 5. The Wyckoff sites, fractional coordinates, and occupations as reported in an earlier study [5] and the thermal parameter U as obtained from Rietveld refinement of LiInCr$_4$O$_8$ in the ambient structure at ~14 GPa. Space group F$\bar{4}$3m, Z = 4, a = 8.2624(13) Å and V = 564.05(26) Å^3.

Atom	Wyckoff	x	y	z	Occupancy	U
Li1	4a	0.0000	0.0000	0.0000	0.9920	0.0610
In1	4a	0.0000	0.0000	0.0000	0.0080	0.0900
Li2	4d	0.7500	0.7500	0.7500	0.0080	0.8000
In2	4d	0.7500	0.7500	0.7500	0.9920	0.0636
Cr	16e	0.3719	0.3719	0.3719	1.0000	0.0064
O1	16e	0.1377	0.1377	0.1377	1.0000	0.0301
O2	16e	0.6107	0.6107	0.6107	1.0000	0.0671

Table 6. The Wyckoff sites, fractional coordinates, and occupations as reported in an earlier study [8] and the thermal parameter U as obtained from Rietveld refinement of LiInCr$_4$O$_8$ in the HP-1 phase at ~14 GPa. Space group I$\bar{4}$m2, Z = 2, a = 5.9030(26) Å, c = 8.0681(60) Å, and V = 281.13(22) Å^3.

Atoms	Wyckoff Sites	x	y	z	Occupancy	U
Li	2a	0.0000	0.0000	0.0000	1.0000	0.0224
In	2d	0.0000	0.5000	0.7500	1.0000	0.0875
Cr	8i	0.2607	0.0000	0.6272	1.0000	0.0067
O1	8i	0.2840	0.0000	0.6340	1.0000	0.0029
O2	8i	0.2510	0.0000	0.1080	1.0000	0.0069

The low-temperature XRD studies on LiInCr$_4$O$_8$ reported the tetragonal phase coexisting with the ambient cubic phase down to the lowest recorded temperature of ~2 K [7,8]. Consistently, our high-pressure studies also show that the sample does not completely transform to the tetragonal phase with further compression. Instead, before this cubic to tetragonal transition is completed, the reflections from the (111) set of planes undergo a broadening at the next recorded pressure (~16 GPa) and then split into two peaks at higher pressures. The XRD pattern recorded at ~19.7 GPa shows clear changes, indicating the existence of a new high-pressure phase (HP-2 phase). The transition from the HP-1 to HP-2 phase is a slow and sluggish first-order structural phase transition where the HP-1 coexists with the evolving HP-2 phase. This new high-pressure phase (HP-2 phase) remains stable up to ~35 GPa. With further compression, the diffraction pattern recorded above ~35 GPa shows the emergence of some new peaks, suggesting a possible transition to a lower-symmetry structure (HP-3 phase). The transition to the HP-3 phase was not completed at the highest pressure recorded in this experiment (i.e., ~36.7 GPa). Due to the broadness of the diffraction peaks, overlapping with the strong reflections from the pressure marker (Au) and the gasket (W) used in these experiments, the high-pressure phases could not be identified. During decompression, the HP-1 and HP-2 phases coexist down up to 12 GPa. Below 8 GPa, the sample slowly transforms to the parent phase and reverts to the ambient pressure crystal structure on complete release of pressure.

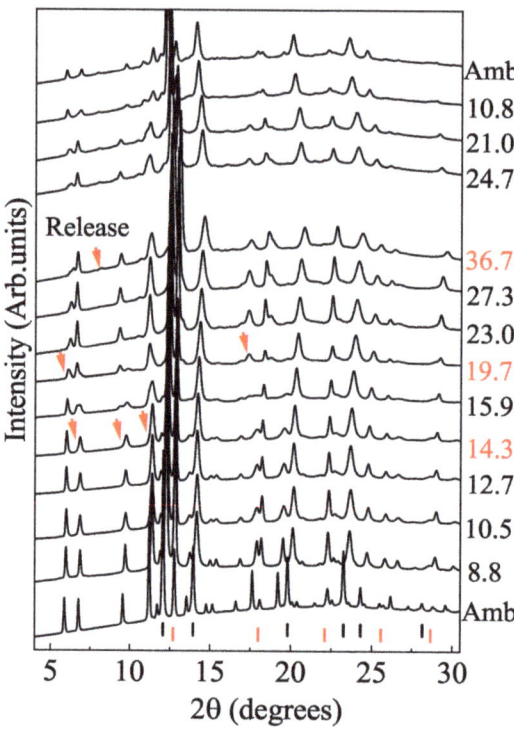

Figure 5. Pressure-dependent XRD patterns at few selected pressures. The pressure values on the right are in GPa. The red arrows indicate the emerging new reflections. The red and black ticks at the bottom are from Au (pressure marker) and W (gasket), respectively.

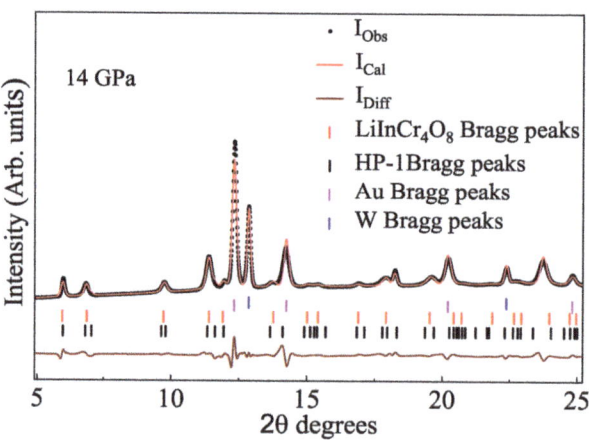

Figure 6. Rietveld refinement of the XRD pattern at ~14 GPa, demonstrating the coexistence of the HP-1 (space group $I\bar{4}m2$) phase with the ambient pressure cubic phase.

It is well known that the diffraction peak width has its origins from instruments, particle size, and stress-induced broadening. In this study, the XRD pattern from standard CeO_2 inside the DAC recorded at ambient conditions was used to characterize the instrumental broadening. Changes in the peak width under compression observed in the present study can be attributed to inhomogeneous strain and change in crystallinity due to phase transformation. For a better understanding of the deformation at high pressures, the behaviour of the diffraction peaks and X-ray peak broadening were studied by plotting the variation of the peak position and the full-width-at-half-maximum (FWHM) value for the reflections from the (111), (200), and (220) planes as a function of pressure in Figure 7a,b, respectively. Structural phase transitions to the HP-1, HP-2, and HP-3 phases at ~14 GPa, ~19 GPa, ~35 GPa, respectively, can be clearly observed from Figure 7b. The FWHM value of all peaks remains almost constant up to ~10 GPa. Above ~10 GPa, all the peaks show a slight discontinuity and broadening, which can be attributed to the non-hydrostatic stresses [38]. However, the (hk0) and (h00) peaks show a significant change in the FWHM as compared to the (111) set of planes. This sudden change is observed in FWHM just before the structural transition at ~14 GPa. Immediately after the phase transition, i.e., above ~14 GPa, both (220) and (200) undergo a clear splitting due to the transition from the cubic to tetragonal phase. Above ~16 GPa, the (111) reflection also shows a sudden change in FWHM, which is an indication of beginning of another structural transition. With further compression, the (111) reflection exhibits a clear splitting (see Figure 7a).

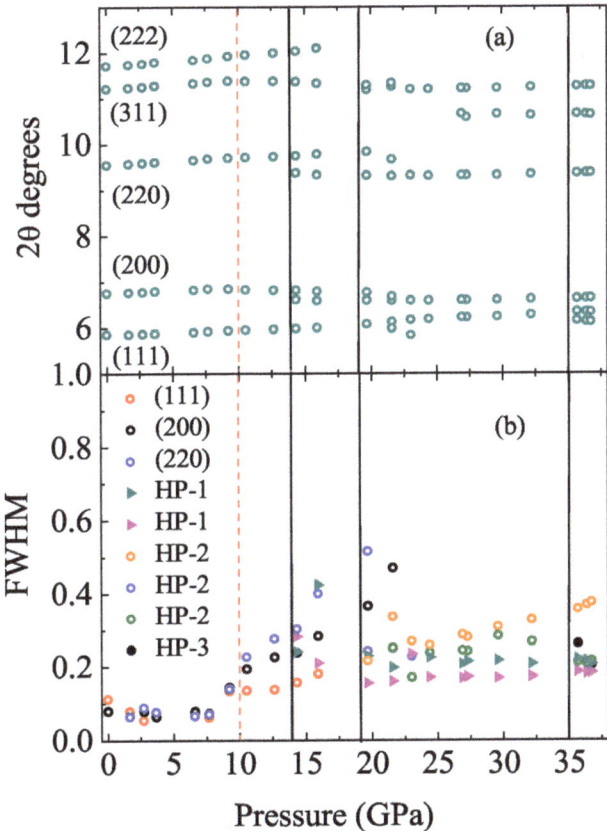

Figure 7. Behaviour of the (**a**) Bragg peaks and (**b**) FWHM with pressure. The vertical red dashed line indicates the hydrostatic limit of the PTM used. Solid black lines mark the transition pressures.

To determine the structural stability of LiInCr$_4$O$_8$, the lattice parameter a and the volume per unit cell V were determined for each recorded pressure from the structural refinements. Both parameters are plotted in Figure 8, where the abrupt decrease in compressibility is notable above ~10 GPa due to nonhydrostatic conditions [39,40]. The P–V data up to ~10 GPa (hydrostatic limit of PTM used) was fitted with a Murnaghan equation of state according to $V(p) = V_0 \cdot [(B'_0/B_0) \cdot p + 1]^{-1/B'_0}$ [41], where the first-order derivative of the bulk modulus B'_0 was fixed to 4 (see Figure 8). The bulk modulus B_0 of the ambient phase was found to be 186.7 ± 6.1 GPa. From the DFT calculations, the ambient pressure crystal structure was relaxed at different target pressures. The energy per formula unit and the corresponding volumes were fitted with the Murnaghan equation of state to compare with the experimental findings. The calculated B_0 and B'_0 amount to 176.8 ± 0.1 GPa and 4.63 ± 0.02, respectively. The difference in the experimental and calculated ambient volume and the non-magnetic configuration of the calculated system could explain the discrepancy between the experimental and theoretical values of the bulk modulus.

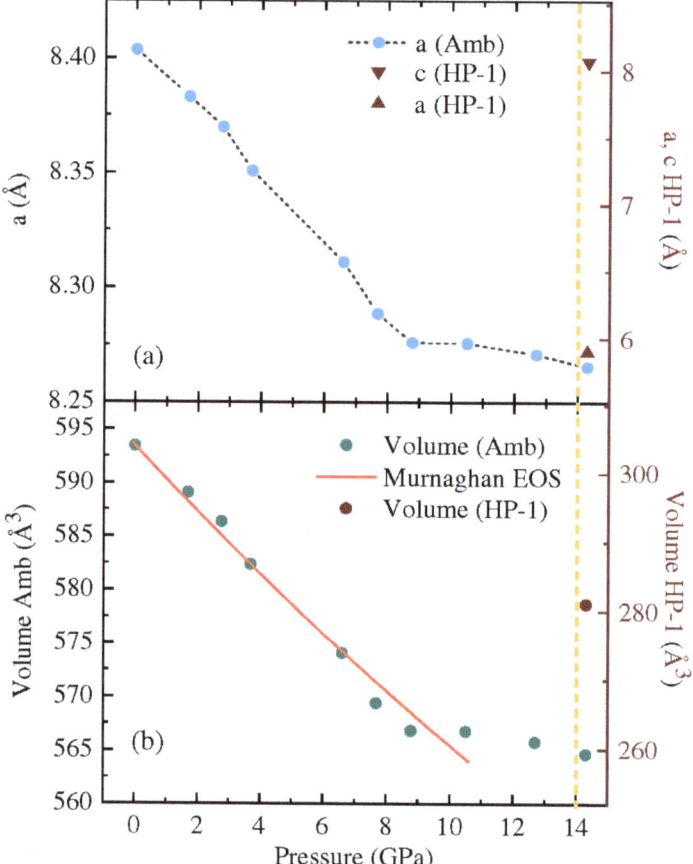

Figure 8. (**a**) Experimental lattice parameters a and c and (**b**) volume per unit cell V of the ambient phase and high-pressure HP-1 phase (scale on the right side of the graph) as a function of pressure. The pressure dependence of the volume is fitted with a Murnaghan equation of state (EOS), as defined in the text.

LiInCr$_4$O$_8$ with two different Cr–Cr distances (d = 2.90 Å and d′ = 3.05 Å) has values that lie in the range of conventional spinels such as ZnCr$_2$O$_4$ with a uniform Cr–Cr distance of 2.944 Å, and that of $CdCr_2O_4$ is 3.041 Å. A theoretical study on $MgCr_2O_4$, $MnCr_2O_4$, and $ZnCr_2O_4$ reports their bulk moduli to be 197.3 GPa, 205.8 GPa, and 215 GPa, respectively, [42], whereas experimentally, $MgCr_2O_4$ is reported to have a bulk modulus of 189 GPa [43,44] and that of $ZnCr_2O_4$ is 183.1 GPa [45]. $LiCrO_2$ is reported to have a bulk modulus of 161 GPa [46]. The experimentally obtained value of B_0 for $LiInCr_4O_8$ from the present study is comparable with that of other chromium spinels.

3.2.2. Raman Spectroscopy Measurements

To investigate the pressure-induced changes in the vibrational properties of $LiInCr_4O_8$, high-pressure Raman spectra were recorded in the spectral range ∼120–850 cm^{-1}. Figure 9 depicts Raman spectra at selected pressures. At the lowest recorded pressure of ∼0.4 GPa, the spectrum shows a splitting of a few Raman modes in the range of 450–600 cm^{-1}. Because the high-pressure XRD measurements do not provide any indication of a structural transition at such low pressures, the splitting of the Raman modes can be attributed to a loss of accidental degeneracy. All observed Raman modes shift to higher frequencies under pressure due to the pressure-induced stiffening of the lattice. The intensity of the Raman mode observed at ∼164 cm^{-1} drops drastically with compression, and this mode vanishes above ∼2.5 GPa. Mode M_2 at ∼310 cm^{-1} also undergoes a significant decrease in the intensity with pressure. Both these modes are associated with the translational motion of the A-site cations. The modes M_4 and M_6, which are related to the O–Li–O bending and O–In stretching vibrations, respectively, undergo a splitting at ∼ 0.4 GPa. The modes M_3 and M_4 combine into a single mode at ∼9 GPa. At ∼2.5 GPa, a further splitting of mode M_6 is observed, and this new mode can be traced up to ∼10 GPa. The splitting of modes associated with internal vibrations of polyhedral units can be attributed to the distortion under pressure. No significant changes in the Raman spectrum are observed with further compression up to ∼14 GPa. Above 14 GPa, the splitting of the mode at ∼545 cm^{-1} can be associated with the structural phase transition as observed from XRD measurements. The most intense mode (M_7), associated with the O–Cr–O bending vibrations, remains the most intense peak up to the highest recorded pressure of ∼ 18 GPa.

The frequency shift of the Raman modes with pressure is summarized in Figure 10, where the emerging new modes can be clearly seen. During pressure release (see Figure 9), the sample slowly reverts to its ambient pressure phase, where the modes at ∼310 cm^{-1} reappear at around 7 GPa, and the mode at ∼164 cm^{-1} can be observed at ∼3 GPa. According to the decompressed spectrum at ambient conditions, the sample has transformed to its ambient phase, consistent with our pressure-dependent XRD results.

The calculated pressure coefficients of Raman modes and the corresponding Grüneisen parameters ($\gamma_i = (B_0/\omega_i)(d\omega_i/dP)$) are tabulated in Table 7. B_0 is the experimentally obtained bulk modulus which was used to calculate the mode Grüneisen parameters for the ambient phase. The modes M_8 and M_9 show relatively larger value of the pressure coefficient, indicating an increase in the force constant under pressure. The tabulated mode Grüneisen parameter relates the vibrational properties to the crystal deformations. The macroscopic Grüneisen parameter γ is a weighted sum of individual γ_i's. The relation between the two is given as [$\gamma = (\sum_i \gamma_i C_i)/(\sum_i C_i)$], where C_i's are the mode contributions to the material's specific heat [47]. Using Einstein's specific heat relation, $C_i = R[x_i^2 exp(x_i)]/[exp(x_i) - 1]^2$ (where $x_i = \hbar\omega_i/k_B T$ and R is the universal gas constant) [48], the macroscopic Grüneisen parameter was calculated using the observed modes from M_1 to M_9 at ambient pressure and its value was found to be $\gamma = 0.99$. A negative contribution to thermal expansion can be associated with a negative mode Grüneisen parameter. In the ambient phase, all the modes are notably contributing to a positive thermal expansion, consistent with recent studies [27,49] stating a positive expansion observed in $LiInCr_4O_8$, whereas other isostructural compounds $LiGaCr_4O_8$ and $LiInCr_4S_8$ are reported to exhibit negative thermal expansion. A softening of the M_{1-HP} mode is, however, observed in the high-pressure phase. All the other modes observed above 14 GPa still show a positive shift with compression.

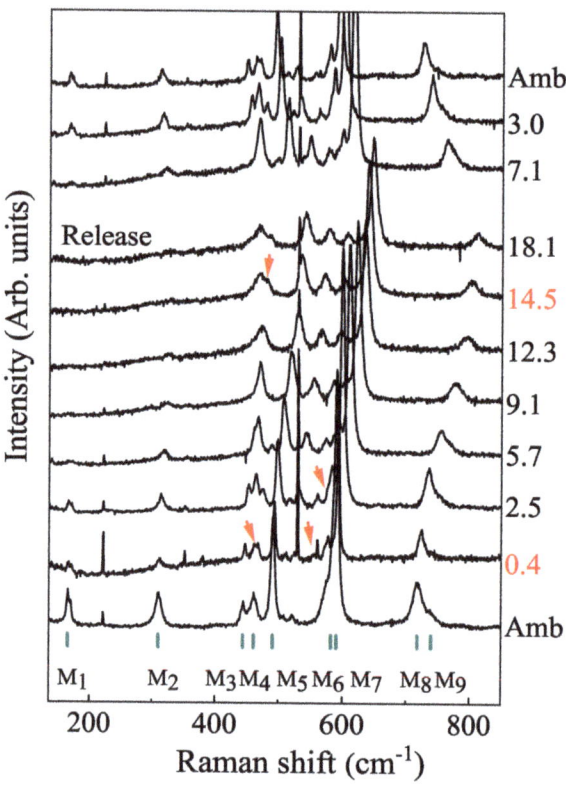

Figure 9. Raman spectra of LiInCr$_4$O$_8$ at selected pressures. The numbers on the right are in GPa scale. Red arrows indicate the emergence of new Raman modes.

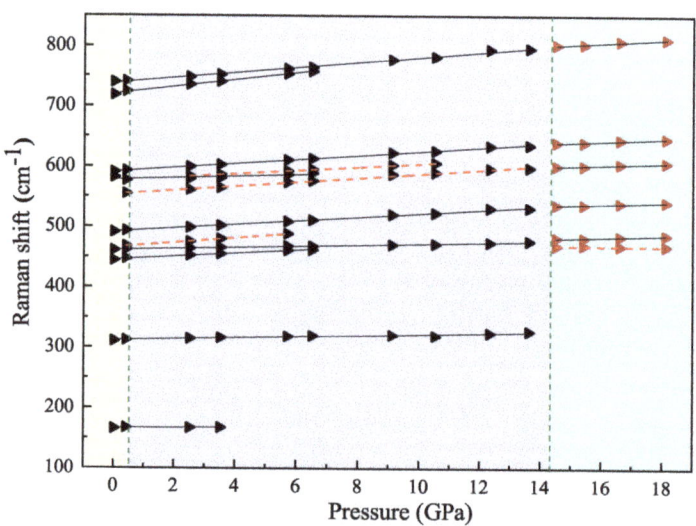

Figure 10. Frequency shift of Raman modes with increasing pressure. The black and red symbols represent the Raman modes of the ambient phase and those observed above 14 GPa, respectively. Different shadings illustrate the changes in the Raman spectrum. The red dashed lines highlight new modes emerging under pressure.

Table 7. Pressure at which the modes are observed, frequency ω_i of observed Raman modes, pressure coefficients $d\omega_i/dP$ of Raman modes, and mode Grüneisen parameter γ_i (see text).

Observed Pressure GPa	Modes	ω_i (cm^{-1})	$(d\omega_i/dP)$ (cm^{-1}GPa^{-1})	γ_i
Amb	M_1	165	0.21	0.23
"	M_2	310	0.90	0.51
"	M_3	444	2.67	1.06
"	M_4	461	0.04	0.40
"	M_5	491	3.02	1.09
"	M_6	582	1.36	0.41
"	M_7	591	3.35	1.00
"	M_8	718	5.82	1.43
"	M_9	739	4.28	1.02
0.4	M_{10}	468	3.92	1.48
0.4	M_{11}	555	3.45	1.10
2.5	M_{12}	584	2.81	0.85
14.5	M_{1-HP}	469	−0.25	−0.09
"	M_{2-HP}	482	1.47	0.54
"	M_{3-HP}	536	1.62	0.53
"	M_{4-HP}	601	1.59	0.47
"	M_{5-HP}	640	2.09	0.58
"	M_{6-HP}	802	2.79	0.62

3.2.3. Infrared Spectroscopy Measurements

The pressure dependence of the absorbance spectrum of LiInCr$_4$O$_8$ is depicted in Figure 11a for pressures up to ~17.4 GPa. The features observed close to 2000 cm^{-1} are due to the multi-phonon absorptions in diamond. In the recorded range, we observed four phonon modes at low frequencies followed by the onset of electronic transitions at around 1000 cm^{-1}. In comparison to the ambient pressure optical conductivity (Figure 4), the onset of electronic transitions is spread out in the absorption spectrum, and the crystal field excitation appears as a broad peak in the spectrum around 16,800 cm^{-1} (2.05 eV), corresponding to the spin-allowed intra-atomic d-d transition between the ground state $^4A_{2g}$ and the excited $^4T_{1g}$ and $^4T_{2g}$ states [28–32]. The pressure dependence of this onset is extracted by extrapolation of a linear fit for each pressure value up to 10 GPa (Figure 12b). Beyond this pressure, the change in the spectral slope leads to unrealistic values for linear extrapolation. The onset of electronic transitions shifts to higher energies with increasing pressure indicating an increase in the band gap under compression, in agreement with the pressure-induced blue-shift of the crystal field excitation (Figure 12b). The observed crystal field excitations were fitted to a Lorentzian profile to obtain the corresponding energy position. This result is consistent with an earlier reported infrared absorption study on CdCr$_2$O$_4$, where a similar behaviour of the crystal field excitation under pressure was observed [32]. Weak features slowly emerging close to ~14,500 cm^{-1} at 4.6 GPa (marked by black arrows in Figure 11a) can be ascribed to the spin-forbidden crystal field transitions that may become infrared-active as a result of lattice vibrations which locally break the centre of symmetry. The absorption spectrum at the highest pressure of ~17.4 GPa shows a changed crystal field excitation profile, which we relate to the pressure-induced structural phase transition observed at ~19 GPa in our XRD measurements. The slightly lower critical pressure extracted from our infrared studies can be explained by the solid PTM used, which is less hydrostatic than the PTM (alcohol mixture) of the XRD experiment [50,51].

The pressure-dependent frequencies of the infrared-active phonon modes are presented in Figure 12a. In this spectral region, the observed modes belong to the internal bending vibrations of the CrO$_6$ octahedral unit and Li–O stretching vibrations. The observed infrared-active modes show a blueshift under pressure, consistent with the pressure behaviour of the Raman modes. The phonon modes P_1 and P_2 merge at 14 GPa, whereas the mode P_4 shows a deviation from a linear fit at the same pressure. At the same pressure, one observes a sudden change in the slope of the pressure-induced shift of the crystal field excitation (Figure 12b). All these changes can be associated with the structural phase transition observed from XRD measurements (discussed earlier). In Figure 12b, a sudden jump in the onset energy is also observed at 7.4 GPa; this, however, cannot be associated with any structural phase transition. The pressure coefficients ($d\omega/dP$) of the observed (P_1–P_4) modes are 5.1, 5.24, 4.28, and 3.69 cm^{-1}GPa^{-1}, respectively. These values are comparable and close to the pressure coefficients of the high-energy Raman modes (M_5, M_7, M_8, and M_9) (see Table 7).

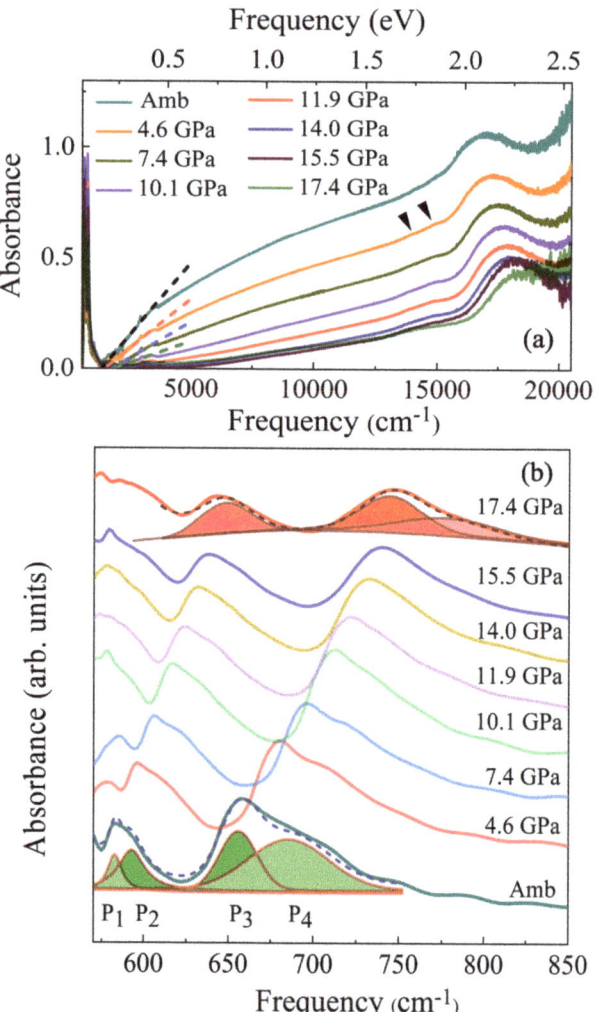

Figure 11. (a) Pressure-dependent infrared absorbance spectra in a broad frequency range 500–20,500 cm^{-1}. The two weak features indicated by black arrows are due to spin-forbidden crystal field excitations. (b) Low-frequency absorbance spectra with phonon modes (labelled P1–P4) as a function of pressure, fitted with Lorentz oscillators.

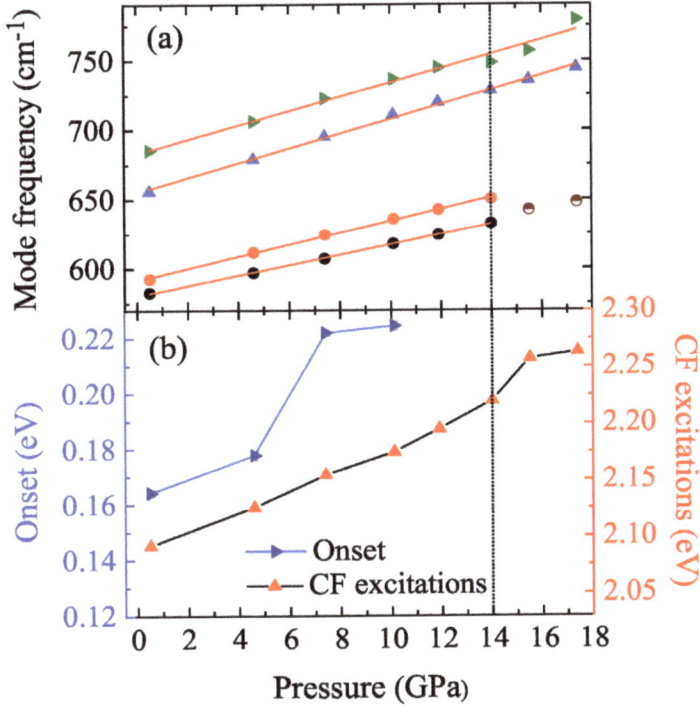

Figure 12. (a) Behaviour of infrared-active phonon mode frequencies with pressure. (b) Onset of electronic transitions and energy of crystal field (CF) excitation as a function of pressure. The vertical line at ∼14 GPa indicates the structural phase transition.

4. Conclusions

In summary, we report multiple pressure-induced phase transitions in the chromium spinel $LiInCr_4O_8$. From the high-pressure XRD measurements, the HP-1 phase with tetragonal structure appearing at ∼14 GPa seems to be an intermediate metastable phase. The system does not completely transform to this structure and always coexists with the ambient pressure cubic phase. The high-pressure infrared and Raman spectroscopy measurements confirm the structural phase transition at ∼14 GPa. All the observed active modes show a pressure-induced blueshift indicative of the stiffening of the lattice. The estimated onset of the electronic transitions from the high-pressure infrared measurements indicates an increase in the band gap under compression. Although the breathing pyrochlore structure of $LiInCr_4O_8$ is known to be a geometrically frustrated system, it appears to be highly resilient when the system reverts to its ambient phase even after undergoing multiple pressure-induced structural changes.

Author Contributions: Investigation and formal analysis, M.V. and M.K.; Visualization, data curation and writing—original draft, M.V.; Writing—review, M.K.; Supervision, resources, writing—review and editing, H.K.P. and C.A.K. All authors have read and agreed to the published version of the manuscript.

Funding: This research received no external funding.

Institutional Review Board Statement: Not applicable.

Informed Consent Statement: Not applicable.

Data Availability Statement: Data is contained within the article.

Acknowledgments: We thank Matthias Schreck for providing technical support for the Raman measurements. We acknowledge Srihari Velaga, ECXRD beamline, Indus-2, RRCAT, India, and Boby Joseph, XPRESS beamline, Elettra Synchrotron, Italy, for their support in using the beamline facilities. We acknowledge the Indus-2, RRCAT, India, and Elettra Synchrotron, Italy, for the provision of beamtime. We also thank J. Ebad-Allah, F. Meggle, S. Rojewski, M. Köpf, and G. Eickerling for the help and fruitful discussions.

Conflicts of Interest: The authors declare no conflict of interest.

References

1. Zhao, Q.; Yan, Z.; Chen, C.; Chen, J. Spinels: Controlled Preparation, Oxygen Reduction/Evolution Reaction Application, and Beyond. *Chem. Rev.* **2017**, *117*, 10121–10211. [CrossRef] [PubMed]
2. Stefan, E.; Irvine, J.T.S. Synthesis and characterization of chromium spinels as potential electrode support materials for intermediate temperature solid oxide fuel cells. *J. Mater. Sci.* **2011**, *46*, 7191–7197. [CrossRef]
3. Kim, D.C.; Ihm, S.K. Application of Spinel-Type Cobalt Chromite as a Novel Catalyst for Combustion of Chlorinated Organic Pollutants. *Environ. Sci. Technol.* **2000**, *35*, 222–226. [CrossRef] [PubMed]
4. Joubert, J.C.; Durif, A. Étude de quelques composés spinelles nouveaux possédant un ordre des cations du type 1/1 sur les sites tétraédriques. *Bull. Soc. Française Min. Cristallogr.* **1966**, *89*, 26. [CrossRef]
5. Okamoto, Y.; Nilsen, G.J.; Attfield, J.P.; Hiroi, Z. Breathing Pyrochlore Lattice Realized in A-Site Ordered Spinel Oxides LiGaCr$_4$O$_8$ and LiInCr$_4$O$_8$. *Phys. Rev. Lett.* **2013**, *110*, 097203. [CrossRef] [PubMed]
6. Tanaka, Y.; Yoshida, M.; Takigawa, M.; Okamoto, Y.; Hiroi, Z. Novel Phase Transitions in the Breathing Pyrochlore Lattice: Li-NMR7 on LiInCr$_4$O$_8$ and LiGaCr$_4$O$_8$. *Phys. Rev. Lett.* **2014**, *113*, 227204. [CrossRef]
7. Nilsen, G.J.; Okamoto, Y.; Masuda, T.; Carvajal, J.R.; Mutka, H.; Hansen, T.; Hiroi, Z. Complex magnetostructural order in the frustrated spinel LiInCr$_4$O$_8$. *Phys. Rev. B* **2015**, *91*, 174435. [CrossRef]
8. Saha, R.; Fauth, F.; Avdeev, M.; Kayser, P.; Kennedy, B.J.; Sundaresan, A. Magnetodielectric effects in A-site cation-ordered chromate spinels LiMCr$_4$O$_8$ (M=Ga and In). *Phys. Rev. B* **2016**, *94*, 064420. [CrossRef]
9. Saha, R.; Dhanya, R.; Bellin, C.; Béneut, K.; Bhattacharyya, A.; Shukla, A.; Narayana, C.; Suard, E.; Carvajal, J.R.; Sundaresan, A. Magnetostructural coupling and magnetodielectric effects in the A -site cation-ordered spinel LiFeCr$_4$O$_8$. *Phys. Rev. B* **2017**, *96*, 214439. [CrossRef]
10. Dewaele, A.; Loubeyre, P.; Mezouar, M. Equations of state of six metals above 94 GPa. *Phys. Rev. B* **2004**, *70*, 094112. [CrossRef]
11. Takemura, K.; Dewaele, A. Isothermal equation of state for gold with a He-pressure medium. *Phys. Rev. B* **2008**, *78*, 104119. [CrossRef]
12. Klotz, S.; Chervin, J.C.; Munsch, P.; Marchand, G.L. Hydrostatic limits of 11 pressure transmitting media. *J. Phys. D Appl. Phys* **2009**, *42*, 075413. [CrossRef]
13. Hammersley, A.P.; Svensson, S.O.; Hanfland, M.; Fitch, A.N.; Hausermann, D. Two-dimensional detector software: From real detector to idealised image or two-theta scan. *High Press. Res.* **1996**, *14*, 235. [CrossRef]
14. Toby, B.H. EXPGUI, a graphical user interface for GSAS. *J. Appl. Crystallogr.* **2001**, *34*, 210. [CrossRef]
15. Mao, H.K.; Xu, J.; Bell, P.M. Calibration of the ruby pressure gauge to 800 kbar under quasi-hydrostatic conditions. *J. Geophys. Res.* **1986**, *91*, 4673. [CrossRef]
16. Celeste, A.; Borondics, F.; Capitani, F. Hydrostaticity of pressure-transmitting media for high pressure infrared spectroscopy. *High Press. Res.* **2019**, *39*, 608. [CrossRef]
17. Blöchl, P.E. Projector augmented-wave method. *Phys. Rev. B* **1994**, *50*, 17953. [CrossRef]
18. Giannozzi, P.; Andreussi, O.; Brumme, T.; Bunau, O.; Nardelli, M.B.; Calandra, M.; Car, R.; Cavazzoni, C.; Ceresoli, D.; Cococcioni, M.; et al. Advanced capabilities for materials modelling with Quantum ESPRESSO. *J. Condens. Matter Phys.* **2017**, *29*, 465901. [CrossRef]
19. Perdew, J.P.; Burke, K.; Ernzerhof, M. Generalized Gradient Approximation Made Simple. *Phys. Rev. Lett.* **1996**, *77*, 3865. [CrossRef]
20. Monkhorst, H.J.; Pack, J.D. Special points for Brillouin-zone integrations. *Phys. Rev. B* **1976**, *13*, 5188. [CrossRef]
21. Lazzeri, M.; Mauri, F. First-Principles Calculation of Vibrational Raman Spectra in Large Systems: Signature of Small Rings in CrystallineSiO$_2$. *Phys. Rev. Lett.* **2003**, *90*, 036401. [CrossRef] [PubMed]
22. Troullier, N.; Martins, J.L. Efficient pseudopotentials for plane-wave calculations. *Phys. Rev. B* **1991**, *43*, 1993. [CrossRef] [PubMed]
23. Mougin, J.; LeBihan, T.; Lucazeau, G. High-pressure study of Cr2O3 obtained by high-temperature oxidation by X-ray diffraction and Raman spectroscopy. *J. Phys. Chem. Solids* **2001**, *62*, 553. [CrossRef]
24. Todorov, N.D.; Abrashev, M.V.; Russev, S.C.; Marinova, V.; Nikolova, R.P.; Shivachev, B.L. Raman spectroscopy and lattice-dynamical calculations of Sc$_3$CrO$_6$ single crystals. *Phys. Rev. B* **2012**, *85*, 214301. [CrossRef]
25. Feng, Y.; Liu, H.; Bian, J.; Xiong, W.; Zhu, S.; Zong, B.; Shi, B.; Fang, B. Structural and Magnetic Properties of the Breathing Pyrochlore LiInCr$_{4-x}$Fe$_x$O$_8$. *Phys. Status Solidi B* **2020**, *257*, 1900685. [CrossRef]

26. Schaftenaar, G.; Noordik, J.H. Molden: A pre- and post-processing program for molecular and electronic structures. *J. Comput. Aided Mol. Des.* **2000**, *14*, 123. [CrossRef] [PubMed]
27. Pokharel, G.; May, A.F.; Parker, D.S.; Calder, S.; Ehlers, G.; Huq, A.; Kimber, S.A.J.; Arachchige, H.S.; Poudel, L.; McGuire, M.A.; et al. Negative thermal expansion and magnetoelastic coupling in the breathing pyrochlore lattice material LiGaCr$_4$S$_8$. *Phys. Rev. B* **2018**, *97*, 134117. [CrossRef]
28. Ohgushi, K.; Okimoto, Y.; Ogasawara, T.; Miyasaka, S.; Tokura, Y. Magnetic, Optical, and Magnetooptical Properties of Spinel-Type ACr$_2$X$_4$ (A = Mn, Fe, Co, Cu, Zn, Cd : X = O, S, Se). *J. Phys. Soc. Japan* **2008**, *77*, 034713. [CrossRef]
29. Brik, M.; Avram, N.M.; Avram, C.N. Crystal field analysis of energy level structure of the Cr$_2$O$_3$ antiferromagnet. *Solid State Commun.* **2004**, *132*, 831. [CrossRef]
30. Brik, M. Crystal Field Analysis, Electron-Phonon Coupling and Spectral Band Shape Modeling in MgO:Cr$_3$. *Z. Naturforschung A* **2005**, *60*, 437. [CrossRef]
31. Larsen, P.K.; Wittekoek, S. Photoconductivity and Luminescence Caused by Band-Band and by Cr$_3$ Crystal Field Absorptions in CdCr$_2$S$_4$. *Phys. Rev. Lett.* **1972**, *29*, 1597. [CrossRef]
32. Rabia, K.; Baldassarre, L.; Deisenhofer, J.; Tsurkan, V.; Kuntscher, C.A. Evolution of the optical properties of chromium spinels CdCr$_2$O$_4$, HgCr$_2$S$_4$ and ZnCr$_2$Se$_4$ under high pressure. *Phys. Rev. B* **2014**, *89*, 125107. [CrossRef]
33. Schmidt, M.; Wang, Z.; Kant, C.; Mayr, F.; Toth, S.; Islam, A.T.M.N.; Lake, B.; Tsurkan, V.; Loidl, A.; Deisenhofer, J. Exciton-magnon transitions in the frustrated chromium antiferromagnets CuCrO$_2$, α-CaCr$_2$O$_4$, CdCr$_2$O$_4$, and ZnCr$_2$O$_4$. *Phys. Rev. B* **2013**, *87*, 224424. [CrossRef]
34. Rudolf, T.; Kant, C.; Mayr, F.; Schmidt, M.; Tsurkan, V.; Deisenhofer, J.; Loidl, A. Optical properties of ZnCr$_2$Se$_4$. *Eur. Phys. J. B* **2009**, *68*, 153. [CrossRef]
35. Figgis, B.N.; Hitchman, M.A. *Ligand Field Theory*; John Wiley & Sons: Hoboken, NJ, USA, 1999.
36. Jórgensen, C.K. Spectra and electronic structure of complexes with sulphur-containing ligands. *Inorg. Chim. Acta. Rev.* **1968**, *2*, 65. [CrossRef]
37. Tarte, P. Effet isotopique ^{6}Li-^{7}Li dans le spectre infra-rouge de composes inorganiques du lithium-I. Carbonate, chromo-ferrite, tungstate, molybdate et nitrate de lithium. *Spectrochim. Acta* **1965**, *21*, 313. [CrossRef]
38. Garg, A.B.; Errandonea, D.; Rodríguez-Hernández, P.; Muñoz, A. ScVO$_4$ under non-hydrostatic compression: A new metastable polymorph. *J. Phys. Condens. Matter* **2016**, *29*, 055401. [CrossRef] [PubMed]
39. Garg, A.B.; Errandonea, D.; Rodríguez-Hernández, P.; López-Moreno, S.; Muñoz, A.; Popescu, C. High-pressure structural behaviour of HoVO$_4$: Combined XRD experiments and *abinitio* calculations. *J. Phys. Condens. Matter* **2014**, *26*, 265402. [CrossRef] [PubMed]
40. Errandonea, D.; Muñoz, A.; Gonzalez-Platas, J. Comment on "High-pressure x-ray diffraction study of YBO$_3$/Eu^{3+}, GdBO$_3$ and EuBO$_3$" [J. Appl. Phys. 115, 043507 (2014)]. *J. Appl. Phys.* **2014**, *115*, 216101. [CrossRef]
41. Murnaghan, F.D. The Compressibility of Media under Extreme Pressures. *Proc. Natl. Acad. Sci. USA* **1944**, *30*, 244. [CrossRef]
42. Catti, M.; Fava, F.F.; Zicovich, C.; Dovesi, R. High-pressure decomposition of MCr$_2$O$_4$ spinels (M = Mg, Mn, Zn) by ab initio methods. *Phys. Chem. Miner.* **1999**, *26*, 389. [CrossRef]
43. Yong, W.; Botis, S.; Shieh, S.R.; Shi, W.; Withers, A.C. Pressure-induced phase transition study of magnesiochromite (MgCr$_2$O$_4$) by Raman spectroscopy and X-ray diffraction. *Phys. Earth Planet. Inter.* **2012**, *196–197*, 75. [CrossRef]
44. Wang, Z.; O'Neill, H.; Lazor, P.; Saxena, S.K. High pressure Raman spectroscopic study of spinel MgCr$_2$O$_4$. *J. Phys. Chem. Solids* **2002**, *63*, 2057. [CrossRef]
45. Wang, Z.; Lazor, P.; Saxena, S.K.; Artioli, G. High-Pressure Raman Spectroscopic Study of Spinel (ZnCr$_2$O$_4$). *J. Solid State Chem.* **2002**, *165*, 165. [CrossRef]
46. Garg, A.; Errandonea, D.; Pellicer-Porres, J.; Martinez-Garcia, D.; Kesari, S.; Rao, R.; Popescu, C.; Bettinelli, M. LiCrO$_2$ Under Pressure: In-Situ Structural and Vibrational Studies. *Crystals* **2018**, *9*, 2. [CrossRef]
47. Liu, G.; Gao, Z.; Ren, J. Anisotropic thermal expansion and thermodynamic properties of monolayer β-Te. *Phys. Rev. B* **2019**, *99*, 195436. [CrossRef]
48. Kamali, K.; Ravindran, T.R.; Ravi, C.; Sorb, Y.; Subramanian, N.; Arora, A.K. Anharmonic phonons of NaZr$_2$(PO$_4$)$_3$ studied by Raman spectroscopy, first-principles calculations, and x-ray diffraction. *Phys. Rev. B* **2012**, *86*, 144301. [CrossRef]
49. Kanematsu, T.; Mori, M.; Okamoto, Y.; Yajima, T.; Takenaka, K. Thermal Expansion and Volume Magnetostriction in Breathing Pyrochlore Magnets LiACr$_4$X$_8$ (A = Ga, In, X = O, S). *J. Phys. Soc. Japan* **2020**, *89*, 073708. [CrossRef]
50. Liang, A.; Turnbull, R.; Bandiello, E.; Yousef, I.; Popescu, C.; Hebboul, Z.; Errandonea, D. High-Pressure Spectroscopy Study of Zn(IO$_3$)$_2$ Using Far-Infrared Synchrotron Radiation. *Crystals* **2020**, *11*, 34. [CrossRef]
51. Kuntscher, C.A.; Pashkin, A.; Hoffmann, H.; Frank, S.; Klemm, M.; Horn, S.; Schönleber, A.; van Smaalen, S.; Hanfland, M.; Glawion, S.; et al. Mott-Hubbard gap closure and structural phase transition in the oxyhalides TiOBr and TiOCl under pressure. *Phys. Rev. B* **2008**, *78*, 035106. [CrossRef]

Disclaimer/Publisher's Note: The statements, opinions and data contained in all publications are solely those of the individual author(s) and contributor(s) and not of MDPI and/or the editor(s). MDPI and/or the editor(s) disclaim responsibility for any injury to people or property resulting from any ideas, methods, instructions or products referred to in the content.

Article

Phase Transformation Pathway of DyPO$_4$ to 21.5 GPa

Jai Sharma, Henry Q. Afful and Corinne E. Packard *

Department of Metallurgical and Materials Engineering, Colorado School of Mines, Golden, CO 80401, USA
* Correspondence: cpackard@mines.edu

Abstract: Interest in the deformation behavior and phase transformations of rare earth orthophosphates (REPO$_4$s) spans several fields of science—from geological impact analysis to ceramic matrix composite engineering. In this study, the phase behavior of polycrystalline, xenotime DyPO$_4$ is studied up to 21.5(16) GPa at ambient temperature using in situ diamond anvil cell synchrotron X-ray diffraction. This experiment reveals a large xenotime–monazite phase coexistence pressure range of 7.6(15) GPa and evidence for the onset of a post-monazite transformation at 13.9(10) GPa to scheelite. The identification of scheelite as the post-monazite phase of DyPO$_4$, though not definitive, is consistent with REPO$_4$ phase transformation pathways reported in both the experimental and the computational literature.

Keywords: high pressure; rare earth orthophosphate; phase transformation; X-ray diffraction

1. Introduction

Rare earth orthophosphates (REPO$_4$s) are highly refractory and insoluble ceramics relevant to various research areas ranging from geoscience to structural ceramics [1]. Much of the world's rare earth element supply comes from naturally occurring xenotime and monazite minerals, whose properties and formation are of great relevance in geochronology and geothermobarometry [2,3] in addition to mineral extraction. The pressure-induced phase transformations of certain REPO$_4$ compositions have also spurred research toward their use as fiber coatings, where they can confer additional plasticity and toughening mechanisms to oxide–oxide ceramic matrix composites [4–6]. At ambient pressure (~10^{-4} GPa), REPO$_4$s adopt either the xenotime (tetragonal, I4$_1$/amd) or monazite (monoclinic, P2$_1$/n) structure. The xenotime structure is also referred to as "zircon" (based on ZrSiO$_4$); however, this study employs the former name because it specifically originates from YPO$_4$ minerals [1]. At high pressures, xenotime compositions transform into the monazite or scheelite (tetragonal, I4$_1$/a) structures (see Figure 1) [7]. The scheelite-type structure has also been observed in other ABO$_4$ materials (e.g., tungstates, molybdates, vanadates, and arsenates) [8–10]. The REPO$_4$ xenotime, monazite, and scheelite structures feature chains of alternating PO$_4$ tetrahedra (shown in gray) and RE-O polyhedra (shown in violet) with RE-O coordination numbers of 8, 9, and 8, respectively.

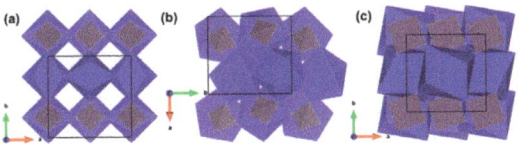

Figure 1. [001] views of REPO$_4$ structure in the (**a**) xenotime, (**b**) monazite, and (**c**) scheelite phases. RE-O polyhedra are shown in violet, PO$_4$ tetrahedra are shown in grey, and the unit cell boundaries are shown as thin black boxes. The 90° rotation of monazite axes with respect to those of xenotime and scheelite is a result of the monoclinic cell setting of monazite, as shown in detailed transformation schemes reported by Hay et al. [5]. Structures are visualized using the VESTA software [11].

Figure 1 shows that the REPO$_4$ structure becomes increasingly compact when transforming from xenotime to monazite to scheelite. This compaction can be attributed to increasing rotation and displacement of the RE-O polyhedra and resultant changes in the phosphate chain linkages (edge-sharing → corner-sharing) [10]. Although the exact unit cell volume losses during these transformations are composition-dependent, the loss associated with the xenotime → monazite transformation is significantly lower than that of the monazite → scheelite transformation [12]. This disparity likely emerges from the fact that the former transformation involves an increase in RE-O coordination (8 → 9), while the latter involves a decrease (9 → 8) [10]. An intermediate anhydrite (orthorhombic, Amma) phase has also been reported in certain xenotime compositions prior to the emergence of monazite when the xenotime composition is subject to high deviatoric stresses (e.g., TbPO$_4$ [13]) or has a composition that is extremely close to the 1 atm (~10^{-4} GPa) xenotime–monazite phase boundary (e.g., Gd$_x$Tb$_{1-x}$PO$_4$ [14]).

Prior studies and reviews have reported phase diagrams showing REPO$_4$ transformation pressures based on a variety of computational and experimental techniques [5,15–18]. Recent advancements in in situ diamond anvil cell (DAC) X-ray diffraction (XRD) experiments require updating the high-pressure REPO$_4$ phase map [12,18–25]. In contrast to Raman spectroscopy and ab initio calculations, XRD provides more direct, crystallographic proof of the existence of REPO$_4$ phases and phase transformations. Figure 2 compiles experimentally observed phase data from DAC XRD studies for all non-radioactive, single-RE compositions except PrPO$_4$, which has only been studied thus far using DAC Raman spectroscopy [26].

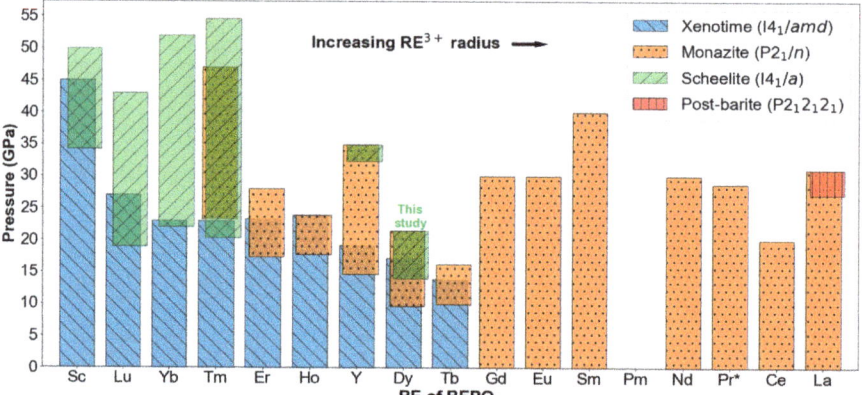

Figure 2. High-pressure phase map of the experimentally observed phases of all single-RE REPO$_4$s except PmPO$_4$ because Pm is both radioactive and extremely rare. The legend lists the phases with their corresponding space groups. For each composition, the upper limit of the highest bar(s) represents the highest pressure at which data are reported and does not represent a phase boundary. See text for references to the sources of phase data. All phase data are compiled from XRD studies except for PrPO$_4$, which has only been characterized via Raman spectroscopy * [26].

Under hydrostatic conditions, the xenotime → monazite transformation has been reported in ErPO$_4$, HoPO$_4$, YPO$_4$, DyPO$_4$, and TbPO$_4$ with onset pressures (P_{onset}) of 17.3 GPa, 17.7 GPa, 14.6 GPa, 9.1(1) GPa, and 9.9 GPa, respectively [12,20–23]. We note that for any number followed by a number in parentheses, the number in parentheses represents the standard deviation of the last digit of the number before the parentheses. In the REPO$_4$ phase transformation literature, there has long been an assumption that the xenotime → monazite P_{onset} varies linearly with RE^{3+} radius, such as many other properties of REPO$_4$s [5]. However, the DyPO$_4$ P_{onset} of 9.1(1) GPa from our 2021 study [22] disrupts this trend, suggesting instead that xenotime → monazite P_{onset} values fall into two clusters:

a high-pressure one around ~16 GPa (ErPO$_4$, HoPO$_4$, and YPO$_4$) and another that is <10 GPa (DyPO$_4$ and TbPO$_4$). Alloyed compositions of Gd$_x$Dy$_{(1-x)}$PO$_4$ and Gd$_x$Tb$_{(1-x)}$PO$_4$ also have transformation pressures that fall within the lower-pressure group, without following a trend with average RE^{3+} radius [5,14,27]. Neither thermodynamic properties (e.g., enthalpies of formation) nor structural properties (e.g., bond lengths and angles) show significant discontinuities between YPO$_4$ and DyPO$_4$, yet these compositions' P$_{onset}$ values are known to differ by at least 5.5 GPa [7,28–31].

The xenotime → monazite transformation in REPO$_4$s has also been described as sluggish and kinetically limited due to the experimentally observed xenotime–monazite phase coexistence being inconsistent with thermodynamic expectations (i.e., Gibbs phase rule) [12,22,23]; this is shown in Figure 2 as the regions where blue and orange bars overlap. The xenotime–monazite phase coexistence ranges for ErPO$_4$, YPO$_4$, and TbPO$_4$ are 6 GPa, 4.6 GPa, and 3.9 GPa, respectively [12,20,23]. In prior experiments, HoPO$_4$ and DyPO$_4$ were not taken to high enough pressures to capture the full xenotime–monazite coexistence range [21,22].

Other xenotime compositions with smaller RE^{3+} radii undergo the xenotime → scheelite transformation, which involves no change in RE-O coordination number. ScPO$_4$, LuPO$_4$, and YbPO$_4$ transform directly to the scheelite structure at P$_{onset}$ values of 34.2 GPa, 19 GPa, and 22 GPa, respectively [12,19]. Their respective xenotime–scheelite phase coexistence ranges (represented by overlap of green and blue bars in Figure 2) are 10.8 GPa, 8 GPa, and 1 GPa, indicating kinetic limitations similar to the xenotime → monazite transformation [12,19].

TmPO$_4$, as described by Stavrou et al., represents a "borderline case" between xenotime compositions that transform to monazite and those that transform to scheelite [18]. In TmPO$_4$, the xenotime → scheelite transformation begins at 20.3 GPa, and these two phases coexist over a 2.7 GPa range. Immediately after the disappearance of xenotime at 23 GPa, monazite emerges at 23.3 GPa and coexists with scheelite until 47 GPa. Then, scheelite persists through the end of the experiment. Stavrou et al. characterize monazite TmPO$_4$ as a "metastable minority phase" and attribute the long monazite–scheelite coexistence to the stabilization of monazite grains "when embedded in a scheelite matrix" [14].

Among the compositions, which adopt the monazite structure at 1 atm (~10^{-4} GPa), only LaPO$_4$ has been shown to undergo a pressure-induced phase transformation. Lacomba-Perales et al. proposed that LaPO$_4$ transforms to barite (orthorhombic, Pnma) based on powder XRD but could not confirm the barite structure due to significant peak overlap [20]. Ruiz-Fuertes et al. used single crystal XRD and second harmonic generation analysis to confirm the post-monazite structure as non-centrosymmetric "post-barite" (orthorhombic, P2$_1$2$_1$2$_1$) [32]. Post-barite first emerges at 27.1 GPa and coexists with monazite through the end of the experiment at 31 GPa. This experimental data conflicts a bit with their ab initio calculations, which show a pressure range where barite is energetically preferred before the emergence of post-barite, but the authors argue large kinetic barriers may explain the lack of barite in their LaPO$_4$ experiments and may hinder barite formation in other monazite REPO$_4$ compositions [32]. Ruiz-Fuertes et al. project (based on ab initio calculations) a post-barite transformation to occur at 45 GPa and 35 GPa in GdPO$_4$ and NdPO$_4$, respectively, with barite as a possible, but unlikely, transition phase.

Our 2021 XRD study [22] reported a DyPO$_4$ xenotime → monazite P$_{onset}$ at 9.1(1) GPa under a quasi-hydrostatic loading rate but did not go to high enough pressures to resolve the xenotime–monazite phase coexistence range [22]. This XRD study aims to identify the end of the xenotime–monazite phase coexistence range by going to higher pressures. Results reveal a xenotime–monazite phase coexistence range of 7.6(15) GPa and a previously unreported phase transformation to a post-monazite phase at 13.9(10) GPa. Comparison to the experimental and the computational literature strongly suggests this post-monazite phase adopts the scheelite (tetragonal, I4$_1$/a) structure.

2. Materials and Methods

Phase-pure xenotime DyPO$_4$ powder was obtained via precipitation reaction involving Dy(NO$_3$)$_3$ · 5H$_2$O ($\geq$99.9% RE oxide basis, Alfa Aesar) precursor and H$_3$PO$_4$ (85% w/w aqueous solution, Alfa Aesar) and subsequent calcination. These two steps are detailed elsewhere [33]. The sample powder consists of sub-micron grains, which exhibit the anisotropic, elongated crystal habit expected of tetragonal materials (see the scanning electron micrograph in Figure S1, see supplementary materials). An energy-dispersive X-ray spectrum of the powder (shown in Figure S2) shows no elemental impurities. In situ DAC XRD was conducted at room temperature at beamline 16-ID-B, HPCAT, Advanced Photon Source, Argonne National Laboratory. Two-dimensional diffraction patterns were collected with the PILATUS 1M-F detector. The X-ray wavelength was 0.42459 Å, and the beam spot size (full width at half maximum) was $\sim$2 µm by $\sim$4 µm. We used a Diacell Helios DAC with a membrane (both from Almax easyLab Inc., Cambridge, MA, USA) driven by a Druck PACE 6000 pressure controller [34]. DAC preparation involved successively loading DyPO$_4$ powder, gold powder (>99.96% metals basis, Alfa Aesar, Ward Hill, MA, USA), ruby chips (Almax easyLab Inc., Cambridge, MA, USA), and 16:3:1 methanol–ethanol–water mixture (MEW) pressure medium into the hole of the 301 stainless steel gasket. The gasket hole diameter and indented thickness were 220 µm and 80 µm, respectively. For pressure marking during initial membrane engagement, ruby was used (R1 fluorescence calibration [35]), while gold was used (third order Birch-Murnaghan EoS [36]) during diffraction data collection. Data collection started at 3.1(2) GPa due to some initial compression required to confirm membrane engagement. There are no reported DyPO$_4$ phase transitions below this starting pressure (as corroborated by our 2021 XRD study) [22]; therefore, the initial jump does not preclude any material insight.

XRD pattern integration, masking, and background subtraction were performed using Dioptas [37]. Pattern fitting was then performed using X'Pert HighScore Plus [38]. This software fits monazite using the P2$_1$/c cell setting as a default. Although both the P2$_1$/c and P2$_1$/n cell settings are valid descriptions of monazite (space group No. 14), fitted lattice parameters were converted to the P2$_1$/n cell setting to facilitate comparison to the literature. The LeBail fitting approach [39] was used instead of traditional Rietveld structural refinement to accommodate the significant preferred orientation present in all scans. This apparent preferred orientation appears due to the small spot size of the beam with respect to the grain size of the sample (effectively sampling a finite number of grains) rather than any inherent orientation of the sample grains. The atmospheric-pressure volume of xenotime DyPO$_4$ (289.39(2) Å^3) was derived from a prior synchrotron XRD pattern of a sample from the same batch as the sample in this study [22]. The following reference structures were used in this study: xenotime DyPO$_4$ from Milligan et al. [40], monazite DyPO$_4$ from Heuser et al. [41], gold from Couderc et al. [42], ruby from Jephcoat et al. [43], and calculated scheelite TbPO$_4$ from López-Solano et al. [23]. Importantly, the scheelite TbPO$_4$ structure file was not employed in LeBail fitting—only in peak position comparison. For the computation involving derived data (e.g., unit cell volume, gold-based pressure, cell setting conversion), Python was used to propagate error with an assumed covariance of zero.

3. Results

During the experiment, the gold lattice parameter decreases steadily as shown in Figure 3a. In Figure 3b, the pressure increases steadily with time and yields an effective sample loading rate of $\sim$20 MPa/s, an order of magnitude faster than that of our 2021 XRD study [22].

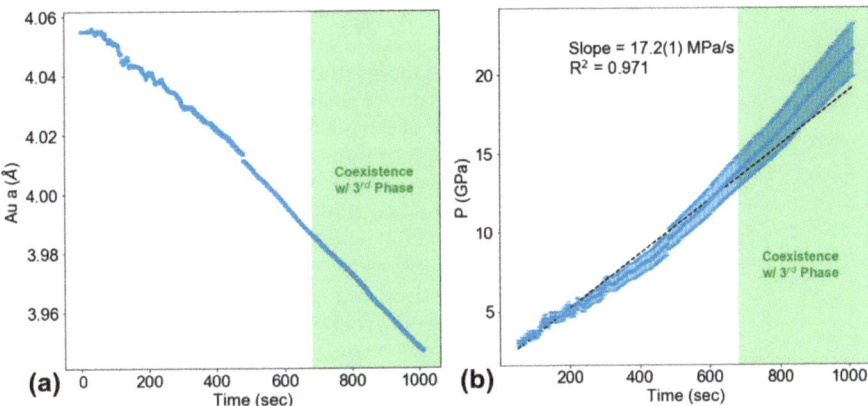

Figure 3. Plots showing (**a**) gold lattice parameter and (**b**) pressure against time during the experiment. Error bars represent standard deviation. The linear fit of the pressure data yields an effective loading rate of 17.2(1) MPa/s. Green, shaded areas represent the pressure range in which a third (post-monazite) phase of $DyPO_4$ appears in the patterns.

Figure 4 shows the pressure evolution of integrated background-subtracted XRD patterns. The square root of intensity is plotted against Q to show weak peaks more clearly.

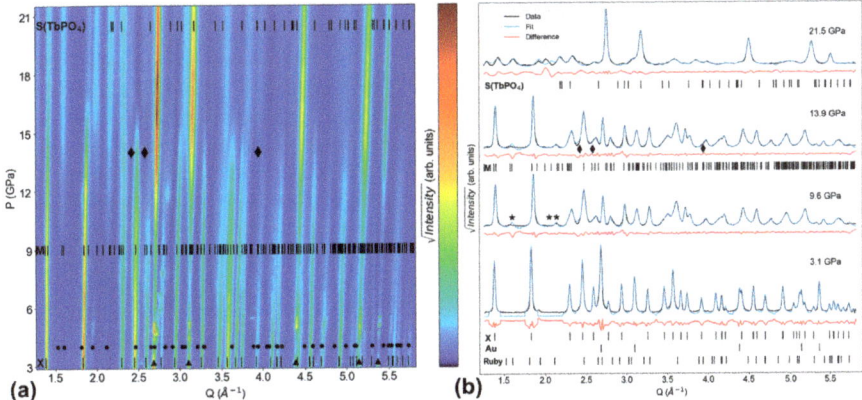

Figure 4. Synchrotron XRD patterns. X, M, and S($TbPO_4$) ticks show Bragg reflections of xenotime $DyPO_4$, monazite $DyPO_4$, and scheelite $TbPO_4$, the last based on ab initio structural data reported by Lopez-Solano et al. [23]. Star symbols mark emerging monazite peaks at 9.6(7) GPa, and diamond symbols mark emerging peaks of the unidentified phase at 13.9(10) GPa. (**a**) A contour plot showing all XRD patterns. Peak positions of gold and ruby are marked with triangles and circles, respectively. (**b**) The LeBail fits of key patterns (initial scan, onset of monazite, onset of unidentified phase, and final scan).

In Figure 4a, sample XRD peaks drift to higher Q and broaden with increasing pressure due to uniform and non-uniform strain, respectively. The first scan shows peaks from xenotime $DyPO_4$, gold (triangles), and ruby (circles). Figure 4 also illustrates the relative loss of sample signal compared to the strong gold signal as pressure increases. P_{onset} values are determined by visual inspection of individual XRD patterns as shown in Figure 4b, not by the coloring in Figure 4a. Figure 4b shows the LeBail fits of key patterns (initial scan, onset of monazite, onset of new phase, and final scan). The first discernible monazite peaks emerge at 9.6(7) GPa at Q = 1.60, 2.05, and 2.14 Å^{-1}; these are the (110), (002), and

($02\bar{1}$) reflections. At P_{onset}, the monazite lattice parameters are a = 6.134(1) Å, b = 6.695(1) Å, c = 6.276(1) Å, and β = 99.79(1)°, and the unit cell volume loss during transformation is 7.90%. Although these lattice parameter uncertainties appear quite small, several checks on the LeBail fits do not change the outcome. The xenotime → monazite P_{onset} is below the hydrostatic limit of the MEW pressure medium (10.5(5) GPa), meaning non-hydrostatic stresses likely do not influence the onset of this transformation. At 10.5(5) GPa, MEW undergoes a glass transition into an amorphous phase that contributes no XRD peaks [44]. Above this hydrostatic limit, the sample stress state is understood to be non-hydrostatic. Starting at 17.2(13) GPa, there are no longer any peaks uniquely attributable to xenotime (based on visual inspection of individual patterns). The disappearance of xenotime peaks by this pressure is also apparent in Figure 4a and yields a xenotime–monazite phase coexistence range of 7.6(15) GPa.

Interestingly, a new set of previously unidentified XRD peaks emerge at 13.9(10) GPa at Q = 2.42, 2.58, and 3.94 Å^{-1} and persist as the pressure increases (see diamonds in Figure 4). These peak positions are inconsistent with xenotime, anhydrite, monazite, ruby, gasket material, gold, or even a "monazite II" phase reported in CeVO$_4$ [45]. Peak positions for the post-monazite phase of TbPO$_4$, scheelite (derived from ab initio calculations at 20.5 GPa), are shown in Figure 4 as there are no available structural data on any post-monazite phases of DyPO$_4$. The final scan at 21.5(16) GPa contains peaks corresponding to monazite, gold, ruby, and the unidentified phase.

Next, we examine the pressure-dependence of lattice parameters more closely, finding consistent lattice parameter deviations around the pressure at which the post-monazite phase emerges. Figure 5 shows the pressure evolution of DyPO$_4$ lattice parameters for the xenotime (a_x and c_x) and monazite (a_m, b_m, c_m, and $β_m$) phases.

Figure 5a shows a_x decreasing fairly steadily and monotonically with two slight disruptions at pressures consistent with the xenotime → monazite P_{onset} (9.6(7) GPa) and with the hydrostatic limit of MEW (10.5(5) GPa). Starting at ~14 GPa, however, a_x stops decreasing, and the lattice parameter uncertainty expands significantly; this change in behavior coincides with the emergence of the unidentified peaks. Figure 5b shows similar behavior in c_x, except this parameter starts decreasing rapidly at ~14 GPa. The axial ratio of xenotime (shown in Figure S3) also shows a dramatic trend change at ~14 GPa. Figure 5c shows a_m decreasing monotonically with slight disruptions at the hydrostatic limit and ~14 GPa. After ~15 GPa, a_m becomes non-monotonic with pressure, and its error bar expands significantly. Figure 5d,e does not show any anomalies in b_m and c_m around 14 or 15 GPa—only minor disruptions around the hydrostatic limit. Figure 5f shows $β_m$ changes monotonicity around the hydrostatic limit, steadily increases with smaller error bars after 12 GPa, then remains almost constant after ~15 GPa with larger error bars. Figure 6 shows all DyPO$_4$ lattice parameters plotted together to better illustrate their relative values and relative compressibilities. The cx, am, bm, and $β_m$ parameters stand out with the most significant changes in behavior beginning at ~14 GPa, coinciding with the emergence of the third (post-monazite) phase of DyPO$_4$.

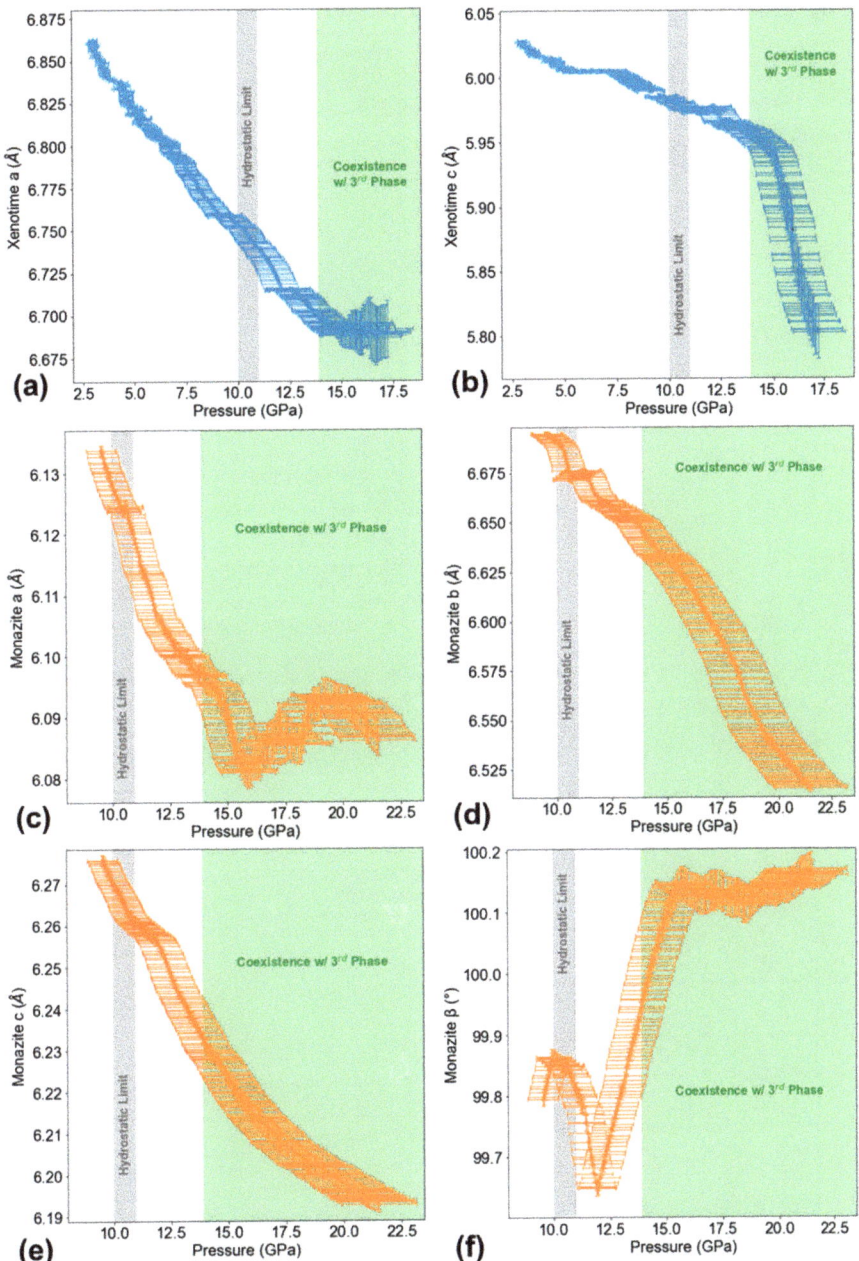

Figure 5. Pressure dependence of xenotime (x) and monazite (m) DyPO$_4$ lattice parameters. Green, shaded areas represent the pressure range in which a third (post-monazite) phase of DyPO$_4$ exists. The thin gray shaded area represents the hydrostatic limit of the pressure medium: (**a**) a_x; (**b**) c_x; (**c**) a_m; (**d**) b_m; (**e**) c_m; (**f**) β_m. All parameters show some irregularity around the hydrostatic limit (10.5(5) GPa). The a_x, c_x, a_m, and β_m parameters exhibit notable changes in behavior after ~14 GPa.

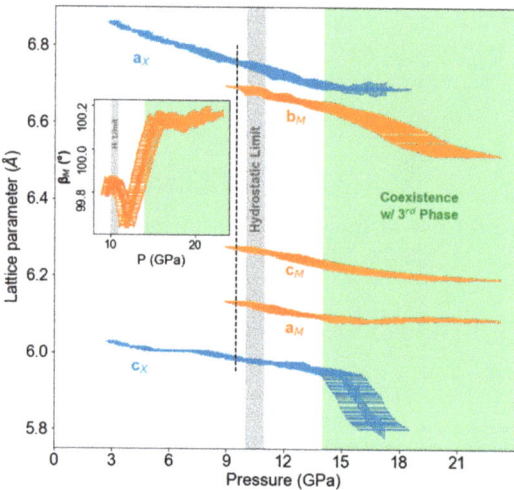

Figure 6. Pressure evolution of all DyPO$_4$ lattice parameters for the xenotime (a_x and c_x) and monazite (a_m, b_m, c_m, and β_m) phases. Error bars represent standard deviation. The vertical dashed line indicates the xenotime → monazite P$_{onset}$. Green, shaded areas represent the pressure range in which a third (post-monazite) phase of DyPO$_4$ exists. The thin gray shaded area represents the hydrostatic limit of the pressure medium. The inset shows the monazite beta angle.

Axial compressibilities of the xenotime and monazite phases are obtained by linearly fitting lattice parameter data from pressures below the hydrostatic limit (10.5 GPa). These values are summarized in Table 1. The monazite beta angle (β_m) was not analyzed as this parameter's non-monotonic behavior precludes a meaningful linear fit.

Table 1. DyPO$_4$ axial compressibilities derived from linear fits of lattice parameter data at pressures below the hydrostatic limit. Negative values indicate compression.

Lattice Parameter	Axial Compressibility (Å,°/GPa) × 10^3	Intercept at 0 GPa (Å,°)	R^2 of Linear Fit
a_x	−16.09 (12)	6.907 (1)	0.9871
c_x	−6.238 (127)	6.045 (1)	0.8876
a_m	−10.67 (22)	6.236 (2)	0.9921
b_m	−11.64 (72)	6.807 (7)	0.9326
c_m	−11.10 (21)	6.382 (2)	0.9934
β_m	–	–	–

4. Discussion

This experiment shows the P$_{onset}$ of DyPO$_4$ xenotime → monazite phase transformation is 9.6(7) GPa when loading at ~20 MPa/s. This pressure is nominally higher than the 9.1(1) GPa P$_{onset}$ observed in our 2021 DyPO$_4$ study under quasi-static loading (~2 MPa/s) [22]; however, the magnitude of the P$_{onset}$ error unfortunately precludes any conclusions regarding rate-dependence of the xenotime → monazite transformation (e.g., thermal activation, mechanism). The xenotime axial compressibilities (see Table 1) and the monazite lattice parameters at P$_{onset}$ are not notably different from those reported in our 2021 DyPO$_4$ study [22], while the monazite axial compressibilities differ significantly in their absolute and relative values. The discrepancy in compressibilities is likely due to this study having a much smaller quasi-hydrostatic pressure regime in which monazite exists (~1 GPa); therefore, there is a much smaller range and significantly fewer monazite datapoints suitable for fitting in this study than in our previous study.

This work also provides, for the first time, an estimate of the full $DyPO_4$ xenotime–monazite phase coexistence range. Figure 4 shows that the xenotime phase is present up to 17.2(13) GPa (also reflected in Figure 2 as the upper bound of the blue bar). This value establishes that, while there is not a trend in the experimentally observed monazite P_{onset} with rare-earth radius, there is a general compositional trend in which the upper pressure bound of xenotime decreases with increasing RE^{3+} radius (see Figure 2). The $DyPO_4$ xenotime–monazite coexistence range is then 7.6(15) GPa, which is significantly larger than that of both neighboring compositions (4.8 GPa for YPO_4 and 4.6 GPa for $TbPO_4$) and slightly larger than that of $ErPO_4$ (6.0 GPa) [12,20,23]. Comparison to $HoPO_4$ is precluded by incomplete xenotime phase transition [13].

Beyond characterizing the xenotime → monazite transformation, this study provides proof of the existence of a new, post-monazite phase of $DyPO_4$. As a reminder, it is crucial to note that no $DyPO_4$ structures other than xenotime and monazite were used during the LeBail fitting (as no other experiment-based $DyPO_4$ structures have been reported). As a result, the fitting process attempted to accommodate the unidentified peaks at Q = 2.42, 2.58, and 3.94 Å^{-1} (emerging at 13.9(10) GPa as seen in Figure 4) with the xenotime and monazite structures. This accommodation explains the anomalies in a_x, c_x, a_m, and β_m after 13.9(10) GPa as shown in Figure 5. The gold lattice parameter variation with time (and its corresponding pressure profile) is smooth and has no interruption at the post-monazite transition pressure (see Figure 3), showing that anomalies in xenotime and monazite lattice parameters do not result from experiment instabilities. Indexing the unidentified peaks to a certain structure or space group is extremely difficult because of a weak sample signal at pressures >14 GPa as well as monazite peaks covering most of the Q range.

Comparison to the literature strongly suggests the post-monazite phase is scheelite (tetragonal, $I4_1/a$). Based on preliminary Raman spectroscopy experiments, Stavrou et al. deduce a xenotime → monazite → scheelite transformation pathway in $DyPO_4$ with a monazite → scheelite P_{onset} at ~33 GPa [18]. This P_{onset} value may be a significant overestimation, as the Raman spectroscopy-based xenotime → monazite P_{onset} has also proven to be a significant overestimation when compared to XRD work [22]. Further analysis of this Raman spectroscopy-based monazite → scheelite P_{onset} is complicated by the fact that the underlying Raman spectra have yet to be published. Nevertheless, the pressure-induced phase transformation pathways of other xenotime $REPO_4$s support the existence of a monazite → scheelite transformation in $DyPO_4$. Figure 2 shows scheelite evolves from monazite (in YPO_4) or directly from xenotime (in $ScPO_4$, $LuPO_4$, $YbPO_4$, and $TmPO_4$) with increasing pressure. Experimental studies of $ErPO_4$, $HoPO_4$, and $TbPO_4$ neither confirm nor deny transformation to scheelite due to limited experimental pressure ranges [20,21,23]. In the case of $TbPO_4$, Lopez-Solano et al. point to "kinetic energy barriers" possibly hindering transformation to scheelite [23]. However, ab initio calculations performed by Bose et al. and Lopez-Solano et al. show $ErPO_4$, $HoPO_4$, and $TbPO_4$ are expected to follow the xenotime → monazite → scheelite phase transformation pathway. Bose et al. predict monazite → scheelite P_{onset} values of ~11 GPa, ~12 GPa, and ~14 GPa for $ErPO_4$, $HoPO_4$, and $TbPO_4$, respectively [17]. Lopez-Solano et al. predict a slightly higher $TbPO_4$ monazite → scheelite P_{onset} of 15.5 GPa [23]. Barite and post-barite also bear consideration for the post-monazite phase, but both seem unlikely given experimental and computational data on $LaPO_4$ put the transition for this and other compositions at pressures above 26 GPa [20,32]. Additionally, Lopez-Solano et al. found the scheelite structure to be energetically favorable to the barite structure [23]. No studies to date report a similar comparison between the scheelite and post-barite structures. Given that the unidentified XRD peaks in this study emerge at a pressure consistent with the expected monazite → scheelite transformation in neighboring compositions, it is likely that these peaks belong to a scheelite $DyPO_4$ phase.

In the absence of scheelite unit cell data for $DyPO_4$, we use $TbPO_4$ for comparison. Lopez-Solano et al. predicted unit cell data of scheelite $TbPO_4$ at 20.5 GPa, which are added to the top of Figure 4 as "S($TbPO_4$)" peak position ticks. Although the new peaks from

DyPO$_4$ do not exactly match these ticks in scans around 20.5 GPa, the peaks are reasonably close to the ticks given the differences in methods, RE, temperature, kinetics, and stress state, which together have confounding effects on the unit cell. If this post-monazite DyPO$_4$ phase is further confirmed to be scheelite, 13.9(10) GPa would be the lowest pressure at which scheelite has been reported in REPO$_4$s to date.

5. Conclusions

This work significantly extends the characterization of the high-pressure phase behavior of DyPO$_4$, which has been limited, particularly at pressures above 15 GPa. Our 2021 XRD study reported a xenotime → monazite P$_{onset}$ at ~9 GPa but did not go to high enough pressures to reveal the xenotime–monazite phase coexistence range. This XRD study goes to higher pressures, showing a xenotime–monazite phase coexistence range of 7.6(15) GPa as well as the emergence of new peaks at 13.9(10) GPa. Contextualizing these new peaks within the experimental and the computational literature provides compelling evidence that monazite DyPO$_4$ undergoes a pressure-induced phase transformation to the scheelite structure. Our results also motivate further XRD studies of other REPO$_4$s (e.g., ErPO$_4$, HoPO$_4$, and TbPO$_4$) at higher pressures to explore possible monazite → scheelite transformations and to elucidate high-pressure phase transformation pathways more broadly.

Supplementary Materials: The following supporting information can be downloaded at: https://www.mdpi.com/article/10.3390/cryst13020249/s1, Figure S1: Scanning electron micrograph showing grain size and morphology of sample powder; Figure S2: Energy-dispersive X-ray spectrum of sample powder; Figure S3: Pressure dependence of the axial ratio (c_x/a_x) of the xenotime unit cell. A file containing XRD pattern fit data shown in Figure 4b is also included in the Supplementary Materials.

Author Contributions: Conceptualization, J.S. and C.E.P.; data curation, J.S., H.Q.A. and C.E.P.; formal analysis, J.S.; writing—original draft preparation, J.S. and H.Q.A.; writing—review and editing, J.S., H.Q.A. and C.E.P.; supervision, C.E.P. All authors have read and agreed to the published version of the manuscript.

Funding: This research was supported by the National Science Foundation under Award No. DMR-1352499. J.S. was supported by the Department of Defense through the National Defense Science & Engineering Graduate Fellowship Program. This work was performed at HPCAT (Sector 16), Advanced Photon Source, Argonne National Laboratory. HPCAT operations are supported by DOE National Nuclear Security Administration under Award No. DE-NA0001974 and DOE Office of Basic Energy Sciences (BES) under Award No. DE-FG02-99ER45775, with partial instrumentation funding by the National Science Foundation (NSF). A.P.S. is supported by DOE-BES, under Contract No. DE-AC02- 06CH11357.

Institutional Review Board Statement: Not applicable.

Informed Consent Statement: Not applicable.

Data Availability Statement: All relevant data that support the findings of this study are available from the corresponding authors upon request.

Acknowledgments: The authors thank Ivar Reimanis for lending us his Diacell Helios DAC, Matthew Musselman for synthesizing the sample used in the experiment, and Jesse Smith at HPCAT for performing initial alignment and calibration. The authors acknowledge the APS synchrotron facilities for provision of beamtime on beamline 16-ID-B.

Conflicts of Interest: The authors declare no conflict of interest.

References

1. Boatner, L.A. Synthesis, Structure, and Properties of Monazite, Pretulite, and Xenotime. *Rev. Mineral. Geochem.* **2002**, *48*, 87–121. [CrossRef]
2. Vielreicher, N.M.; Groves, D.I.; Fletcher, I.R.; McNaughton, N.J.; Rasmussen, B. Hydrothermal Monazite and Xenotime Geochronology: A New Direction for Precise Dating of Orogenic Gold Mineralization. *SEG Discov.* **2003**, *70*, 1–16. [CrossRef]

3. Cox, M.A.; Cavosie, A.J.; Poelchau, M.; Kenkmann, T.; Bland, P.A.; Miljković, K. Shock Deformation Microstructures in Xenotime from the Spider Impact Structure, Western Australia. In *Large Meteorite Impacts and Planetary Evolution VI*; Geological Society of America: Boulder, CO, USA, 2021; pp. 449–464.
4. Morgan, P.E.D.; Marshall, D.B. Ceramic Composites of Monazite and Alumina. *J. Am. Ceram. Soc.* **1995**, *78*, 1553–1563. [CrossRef]
5. Hay, R.S.; Mogilevsky, P.; Boakye, E. Phase Transformations in Xenotime Rare-Earth Orthophosphates. *Acta Mater.* **2013**, *61*, 6933–6947. [CrossRef]
6. Hay, R.S.; Boakye, E.E.; Mogilevsky, P.; Fair, G.E.; Parthasarathy, T.A.; Davis, J.E. Transformation Plasticity in (Gdx Dy1-x) PO4 Fiber Coatings during Fiber Push Out. *J. Am. Ceram. Soc.* **2013**, *96*, 1586–1595. [CrossRef]
7. Ni, Y.; Hughes, J.M.; Mariano, A.N.; Et, N.I.; Crystal, A.L. Crystal Chemistry of the Monazite and Xenotime Structures. *Am. Mineral.* **1995**, *80*, 21–26. [CrossRef]
8. Errandonea, D.; Manjón, F.J. Pressure Effects on the Structural and Electronic Properties of ABX4 Scintillating Crystals. *Prog. Mater. Sci.* **2008**, *53*, 711–773. [CrossRef]
9. Minakshi, M.; Mitchell, D.R.G.; Baur, C.; Chable, J.; Barlow, A.J.; Fichtner, M.; Banerjee, A.; Chakraborty, S.; Ahuja, R. Phase Evolution in Calcium Molybdate Nanoparticles as a Function of Synthesis Temperature and Its Electrochemical Effect on Energy Storage. *Nanoscale Adv.* **2019**, *1*, 565–580. [CrossRef]
10. Macey, B.J. *The Crystal Chemistry of MTO4 Compounds with the Zicron, Scheelite, and Monazite Structure Types*; Virginia Tech: Blackburg, VA, USA, 2009.
11. Momma, K.; Izumi, F. VESTA 3 for Three-Dimensional Visualization of Crystal, Volumetric and Morphology Data. *J. Appl. Crystallogr.* **2011**, *44*, 1272–1276. [CrossRef]
12. Zhang, F.X.; Wang, J.W.; Lang, M.; Zhang, J.M.; Ewing, R.C.; Boatner, L.A. High-Pressure Phase Transitions of ScPO4 and YPO4. *Phys. Rev. B Cover. Condens. Matter Mater. Phys.* **2009**, *80*, 184114. [CrossRef]
13. Lösch, H.; Hirsch, A.; Holthausen, J.; Peters, L.; Xiao, B.; Neumeier, S.; Schmidt, M.; Huittinen, N. A Spectroscopic Investigation of Eu3+ Incorporation in LnPO4 (Ln = Tb, Gd1-XLux, X = 0.3, 0.5, 0.7, 1) Ceramics. *Front. Chem.* **2019**, *7*, 94. [CrossRef] [PubMed]
14. Tschauner, O.; Ushakov, S.V.; Navrotsky, A.; Boatner, L.A. Phase Transformations and Indications for Acoustic Mode Softening in Tb-Gd Orthophosphate. *J. Phys. Condens. Matter* **2016**, *28*, 035403. [CrossRef] [PubMed]
15. Kolitsch, U.; Holtstam, D. Crystal Chemistry of REEXO4 Compounds (X = P,As, V). II. Review of REEXO4 Compounds and Their Stability Fields. *Eur. J. Mineral.* **2004**, *16*, 117–126. [CrossRef]
16. Finch, R.J. Structure and Chemistry of Zircon and Zircon-Group Minerals. *Rev. Mineral. Geochem.* **2003**, *53*, 1–25. [CrossRef]
17. Bose, P.P.; Mittal, R.; Chaplot, S.L.; Loong, C.K.; Boatner, L.A. Inelastic Neutron Scattering, Lattice Dynamics, and High-Pressure Phase Stability of Zircon-Structured Lanthanide Orthophosphates. *Phys. Rev. B Cover. Condens. Matter Mater. Phys.* **2010**, *82*, 094309. [CrossRef]
18. Stavrou, E.; Tatsi, A.; Raptis, C.; Efthimiopoulos, I.; Syassen, K.; Muñoz, A.; Rodríguez-Hernández, P.; López-Solano, J.; Hanfland, M.; Muñoz, A.; et al. Effects of Pressure on the Structure and Lattice Dynamics of TmPO$_4$: Experiments and Calculations. *Phys. Rev. B Condens. Matter Mater. Phys.* **2012**, *85*, 24117. [CrossRef]
19. Zhang, F.X.; Lang, M.; Ewing, R.C.; Lian, J.; Wang, Z.W.; Hu, J.; Boatner, L.A. Pressure-Induced Zircon-Type to Scheelite-Type Phase Transitions in YbPO4 and LuPO4. *J. Solid State Chem.* **2008**, *181*, 2633–2638. [CrossRef]
20. Lacomba-Perales, R.; Errandonea, D.; Meng, Y.; Bettinelli, M. High-Pressure Stability and Compressibility of A PO4 (A=La, Nd, Eu, Gd, Er, and Y) Orthophosphates: An X-ray Diffraction Study Using Synchrotron Radiation. *Phys. Rev. B Cover. Condens. Matter Mater. Phys.* **2010**, *81*, 064113. [CrossRef]
21. Gomis, O.; Lavina, B.; Rodríguez-Hernández, P.; Muñoz, A.; Errandonea, R.; Errandonea, D.; Bettinelli, M. High-Pressure Structural, Elastic, and Thermodynamic Properties of Zircon-Type HoPO4 and TmPO4. *J. Phys. Condens. Matter* **2017**, *29*, 095401. [CrossRef]
22. Sharma, J.; Musselman, M.; Haberl, B.; Packard, C.E. In Situ Synchrotron Diffraction of Pressure-Induced Phase Transition in DyPO4 under Variable Hydrostaticity. *Phys. Rev. B* **2021**, *103*, 184105. [CrossRef]
23. López-Solano, J.; Rodríguez-Hernández, P.; Muñoz, A.; Gomis, O.; Santamaría-Perez, D.; Errandonea, D.; Manjón, F.J.; Kumar, R.S.; Stavrou, E.; Raptis, C. Theoretical and Experimental Study of the Structural Stability of TbPO4 at High Pressures. *Phys. Rev. B Cover. Condens. Matter Mater. Phys.* **2010**, *81*, 144126. [CrossRef]
24. Heuser, J.M.; Palomares, R.I.; Bauer, J.D.; Rodriguez, M.J.L.; Cooper, J.; Lang, M.; Scheinost, A.C.; Schlenz, H.; Winkler, B.; Bosbach, D.; et al. Journal of the European Ceramic Society Structural Characterization of (Sm, Tb) PO4 Solid Solutions and Pressure-Induced Phase Transitions. *J. Eur. Ceram. Soc.* **2018**, *38*, 4070–4081. [CrossRef]
25. Huang, T.; Lee, J.S.; Kung, J.; Lin, C.M. Study of Monazite under High Pressure. *Solid State Commun.* **2010**, *150*, 1845–1850. [CrossRef]
26. Errandonea, D.; Gomis, O.; Rodriguez-Hernández, P.; Munz, A.; Ruiz-Fuertes, J.; Gupta, M.; Achary, S.N.; Hirsch, A.; Manjon, F.J.; Peters, L.; et al. High-Pressure Structural and Vibrational Properties of Monazite-Type BiPO4, LaPO4, CePO4, and PrPO4. *J. Phys. Condens. Matter* **2018**, *30*, 065401. [CrossRef] [PubMed]
27. Musselman, M.A.; Wilkinson, T.M.; Haberl, B.; Packard, C.E. In Situ Raman Spectroscopy of Pressure-Induced Phase Transformations in Polycrystalline TbPO4, DyPO4, and GdxDy(1−x)PO4. *J. Am. Ceram. Soc.* **2018**, *101*, 2562–2570. [CrossRef]
28. Rustad, J.R. Density Functional Calculations of the Enthalpies of Formation of Rare-Earth Orthophosphates. *Am. Mineral.* **2012**, *97*, 791–799. [CrossRef]

29. Kowalski, P.M.; Li, Y. Relationship between the Thermodynamic Excess Properties of Mixing and the Elastic Moduli in the Monazite-Type Ceramics. *J. Eur. Ceram. Soc.* **2016**, *36*, 2093–2096. [CrossRef]
30. Ushakov, S.V.; Helean, K.B.; Navrotsky, A.; Boatner, L.A. Thermochemistry of Rare-Earth Orthophosphates. *J. Mater. Res.* **2001**, *16*, 2623–2633. [CrossRef]
31. Blanca Romero, A.; Kowalski, P.M.; Beridze, G.; Schlenz, H.; Bosbach, D. Performance of DFT+U Method for Prediction of Structural and Thermodynamic Parameters of Monazite-Type Ceramics. *J. Comput. Chem.* **2014**, *35*, 1339–1346. [CrossRef]
32. Ruiz-Fuertes, J.; Hirsch, A.; Friedrich, A.; Winkler, B.; Bayarjargal, L.; Morgenroth, W.; Peters, L.; Roth, G.; Milman, V. High-Pressure Phase of $LaPO_4$ Studied by X-ray Diffraction and Second Harmonic Generation. *Phys. Rev. B Cover. Condens. Matter Mater. Phys.* **2016**, *94*, 134109. [CrossRef]
33. Musselman, M.A. In Situ Raman Spectroscopy of Pressure-Induced Phase Transformations in $DyPO_4$ and $Gd_xDy(1-x)PO_4$. Master's Thesis, Colorado School of Mines, Golden, CO, USA, 2017.
34. Sinogeikin, S.V.; Smith, J.S.; Rod, E.; Lin, C.; Kenney-Benson, C.; Shen, G. Online Remote Control Systems for Static and Dynamic Compression and Decompression Using Diamond Anvil Cells. *Rev. Sci. Instrum.* **2015**, *86*, 072209. [CrossRef] [PubMed]
35. Shen, G.; Wang, Y.; Dewaele, A.; Wu, C.; Fratanduono, D.E.; Eggert, J.; Klotz, S.; Dziubek, K.F.; Loubeyre, P.; Fat'yanov, O.V.; et al. Toward an International Practical Pressure Scale: A Proposal for an IPPS Ruby Gauge (IPPS-Ruby2020). *High Press. Res.* **2020**, *40*, 299–314. [CrossRef]
36. Anderson, O.L.; Isaak, D.G.; Yamamoto, S. Anharmonicity and the Equation of State for Gold. *J. Appl. Phys.* **1989**, *65*, 1534–1543. [CrossRef]
37. Prescher, C.; Prakapenka, V.B. DIOPTAS: A Program for Reduction of Two-Dimensional X-ray Diffraction Data and Data Exploration. *High Press. Res.* **2015**, *35*, 223–230. [CrossRef]
38. Degen, T.; Sadki, M.; Bron, E.; König, U.; Nénert, G. The High Score Suite. *Powder Diffr.* **2014**, *29*, S13–S18. [CrossRef]
39. Bail, A. Le Whole Powder Pattern Decomposition Methods and Applications: A Retrospective. *Powder Diffr.* **2005**, *20*, 316–326. [CrossRef]
40. Milligan, W.O.; Mullica, D.F.; Beall, G.W.; Boatner, L.A. The Structures of Three Lanthanide Orthophosphates. *Inorganica Chim. Acta* **1983**, *70*, 133–136. [CrossRef]
41. Heuser, J.M.; Neumeier, S.; Peters, L.; Schlenz, H.; Bosbach, D.; Deissmann, G. Structural Characterisation of Metastable Tb- and Dy-Monazites. *J. Solid State Chem.* **2019**, *273*, 45–52. [CrossRef]
42. Couderc, J.J.; Garigue, G.; Lafourcade, L.; Nguyen, Q.T. Standard X-ray Diffraction Powder Patterns. *Z. Met.* **1959**, *50*, 708–716.
43. Jephcoat, A.P.; Hemley, R.J.; Mao, H.K. X-ray Diffraction of Ruby (Al_2O_3:Cr_3^+) to 175 GPa. *Phys. B+C* **1988**, *150*, 115–121. [CrossRef]
44. Klotz, S.; Chervin, J.-C.C.; Munsch, P.; Le Marchand, G. Hydrostatic Limits of 11 Pressure Transmitting Media. *J. Phys. D Appl. Phys.* **2009**, *42*, 075413. [CrossRef]
45. Garg, A.B.; Shanavas, K.V.; Wani, B.N.; Sharma, S.M. Phase Transition and Possible Metallization in $CeVO_4$ under Pressure. *J. Solid State Chem.* **2013**, *203*, 273–280. [CrossRef]

Disclaimer/Publisher's Note: The statements, opinions and data contained in all publications are solely those of the individual author(s) and contributor(s) and not of MDPI and/or the editor(s). MDPI and/or the editor(s) disclaim responsibility for any injury to people or property resulting from any ideas, methods, instructions or products referred to in the content.

Article

Strain-Rate Dependence of Plasticity and Phase Transition in [001]-Oriented Single-Crystal Iron

Nourou Amadou [1,2,*], Abdoul Razak Ayouba Abdoulaye [2], Thibaut De Rességuier [1,*] and André Dragon [1]

[1] Institut Pprime, CNRS, ENSMA. Université de Poitiers, F-86961 Futuroscope, France
[2] Département de Physique, Université Abdou Moumouni de Niamey, Niamey P.O. Box 10662, Niger
* Correspondence: nourou.amadou@ensma.fr (N.A.); resseguier@ensma.fr (T.D.R.)

Abstract: Non-equilibrium molecular dynamics simulations have been used to investigate strain-rate dependence of plasticity and phase transition in [001]-oriented single-crystal iron under ramp compression. Here, plasticity is governed by deformation twinning, in which kinetics is tightly correlated with the loading rate. Over the investigated range of strain rates, a hardening-like effect is found to shift the onset of the structural bcc-to-hcp phase transformation to a high, almost constant stress during the ramp compression regime. However, when the ramp evolves into a shock wave, the bcc–hcp transition is triggered whenever the strain rate associated with the plastic deformation reaches some critical value, which depends on the loading rate, leading to a constitutive functional dependence of the transition onset stress on the plastic deformation rate, which is in overall consistence with the experimental data under laser compression.

Keywords: plasticity; iron; alpha–epsilon; phase transition; molecular dynamics simulations; ramp; shock wave; hardening-like effect

1. Introduction

Over the past few decades, there has been an increasing interest in using dynamic compression at extremely high strain rates to investigate both plasticity and pressure-induced phase transition [1–3]. Usually, such studies are performed under the conditions of uniaxial strain, which involve large deviatoric stresses. High strain rates affect material behavior in both the mechanisms of plastic deformation and the kinetics of polymorphic phase transitions [4,5]. They can even affect melting temperatures at high pressures [6]. Shock-loading experiments were classically realized under planar plate impacts where the strain rates are typically about 10^4 to 10^6 s^{-1} [7–9]. Then, iron was found to plastically yield at a stress of 0.92 to 1.3 GPa, while the ground-state body-centered cubic (bcc) structure (α phase) was found to transform into the high-pressure hexagonal close-packed (hcp) structure (ε phase) at a stress σ_T of about 12.88 to 14.26 GPa, which is higher for single-crystal than polycrystalline iron [7,9,10]. Higher strain rates can be achieved under laser ramp compression. Indeed, strain rates from 3 to ~9 × 10^7 s^{-1} were reported by Amadou et al. [11] under nanosecond laser compression, where the phase transformation onset stress, σ_T, was found to vary from 11 to 25 GPa with a constant completion time of 1 ns, suggesting the existence of an isokinetic regime over the explored range of strain rates. Furthermore, an extended analysis of strain-rate effects, up to 10^8 s^{-1}, on both plastic flow and phase transition kinetics in iron was reported by Smith et al. [12,13], where the elastic limit (σ_E) in polycrystalline iron was found to vary from ~1 to ~5.5 GPa, while σ_T ranged from ~14 to ~40 GPa. Moreover, a constitutive functional dependence on the plastic deformation strain rate, $\dot{\varepsilon}_P$, was evidenced for both σ_E and σ_T. At strain rates higher than about 5 × 10^6 s^{-1}, a sharp increase in σ_E with increasing $\dot{\varepsilon}_P$ was interpreted as a transition in the plastic flow regime from thermally activated to phonon drag dislocation dynamics, which affects the structural phase transition kinetics by limiting the new phase growth

through energy dissipation [13]. In this regime, σ_T was found to scale as $\dot{\varepsilon}_P^{0.18}$. A strain rate on the order of about 10^8 s^{-1} was also reported by Hawreliak et al. [14]. However, σ_T was found to be much lower, at about 15 GPa, with no obvious correlation with the strain rate. Even higher strain rates up to about 10^9 s^{-1} are currently available under picosecond laser compression [15,16]. In this ultra-fast deformation regime, Crowhurst et al. [17] and Hwang et al. [18] found σ_T values of 25 and 34 GPa, respectively.

These experimental ultra-high strain rates are comparable to those currently available in molecular dynamic simulations. Thus, Gunkelmann et al. [19], using the modified version of the Ackland EAM potential [20] to study the behavior of polycrystalline iron at a strain rate of about 10^9 s^{-1}, showed that yielding occurs through dislocation activities around the grain boundary at 10 GPa with the onset stress of a structural bcc-to-hcp phase transformation σ_T of about 23 GPa. A strain-rate regime ranging from 10^9 s^{-1} to 10^{11} s^{-1} was explored by Wang et al. [21] using a modified analytic embedded atom model [22], and they found that σ_T varied from 25 to 38 GPa with an exponential dependence in $\dot{\varepsilon}_P$, $\sigma_T \propto \dot{\varepsilon}_P^{0.196}$. Finally, using Voter–Chen potential [23], singularized by a lack of plasticity before the structural phase transformation [24], Shao et al. reported a linear dependence of σ_T on $\dot{\varepsilon}_P$ [25] over a strain rate range from 10^{10} s^{-1} to 10^{11} s^{-1} with σ_T ranging from 15 to 25 GPa. Thus, despite extensive research, the dynamic response of iron, including the bcc–hcp phase transition, its kinetics, and its strain-rate dependence under high strain rates still remains an open issue. In this context, using the modified version of the Ackland iron potential [20], we have reported in previous papers that defect-free single-crystal iron at a 50 K initial temperature subjected to ramp compression along the [001] direction exhibits a hardening-like effect, which has been shown to inhibit the nucleation of the hcp phase so that the onset of the phase transformation is shifted to very high pressures (on the order of 100 GPa) [5,26–28]. This remarkable effect, never reported before, deserves extensive study because it may contribute to the strong variations in the transition onset pressure with the loading rate classically observed experimentally. Here, we go further to explore the strain-rate dependence of both plasticity and phase transition and to investigate the influence of the hardening-like effect on the $\sigma_T - \dot{\varepsilon}_P$ relationship. Because we reported that this hardening-like effect is more important when the ramp wave propagates along the [001] direction than along other low-index crystallographic directions [26], we choose to simulate the response of a [001]-oriented single-crystal iron.

2. Method and Computational Details

Samples with up to 28 million atoms, $100a_0 \times 100a_0$ cross-section, and a length up $1400a_0$ (with a_0 = 2.87 Angstrom, the lattice constant) were simulated using the Large-scale Atomic/Molecular Massively Parallel Simulator (LAMMPS) molecular dynamics code [29]. The interactions between atoms were modeled through the embedded atom model (EAM) formalism [30,31], where the iron model was a modified version of the Ackland potential [20], which was successfully used to address both plasticity and phase transition in iron [5,19,26]. The sample was thermalized at a 50 K initial temperature, then dynamic ramp-wave compression was realized by driving an effective infinite-mass wall piston with an imposed velocity $v(t)$ along the z-axis, oriented along the [001] crystallographic direction, while periodic boundary conditions were used for the transverse directions. In order to investigate strain-rate effects, the piston velocity was increased linearly from 0 to 1600 m·s^{-1} with a rise time varying from 15 to up to 150 ps. The corresponding loading rates were about $1–10 \times 10^9$ s^{-1}, which is comparable with that reported under picosecond laser dynamic compression [15–18]. The local thermodynamic and mechanical properties, such as longitudinal stress σ_z, shear stress, temperature, etc., were evaluated within a spatial planar bin (of 3-lattice constant width) perpendicular to the wave propagation direction in the same manner as our previous work [5,26,27]. Finally, local structural analysis was performed by adaptative common neighbor analysis (CNA), centrosymmetry, and DXA, as implemented in OVITO software (https://www.ovito.org, accessed on 18 January 2023) [32].

3. Results and Discussion

Figure 1 shows snapshots of atom configurations for simulations where the ramp rise time was 30, 45, and 60 ps. At each loading rate, the compressed material exhibits four distinct regimes: an elastic compression of bcc-iron (yellow zone); a regime where defects associated with plastic deformation can be observed (green zone with brown micro-features identified as twins); a regime where these defects associated with plastic deformation progressively disappear (blue zone); and a regime where iron is transformed from bcc to hcp phase (violet zone). More details on wave propagation and splitting of the compression fronts can be found in ref. [5]. At each loading rate, the sample is found to yield at a stress σ_z of about 12 GPa via the generation of micro-twins with no sizeable influence on the loading rate. The fact that twinning deformation is independent of the loading rate is consistent with the well-known existence of the twinning threshold in bcc materials, such as tantalum, reported under both experiments [1,33] and molecular dynamics simulations [34]. Thus, regardless of the loading rate, twinning is found to be the main deformation mechanism for the defect-free single-crystal iron ramp compressed along the [001] crystallographic direction.

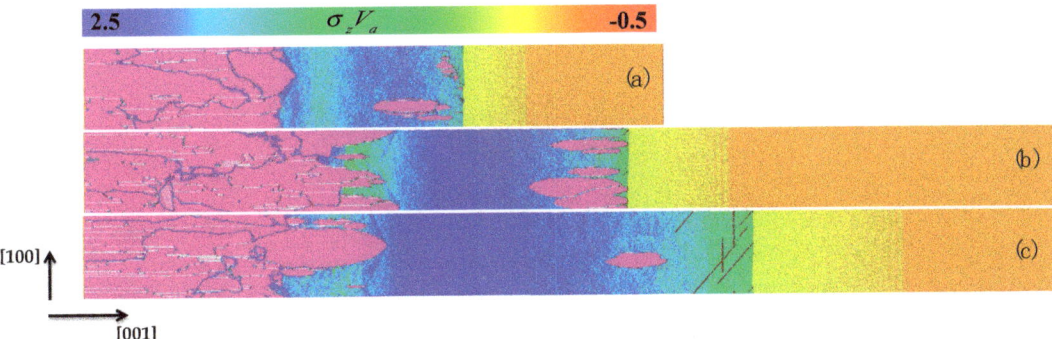

Figure 1. Snapshots of the atom configurations at times 33 ps (**a**), 48 ps (**b**), and 60 ps (**c**) under ramp compression rise times of 30 ps (**a**), 45 ps (**b**), and 60 ps (**c**). The bcc atoms are colored according to their σ_z Va value, where Va is the atomic volume, and σ_z is the stress component along the wave propagation z-axis. The hcp and fcc atoms, detected by the adaptative CNA analysis, are colored in magenta and white, while twins (detected by centrosymmetry analysis) are colored in brown (regardless of the stress). A double nucleation front of the hcp phase (violet) can be observed since the transformation occurs (i) at the top of the pressure ramp propagating from left to right and (ii) where the ramp wave steepens into a shock (see text for more details). Note that a longer ramp rise time requires longer propagation distance. Thus, the sample length in the wave propagation direction was 800 a_0 for (**a**), while it was 1400 a_0 for both (**b**) and (**c**).

Figure 2 shows the evolution of the twin fraction during the compression for a ramp rise time of 15, 30, 45, and 60 ps. After the nucleation period, the twin fraction grows rapidly, reaching its maximum within 3–4 ps. Then, the twin fraction starts to decrease, i.e., twins formed under moderate pressure are removed upon further compression, which we refer to as a hardening-like effect (see Figure 1, blue zone). The longer the ramp rise time, the more important the twins' fraction and the duration of the subsequent receding phase. Thus, upon increasing the ramp rise time, the maximum twins' fraction increases from 28% to a value as high as ≈67%, in consistence with the 50% twins' fraction usually assumed under dynamic loading [1,33]. Furthermore, this increase in the twins' fraction upon increasing the ramp duration is consistent with both experimental observations [35] and thermo-mechanical simulations, including a model of twins' nucleation and growth [36]. In the region affected by the hardening-like effect (blue zone), where twins are almost fully removed and no dislocation activities can be detected, the shear stress increases with compression, as can be seen in Figure 3. Such elastic stiffening of the bcc matrix leads to a confinement effect inhibiting the nucleation of the hcp phase. This confinement effect has been interpreted in the context of classical nucleation and growth theory, and it has been shown that in order to activate new phase nucleation two different scenarios are possible [5]: (i) ramp compressing must be kept at a much higher pressure by increasing the piston maximum velocity or (ii) the ramp wave, which steepens during its propagation due to the increase in sound velocity with pressure in the bcc phase, should be allowed to evolve into a shock wave. Thus, for the simulations shown in Figure 1, both piston maximum velocity and propagation distance are enough that two nucleation fronts can be observed. The first front, on the left part of the sample, corresponds to the top of the ramp, where pressure is sufficiently high for the hcp nuclei to grow and reach a critical size, despite a rigid, confining bcc matrix due to the hardening-like effect. The second nucleation front, in the middle of the sample, corresponds to the second scenario above (ii), where the so-called P1 wave evolves into a shock wave during its propagation.

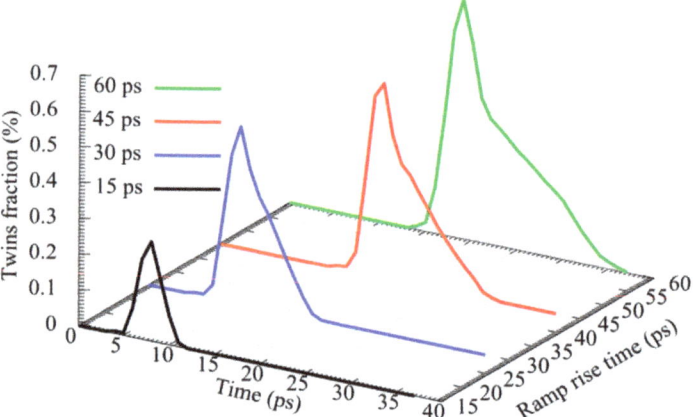

Figure 2. Temporal evolution of the twin fraction for various ramp rise times.

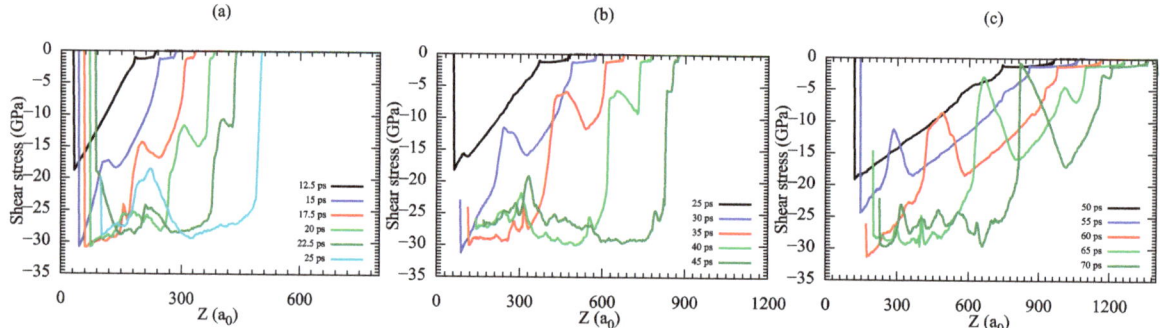

Figure 3. Shear stress profiles along the loading direction Z for ramp rise times of 15 ps (**a**), 30 ps (**b**), and 60 ps (**c**).

Thus, the resulting instantaneous thermodynamic driving force is high enough for hcp embryos to grow and reach a critical size almost immediately behind the shock front [37]. If the ramp rise time is increased to a sufficiently high value, the P1 wave does not form a shock upon the simulation duration. Thus, the second nucleation front can no longer be observed for a ramp rise time above 150 ps.

Figure 4 shows the structural phase transition onset stress, σ_T, as a function of the ramp rise time for both scenarios (i) and (ii) mentioned above. In the first scenario, upon increasing the ramp rise time, i.e., decreasing the loading strain rate, the phase transition onset stress can be observed to remain almost constant, at about 100 GPa (red triangles), so that no loading rate dependence can be observed. This observation can be interpreted as follows: As the hardening-like effect confines the hcp embryos, only the material strength determines the maximum pressure supported by the bcc matrix. Yet, it has been shown that above a critical strain rate of 10^5 s^{-1}, the material strength becomes strain-rate independent [38,39]. This result highlights the crucial role of the strength of the parent phase matrix in the dynamics of structural phase transformation due to the generation of the elastic strains necessary to accommodate the difference between the parent and daughter crystalline structures.

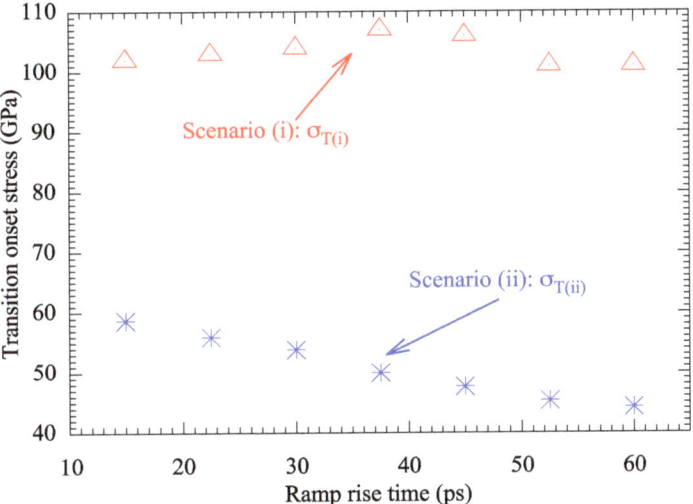

Figure 4. The structural α to ε phase transformation onset stress as a function of the ramp rise time according to scenario (i) where high compression overcomes the hardening-like effect (red triangles) and scenario (ii) where the ramp wave steepens into a shock (blue stars).

Although the transition onset stress in scenario (i) remains almost constant after the hardening-like effect, the pressure relaxation accompanying the structural phase transformation is found to be rate-dependent (Figure 5). Indeed, this relaxation is due to the coexistence of both bcc and hcp phases in the sample, which causes a drop in the sound velocity that evolves as a result of the balance between the hcp nucleation and loading rates [13]. Moreover, increasing the ramp compression duration seems to enhance both the size and the ovaloid shape of the hcp nuclei (see Figure 6).

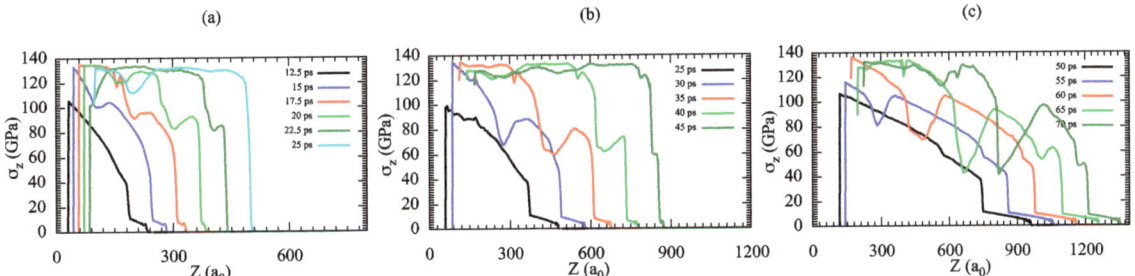

Figure 5. Stress profiles along the loading direction, Z, for the ramp rise time of 15 ps (**a**), 30 ps (**b**), and 60 ps (**c**), corresponding to Figure 3.

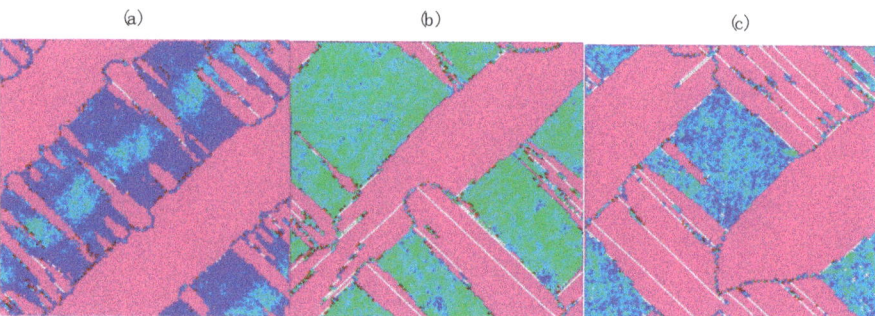

Figure 6. Snapshots of atom configurations showing a cross-section of the phase transformation front once compression overcomes the hardening-like effect scenario (i) for a ramp compression rise time of 30 ps (**a**), 45 ps (**b**), and 60 ps (**c**).

Ramp compression can be seen as a succession of infinitesimal compression waves, which all propagate at the speed of sound in the upstream medium. Because this speed usually increases with pressure (above the elastic limit), these elementary waves catch up with each other so that the ramp becomes steeper with increasing propagation distance and evolves naturally into shock waves. Its rise time and the associated plastic deformation strain rate, $\dot{\varepsilon}_P$, is thought to be associated with the microscopic processes of dislocation multiplication and motion, twinning, vacancy production, precipitate alteration, etc. [40,41]. Figure 5 clearly shows such gradual steepening of the compressive wave beyond the elastic limit (Pz > 12 GPa). This wave, usually referred to as the P1 wave, is associated to the plastic compression of the bcc phase, up to the onset of the transition to hcp. The average plastic deformation strain rate, $\dot{\varepsilon}_P$, associated with the P1 wave is calculated as follows [41,42]:

$$\dot{\varepsilon}_P = \frac{1}{\Delta z_{P_1}} \int_{\Delta z_{P_1}} \dot{\varepsilon}(z) dz \qquad (1)$$

where Δz_{P_1} is the width of the P$_1$ wave between the elastic front and the bcc–hcp phase transition front in the scenario (i), as can be seen in Figure 7. The evolution of $\dot{\varepsilon}_P$ as a function of time for various ramp rise times is presented in Figure 8. It ranges from a few times 10^9 s^{-1} to about 10^{11} s^{-1}, in consistence with the values reported under the picosecond laser compression experiment [16,17]. As expected, when the ramp steepens, $\dot{\varepsilon}_P$ increases as a function of time more rapidly for lower ramp rise time than for higher ramp rise time [43]. For each rise time, $\dot{\varepsilon}_P$ shows a vertical asymptotic upward variation corresponding to the shock formation during the wave propagation through the sample. The lower the loading ramp rise time, the earlier the shock formation. Due to the complexity of the plastic response involved here, i.e., twins' formation, hardening-like effect, and phase transition, including hcp phase nucleation and growth, the time taken for steepening is much lower than the ratio of ~1.4 to 1.6 times the rise time of the ramp found from the analytical models and MD simulations of simple materials such as Al and Cu [43,44].

The bcc-to-hcp phase transition is triggered in scenario (ii) whenever the plastic strain rate, $\dot{\varepsilon}_P$, reaches a certain critical value that depends on the loading conditions (see Figure 8). This critical $\dot{\varepsilon}_P^c$ ranges from 6.71×10^{10} s^{-1} to 0.88×10^{10} s^{-1} when the ramp loading rise time increases from 15 to 60 ps. Indeed, when these critical values are reached, the dynamic loading switches from ramp to shock compression, where the material is carried almost instantaneously from the thermodynamic stability conditions of the bcc phase to those of the hcp phase. The resulting driving force is high enough for hcp embryos to reach a stable critical volume almost instantly behind the shock front. Thus, the hcp phase starts to nucleate behind the elastic–plastic transformation front leading to a second phase transition front in the sample, referenced above as scenario (ii), as can be seen in Figure 1. In this scenario, the transition onset stress, σ_T, can be observed to decrease from 60 to 42 GPa upon increasing the ramp rise time from 15 to 60 ps. Thus, in contrast with scenario (i), σ_T is rate-dependent. It increases with the strain rate, which is due to the kinetics of the transformation, and is consistent with experimental observations [11,13,45].

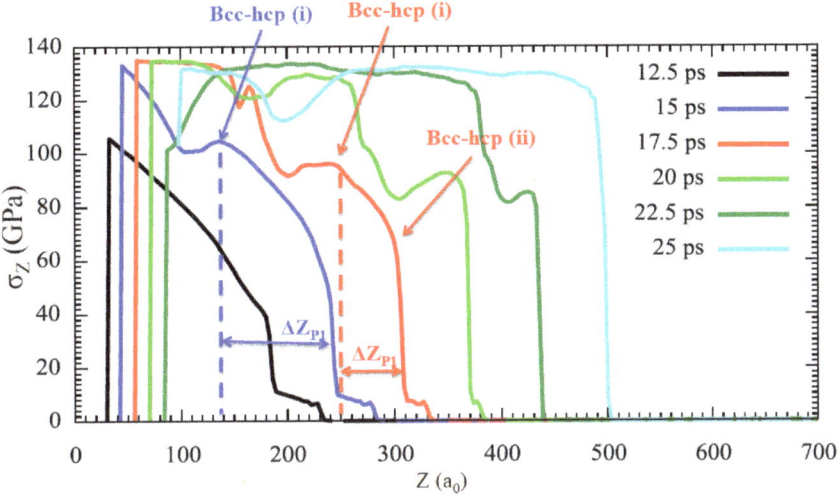

Figure 7. Stress profiles at successive times showing the evolution of the ramp compression, which steepens into a shock wave during its propagation. The width of the P1 wave is used to evaluate the plastic deformation strain rate. The bcc–hcp phase transformation occurs either under ramp compression on a ~100 GPa order scenario (i) or upon an increase in the strain-rate scenario (ii).

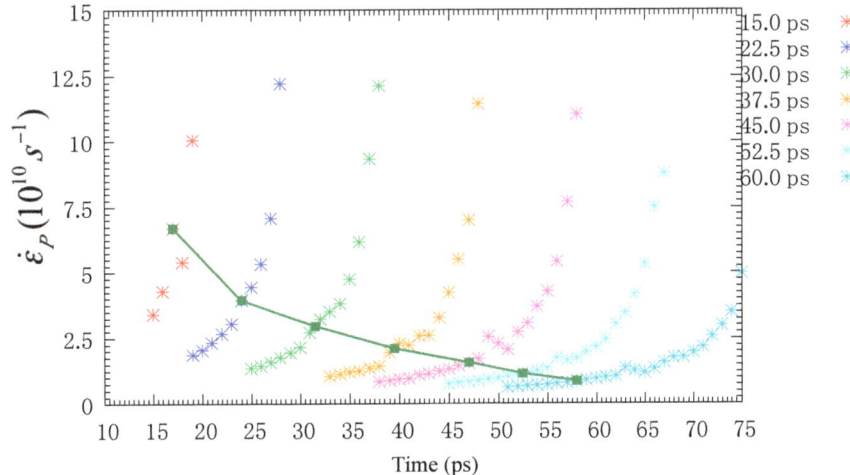

Figure 8. Time evolution of the plastic strain rate (stars) during ramp wave propagation through the sample with the critical plastic strain rate, $\dot{\varepsilon}_P^c$, (forest-green line) evaluated at the beginning of the hcp phase nucleation behind the compression front, once the ramp wave has steepened into a shock.

Figure 9 shows the variation in σ_T, as a function of $\dot{\varepsilon}_P^c$ (red stars), in comparison with various data reported in the literature under both experiments and MD simulations. The black line is the prediction of the Swegle–Grady (SG) law $\sigma_T = \alpha \dot{\varepsilon}_P^\beta$ [40], using the parameters reported in ref. [14]. The red curve presents the fit of our data using a similar law where $\alpha = 44.88 \pm 0.47$ and $\beta = 0.15 \pm 0.01$. MD simulations are from Wang et al. [21], using a MEAM potential where the plasticity is dominated by twins; Shao et al. [25], using the Voter–Chen EAM potential [23], which does not predict plastic deformation before the structural phase transformation; and Gunkelmann et al. [19], using the same potential as we did.

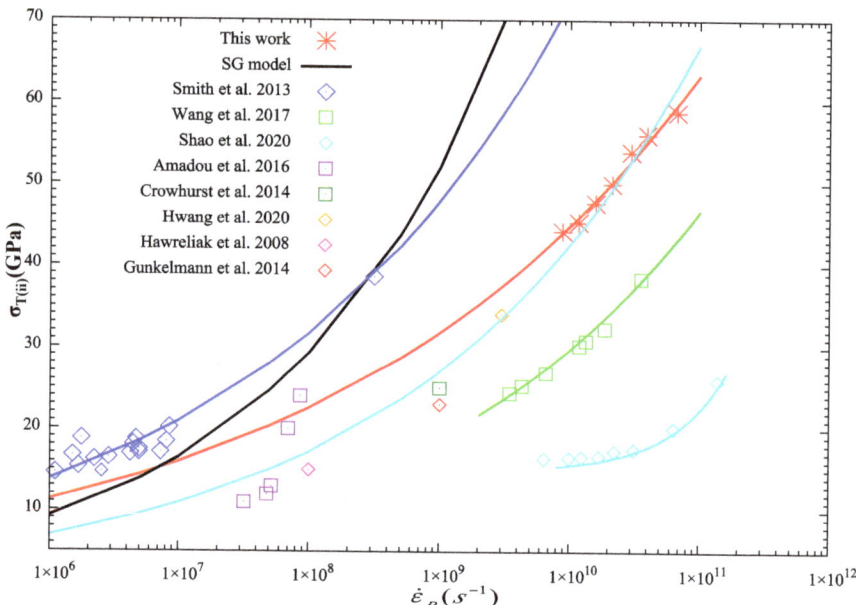

Figure 9. Variation in the onset stress, σ_T (in scenario (ii)), with the plastic strain rate, $\dot{\varepsilon}_P^c$ (red stars), compared with various data reported in the literature in both experiments and MD simulations, see text for more details.

Experimental data come from Smith et al. [13], where plastic deformation was thought to be dominated by phonon scattering from defects, nanosecond laser ramp compression iron by Amadou et al. [11] and Hawreliak et al. [14], or picosecond laser compression by Crowhurst et al. [17] and Hwang et al. [18]. Although our data deviate from the predictions by both the original SG model and Smith et al.'s constitutive relationship [13] at an ultra-high strain rate, their extrapolation to lower strain rates using this updated SG fit is consistent with the experimental data, including those by Amadou et al. [11], Smith et al. [13], and Hwang et al. [18]. On the other hand, the magnitude of the power law exponent in our simulations is slightly different from the 0.196 value reported by Wang et al. [21] in single-crystal iron for the same crystallographic orientation under MD simulations using the MEAM interatomic potential. This discrepancy is thought to be due to the difference in the underlying plastic deformation micro-process predicted by the different potentials. Indeed, in our simulations, twinning is followed by a hardening-like effect under further compression, while such a post-twinning effect is not predicted by the MEAM potential used by Wang et al.

4. Conclusions

NEMD simulations were used to investigate the strain-rate dependence of plasticity and phase transition in the [001]-oriented single-crystal iron at 50 K under dynamic ramp compression between 10^9 s^{-1} and 10^{10} s^{-1}. Iron was found to yield at 12 GPa, regardless of the loading rate, through the generation of micro-twins, which grew rapidly to a peak fraction before receding upon further compression. The longer the ramp rise time, the higher the maximum twins' fraction and the longer the subsequent receding regime, in consistence with both experimental and theoretical observations in iron. As reported previously, twin recession in the absence of any dislocation slip induced a hardening-like effect, which shifted the phase transition onset stress to a very high value of ≈100 GPa scenario (i) independently of the loading rate. On the other hand, the phase transition could be triggered at lower stresses when the ramp evolved into shock wave scenario (ii). Then, the onset stress σ_T was evidenced to be strain-rate-dependent. Indeed, the transition was triggered whenever the strain rate associated with the plastic deformation reached some critical value, $\dot{\varepsilon}_P^c$. The higher the ramp compression time, the lower both $\dot{\varepsilon}_P^c$ and σ_T. Thus, the onset stress, σ_T, in this scenario has been shown to follow a Swegle–Grady power law type [40] in $\dot{\varepsilon}_P^c$ with an exponent of 0.15, i.e., $\sigma_T \propto \dot{\varepsilon}_P^{c\,0.15}$, which is in overall consistence with some experimental data under laser compression.

Author Contributions: Conceptualization, N.A.; methodology, N.A. and T.D.R.; software, N.A.; validation, N.A. and T.D.R.; formal analysis, N.A. and A.R.A.A.; investigation, N.A.; resources, T.D.R.; data curation, N.A. and A.R.A.A.; writing—original draft preparation, N.A.; writing—review and editing, N.A., T.D.R., and A.D.; visualization, N.A. and A.R.A.A.; supervision, N.A.; project administration, N.A.; funding acquisition, T.D.R. and A.D. All authors have read and agreed to the published version of the manuscript.

Funding: This research was funded by a grant allocated by the *Conseil Scientifique de l'ENSMA* on 1 December 2021.

Data Availability Statement: The raw/processed data required to reproduce these findings cannot be shared at this time as the data also form part of an ongoing study.

Acknowledgments: Computations were performed on the supercomputer facilities of the Mésocentre de calcul de Poitou Charentes (France). We thank Mikaël Gueguen and Gérald Sailly (Institut Pprime) for their kind help to set up work stations and access.

Conflicts of Interest: The authors declare no conflict of interest.

References

1. Meyers, M.A. *Dynamic Behavior of Materials*; John Wiley & Sons: Hoboken, NJ, USA, 1994; p. 1.
2. Davidson, R.C.; Arnett, D.; Dahlburg, J.; Dimotakis, P.; Dubin, D.; Gabrielse, G.; Hammer, D.; Katsouleas, T.; Kruer, W.; Lovelas, R.; et al. *Frontiers in High Energy Density Physics: The X-Games of Contemporary Science*; The National Academies Press: Washington, DC, USA, 2003.
3. Lorenzana, H.E.; Belak, J.F.; Bradley, K.S.; Bringa, E.M.; Budil, K.S.; Cazamias, J.U.; El-Dasher, B.; Hawreliak, J.A.; Hessler, J.; Kadau, K.; et al. Shocked materials at the intersection of experiment and simulation. *Sci. Model. Simul.* **2008**, *15*, 159–186. [CrossRef]
4. Rudd, R.E.; Germann, T.C.; Remington, B.A.; Wark, J.S. Metal deformation and phase transitions at extremely high strain rates. *MRS Bull.* **2010**, *35*, 999–1006. [CrossRef]
5. Amadou, N.; de Resseguier, T.; Dragon, A.; Brambrink, E. Coupling between plasticity and phase transition in shock- and ramp-compressed single-crystal iron. *Phys. Rev. B* **2018**, *98*, 024104. [CrossRef]
6. Baty, S.R.; Burakovsky, L.; Errandonea, D. Ab Initio Phase Diagram of Copper. *Crystals* **2021**, *11*, 537. [CrossRef]
7. Barker, L.M.; Hollenbach, R.E. Shock wave study of the $\alpha \rightleftharpoons \varepsilon$ phase transition in iron. *J. Appl. Phys.* **1974**, *45*, 4872. [CrossRef]
8. Arnold, W. *Dynamisches Werkstoffverhalten von Armco-Eisen bei Stosswellenbelastung*; VDI-Verlag: Duesseldorf, Germany, 1992.
9. Jensen, B.; Gray, G.T., III; Hixson, R.S. Direct measurements of the $\alpha-\varepsilon$ transition stress and kinetics for shocked iron. *J. Appl. Phys.* **2009**, *105*, 103502. [CrossRef]
10. Boettger, J.C.; Wallace, D.C. Metastability and dynamics of the shock-induced phase transition in iron. *Phys. Rev. B* **1997**, *55*, 2840–2849. [CrossRef]
11. Amadou, N.; de Resseguier, T.; Brambrink, E.; Vinci, T.; Benuzzi-Mounaix, A.; Huser, G.; Morard, G.; Guyot, F.; Miyanishi, K.; Ozaki, N.; et al. Kinetics of the iron $\alpha-\varepsilon$ phase transition at high-strain rates: Experiment and model. *Phys. Rev. B* **2016**, *93*, 214108. [CrossRef]
12. Smith, R.F.; Eggert, J.H.; Rudd, R.E.; Swift, D.C.; Bolme, C.A.; Collins, G.W. High strain-rate plastic flow in Al and Fe. *J. Appl. Phys.* **2011**, *110*, 123515. [CrossRef]
13. Smith, R.F.; Eggert, J.H.; Swift, D.C.; Wang, J.; Duffy, T.S.; Braun, D.G.; Rudd, R.E.; Reisman, D.B.; Davis, J.-P.; Knudson, M.D.; et al. Time-dependence of the alpha to epsilon phase transformation in iron. *J. Appl. Phys.* **2013**, *114*, 223507. [CrossRef]

14. Hawreliak, J.A.; El-Dasher, B.; Lorenzana, H.; Kimminau, G.; Higginbotham, A.; Nagler, B.; Vinko, S.M.; Murphy, W.J.; Whitcher, T.; Wark, J.S.; et al. In situ X-ray diffraction measurements of the c/a ratio in the high-pressure ε phase of shock-compressed polycrystalline iron. *Phys. Rev. B* **2011**, *83*, 144114. [CrossRef]
15. Ashitkov, S.; Komarov, P.; Romashevskiy, S.; Struleva, E.; Evlashin, S. Shock compression of magnesium alloy by ultrashort loads driven by sub-picosecond laser pulses. *J. Appl. Phys.* **2022**, *132*, 175104. [CrossRef]
16. Crowhurst, J.C.; Armstrong, M.R.; Knight, K.B.; Zaug, J.M.; Behymer, E.M. Invariance of the Dissipative Action at Ultrahigh Strain Rates Above the Strong Shock Threshold. *Phys. Rev. Lett.* **2011**, *107*, 144302. [CrossRef] [PubMed]
17. Crowhurst, J.C.; Reed, B.W.; Armstrong, M.R.; Radousky, H.B.; Carter, J.A.; Swift, D.C.; Zaug, J.M.; Minich, R.W.; Teslich, N.E.; Kumar, M. The phase transition in iron at strain rates up to 10^9 s^{-1}. *J. Appl. Phys.* **2014**, *115*, 113506. [CrossRef]
18. Hwang, H.; Galtier, E.; Cynn, H.; Eom, I.; Chun, S.H.; Bang, Y.; Hwang, G.C.; Choi, J.; Kim, T.; Kong, M.; et al. Subnanosecond phase transition dynamics in laser-shocked iron. *Sci. Adv.* **2020**, *6*, eaaz5132. [CrossRef]
19. Gunkelmann, N.; Bringa, E.M.; Tramontina, D.R.; Ruestes, C.J.; Suggit, M.J.; Higginbotham, A.; Wark, J.S.; Urbassek, H.M. Shock waves in polycrystalline iron: Plasticity and phase transitions. *Phys. Rev. B* **2014**, *89*, 140102. [CrossRef]
20. Gunkelmann, N.; Bringa, E.M.; KKang, K.; Ackland, G.J.; Ruestes, C.J.; Urbassek, H.M. Polycrystalline iron under compression: Plasticity and phase transitions. *Phys. Rev. B* **2012**, *86*, 144111. [CrossRef]
21. Wang, K.; Chen, J.; Zhu, W.; Hu, W.; Xiang, M. Phase transition of iron-based single crystals under ramp compressions with extreme strain rates. *Int. J. Plast.* **2017**, *96*, 56–80. [CrossRef]
22. Wang, K.; Xiao, S.; Deng, H.; Zhu, W.; Hu, W. An atomic study on the shock-induced plasticity and phase transition for iron-based single crystals. *Int. J. Plast.* **2014**, *59*, 180–198. [CrossRef]
23. Harrison, R.; Voter, A.F.; Chen, S.-P. *Atomistic Simulation of Material*; Vitek, V., Srolovitz, D.J., Eds.; Plenum: New York, NY, USA, 1989; p. 219.
24. Kadau, K.; Germann, T.C.; Lomdahl, P.S.; Holian, B.L. Atomistic simulations of shock-induced transformations and their orientation dependence in bcc Fe single crystals. *Phys. Rev. B* **2005**, *72*, 064120. [CrossRef]
25. Shao, J.-L.; He, W.; Xi, T.; Xin, J. Microscopic insight into the structural transition of single crystal iron under the ramp wave loading. *Comput. Mater. Sci.* **2020**, *182*, 109772. [CrossRef]
26. Amadou, N.; de Resseguier, T.; Dragon, A.; Brambrink, E. Effects of orientation, lattice defects and temperature on plasticity and phase transition in ramp-compressed single crystal iron. *Comput. Mater. Sci.* **2020**, *172*, 109318. [CrossRef]
27. Amadou, N.; de Resseguier, T.; Dragon, A. Coupling between plasticity and phase transition in single crystal iron at ultra-high strain rate. *AIP Conf. Proc.* **2020**, *2272*, 070001.
28. Amadou, N.; de Resseguier, T.; Dragon, A. Influence of point defects and grain boundaries on plasticity and phase transition in uniaxially-compressed iron. *Comput. Condens. Matter* **2021**, *27*, e00560. [CrossRef]
29. Plimpton, S. Fast Parallel Algorithms for Short-Range Molecular Dynamics. *J. Comp. Phys.* **1995**, *117*, 1–19. [CrossRef]
30. Daw, M.S.; Baskes, M.I. Semiempirical, Quantum Mechanical Calculation of Hydrogen Embrittlement in Metals. *Phys. Rev. Lett.* **1983**, *50*, 1285–1288. [CrossRef]
31. Foiles, S.M.; Baskes, M.I.; Daw, M.S. Embedded-atom-method functions for the fcc metals Cu, Ag, Au, Ni, Pd, Pt, and their alloys. *Phys. Rev. B* **1986**, *33*, 7983–7991. [CrossRef] [PubMed]
32. Stukowski, A. Visualization and analysis of atomistic simulation data with OVITO–the Open Visualization Tool, Modelling Simul. *Mater. Sci. Eng.* **2010**, *18*, 015012.
33. Wehrenberg, C.E.; McGonegle, D.; Bolme, C.; Higginbotham, A.; Lazicki, A.; Lee, H.J.; Nagler, B.; Park, H.-S.; Remington, B.A.; Rudd, R.E.; et al. In situ X-ray diffraction measurement of shock-wave-driven twinning and lattice dynamics. *Nature* **2017**, *550*, 496. [CrossRef]
34. Zepeda-Ruiz, L.A.; Stukowski, A.; Oppelstrup, T.; Bulatov, V.V. Probingthe limits of metal plasticity with molecular dynamics simulations. *Nature* **2017**, *550*, 492. [CrossRef]
35. Stone, G.A.; Orava, R.N.; Gray, G.T.; Pelton, A.R. An Investigation of the Influence of Shock-Wave Profile on the Mechanical and Thermal Responses of Polycrystalline Iron. U. S. Army Research Office, Report Number SMTJ-78 1978. p. 30. Available online: https://archive.org/details/DTIC_ADA049764 (accessed on 23 January 2022).
36. de Resseguier, T.; Hallouin, M. Stress relaxation and precursor decay in laser shock-loaded iron. *J. Appl. Phys.* **1998**, *84*, 1932–1938. [CrossRef]
37. Knudson, M.D.; Gupta, Y.M. Transformation kinetics for the shock wave induced phase transition in cadmium sulfide crystals. *J. Appl. Phys.* **2002**, *91*, 9561–9571. [CrossRef]
38. Steinberg, D.J.; Cochran, S.G.; Guinan, M.W. A constitutive model for metals applicable at high-strain rate. *J. Appl. Phys.* **1980**, *51*, 1498–1504. [CrossRef]
39. Remington, B.A.; Allen, P.; Bringa, E.M.; Hawreliak, J.; Ho, D.; Lorenz, K.T.; Lorenzana, H.; McNaney, J.M.; Meyers, M.A.; Pollaine, S.W.; et al. Material dynamics under extreme conditions of pressure and strain rate. *Mater. Sci. Technol.* **2006**, *22*, 474–488. [CrossRef]
40. Swegle, J.W.; Grady, D.E. Shock viscosity and the prediction of shock wave rise times. *J. Appl. Phys.* **1985**, *58*, 692–701. [CrossRef]
41. Grady, D.E. Structured shock waves and the fourth-power law. *J. Appl. Phys.* **2010**, *107*, 013506. [CrossRef]
42. Ravelo, R.; Germann, T.C.; Guerrero, O.; An, Q.; Holian, B.L. Shock-induced plasticity in tantalum single crystals: Interatomic potentials and large-scale molecular-dynamics simulations. *Phys. Rev. B* **2013**, *88*, 134101. [CrossRef]

43. Swift, D.C.; Kraus, R.G.; Loomis, E.N.; Hicks, D.G.; McNaney, J.M.; Johnson, R.P. Shock formation and the ideal shape of ramp compression waves. *Phys. Rev. E* **2008**, *78*, 066115. [CrossRef]
44. Higginbotham, A.; Hawreliak, J.; Bringa, E.M.; Kimminau, G.; Park, N.; Reed, E.; Remington, B.A.; Wark, J.S. Molecular dynamics simulations of ramp-compressed copper. *Phys. Rev. B* **2012**, *85*, 024112. [CrossRef]
45. Amadou, N.; Brambrink, E.; de Rességuier, T.; Manga, A.O.; Aboubacar, A.; Borm, B.; Molineri, A. Laser-Driven Ramp Compression to Investigate and Model Dynamic Response of Iron at High Strain Rates. *Metals* **2016**, *6*, 320. [CrossRef]

Disclaimer/Publisher's Note: The statements, opinions and data contained in all publications are solely those of the individual author(s) and contributor(s) and not of MDPI and/or the editor(s). MDPI and/or the editor(s) disclaim responsibility for any injury to people or property resulting from any ideas, methods, instructions or products referred to in the content.

Article

Pressure-Induced Monoclinic to Tetragonal Phase Transition in $RTaO_4$ (R = Nd, Sm): DFT-Based First Principles Studies

Saheli Banerjee [1,2], Amit Tyagi [1] and Alka B. Garg [1,2,*]

1. High Pressure and Synchrotron Radiation Physics Division, Bhabha Atomic Research Centre, Mumbai 400085, India
2. Homi Bhabha National Institute, Anushakti Nagar, Mumbai 400094, India
* Correspondence: alkagarg@barc.gov.in

Abstract: In this manuscript, we report the density functional theory-based first principles study of the structural and vibrational properties of technologically relevant M′ fergusonite ($P2/c$)-structured $NdTaO_4$ and $SmTaO_4$ under compression. For $NdTaO_4$ and $SmTaO_4$, ambient unit cell parameters, along with constituent polyhedral volume and bond lengths, have been compared with earlier reported parameters for $EuTaO_4$ and $GdTaO_4$ for a better understanding of the role of lanthanide radii on the primitive unit cell. For both the compounds, our calculations show the presence of first-order monoclinic to tetragonal phase transition accompanied by nearly a 1.3% volume collapse and an increase in oxygen coordination around the tantalum (Ta) cation from ambient six to eight at phase transition. A lower bulk modulus obtained in the high-pressure tetragonal phase when compared to the ambient monoclinic phase is indicative of the more compressible unit cell under pressure. Phonon modes are calculated for the ambient and high-pressure phases with compression for both the compounds along with their pressure coefficients. One particular IR mode has been observed to show red shift in the ambient monoclinic phase, possibly leading to the instability in the compounds under compression.

Keywords: high pressure; rare earth tantalates; first principles calculations; Raman modes; phase transition

1. Introduction

RBO_4 (R: rare earth; B: a pentavalent cation such as V, W, Mo, Nb, Ta, As, and P) compounds are the subject of extensive research due to their promising applications in areas such as proton-conducting solid oxide fuel cells [1], and as a host for nuclear radioactive waste immobilization [2,3]. It has been well established that the B cation plays a crucial role in deciding the stable structure of RBO_4 compounds. Depending on the ratio of B cationic radii to lanthanide radii, RBO_4 compounds are reported to crystallize in either tetragonal (zircon, scheelite type) or monoclinic (fergusonite, monazite, wulframite type) structures [4]. Rare earth orthovanadate (RVO_4) are generally synthesized in a zircon (tetragonal) structure with a $I4_1/amd$ space group, while rare earth orthotungstates (RWO_4) and molybates ($RMoO_4$) have been reported to crystallize in scheelite (tetragonal) structure with $I4_1/a$ crystal symmetry [5–7]. A monoclinic M fergusonite structure with space group $I2/a$ has been established as a stable structure for rare earth orthoniobates ($RNbO_4$) at ambient temperature and pressure [8]. Depending upon the atomic radii of the lanthanide cation, the crystal structure of rare earth orthophosphate is either zircon (R < Gd) or monazite (R ≥ Gd) [9]. All the structures are closely related to each other by group–subgroup relations. A tetragonal scheelite structure ($I4_1/a$) is a subgroup of a tetragonal zircon ($I4_1/amd$) structure and the transformation between these two structures is generally of the first-order reconstructive type. A scheelite structure is transformed to a fergusonite structure by means of another translationgleiche, which involves lowering of point group symmetry from 4/m to 2/m [10]. Among all RBO_4 compounds, the $RTaO_4$ family of compounds exhibit polymorphism at ambient pressure and temperature conditions, which makes them of great interest from theoretical and technological point of views [11,12]. In $RTaO_4$ compounds, the final stable structure not only depends on the lanthanide radii but also depends on the heating temperature. Most of the compounds belonging to the $RTaO_4$ family stabilize either in a M fergusonite or M′ fergusonite structure at room temperature depending on the processing parameters. The main difference between the M fergusonite and M′ fergusonite structures is the oxygen coordination around the Ta atom. In a M fergusonite structure, the Ta atom is surrounded by an oxygen tetrahedra, whereas an oxygen octahedra is formed in a M′ fergusonite structure. In the M fergusonite phase, doubling of the b axis has been observed compared to M′ structure, while the other two unit cell axes have almost the same value in both structures. In recent times, $RTaO_4$ family compounds are being investigated extensively due to their potential applications in the field of scintillators and laser materials, owing to their high thermal stability and good chemical stability [13,14]. These compounds are also proposed as excellent alternatives to using Yttria-stabilized zirconia as ceramic thermal barrier coatings (TBCs) due to their lower thermal conductivity and better fracture toughness at high temperatures [15]. As is well known, most often it is the crystal structure that determines the properties of a material, and the structure may be altered by varying the thermodynamic parameters. Recently, by x-ray diffraction, the Raman spectroscopic technique and density functional theory (DFT)-based first principles calculations, we have shown the structural instability of $EuTaO_4$ and $GdTaO_4$ [16,17]. In continuation to this work, to understand the compression behavior of the $RtaO_4$ family of compounds, we have performed density functional theory-based first-principle simulations on technologically important $NdTaO_4$ and $SmTaO_4$. Wenhui Xiao et al. [18] have reported that the M′ fergusonite structure is more stable compared to the M fergusonite structure. Therefore, in the present work, we report the results from DFT-based first principles calculations on the M′ fergusonite structure under compression. We have also calculated the equation of state for both the compounds using the third-order Birch–Murnaghan equation of state (BM-EOS). Compressibility analysis of the simulated volume of constituent polyhedral units RO_8 and TaO_6 indicates that the major contribution to the bulk modulus comes from RO_8 polyhedra. This behavior validates Hazen and Finger's proposed empirical model for predicting the bulk modulus, with contributions from the rare earth polyhedral unit as seen in RVO_4, RWO_4, $RMoO_4$, $RNbO_4$ compounds [6,7,19–25]. We have also calculated the pressure evolution of the Raman and IR modes, which is consistent with earlier reported results for $EuTaO_4$ and $GdTaO_4$ for the same structure [16,17]. Further, we have compared the results from this work with previously investigated $EuTaO_4$ and $GdTaO_4$ to establish the role of lanthanide contraction present in the $RTaO_4$ family.

2. Computational Details

DFT-based first-principle simulations were carried out as implemented in Quantum Espresso [26] for determination of stable structures at ambient pressure as well as to investigate the influence of pressure on structural and vibrational properties. This is based on density functional theory, plane waves, and pseudopotentials. The projector-augmented wave (PAW) scheme [27], which describes electron–ion interactions, was employed as pseudo potential in self-consistency field calculation. Appropriate pseudopotentials are taken from the Pslibrary [28] considering 11 valence electrons for Nd ($5s^2 6s^{1.5} 5p^6 6p^{0.5} 5d^1$) and Sm ($5s^2 6s^{1.5} 5p^6 6p^{0.5} 5d^1$), 27 valence electrons for Ta ($4f^{14} 5d^3 5p^6 6s^2 5s^2$) and 6 valence electrons for oxygen ($2s^2 2p^4$). A prescribed generalized gradient approximation (GGA) based on the parametrization proposed by Perdew, Burke, and Ernzherhof (PBE) [29] has been accounted for the for calculation of exchange and correlation energy for both the compounds with the lowest-energy M′ fergusonite structure. Wave functions in the Kohn–Sham equation are expanded in a plane wave basis set due to the major advantage of orthonormality and since it is easy to control the convergence with respect to the size of the basis with only one parameter E_{cut}. In our calculation, the plane wave basis was extended up to 70 Ry for both the compounds ($NdTaO_4$ and $SmTaO_4$) to achieve highly converged results in the PAW scheme after thoroughly going through the convergence test. A dense Monkhorst pack grid of 8 × 8 × 8 is used for Brillouin zone integrations. Geometric optimization of NdTaO$_4$ and SmTaO$_4$ structures has been achieved using the Broyden–Fletcher–Goldfarb–Shanno (BFGS) minimization algorithm [30], where the structures have been fully optimized to the equilibrium condition at ambient pressure, by minimizing the forces on the atoms and the stress tensor. Helmann–Feynman forces lower than 0.00003 eV/atom on each atom in the unit cell and maximum deviation among the diagonal components of the stress tensor on a unit cell lower than 0.1 GPa ensure a fully relaxed structure. The same steps were followed while calculating from ambient pressure to 50 GPa with an interval of 1 GPa. After obtaining the equilibrium structures at different pressures, phonon frequencies were calculated at the center of the Brillouin zone using density functional perturbation theory (DFPT) as implemented in the Quantum Espresso code [31]. Simulations were performed at zero temperature and under a hydrostatic environment. The stable structures and transition pressures were obtained by analyzing the enthalpy–pressure curve.

3. Results and Discussions

3.1. Ambient Structure

The optimized volume obtained at ambient pressure for both the compounds in the M′ fergusonite structure matches very well with previously reported experimental values [32]. Unit cell parameters obtained from the geometrically relaxed structure with constituent bond lengths along with the previously reported experimental data are given in Table 1. The equilibrium volume obtained for NdTaO$_4$ is 159.1 Å^3 and 155.4 Å^3 for SmTaO$_4$, which are within 1% and 0.3% of the experimentally observed volume. Figure 1 shows the polyhedral representation of NdTaO$_4$ and SmTaO$_4$ belonging to a P2/c (space group no.:13, Z = 2) structure with 2/m point group symmetry. The rare earth cation (Nd/Sm) is surrounded by eight oxygen, forming dodecahedra while oxygen coordination around the Ta cation is six. The formation of oxygen octahedra around tantalum is a distinctive signature of the M′ fergusonite structure that makes it different from the M fergusonite structure with space group I2/a (space group: 15). Each rare earth cation is bonded with eight oxygen with four different R-O bond distances while each Ta cation bonds to six oxygen with three different bond distances, making all the constituent units distorted. These distorted polyhedral units provide structural stability to these compounds against a large range of pressure/temperature, compared to zircon- or scheelite-structured compounds [6,7,19–25]. In the calculated M′ fergusonite structure, the Nd/Sm and Ta atoms occupy the 2e and 2f Wyckoff positions while oxygen atoms O1 and O2 occupy the 4g position. In Figure 2, we have plotted ambient pressure lattice parameters, the unit cell volume along with the constituent polyhedral volume (RO$_8$, TaO$_6$), and R-O and Ta-O bond lengths for both the compounds along with previously reported data for EuTaO$_4$ and GdTaO$_4$ for a better understanding of the role of the rare earth cation. An increasing linear trend has been observed in all the parameters with lanthanum cationic radii except the volume of TaO$_6$ octahedra and its constituent bond lengths. This indicates the major influence of the R cation on structural parameters. In all four RTaO$_4$ compounds, the polyhedral volume of all distorted TaO$_6$ octahedra remains almost the same.

3.2. Structural Behavior under Compression

3.2.1. The Low-Pressure Phase

Simulated unit cell parameters for both the compounds, NdTaO$_4$ and SmTaO$_4$, have been plotted at different pressures in Figure 3a,b. The anisotropic compressibility of the b axis is clearly seen, which is 2 fold as compressible compared to the other two axes. This particular behavior has been observed in almost all RBO$_4$ compounds irrespective of their ambient structure [5–7,9]. The axial compressibility obtained by fitting the calculated lattice parameters to third-order BM-EOS [33] for NdTaO$_4$ is K_a = 1.89 × 10^{-3} GPa^{-1}, K_b = 3.79 × 10^{-3} GPa^{-1} and K_c = 1.75 × 10^{-3} GPa^{-1}; and SmTaO$_4$ are K_a = 1.66 × 10^{-3} GPa^{-1},

$K_b = 3.64 \times 10^{-3}$ GPa^{-1} and $K_c = 1.60 \times 10^{-3}$ GPa^{-1}, respectively. The structural arrangement of relatively more compressible RO_8 polyhedral units along the b axis could be the reason for anisotropic compression as has been reported in other rare earth metal oxides. Similar behavior has also been observed in EuTaO$_4$ and GdTaO$_4$ [16,17]. It is interesting to note that a decreasing trend is observed in the axial compressibility as we go from higher ionic radii to lower ionic radii of the lanthanide cation, indicating the more incompressible behavior of RTaO$_4$ compounds with a lower ionic radii lanthanide cation. Figure 4 shows the pressure evolution of the unit cell volume for both the compounds. The bulk modulus obtained by fitting the simulated pressure–volume data to third-order BM-EOS is 145.1 and 147.2 GPa for NdTaO$_4$ and SmTaO$_4$, respectively, which is in good agreement with the bulk modulus reported for the RTaO$_4$ family of compounds. The bulk modulus reported for NdVO$_4$ [34] in the zircon and scheelite structures is 124.2 and 136 GPa, respectively, which is similar to the obtained bulk modulus of NdTaO$_4$ in the present studies. A similar bulk modulus is observed for NdNbO$_4$ (138.32 GPa) [18]. Similarly, for SmTaO$_4$, the obtained bulk modulus in the present work is similar to that reported for other SmBO4 compounds [35]. This indicates that it is NdO$_8$/SmO$_8$ polyhedral units that mostly contribute to the bulk modulus. A similar bulk modulus for various RBO$_4$ compounds has already been reported and reaffirms that it is indeed the lower valence polyhedral units (RO_8) that mainly contribute to the bulk modulus [16,17]. This can be validated using Hazen and Finger's proposed empirical model $B_0 = N \times Z/(d_{R-O})^3$, where B_0 is the bulk modulus, N is the dimensional-less proportional constant (610 for tantalates and niobates), Z is the formal charge of the R cation and d_{R-O} is the average cation–anion distance, which only considers the rare earth polyhedral unit for predicting the bulk modulus [36]. For NdTaO$_4$ with d_{Nd-O} = 2.460 Å and Z = 3, the calculated bulk modulus using Hazen and Finger's equation is 122.9 GPa, which is similar to the simulated value obtained in the present work. For the SmTaO$_4$ compound, the bulk modulus obtained using d_{Sm-O} = 2.460 Å is 127.4 GPa.

Table 1. Comparison of simulated ambient pressure lattice parameters and constituent bond lengths of NdTaO$_4$ and SmTaO$_4$ with previously reported experimental values.

Lattice Parameters	NdTaO$_4$		SmTaO$_4$	
	Experiment [32]	Calculated	Experiment [32]	Calculated
a (Å)	5.2437(4)	5.257	5.2065(4)	5.206
b (Å)	5.5969(4)	5.638	5.5542(4)	5.571
c (Å)	5.4275(4)	5.451	5.3947(4)	5.397
β (degree)	96.767(9)	96.79	97.721(9)	96.74
Bond length	2×	2×	2×	2×
R-O2 (Å)	2.454	2.459	2.414	2.4179
R-O1(Å)	2.371	2.371	2.346	2.33622
R-O1(Å)	2.608	2.603	2.586	2.57709
R-O2(Å)	2.408	2.409	2.377	2.37666
Ta-O1(Å)	1.864	1.879	1.874	1.88026
Ta-O2(Å)	2.201	2.230	2.211	2.22584
Ta-O2(Å)	1.996	2.003	1.990	1.99934

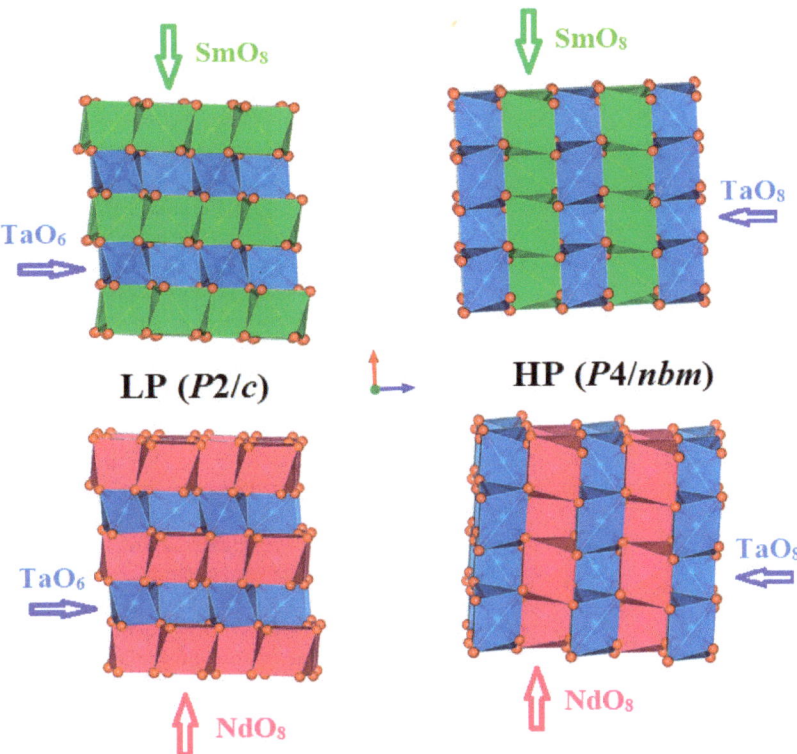

Figure 1. Polyhedral representation of SmTaO$_4$ (upper) and NdTaO$_4$ (lower) in both LP (P2/c) and HP (P4/nbm) phases. Sm, Nd, Ta, and O atoms are green, pink, blue and red, respectively.

3.2.2. The High-Pressure Phase

Earlier, based on our X-ray diffraction and Raman spectroscopic measurements along with DFT-based first principles calculations on EuTaO$_4$ and GdTaO$_4$, an isostructural transition was reported with a sudden drop in the monoclinic angle from 96° to 90° at the transition. This isostructural monoclinic phase could also be described as the pseudo orthorhombic structure due to all unit cell angles of 90°. Therefore, orthorhombic Pcna was tested as an alternative description against the isostructural monoclinic phase and our calculation reveals that, at high pressure, both the structures are energetically favorable. Since NdTaO$_4$ and SmTaO$_4$ are found to be synthesized in the same space group (P2/c) as EuTaO$_4$ and GdTaO$_4$, the same orthorhombic space group has been tested as the possible high-pressure phase for both the compounds in the present work. For NdTaO$_4$, the orthorhombic structure with the *Pcna* space group becomes energetically favorable at approximately 40 GPa as seen in Figure 5a, which depicts enthalpy difference as a function of pressure (the monoclinic phase P2/c has been taken as a reference). Similar behavior has been observed at approximately 33.5 GPa for SmTaO$_4$ (Figure 5b). This orthorhombic structure can also be alternatively described by a pseudo tetragonal structure due to almost the same value of lattice parameter a and c as seen in calculation. Therefore, the higher-symmetry tetragonal structure P4/nbm with 4/mmm point group symmetry was also tested against the Pcna structure and indeed tetragonal P4/nbm is a lower-energy structure at approximately 33 GPa for SmTaO$_4$ and approximately 40 GPa for NdTaO$_4$ as seen in Figure 5a,b. The phase transition in both the compounds is associated with a nearly 1.3% volume collapse at transition pressure along with change in oxygen coordination around the Ta cation from ambient six to eight at phase transition. In Figure 3, we have plotted lattice parameters of the high-pressure tetragonal phase (HP) along with unit cell parameters corresponding to the low-pressure monoclinic phase (LP). The axial compressibility obtained by fitting the calculated lattice parameters to third-order BM-EOS for NdTaO$_4$ and SmTaO$_4$ in the HP phases is $K_a = 3.75 \times 10^{-3}$ GPa^{-1}, $K_c = 1.02 \times 10^{-3}$ GPa^{-1} and $K_a = 3.4 \times 10^{-3}$ GPa^{-1}, $K_c = 1.04 \times 10^{-3}$ GPa^{-1}, respectively. Unit cells for both the compounds also undergo anisotropic compression in the high-pressure tetragonal structure as observed in the low-pressure monoclinic structure. The simulated pressure volume data fitted with the third-order Birch–Murnaghan (BM) equation of state (EOS) yields a bulk modulus of 123.74 GPa for NdTaO$_4$ and 130.60 GPa for SmTaO$_4$. It is interesting to note that the high-pressure phase has a lower bulk modulus than the low-pressure phase, indicating a more compressible high-pressure phase in spite of volume collapse at phase transition. This can be understood by analyzing the compressional behavior of the constituent polyhedral unit in both the LP phase and the HP phase. In low-pressure phase, the compounds NdTaO$_4$ and SmTaO$_4$ are made up of highly compressible NdO$_8$ and SmO$_8$ polyhedra as evident from their value of bulk modulus being 117.3 and 115 GPa respectively, while TaO$_6$ octahedra in both the compounds is highly incompressible with incompressibility modulus being 262.6 and 258.7 GPa for NdTaO$_4$ and SmTaO$_4$ respectively. The modulus of incompressibility has been obtained using calculated pressure and polyhedral volume fitted to third-order BM-EOS. A large difference in the bulk modulus at the LP phase affirms that indeed lower valence rare earth polyhedra significantly contribute to the compressibility of the compound in the low-pressure monoclinic phase, which validates Hazen and Finger's proposed empirical model for predicting the bulk modulus, taking contribution from rare earth polyhedral unit as seen in RVO$_4$, RWO$_4$, RMoO$_4$, RNbO$_4$ compounds [6,7,19–25]. In Figure 6a,b, we have plotted the distortion index of bond lengths for the polyhedral units of both NdTaO$_4$ and SmTaO$_4$ compounds, which show that RO_8 polyhedra become fully symmetric under compression, while that of TaO$_6$ more distorted and distortion in the bond length increases sharply at phase transition with the formation of a TaO$_8$ polyhedral unit. The distortion index of the bond length as defined by Baur [37] has been computed using VESTA software [38]. For the NdTaO$_4$ compound, the bulk modulus obtained for constituent polyhedral NdO$_8$ and TaO$_8$ in the HP phase is 105.9 and 99.8 GPa, respectively, whereas the bulk modulus for the SmTaO$_4$ compound obtained for constituent polyhedral SmO$_8$ and TaO$_8$ in the HP phase is 109.4 and 111.42 GPa, respectively. The similar bulk modulus indicates equal contribution to the compressibility of the high-pressure unit cell unlike the low-pressure phase. Incompressible TaO$_6$ octahedra change to very compressible TaO$_8$ at phase transition due to the increase in oxygen coordination under pressure and can be cited as among the reasons for the lower bulk modulus of the high-pressure phase. It should also be noted that the volume collapse at transition pressure is much smaller (~1.3%), which also supports the more compressible high-pressure unit cell. In Table 2, we have tabulated the atom positions along with their unit cell lattice parameters at ambient pressure and phase transition pressure for both the compounds. It can be clearly seen that all the constituent atoms show significant atomic rearrangement in position when compared to their ambient pressure. A more packed high-pressure unit cell is a consequence of the more effective packing of oxygen anions surrounding the Ta cation. To analyze pressure effects from this perspective, we calculated the pressure evolution of R-O and Ta-O bond distances for both the LP and HP phases. Results are summarized in Figure 7a,b, which shows calculated bond distances plotted against pressure. Figure 7a,b clearly shows that for both the compounds, at LP phase, the two largest bond distances between Nd-O decrease at a faster rate than the other six Nd-O bond lengths, whereas the Ta-O bond lengths show lower compressibility than the Nd-O bond lengths. In the high-pressure tetragonal phase, NdO$_8$/SmO$_8$ polyhedra become fully symmetric and the largest Ta-O bond shows most compressibility among all the constituent bonds, indicating the contribution of TaO$_8$ towards the compressibility of the unit cell.

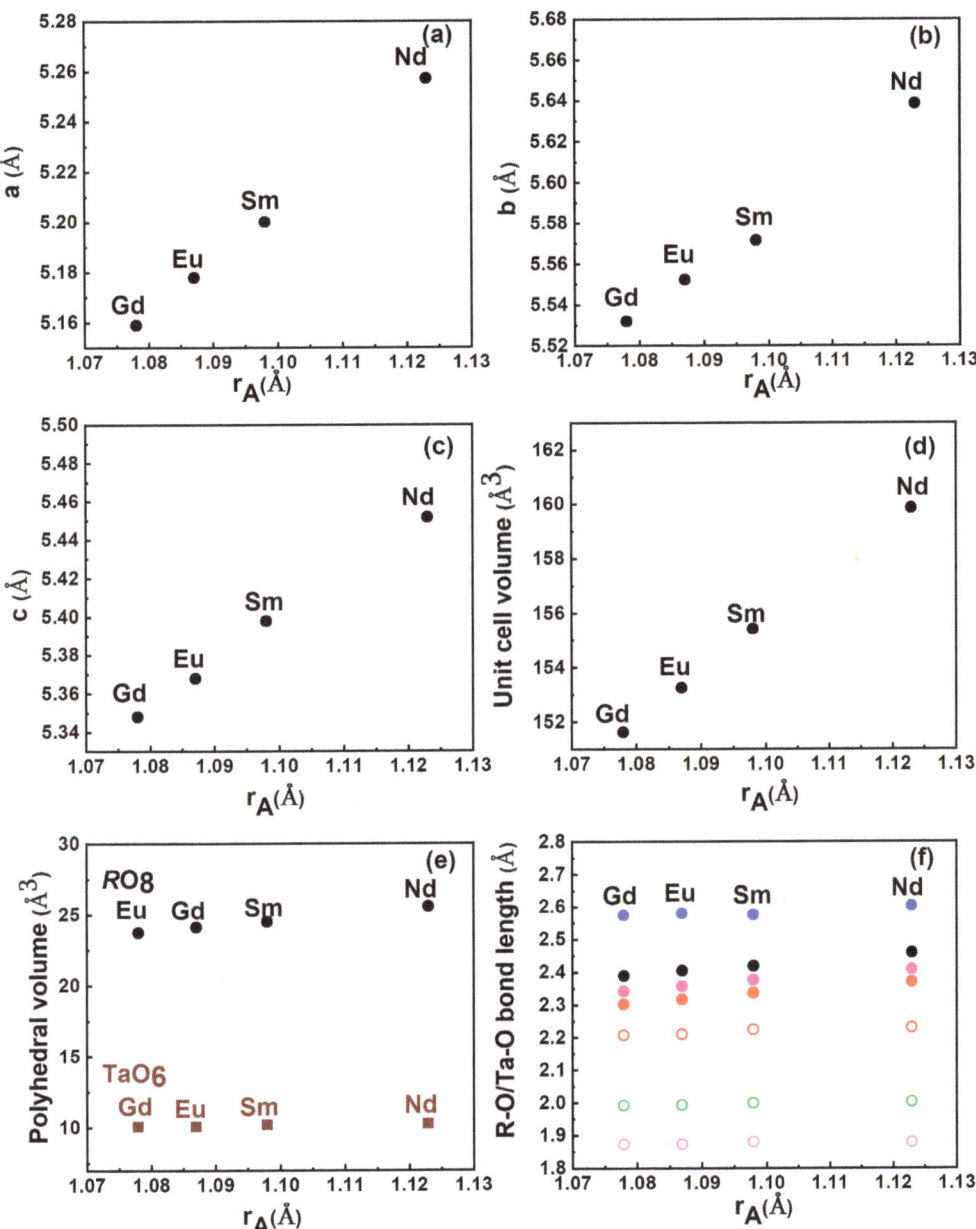

Figure 2. Dependence of lattice parameters: (**a**) a, (**b**) b, (**c**) c, (**d**) unit cell volume V, (**e**) constituent polyhedral volume and (**f**) interatomic distances with the ionic radii of the lanthanide cation (Gd, Eu, Sm, and Nd).

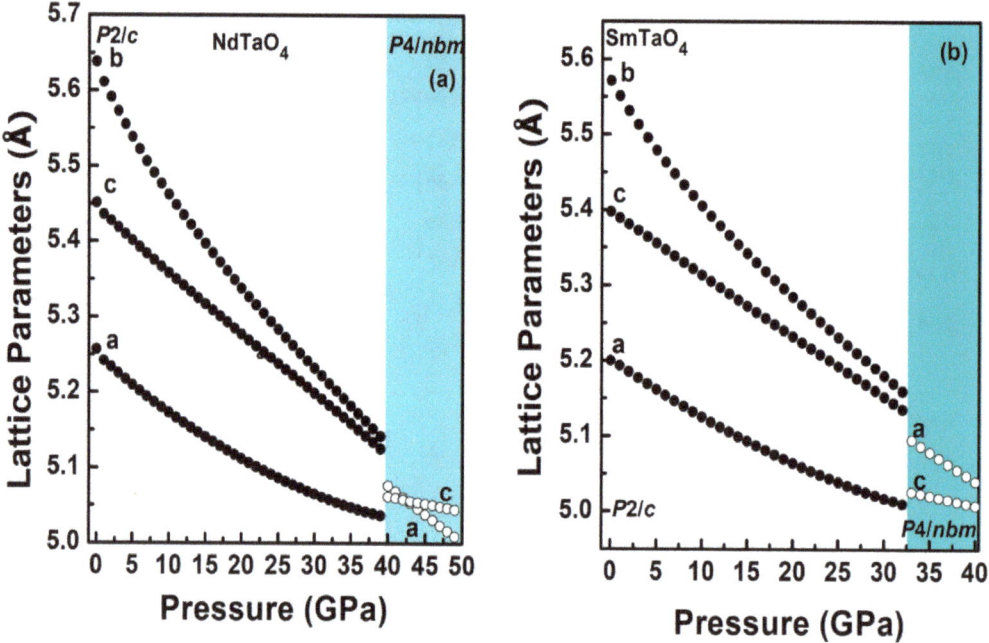

Figure 3. Pressure dependence of unit cell lattice parameters for the compounds (**a**) NdTaO$_4$ and (**b**) SmTaO$_4$ in the low-pressure phase (solid circle) and the high-pressure phase (empty circle). The colored region describes the high-pressure tetragonal phase.

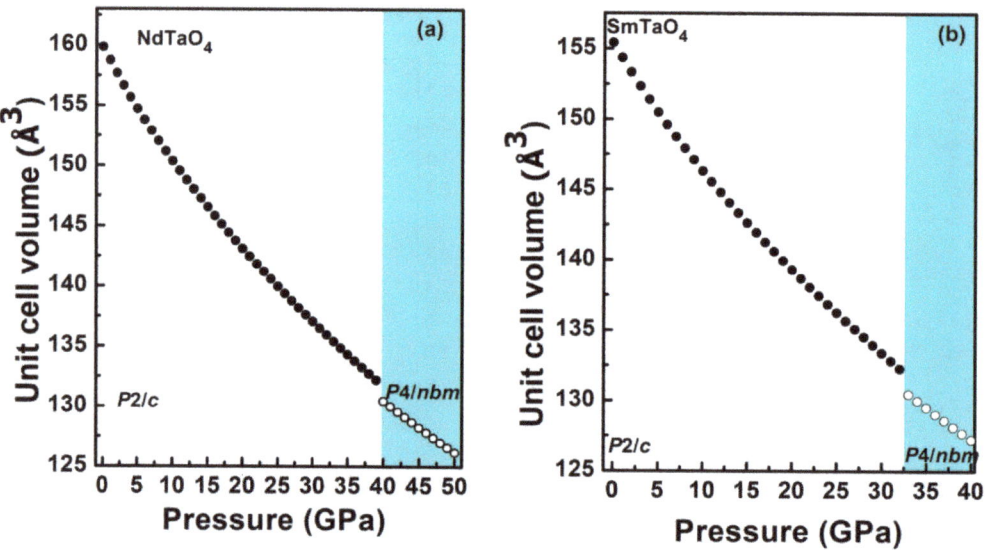

Figure 4. Pressure dependence of the unit cell lattice volume for the compounds (**a**) NdTaO$_4$ and (**b**) SmTaO$_4$, of the low-pressure phase (solid circle) and the high-pressure phase (empty circle). The colored region describes the high-pressure tetragonal phase.

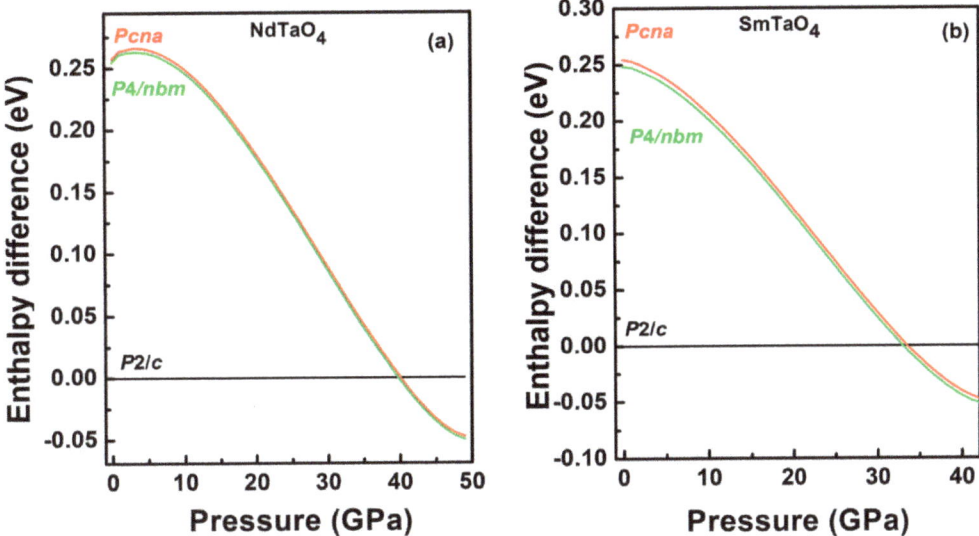

Figure 5. Enthalpy difference (eV) versus pressure for (**a**) NdTaO$_4$ and (**b**) SmTaO$_4$. The ambient pressure monoclinic $P2/c$ phase (black) has been taken as a reference for the both compounds. Red and green lines correspond to the *Pcna* (orthorhombic) and $P4/nbm$ (tetragonal) structures, respectively.

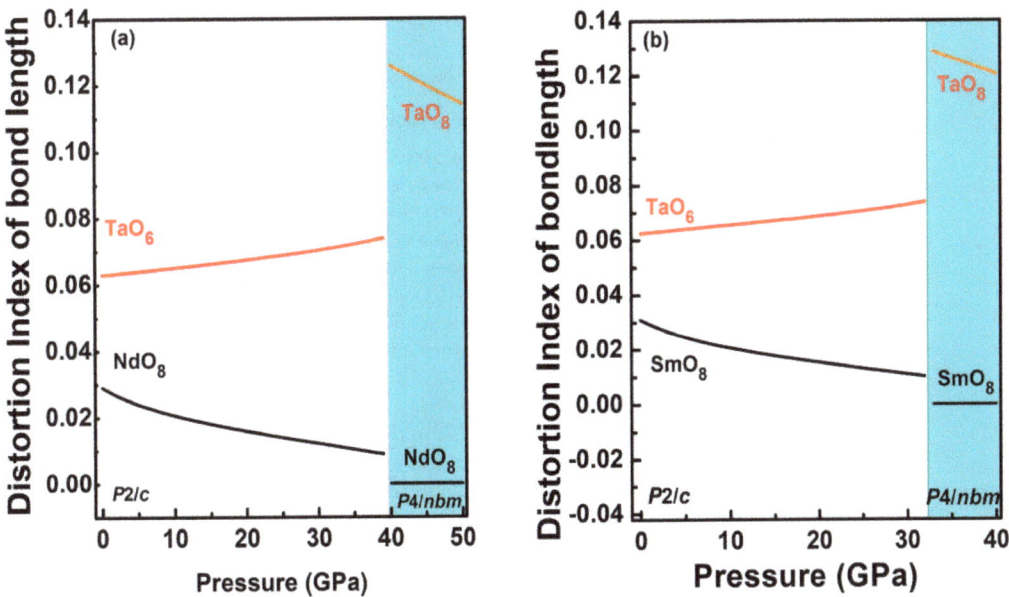

Figure 6. Pressure dependence of the distortion index of the bond length of NdO$_8$ and SmO$_8$ (black), TaO$_6$ (red) and TaO$_8$ (orange) for the compounds (**a**) NdTaO$_4$ and (**b**) SmTaO$_4$, of the low-pressure phase (white region) and the high-pressure phase (colored region).

Table 2. Calculated structural parameters along with the atomic positions of NdTaO$_4$ and SmTaO$_4$ at ambient pressure and transition pressure.

	NdTaO$_4$: LP Monoclinic Phase (P2/c) @ Ambient Pressure a = 5.2504 Å; b = 5.6312 Å; c = 5.4446 Å; β = 96.79°		
Nd (2e)	0.0000	0.23488	0.2500
Ta (2f)	0.5000	0.31247	0.7500
O1 (4g)	0.74508	0.90645	0.39125
O2 (4g)	0.27392	0.56319	0.49246
	NdTaO$_4$: HP Tetragonal Phase (P4/nbm) @ 40 GPa a = 5.0765 Å; c = 5.0616 Å		
Nd (2b)	0.75000	0.75000	0.500
Ta (2c)	0.75000	0.25	0.0
O (8m)	0.45494	0.54506	0.23076
	SmTaO$_4$: LP Monoclinic Phase (P2/c) @ Ambient Pressure a = 5.2025 Å; b = 5.5722 Å; c = 5.3985 Å; β = 96.74°		
Sm (2e)	0.0000	0.23438	0.2500
Ta (2f)	0.5000	0.30982	0.7500
O1 (4g)	0.74726	0.91026	0.39507
O2 (4g)	0.27135	0.56359	0.49281
	SmTaO$_4$: HP Tetragonal Phase (P4/nbm) @ 33 GPa a = 5.0949 Å; c = 5.0262 Å		
Sm (2b)	0.75000	0.75000	0.500
Ta (2c)	0.75000	0.25	0.0
O (8m)	0.45372	0.54628	0.23465

3.3. Vibrational Properties under Compression

The primitive unit cell of the M' fergusonite structure has two formula units, giving rise to a total of 36 phonon modes for both the compounds belonging to the P2/c space group with 2/m point group symmetry. Out of a total of 36 phonon modes, 18 modes are Raman active (8A$_g$ + 10B$_g$), 15 modes are IR active (7A$_u$ + 8B$_u$) and 3 are low-frequency acoustic modes. The A and B modes are one-dimensional irreducible representations which are symmetric and antisymmetric with respect to the principle axis of symmetry. The assignment of Raman and IR modes is performed in accordance with DFPT as implemented in Quantum ESPRESSO. Ambient pressure Raman and IR modes have been tabulated in Tables 3 and 4, respectively. We have also included previously published Raman mode frequencies from theoretical calculations for the EuTaO$_4$ and GdTaO$_4$ compounds [16,17]. The frequency distribution of the Raman modes is quite similar to the Raman mode distribution in the wolframite structure, which belongs to the same space group P2/c as M' fergusonite [39,40]. A total of twelve low-frequency Raman modes are present in the frequency region 100–400 cm^{-1}, two Raman modes are in the 400–600 cm^{-1} range and four are in the higher-frequency side, 600–800 cm^{-1}. It is interesting to note that we have not observed any frequency gap as observed in the Raman spectrum of the zircon or scheelite structure [41]. This can be understood by group–subgroup relationships among zircon–scheelite–fergusonite structures by virtue of the reduction in point group symmetry from 4/mmm to 4/m to 2/m, which in turn increases the allowed numbers of Raman modes and hence fills the frequency gap. All the calculated Raman modes can be categorized as internal modes or external modes. Internal modes, lying in the higher frequency region correspond to TaO$_6$ octahedra while, modes at lower frequency are external modes describing the movement of rigid TaO$_6$ unit against the lanthanide cation [42]. Out of a total of six internal modes of TaO$_6$ octahedra, the 2 A$_g$ and 2 B$_g$ modes lie in the higher–frequency region that is from 600 to 800 cm^{-1}, and two with A_g symmetry appear in the 390–500 cm^{-1} range [42]. Identified internal modes are marked by an asterisk in Table 3. The highest-frequency A_g mode, which appears at approximately 762 cm^{-1} for NdTaO$_4$ and 771 cm^{-1} for SmTaO$_4$, describes the symmetric stretching mode of TaO$_6$ octahedra. The frequency of the majority of the Raman active modes are observed to increase with a decrease in the lanthanum ionic radii except the two A_g modes (108 cm^{-1} for NdTaO$_4$ and 107cm^{-1} for SmTaO$_4$; 220 cm^{-1} for NdTaO$_4$ and 219 cm^{-1} for SmTaO$_4$) whereas the frequency of the two B_g modes (at 119 and 138 cm^{-1}) remains unaltered by the change in lanthanide cationic radii. This observation is valid when we extend our comparison with the published Raman mode frequencies of EuTaO$_4$ and GdTaO$_4$. Table 3 lists the pressure evolution of all the Raman active modes in P2/c structure obtained by quadratic fitting of data points under pressure. No Raman mode softening has been seen in both the compounds due to the absence of negative pressure coefficients. Hardening of all the Raman active modes has been associated with the monoclinic fergusonite phase of the other compound such as rare earth niobates GdNbO$_4$ and EuNbO$_4$, as well as rare earth tantalates GdTaO$_4$ and EuTaO$_4$ [16,17,22,23]. For rare earth vanadate, mode softening has been observed in the zircon or scheelite phase, but no mode softening has been reported in the fergusonite structure consistent with our current observation in the present work [41,43]. There is crossover between the B_g and A_g modes located at 380–400 cm^{-1} due to a nearly 3-fold higher pressure coefficient of the B_g mode than the A_g mode. No other mode crossover has been observed in spite of the large difference in the pressure coefficient of the Raman modes. In the HP phase, stabilized in the tetragonal structure, calculation predicts the presence of the 11 Raman active modes (5 E_g + 2A_{1g} + 1B_{1g} + 3B_{2g}). The pressure evolution of the Raman active modes are shown in Table 5. In the HP phase, all modes show positive pressure coefficients except three modes, which show nonlinear behavior under compression.

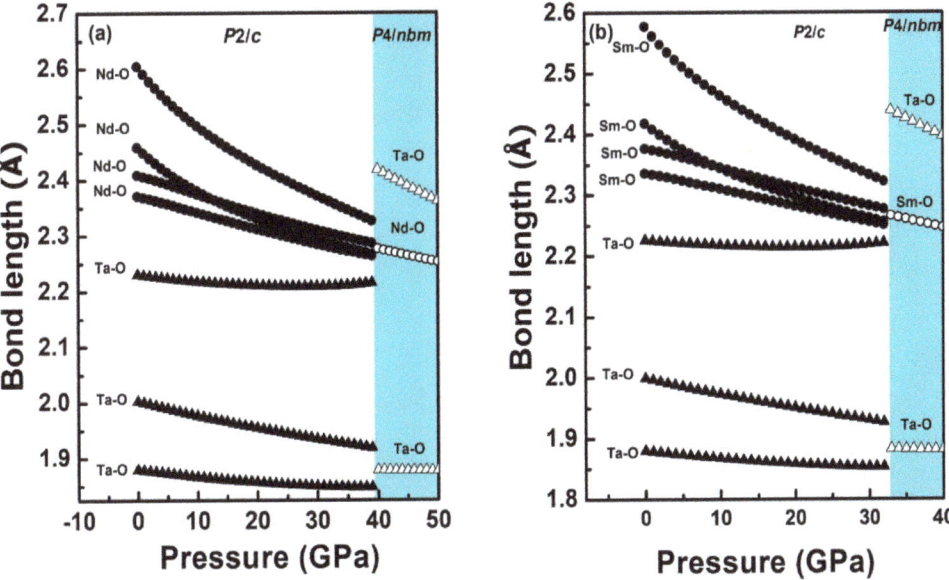

Figure 7. (a) Nd-O bond lengths versus pressure in the LP phase (solid circle) and the HP phase (empty circle), Ta-O bond lengths versus pressure in the LP phase (solid triangle) and the HP phase (empty triangle); (b) Sm-O bond lengths versus pressure in LP phase (solid circle) and HP phase (empty circle), Ta-O bond lengths versus pressure in the LP phase (solid triangle) and the HP phase (empty triangle). The white region describes the LP phase while the colored region is for the HP phase.

Table 3. Calculated Raman modes for NdTaO$_4$ and SmTaO$_4$ at ambient pressure along with their pressure coefficients (present work). Reported data of earlier studies on EuTaO$_4$ and GdTaO$_4$ have been included for comparison. Frequencies marked with an asterisk (*) correspond to internal modes of TaO$_6$ octahedra.

Raman Frequency	NdTaO$_4$		SmTaO$_4$		EuTaO$_4$ [16]		GdTaO$_4$ [17]	
	ω	dω/dP	ω	dω/dP	ω	dω/dP	ω	dω/dP
B_g	99.3	1.42	101.6	1.83	102.3	1.25	100.6	1.64
A_g	108.7	1.40	107.5	2.45	107.1	1.14	103.5	1.84
B_g	119.6	1.28	119.7	1.53	119.8	1.10	119.1	1.27
B_g	138.7	2.22	138.4	2.95	138.5	1.67	138.5	2.49
B_g	164.3	0.88	168.7	1.43	172.0	0.94	171	0.92
A_g	176.7	0.76	177.2	1.05	177.6	0.54	177.9	0.41
A_g	220.6	4.07	219	4.7	216.9	4.17	211.1	4.74
B_g	257.1	2.41	261.9	2.99	263.6	2.65	261.6	2.66
A_g	257.6	2.85	262.5	3.04	264.3	2.70	263.4	2.18
B_g	314.1	4.22	321.9	5.19	325.2	4.62	322	4.78
B_g	383.5	4.20	392.5	5.3	396.0	3.96	395.3	4.80
A_g*	392.6	1.36	397.1	1.8	400.3	1.68	402.6	1.76
B_g	466.9	1.7	475.1	2.3	480.8	1.88	486.7	1.53
A_g*	472.9	1.42	478.7	1.97	482.2	1.84	487.6	1.61
B_g*	600.3	3.65	613	4.11	620.5	3.93	626.3	4.29
A_g*	609.3	3.62	620.8	4.05	627.6	3.77	633.4	4.15
B_g*	627.4	4.78	643.4	4.93	651.1	4.17	661.7	4.76
A_g*	762.2	3.55	771.5	3.58	777.1	3.11	785.9	3.69

Table 4. Calculated IR modes for NdTaO$_4$ and SmTaO$_4$ at ambient pressure along with their pressure coefficients (present work).

IR Frequency	NdTaO$_4$ @ Ambient Pressure		SmTaO$_4$ @ Ambient Pressure	
	ω	dω/dP	ω	dω/dP
A_u	135.5	2.13	134.3	2.88
B_u	138.6	1.58	137	2.92
B_u	156.4	3.85	155.1	5.03
B_u	207.5	−1.44	204.2	−1.54
A_u	249.1	Nonlinear	253.7	Nonlinear
B_u	257.2	2.24	257.2	2.97
B_u	275.8	2.24	281.9	2.39
A_u	311.1	0.87	310.8	1.53
A_u	354.7	3.33	359.4	3.81
B_u	382	3.26	387.3	3.59
B_u	480.7	4.12	491.5	4.65
A_u	503.6	3.74	512.4	4.23
A_u	561.3	4.33	572.8	4.68
B_u	607.5	4.23	616.7	4.66
A_u	743.2	3.77	752.2	3.96

As seen in Tables 4 and 5 out of 15 IR modes show a decreasing trend when we go from the NdTaO$_4$ to SmTaO$_4$ compound, whereas one B_u mode (257 cm^{-1}) remains unaltered. The remaining nine modes show an increase in frequency with lower lanthanide radii. From Table 4, it can be clearly seen that one IR active mode (A_u) shows a nonlinear response in frequency under compression. The presence of the negative pressure coefficient of the IR active mode (B_u) indicates phonon softening in the compound with pressure. Phonon softening was cited as among the crucial trigger points for pressure-induced instability in metal oxides [39]. The other 13 IR modes have a positive pressure coefficient. According to our calculation, the high-pressure phase for both the compounds have seven IR active modes ($4E_u + 3A_{2u}$). IR active modes at transition pressure, along with pressure coefficients, are summarized in Table 6. It can be clearly seen in Table 6 that all the modes show a positive pressure coefficient except the lowest-frequency A_{2u} mode, which shows a nonlinear response under pressure.

Table 5. Calculated Raman modes for NdTaO$_4$ and SmTaO$_4$ at transition pressure along with their pressure coefficients (present work).

Raman Frequency	NdTaO$_4$(HP) @ 40 GPa		SmTaO$_4$(HP) @ 33 GPa	
	ω	$d\omega/dP$	ω	$d\omega/dP$
E_g	103.1	Nonlinear	99.6	Nonlinear
B_{2g}	163.6	Nonlinear	159.4	Nonlinear
E_g	208.7	1.1	199.9	1.2
B_{1g}	357.5	0.92	343.9	1.26
E_g	412.3	1.93	397.2	2.16
A_{1g}	412.9	Nonlinear	408.7	Nonlinear
B_{2g}	513	2.77	501	2.73
E_g	546.6	2.9	533.6	3.09
E_g	741.8	1.55	738.4	1.85
B_{2g}	750	1.85	743.8	2.09
A_{1g}	823.7	1.12	815.7	1.25

Table 6. Calculated IR modes for NdTaO$_4$ and SmTaO$_4$ at 40 and 33 GPa along with their pressure coefficients (present work).

IR Frequency	NdTaO$_4$(HP)40 GPa		SmTaO$_4$(HP)33 GPa	
	ω	$d\omega/dP$	ω	$d\omega/dP$
A_{2u}	109	Nonlinear	110.6	Nonlinear
E_u	136.2	1.92	126.2	2.24
E_u	183.2	1.36	183.8	1.33
A_{2u}	281.2	1.80	264.1	1.89
E_u	509.3	2.26	494.6	2.52
A_{2u}	669.1	0.62	659.6	0.85
E_u	699.9	1.53	687.7	1.72

4. Conclusions

To conclude, the compressional behavior of the M' fergusonite-structured NdTaO$_4$ and SmTaO$_4$, investigated through DFT-based first-principle simulations indicate pressure-induced first-order phase transition from monoclinic to tetragonal structure. The transition is accompanied by an increase in oxygen coordination around the Ta cation from six to eight and a nearly 1.3% volume reduction at transition pressure (40 GPa for NdTaO$_4$ and 33 GPa for SmTaO$_4$). In the low-pressure monoclinic phase, the compressibility of the unit cell has a major contribution from rare earth polyhedra, whereas both the rare earth polyhedra and tantalum polyhedra exhibit a similar contribution towards the compressibility of the unit cell for both the compounds at the HP tetragonal phase, which in turn explains the lower bulk modulus obtained in the HP phase. The pressure evolution of phonon modes has been evaluated in both the LP phase and the HP phase. No Raman mode softening has been seen in both the compounds, although one particular IR mode has been observed to show red shift under pressure, possibly leading to instability in the compounds. Earlier reported experimental high-pressure studies on EuTaO$_4$ and GdTaO$_4$ show isostructural first-order reversible phase transition at approximately 20 GPa, although theoretical calculations predict an equally probable orthorhombic Pcna as an alternative description of the HP phase near 43 and 40 GPa, respectively. The experimental and theoretical phase transition pressures differ due to the crucial dependence of a non-hydrostatic stress environment on phase transition. Therefore, a high-pressure study on NdTaO$_4$ and SmTaO$_4$ by experimental techniques is desirable for a better understanding of structural and vibrational changes under compression, which is out of scope in the present work.

Author Contributions: Conceptualization, methodology, validation, writing—review and editing, visualization, supervision: A.B.G. (corresponding Author); Formal analysis, investigation, writing—original draft preparation, visualization: S.B. (First author); Data curation, visualization: A.T. (second author). All authors have read and agreed to the published version of the manuscript.

Funding: This research received no external funding.

Institutional Review Board Statement: Not applicable.

Informed Consent Statement: Not applicable.

Data Availability Statement: The data are available from the corresponding author on reasonable request.

Acknowledgments: S.B. and A.T. acknowledge T. Sakuntala for providing motivation and encouragement.

Conflicts of Interest: The authors declare no conflict of interest.

References

1. Haugsrud, R.; Norby, T. Proton conduction in rare-earth ortho-niobates and ortho-tantalates. *Nat. Mater.* **2006**, *5*, 193–196. [CrossRef]
2. Yang, H.; Teng, Y.; Ren, X.; Wu, L.; Liu, H.; Wang, S.; Xu, L. Synthesis and crystalline phase of monazite-type $Ce_{1-x}Gd_xPO_4$ solid solutions for immobilization of minor actinide curium. *J. Nucl. Mater.* **2014**, *444*, 39–42. [CrossRef]
3. Heuser, J.; Bukaemskiy, A.; Neumeier, S.; Neumann, A.; Bosbach, D. Raman and infrared spectroscopy of monazite-type ceramics used for nuclear waste conditioning. *Prog. Nucl. Energy* **2014**, *72*, 149–155. [CrossRef]
4. Errandonea, D.; Manjon, F.J. Pressure effects on the structural and electronic properties of $ABX4$ scintillating crystals. *Prog. Mater. Sci.* **2008**, *53*, 711–773. [CrossRef]
5. Errandonea, D.; Garg, A.B. Recent progress on the characterization of the high-pressure behaviour of AVO_4 orthovanadates. *Prog. Mater. Sci.* **2018**, *97*, 123–169. [CrossRef]
6. López-Moreno, S.; Rodríguez-Hernández, P.; Muñoz, A.; Romero, A.H.; Errandonea, D. First-principles calculations of elec-tronic, vibrational, and structural properties of scheelite $EuWO_4$ under pressure. *Phys. Rev. B* **2011**, *84*, 064108. [CrossRef]
7. Errandonea, D.; Santamaria-Perez, D.; Achary, S.N.; Tyagi, A.K.; Gall, P.; Gougeon, P. High-pressure x-ray diffraction study of $CdMoO_4$ and $EuMoO_4$. *J. Appl. Phys.* **2011**, *109*, 043510. [CrossRef]
8. Siqueira, K.P.F.; Moreira, R.L.; Dias, A. Synthesis and Crystal Structure of Lanthanide Orthoniobates Studied by Vibrational Spectroscopy. *Chem. Mater.* **2010**, *22*, 2668–2674. [CrossRef]
9. Lacomba-Perales, R.; Errandonea, D.; Meng, Y.; Bettinelli, M. High-pressure stability and compressibility of APO_4 (A = La, Nd, Eu, Gd, Er, and Y) orthophosphates: An x-ray diffraction study using synchrotron radiation. *Phys. Rev. B* **2010**, *81*, 064113. [CrossRef]
10. Errandonea, D.; Pellicer-Porres, J.; Manjon, F.J.; Segura, A.; Ferrer Roca, C.; Kumar, R.; Tschauner, O.; Lopez-Solano, J.; Ro-driguez, P.; Radescu, S.; et al. Determination of the High-Pressure Crystal Structure of $BaWO_4$ and $PbWO_4$. *Phys. Rev. B* **2006**, *73*, 224103. [CrossRef]
11. Siqueira, K.P.F.; Dias, A. Effect of the processing parameters on the crystalline structure of lanthanide orthotantalates. *Mater. Res.* **2013**, *17*, 167–173. [CrossRef]
12. Siqueira, K.P.F.; Carvalho, G.B.; Dias, A. Influence of the processing conditions and chemical environment on the crystal structures and phonon modes of lanthanide orthotantalates. *Dalton Trans.* **2011**, *40*, 9454–9460. [CrossRef] [PubMed]
13. Kim, C.; Lee, W.; Melis, A.; Elmughrabi, A.; Lee, K.; Park, C.; Yeom, J.-Y. A Review of Inorganic Scintillation Crystals for Extreme Environments. *Crystals* **2021**, *11*, 669. [CrossRef]
14. Yang, H.; Peng, F.; Zhang, Q.; Guo, C.; Shi, C.; Liu, W.; Sun, G.; Zhao, Y.; Zhang, D.; Sun, D.; et al. A promising high-density scintillator of $GdTaO_4$ single crystal. *Crystengcomm* **2014**, *16*, 2480–2485. [CrossRef]
15. Wang, J.; Chong, X.; Zhou, R.; Feng, J. Microstructure and thermal properties of $RETaO_4$ (RE = Nd, Eu, Gd, Dy, Er, Yb, Lu) as promising thermal barrier coating materials. *Scr. Mater.* **2017**, *126*, 24–28. [CrossRef]
16. Banerjee, S.; Garg, A.B.; Poswal, H.K. Pressure driven structural phase transition in $EuTaO_4$: Experimental and first principles investigations. *J. Phys. Condens. Matter* **2022**, *34*, 135401. [CrossRef]
17. Banerjee, S.; Garg, A.B.; Poswal, H.K. Structural and vibrational properties of $GdTaO4$ under compression: An insight from experiment and first principles simulations. *J. Appl. Phys.* **2023**, *133*, 025902. [CrossRef]
18. Xiao, W.; Yang, Y.; Pi, Z.; Zhang, F. Phase Stability and Mechanical Properties of the Monoclinic, Monoclinic-Prime and Tetragonal $REMO_4$ (M = Ta, Nb) from First-Principles Calculations. *Coatings* **2022**, *12*, 73. [CrossRef]
19. Garg, A.B.; Errandonea, D.; Rodríguez-Hernández, P.; López-Moreno, S.; Muñoz, A.; Popescu, C. High-pressure structural behaviour of $HoVO_4$: Combined XRD experiments and ab initio calculations. *J. Phys. Condens. Matter* **2014**, *26*, 265402. [CrossRef]
20. Garg, A.B.; Errandonea, D. High-pressure powder x-ray diffraction study of $EuVO_4$. *J. Solid State Chem.* **2015**, *226*, 147–153. [CrossRef]

21. Wang, X.; Wang, B.; Tan, D.; Xiao, W.; Song, M. Phase transformations of zircon-type DyVO4 at high pressures up to 36.4 GPa: X-ray diffraction measurements. *J. Alloy. Compd.* **2021**, *875*, 159926. [CrossRef]
22. Garg, A.B.; Rao, M.R.; Errandonea, D.; Pellicer-Porres, J.; Martinez-Garcia, D.; Popescu, C. Pressure-induced instability of the fergusonite phase of EuNbO4 studied by in situ Raman spectroscopy, x-ray diffraction, and photoluminescence spectroscopy. *J. Appl. Phys.* **2020**, *127*, 175905. [CrossRef]
23. Pellicer-Porresa, J.; Garg, A.B.; Vázquez-Socorroa, D.; Martínez-García, D.; Popescu, C.; Errandonea, D. Stability of the fer-gusonite phase in GdNbO4 by high pressure XRD and Raman experiments. *J. Solid State Chem.* **2017**, *251*, 14–18. [CrossRef]
24. Garg, A.B.; Liang, A.; Errandonea, D.; Rodríguez-Hernández, P.; Muñoz, A. Monoclinic–triclinic phase transition induced by pressure in fergusonite-type YbNbO4. *J. Phys. Condens. Matter* **2022**, *34*, 174007. [CrossRef]
25. Garg, A.B.; Errandonea, D.; Rodríguez-Hernández, P.; Muñoz, A. High-pressure monoclinic–monoclinic transition in fergu-sonite-type HoNbO4. *J. Phys. Condens. Matter* **2021**, *33*, 195401. [CrossRef] [PubMed]
26. Giannozzi, P.; Baseggio, O.; Bonfà, P.; Brunato, D.; Car, R.; Carnimeo, I.; Cavazzoni, C.; de Gironcoli, S.; Delugas, P.; Ruffino, F.F.; et al. Quantum ESPRESSO toward the exascale. *J. Chem. Phys.* **2020**, *152*, 154105. [CrossRef]
27. Blöchl, P.E. Projector augmented-wave method. *Phys. Rev. B* **1994**, *50*, 17953–17979. [CrossRef]
28. Corso, A.D. Pseudopotentials periodic table: From H to Pu. *Comput. Mater. Sci.* **2014**, *95*, 337–350. [CrossRef]
29. Perdew, J.P.; Burke, K.; Ernzerhof, M. Generalized Gradient Approximation Made Simple. *Phys. Rev. Lett.* **1997**, *78*, 1396. [CrossRef]
30. Head, J.D.; Zerner, M.C. A Broyden—Fletcher—Goldfarb—Shanno optimization procedure for molecular geometries. *Chem. Phys. Lett.* **1985**, *122*, 264–270. [CrossRef]
31. Baroni, S.; de Gironcoli, S.; Corso, A.D.; Giannozzi, P. Phonons and related crystal properties from density-functional per-turbation theory. *Rev. Mod. Phys.* **2001**, *73*, 515. [CrossRef]
32. Hartenbach, I.; Lissner, F.; Nikelski, T.; Meier, S.F.; Müller-Bunz, H.; Schleid, T. About Lanthanide Oxotantalates with the Formula MTaO4 (M = La–Nd, Sm–Lu). *Z. Anorg. Und Allg. Chem.* 2005 *631*, 2377–2382.
33. Birch, F. Finite Elastic Strain of Cubic Crystals. *Phys. Rev.* **1947**, *71*, 809–824. [CrossRef]
34. Panchal, V.; Errandonea, D.; Manjón, F.; Muñoz, A.; Rodríguez-Hernández, P.; Achary, S.; Tyagi, A. High-pressure lattice-dynamics of NdVO. *J. Phys. Chem. Solids* **2017**, *100*, 126–133. [CrossRef]
35. Garg, A.B.; Errandonea, D.; Rodríguez-Hernández, P.; Muñoz, A. ScVO4 under non-hydrostatic compression: A new meta-stable polymorph. *J. Phys. Condens. Matter* **2017**, *29*, 055401. [CrossRef] [PubMed]
36. Hazen, R.M.; Finger, L.W. Bulk modulus-volume relationship for cation-anion polyhedra. *J. Geophys. Res. Solid Earth* **1979**, *84*, 6723–6728. [CrossRef]
37. Baur, W.H. The geometry of polyhedral distortions. Predictive relationships for the phosphate group. *Acta Crystallogr. Sect. B Struct. Crystallogr. Cryst. Chem.* **1974**, *30*, 1195–1215. [CrossRef]
38. Momma, K.; Izumi, F. VESTA 3 for three-dimensional visualization of crystal, volumetric and morphology data. *J. Appl. Crystallogr.* **2011**, *44*, 1272–1276. [CrossRef]
39. Errandonea, D.; Popescu, C.; Garg, A.B.; Botella, P.; Martinez-García, D.; Pellicer-Porres, J.; Rodríguez-Hernández, P.; Muñoz, A.; Cuenca-Gotor, V.; Sans, J.A. Pressure-induced phase transition and band-gap collapse in the wide-band-gap semiconductor InTaO4. *Phys. Rev. B* **2016**, *93*, 035204. [CrossRef]
40. Errandonea, D.; Ruiz-Fuertes, J. A Brief Review of the Effects of Pressure on Wolframite-Type Oxides. *Crystals* **2018**, *8*, 71. [CrossRef]
41. Marqueño, T.; Errandonea, D.; Pellicer-Porres, J.; Martinez-Garcia, D.; Santamaria-Pérez, D.; Muñoz, A.; Rodríguez-Hernández, P.; Mujica, A.; Radescu, S.; Achary, S.N.; et al. High-pressure polymorphs of gadolinium orthovanadate: X-ray diffraction, Raman spectroscopy, and *ab initio* calculations. *Phys. Rev. B* **2019**, *100*, 064106. [CrossRef]
42. Ruiz-Fuertes, J.; Errandonea, D.; López-Moreno, S.; González, J.; Gomis, O.; Vilaplana, R.; Manjón, F.J.; Muñoz, A.; Rodríguez-Hernández, P.; Friedrich, A.; et al. High-pressure Raman spectroscopy and lat-tice-dynamics calculations on scintillating MgWO4: A comparison with isomorphic compounds. *Phys. Rev. B* **2011**, *83*, 214112. [CrossRef]
43. Errandonea, D.; Achary, S.N.; Pellicer-Porres, J.; Tyagi, A.K. Pressure-Induced Transformations in PrVO4 and SmVO4 and Isolation of High-Pressure Metastable Phases. *Inorg. Chem.* **2013**, *52*, 5464–5469. [CrossRef] [PubMed]

Disclaimer/Publisher's Note: The statements, opinions and data contained in all publications are solely those of the individual author(s) and contributor(s) and not of MDPI and/or the editor(s). MDPI and/or the editor(s) disclaim responsibility for any injury to people or property resulting from any ideas, methods, instructions or products referred to in the content.

Article

Theoretical Study of Pressure-Induced Phase Transitions in Sb$_2$S$_3$, Bi$_2$S$_3$, and Sb$_2$Se$_3$

Estelina Lora da Silva [1,†,*], Mario C. Santos [2,3,†], Plácida Rodríguez-Hernández [4,†], Alfonso Muñoz [4,†] and Francisco Javier Manjón [3,†]

1. IFIMUP, Institute of Physics for Advanced Materials, Nanotechnology and Photonics, Department of Physics and Astronomy, Faculty of Sciences, University of Porto, Rua do Campo Alegre, 687, 4169-007 Porto, Portugal
2. Instituto dos Pupilos do Exército, Estrada de Benfica, n° 374, 1549-016 Lisboa, Portugal
3. Instituto de Diseño para la Fabricación y Producción Automatizada, MALTA Consolider Team, Universitat Politècnica de València, 46022 València, Spain
4. Departamento de Física, Instituto de Materiales y Nanotecnología, MALTA Consolider Team, Universidad de La Laguna, 38200 Tenerife, Spain
* Correspondence: estelina.silva@fc.up.pt
† These authors contributed equally to this work.

Abstract: We report an ab initio study of Sb$_2$S$_3$, Sb$_2$Se$_3$, and Bi$_2$S$_3$ sesquichalcogenides at hydrostatic pressures of up to 60 GPa. We explore the possibility that the *C2/m*, *C2/c*, the disordered *Im-3m*, and the *I4/mmm* phases observed in sesquichalcogenides with heavier cations, viz. Bi$_2$Se$_3$, Bi$_2$Te$_3$, and Sb$_2$Te$_3$, could also be formed in Sb$_2$S$_3$, Sb$_2$Se$_3$, and Bi$_2$S$_3$, as suggested from recent experiments. Our calculations show that the *C2/c* phase is not energetically favorable in any of the three compounds, up to 60 GPa. The *C2/m* system is also unfavorable for Sb$_2$S$_3$ and Bi$_2$S$_3$; however, it is energetically favorable with respect to the *Pnma* phase of Sb$_2$Se$_3$ above 10 GPa. Finally, the *I4/mmm* and the disordered body-centered cubic-type *Im-3m* structures are competitive in energy and are energetically more stable than the *C2/m* phase at pressures beyond 30 GPa. The dynamical stabilities of the *Pnma*, *Im-3m*, *C2/m*, and *I4/mmm* structural phases at high pressures are discussed for the three compounds.

Keywords: density functional theory; high-pressure effects; lattice dynamics

Citation: da Silva, E.L.; Santos, M.C.; Rodríguez-Hernández, P.; Muñoz, A; Manjón, F.J. Theoretical Study of Pressure-Induced Phase Transitions in Sb$_2$S$_3$, Bi$_2$S$_3$, and Sb$_2$Se$_3$. *Crystals* **2023**, *13*, 498. https://doi.org/10.3390/cryst13030498

Academic Editors: Andrei Vladimirovich Shevelkov, Daniel Errandonea and Enrico Bandiello

Received: 23 February 2023
Revised: 8 March 2023
Accepted: 12 March 2023
Published: 14 March 2023

Copyright: © 2023 by the authors. Licensee MDPI, Basel, Switzerland. This article is an open access article distributed under the terms and conditions of the Creative Commons Attribution (CC BY) license (https://creativecommons.org/licenses/by/4.0/).

1. Introduction

A great deal of attention has been paid to the family of A$_2$X$_3$ sesquichalcogenides in the last decade since the identification of the trigonal tetradymite-like *R-3m* phases of the group-15 sesquichalcogenides Sb$_2$Te$_3$, Bi$_2$Se$_3$, and Bi$_2$Te$_3$ as 3D topological insulators [1,2]. Such topological insulators are a state of quantum matter where the bulk is a trivial insulator, whereas the surface states are topologically-protected conducting states due to time-reversal symmetry and strong spin-orbit coupling. Respective properties allow for potential applications in the fields of spintronics as well as for quantum computing [3].

These sesquichalcogenides have also been well studied in relation to their thermoelectric properties and as phase change materials [4,5].

Stibnite (Sb$_2$S$_3$), bismuthinite (Bi$_2$S$_3$) and antimonselite (Sb$_2$Se$_3$) minerals belong to group-15 sesquichalcogenides. Such systems do not crystallize into the tetradymite-type structure at room conditions, but instead into the orthorhombic U$_2$S$_3$-type (*Pnma*) structure (Figure 1b). Several technological applications can be considered for such compounds. These include, applications for photovoltaic solar cells, X-ray computed tomography detectors, fuel cells, biomolecules and gas sensors, solid-state batteries, fiber lasers, and photoelectrochemical devices [6–16].

High-pressure (HP) studies of Sb$_2$Te$_3$, Bi$_2$Se$_3$, and Bi$_2$Te$_3$ have shown that they undergo pressure-induced phase transitions (PTs) to monoclinic *C2/m* and *C2/c* structures, to the disordered solid solution *Im-3m* phase, and to the body-centered tetragonal *I4/mmm* structure [17–26]. These HP studies have shown a plethora of interesting properties, such as enhanced thermoelectric properties [? ?] and superconductivity, which could be of topological nature [29–33]. Therefore, there is a fundamental interest in identifying if the above mentioned HP phases can be associated to group-15 sesquichalcogenides with the *Pnma* structure crystallizing at room pressure. Since the *Pnma* structure has also been identified as a possible HP post-perovskite phase of the (Mg,Fe)SiO$_3$ and NaFeN$_3$ compounds [34,35], the study of Sb$_2$S$_3$, Bi$_2$S$_3$ and Sb$_2$Se$_3$ at HPs could also provide useful information about the structures and stability of ABO$_3$ orthorhombic minerals at pressure conditions close to those of the Earth's mantle.

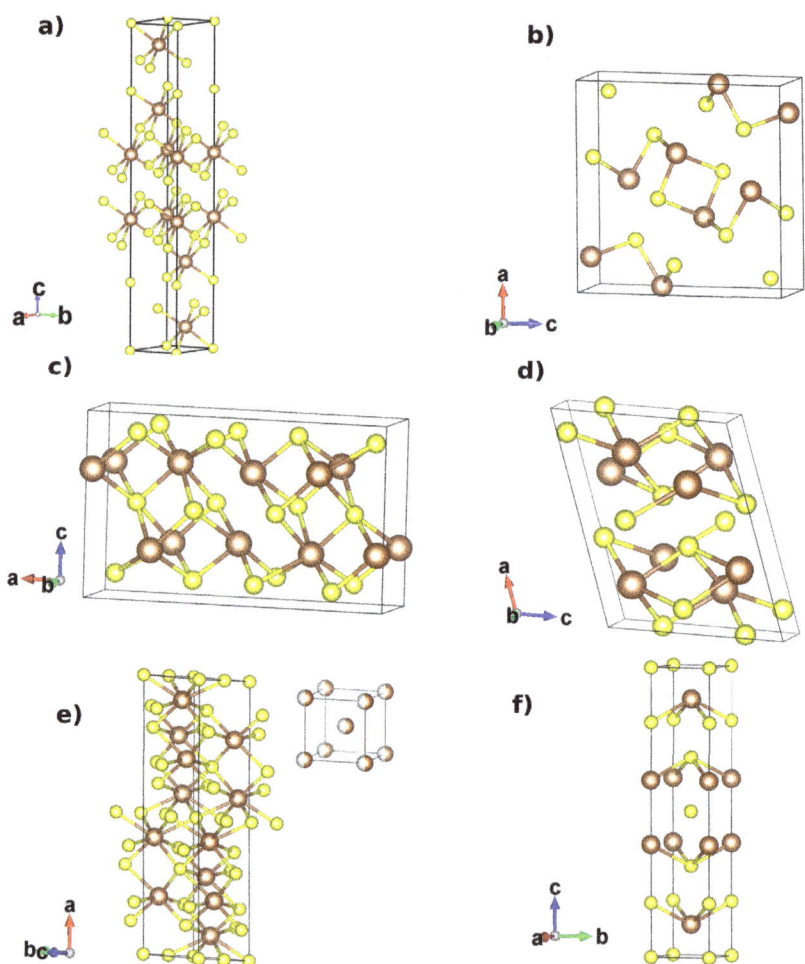

Figure 1. Unit-cell representations of the crystal structures of the *R-3m* (Z = 1) (**a**), *Pnma* (Z = 4) (**b**), *C2/m* (Z = 2) (**c**), *C2/c* (Z = 2) (**d**), the *C2/m* 9-/10-fold structure used to model the disordered bcc-type *Im-3m* phase (Z = 2) (**e**), and the *I4/mmm* (Z = 1) (**f**), and A$_2$X$_3$ sesquichalcogenide structures (A = Sb, Bi; X = S, Se). The A cations and X anions are shown as brown and yellow spheres, respectively.

Initial experimental HP studies of Sb$_2$S$_3$, Bi$_2$S$_3$, and Sb$_2$Se$_3$ have shown that the *Pnma* structure is stable under compression, with first-order PTs occurring at around 50 GPa [36–41]. A comparative study between experimental and theoretical analysis suggested that the *Pnma* structure of the three U$_2$S$_3$-type sesquichalcogenides should actually be stable up until 50 GPa [39]. Moreover, another work has reported that the crystalline *Pnma* phase for Sb_2Se_3 is in fact stable even at 70 GPa [?]. Curiously enough, HP studies have found that *Pnma*-type Sb_2Se_3 becomes a topological superconductor at around 10 GPa and at a low temperature value of 2.5 K [43], exhibiting highly conducting spin-polarized surface states, and similar to what occurs for Bi_2Se_3 [44]. Moreover, HP superconductivity has been recently found in amorphous Sb_2Se_3 and attributed to the crystallization of the material in a phase with possible "metavalent" bonding as that present in tetradymite-like Sb_2Te_3, Bi_2Se_3, Bi_2Te_3 [45,46].

It must be noted, however, that some experimental HP studies have suggested that several first- and second-order PTs occur for Sb_2S_3 up to 50 GPa [47–49]. Furthermore, it has also been suggested that the HP phases of Sb_2S_3 could be similar to those observed for heavier sesquichalcogenides such as Bi_2Se_3, Bi_2Te_3, and Sb_2Te_3. Moreover, a theoretical study of Bi_2S_3 at HP predicts the system to be unstable under compression, and decomposing into another stoichiometric system [50]. Therefore, there remains the question for whether the different structural phases (*C2/m*, *C2/c*, and disordered *Im-3m* and *I4/mmm*) observed for heavier cation sesquichalcogenides could be observed at HP on the three U_2S_3-type minerals, and at different pressure conditions.

In this work, we report theoretical simulations between 0 GPa and 60 GPa of the *Pnma* and hypothetical *R-3m*, *C2/m*, and *C2/c*; and the disordered *Im-3m* and *I4/mmm* phases for Sb_2S_3, Sb_2Se_3, and Bi_2S_3 (Figure 1), with a view to assessing which, if any, are likely to be observed at HP. This work complements a previous study in which we examined the stability of the tetradymite-like (*R-3m*) phase with respect to the *Pnma* phase, up until 10 GPa, for the three U_2S_3-type minerals [51].

2. Theoretical Methodology

The six studied crystalline phases (*Pnma*, *R-3m*, *C2/c*, *C2/m*, disordered *Im-3m*, and *I4/mmm*) for Sb_2S_3, Bi_2S_3, and Sb_2Se_3 were simulated within the framework of density-functional theory (DFT) [52]. The Vienna Ab initio Simulation Package (VASP) package [53] was used within the projector augmented-wave (PAW) scheme. The datasets included six valence electrons for $S[3s^23p^4]$ and $Se[4s^24p^4]$, and 15 valence electrons for $Sb[4d^{10}5s^25p^3]$ and $Bi[5d^{10}6s^26p^3]$. Total energy convergence was achieved with a plane-wave kinetic-energy cut-off of 600 eV. The Perdew-Burke-Ernzerhof parameterization revised for solids (PBEsol), of the generalized-gradient approximation (GGA) exchange-correlation (xc) functional [54], was considered for all the calculations.

The sampling of the Brillouin-zone (BZ) was converged with Γ-centered Monkhorst-Pack [55] grids employing adequate meshes for the different structural phases of the three compounds: *Pnma*-$6 \times 10 \times 6$, *R-3m*-$12 \times 12 \times 12$, *C2/m*-$6 \times 12 \times 6$, *C2/c*-$10 \times 10 \times 8$, disordered *Im-3m* (using a *C2/m* conventional cell)-$6 \times 12 \times 12$, and *I4/mmm*-$12 \times 12 \times 12$.

The *Im-3m* phase is a body-centered cubic (bcc) disordered structure, considered as a disordered solid solution, and it has been theoretically predicted and experimentally found for Bi_2Te_3 [17]. For sesquichacogenides with A_2X_3 stoichiometry, the bcc lattice site (2a Wyckoff position) is randomly occupied by 40% of A cations and 60% of X anions. This means that such a structure is a disordered phase with a mixture of cations and anions randomly sharing the same bcc crystallographic position and forming an A-X substitutional alloy [17]. Due to the theoretical difficulty in simulating the disordered *Im-3m* structure, we have used a 9/10-fold *C2/m* structure (the formation of 9/10 chemical A-X bonds), as was previously employed for Bi_2Te_3 [17] and Bi_2Se_3 [22]. Moreover, it has been observed that the 9/10-fold *C2/m* structure presents a bcc-like structural order, in agreement with the observed XRD patterns [17,49], therefore giving support to employ the calculated intermediate bcc-like monoclinic *C2/m* phase to confirm the experimental presence of the disordered *Im-3m* system.

For the structural relaxations, the atomic positions and the unit-cell parameters were allowed to change during the ionic relaxation, for different volume values. From these relaxations, we obtained the respective external pressure for the specific isotropic volume compression. The pressure-volume (P-V) curves for all the compounds were fitted to a third-order Birch-Murnaghan equation of state [56,57] to obtain the equilibrium volume, bulk modulus, and respective pressure derivative. The enthalpy (H) curves were computed by considering the relation, $H = E + pV$, where E is the total electronic energy of the system, p is pressure, and V is the volume. The analysis and comparison of the H curves for the different polymorphs provides deep insight regarding the thermodynamic stability of each phase for increasing pressure values, up until the studied pressure range (60 GPa).

Lattice dynamics calculations, within the harmonic approximation, were performed at different pressure points, which were found to be energetically favorable. These were considered for the *Pnma*, the disordered *Im-3m*, and the tetragonal phases. We have also calculated the phonon band structure for the *C2/m* system of Sb_2Se_3 to confirm the dynamical stability at the pressure point where the energetic stability is evidenced (20 GPa). The phonon properties were computed by using the supercell finite-displacement method implemented in the Phonopy package [58], with VASP being used as the second-order force calculator. Supercells were expanded up to $2 \times 4 \times 2$ for the *Pnma* systems, and $2 \times 2 \times 2$ for the disordered and tetragonal phases, enabling the exact calculation of frequencies at the zone center (Γ) and inequivalent zone-boundary wavevectors, which were then interpolated to obtain phonon-dispersion curves.

Since our calculations on the *Pnma* phases for the three compounds under study were in good agreement with the overall data found in the literature [51], we have proceeded in carrying out a theoretical study of the hypothetical *R-3m*, *C2/m*, and *C2/c*; and disordered *Im-3m* and *I4/mmm* phases of Sb_2S_3, Bi_2S_3, and Sb_2Se_3, to probe whether such polymorphs could be energetically competitive under hydrostatic pressure.

3. Results and Discussion

3.1. Energetic Stability

Figure 2a–c shows the pressure-dependence of the enthalpy differences relative to the stable phase at ambient pressure between the six above mentioned phases of Sb_2S_3, Bi_2S_3, and Sb_2Se_3, respectively. The values of the predicted transition pressures between the different phases are summarized in Table 1.

Table 1. Theoretical estimation of the pressure-induced phase transitions of *R-3m* → *Pnma*, *Pnma* → disordered *C2/m*, *C2/m* → disordered *Im-3m*, and *Pnma* → disordered *Im-3m* for the Sb_2Se_3, Sb_2S_3, and Bi_2S_3 compounds (presented in units of GPa).

	Sb_2S_3	Bi_2S_3	Sb_2Se_3
R-3m → *Pnma*	–	–	4.8 *
Pnma → *C2/m*	–	–	9.9
C2/m → *Im-3m*	–	–	22.1
Pnma → *Im-3m*	35.1	30.1	–

* Ref. [51].

From the enthalpy plots (Figure 2), we observe that:

1. At 0 GPa, the orthorhombic *Pnma* structure is energetically stable for Bi_2S_3 and Sb_2S_3; however, for Sb_2Se_3, it is the trigonal *R-3m* phase that is the most energetically favorable phase at 0 GPa [51].
2. The two monoclinic *C2/c* and *C2/m* phases do not become energetically competitive with the ground-state system over the range of pressures examined. However, an exception occurs for the *C2/m* phase of the Sb_2Se_3 structure, which competes energetically with the *Pnma* polymorph at 9.9 GPa.
3. The bcc-like disordered *Im-3m* structure is the most energetically stable phase at pressures above 35.1, 30.1, and 22.1 GPa for Sb_2S_3, Bi_2S_3, and Sb_2Se_3, respectively.
4. The body-centered tetragonal *I4/mmm* phase, although being quite close in energy to the bcc-like *Im-3m* structure (namely, for Bi_2S_3 close to 30 GPa, and for Sb2Se3 between

30 and 50 GPa), is never energetically favorable within the studied pressure range, for either of the three studied compounds.

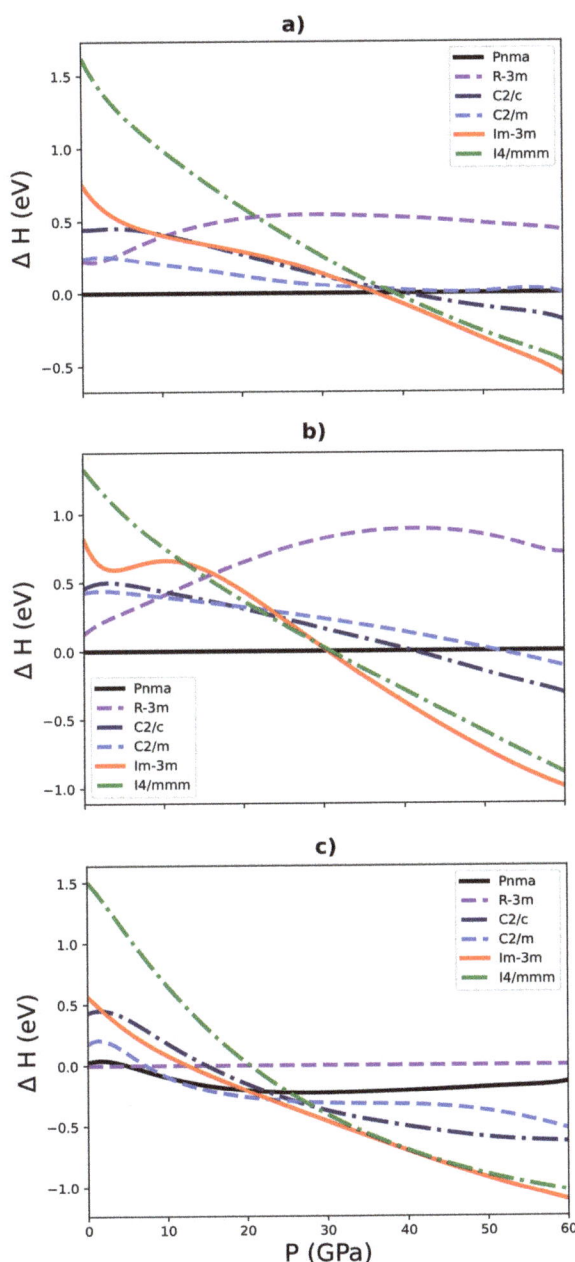

Figure 2. Calculated relative enthalpy vs. pressure curves, for the different possible phases (shown in Figure 1) of Sb_2S_3 (**a**), Bi_2S_3 (**b**), and Sb_2Se_3 (**c**), relative to the lowest-energy phase at ambient pressure: the *Pnma* phase for Sb_2S_3 and Bi_2S_3, and the *R-3m* phase for Sb_2Se_3.

With respect to the first point, referring to Bi_2S_3 and Sb_2S_3, our calculations predict that the *Pnma* structure is energetically the most stable phase, up to 30 GPa, and in good agreement with experimental evidences that show the observation of this phase, both at room and also at HP conditions [30–33,45,46]. Surprisingly, however, our simulations indicate that the Sb_2Se_3 *R-3m* phase is the most stable polymorph at pressures below 4.8 GPa; being both the *Pnma* and *R-3m* phases, energetically competitive between 0 and 4.8 GPa [51].

Regarding the second point, our analysis further shows that the two monoclinic *C2/c* and *C2/m* phases are never energetically competitive in the three compounds throughout the studied pressure range up to 60 GPa; with the exception of the *C2/m* phase of the Sb_2Se_3 close to 10 GPa. These results are consistent with experimental analysis obtained from Refs. [30–33,45,46], in which no PT had been observed for the Sb_2S_3 and Bi_2S_3 systems up until ~50 GPa. However, the respective results are not consistent with three recent studies reporting evidences of PTs in Sb_2S_3 [47–49]. A PT to an unknown phase was claimed to occur at around 15 GPa [47,48], and several transitions were also reported between 10 and 25 GPa, and tentatively proposed to be the *R-3m*, *C2/c*, and *C2/m* structural phases [49]. In this context, it must be stressed that our calculations are performed for pure hydrostatic conditions; therefore, it is not expected that a complete agreement occurs with experiments if non-hydrostatic conditions are considered for the experiments.

As for the third point, from a thermodynamic point of view, our results indicate that the bcc-like disordered *Im-3m* phase, initially identified for Bi_2Se_3, Bi_2Te_3, and Sb_2Te_3 [17,19,22], seems to be energetically favorable at HP for our three materials of interest. These results are consistent with the observation of such a phase at around 50 GPa for Sb_2Se_3 [36] and above 25 GPa for Sb_2S_3 [47–49]. However, our results do not agree with those found for Bi_2S_3 [37,40], for which a disorder, attributed mostly to a pressure-induced amorphization (PIA), has been observed above 50 GPa. Notably, PIA in Bi_2S_3 is consistent with a recent theoretical work that claims that a respective system is unstable above 31.5 GPa, decomposing into a mixture of BiS_2 and BiS compounds [50]. In this context, it must be stressed that the existence of a bcc-like structure at HP is not only expected for A_2X_3 sesquichalcogenides, but also for AX (A=Ge,Sn,Pb) chalcogenides [? ?]. All these compounds have in common the property of being cataloged as evidencing metavalent bonding at room pressure, such as SnTe, PbS, PbSe, PbTe, Sb_2Te_3, Bi_2Se_3, or Bi_2Te_3 [46]; or as being compounds with p-type covalent bonds that develop metavalent bonding at HP when approaching a six-fold coordination, such as GeSe, SnSe, As_2S_3, and also U_2S_3-type sesquichalcogenides [61,62], and ultimately, they will reach eight-fold coordination typical of bcc-like metals at very HP.

With respect to the fourth point, the *I4/mmm* structure was firstly proposed for Bi_2Se_3 by combining experimental and theoretical data [25]. It was found that the structural phase transition pathway for Bi_2Se_3 followed the sequence: *R-3m* → *C2/m* → *C2/c* → *I4/mmm*, when the quasi-hydrostatic pressure was considered. The *C2/c* phase would, however, be suppressed when nonhydrostatic pressure conditions are taken into account [25]. These results are also compatible with Raman analysis and XRD data performed on Bi_2Se_3, where it was evidenced that the stability of the *I4/mmm* phase was up to 81.2 GPa [24]. Moreover, it has also been claimed [26], that the alloying of Bi_2Se_3 with Bi_2Te_3 enables transitions to occur from the monoclinic phases, up to the disordered *Im-3m* at ~19 GPa, which then would be surpassed energetically by the tetragonal *I4/mmm* polymorph at ~23 GPa. Curiously enough, the energetic competition between the *C2/m*, *C2/c*, and *I4/mmm* phases is observed at HP for all three U_2S_3-type sesquichalcogenides (if we discard the disordered *Im-3m* phase), and the *I4/mmm* structure would indeed be energetically the most favorable polymorph at very HP. However, the disordered *Im-3m* phase is always energetically more competitive than *I4/mmm* for the U_2S_3-type sesquichalcogenides, and so this latter phase is, *a priori*, not expected to be observed at HP.

Finally, we have to mention that we have also performed the enthalpy calculations by employing two further xc functionals, in order to confirm the present theoretical data (see Appendix A). The employed functionals were the PBE-D2 method of Grimme (which takes into account the dispersion correction term) [63], and LDA [64]. From the analysis of the plots (Figures A1 and A2) we can infer similar energetic trends of the six polymorphs; however, mild differences are observed between the two competing phases of the disordered *Im-3m* and *I4/mmm*—detailed discussion is found in Appendix A.

In summary, the agreement of our results regarding the observation of the disordered *Im-3m* phase for Sb_2Se_3 and Sb_2S_3, but not for Bi_2S_3, suggests that thermodynamic stability is not sufficient to explain the lack of the HP disordered phase for the latter compound. In the following section, we discuss the dynamical stability of the *Pnma*, *C2/m*, *Im-3m*, and *I4/mmm* phases as a function of pressure, in order to provide a deeper understanding regarding this question.

3.2. Dynamical Stability

Energetic stability is a necessary, but it is not a sufficient condition for a structural phase to be synthetically accessible. One should also probe the dynamical stability of the system, which requires the study of the phonon frequencies. If imaginary frequencies emerge (as represented by negative frequencies in the phonon dispersion curves), we can analyze the results by considering that the system is at a potential-energy maximum (transient state), undergoing a phase transition, and therefore cannot be kinetically stable at the given temperature and/or pressure conditions [65–70].

In this section, we consider the phonon properties of different phases for the three compounds, which were observed to be energetically the most favorable (Figure 2) at different pressure values, namely:

1. The *C2/m* phase of Sb_2Se_3 at 20 GPa.
2. The disordered bcc-type *Im-3m* phase of the three compounds above 30 GPa.
3. The *Pnma* phase of the three compounds at 50 GPa.
4. The *I4/mmm* phase of the three compounds at HP.

3.2.1. The *C2/m* Phase of Sb_2Se_3 at 20 GPa

By analyzing the enthalpies as a function of pressure (Figure 2), we observe that the *C2/m* phase in Sb_2Se_3 is thermodynamically the most stable phase between 10 and 20 GPa. In order to confirm the dynamical stability, we have performed the phonon dispersion curves of respective system at 20 GPa (Figure 3). The dispersion curves do not evidence any imaginary modes; therefore, we may conclude that the *C2/m* phase in Sb_2Se_3 is dynamically stable in this pressure range and could potentially be observed after the *Pnma* phase, although it has not yet been experimentally observed [36].

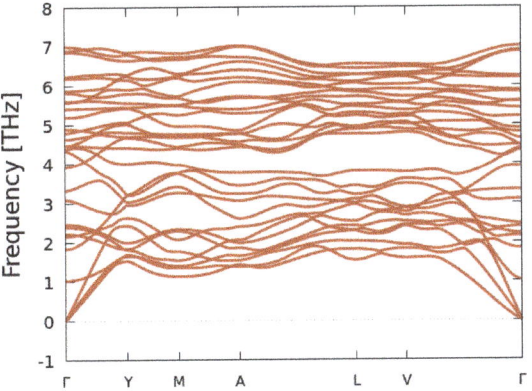

Figure 3. Harmonic phonon dispersion curves of the *C2/m* phase of the Sb_2Se_3 compound at 20 GPa.

3.2.2. The Disordered bcc-Type *Im-3m* Phase of the Three Compounds above 30 GPa

To assess the possibility of dynamical stability for the disordered *Im-3m* phases of the three compounds at HP, we have evaluated the phonon dispersion curves at pressure values of 30 GPa, which is close to the transition pressures observed in Figure 2; and at higher pressures of 50 (Sb_2Se_3) and 60 GPa (Sb_2S_3, Bi_2S_3).

As illustrated in Figure 4, all three disordered structures at 30 GPa show imaginary modes along the dispersion curves, thus indicating that these structures are dynamically unstable at this pressure range.

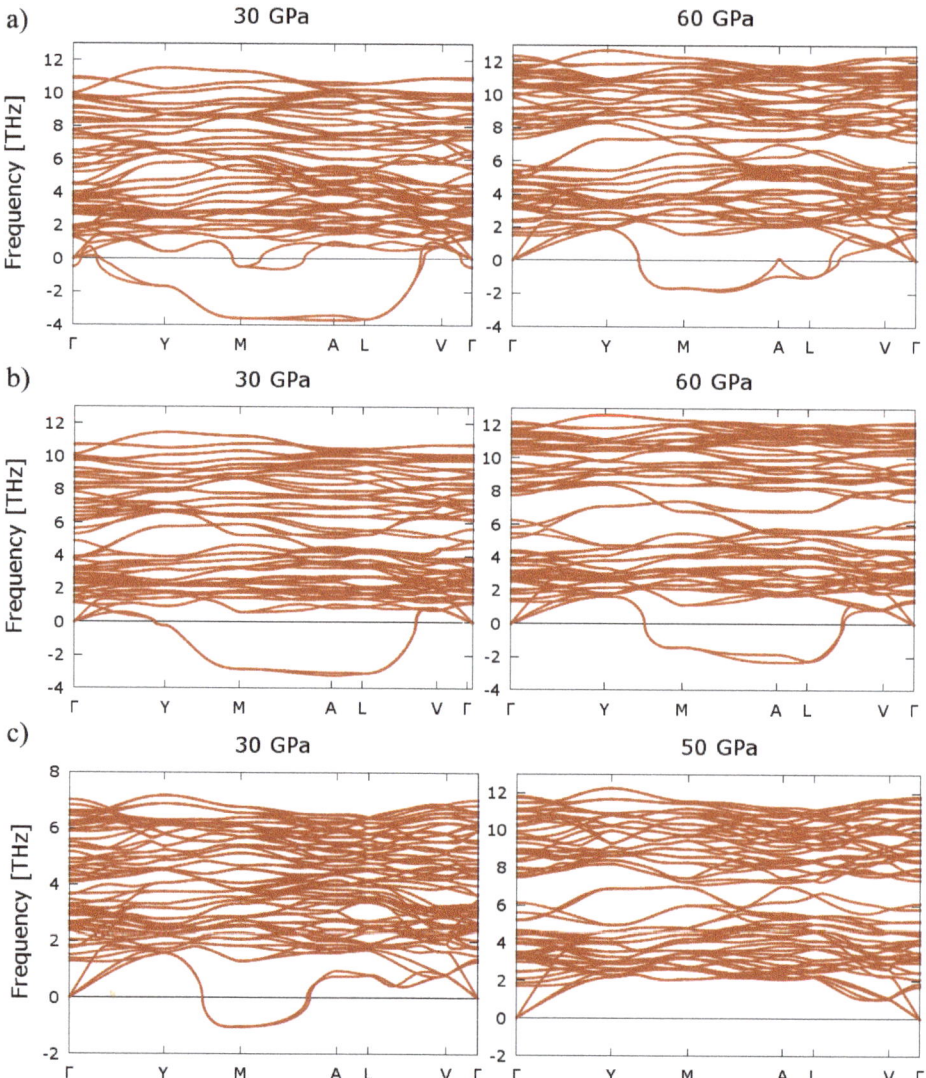

Figure 4. Harmonic phonon dispersion curves of the disordered bcc-like *Im-3m* phases of Sb_2S_3 (**a**), Bi_2S_3 (**b**), and Sb_2Se_3 (**c**); and calculated at 30 GPa (**left**), and 50 (Sb_2Se_3) or 60 GPa (Sb_2S_3 and Bi_2S_3; **right**). The BZ **q**−vector description represents the *C2/m* space−group, according to the symmetry of the employed cell.

The phonon dispersion curves of Sb_2S_3 and Bi_2S_3 still present imaginary modes at 60 GPa (Figure 4), thus indicating that neither compound is likely to adopt this phase up to this pressure range. We note, however, that the dynamical instabilities found for Sb_2S_3 and Bi_2S_3 both decrease (the imaginary, soft modes shift to higher frequency values, towards positive values) with increasing pressure, suggesting that this phase could in principle become stable at pressures above 60 GPa. In this context, we must note that Efthimiopoulos et al. [37] had observed a pressure-induced amorphization above 50 GPa for Bi_2S_3; however, the authors were not able to identify the phase to be the disordered Im-$3m$ structure, even at 65 GPa. On the other hand, the experimental data for Sb_2S_3, suggests that the disordered bcc-like phase exists between 28.2 and 50.2 GPa [49]. However, it must be noted that experimental measurements detailed in Ref. [49] were carried out under non-hydrostatic behavior due to the employed pressure-transmitting medium.

Finally, our calculations suggest that the Im-$3m$ phase becomes dynamically stable in Sb_2Se_3 already at 50 GPa; a result that is in agreement with the experimental observation of this phase at around 50 GPa [36].

In the light of the above results regarding the disordered Im-$3m$ phase of Sb_2S_3, Bi_2S_3, and Sb_2Se_3, we can speculate that the stability of the disordered solid solution in group-15 sesquichalcogenides seems to be related to the sizes of cations and anions. Since the Im-$3m$ phase is consistently being evidenced at HP for sesquichalcogenides with heavier cations and anions (Sb_2Se_3, Sb_2Te_3, Bi_2Se_3, and Bi_2Te_3), the observation of such a HP phase could be related to the radius sizes [71] of Se, Te, Sb, and Bi (atomic radii: $r_{Se} = 117$, $r_{Te} = 137$, $r_{Sb} = 141$, and $r_{Bi} = 182$ pm, respectively). Stemming on these values, we can infer that the disordered solid solutions are energetically favorable in sesquichalcogenides if the atomic radii of the cation and anion differ by less than ~ 65 pm, or if the size ratio between them is smaller than ~ 1.55 (case of Bi_2Se_3). In this context, Sb_2S_3, which shows a radius difference of between r_{Sb} and r_S of 37 pm (141 − 104 = 37 pm) and a size ratio of 1.35, meets the criteria for the observation of the disordered Im-$3m$ phase at HP. However, Bi_2S_3, which evidences a larger radius difference (78 pm) and ratio (1.75) between B and S, does not satisfy the above criteria. Such results would therefore be consistent with the amorphous phase observed for Bi_2S_3 at HP [37,40].

Finally, it must be considered, and as suggested in Ref. [17], that the atomic radii between the anion and cation tend to become approximately equal at HP due to a higher probability of charge transfer from cation to anion. Therefore, HP inherently creates a favorable environment for the disordered solid solutions due to the decrease in the difference between the cation and anion atomic radii. This means that we cannot discard that the disordered solid solution in Bi_2S_3 could occur at very HP values, namely, when the difference between the two radii decreases below 65 pm and when the ratio decreases below 1.55.

3.2.3. The *Pnma* Phase of the Three Compounds at 50 GPa

In order to study the dynamical stability of the well known low-pressure *Pnma* phase at HP, we present in Figure 5 the phonon dispersion curves of the respective phase for Sb_2S_3, Bi_2S_3, and Sb_2Se_3 at 50 GPa.

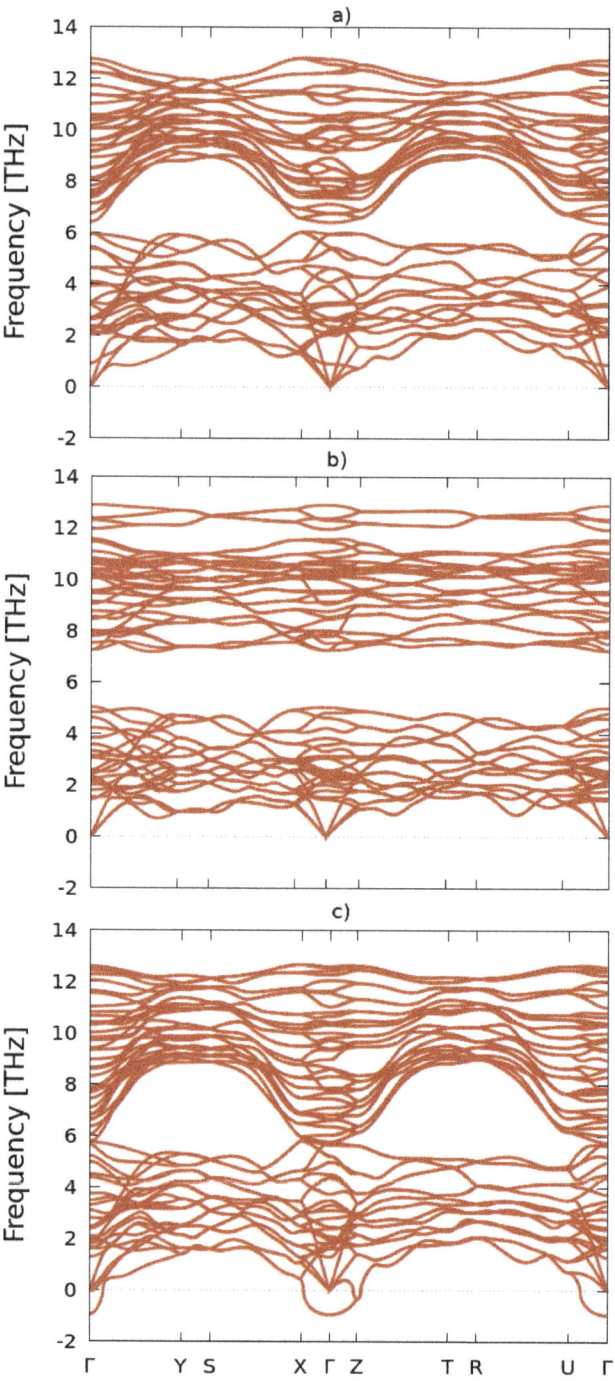

Figure 5. Harmonic phonon dispersion curves of the *Pnma* phase of Sb_2S_3 (**a**), Bi_2S_3 (**b**), and Sb_2Se_3 (**c**), calculated at 50 GPa.

Curiously enough, we find that the *Pnma* phase for Sb$_2$Se$_3$ is already unstable at 50 GPa; therefore, a PT should occur at smaller pressures to the *C2/m* phase, and ultimately, to the disordered *Im-3m* phase, as previously commented. On the other hand, the *Pnma* phase for Sb$_2$S$_3$ and Bi$_2$S$_3$ is still dynamically stable at 50 GPa, although thermodynamically, it is not the most stable phase (Figure 2). These results, together with the dynamical instability observed for the disordered phase of Sb$_2$S$_3$ and Bi$_2$S$_3$ at 50 GPa (Figure 4) and the thermodynamic instability of the *C2/m* and *C2/c* phases, suggest that only the *Pnma* structure should be observed up to 50 GPa for both compounds, unless there is a decomposition, as theoretically predicted for Bi$_2$S$_3$ [50].

3.2.4. The *I4/mmm* Phase of the Three Compounds at HP

By considering the enthalpy curves of the three compounds (Figure 2), we can infer that the *I4/mmm* phase is very close in energy with the disordered *Im-3m* system at pressures above 30 GPa. We therefore have carried out phonon calculations to probe the dynamical stability of this phase, and for all three compounds.

With regard to the Sb$_2$S$_3$ system, we observe that only at 60 GPa is the *I4/mmm* structural phase dynamically stable (Figure 6). Below this pressure value, imaginary modes are observed in the vicinity of the Γ-point, which harden for increasing pressure points. This result, together with the dynamic instability of the disordered *Im-3m* phase still at 60 GPa, suggests that both the *I4/mmm* and *Im-3m* structures are not expected to be observed at pressures below 60 GPa; however, any of these systems could potentially be observed above this pressure range.

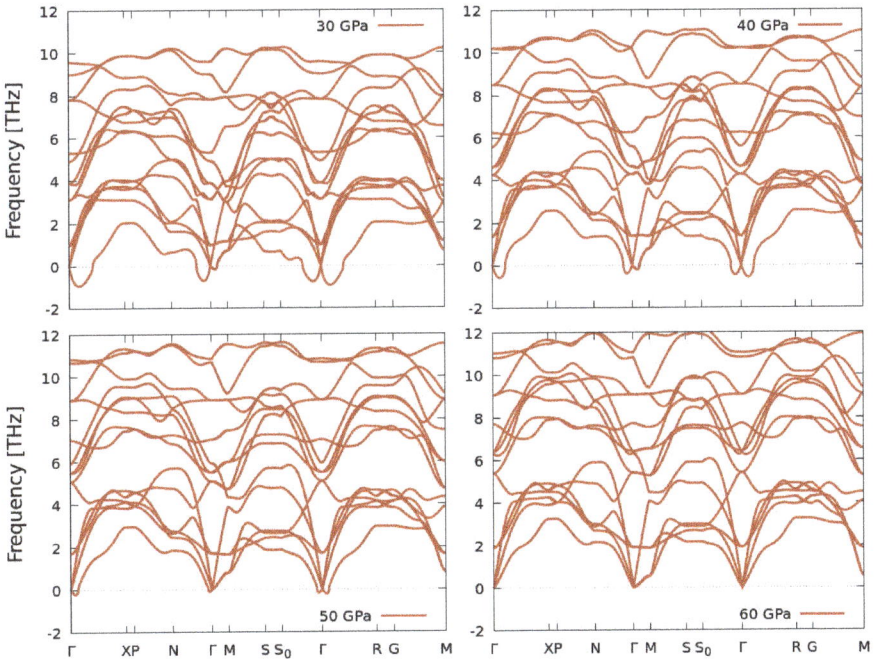

Figure 6. Harmonic phonon dispersion curves of the *I4/mmm* phase of the Sb$_2$S$_3$ compound between 30 and 60 GPa.

With respect to the Bi$_2$S$_3$ compound (Figure 7), we observe that the *I4/mmm* phase is already dynamically stable at 50 GPa, maintaining the stability at 60 GPa. Therefore, the *I4/mmm* phase could in fact be observed at above 40 GPa, if there was no sample decomposition, as theoretically predicted [50].

Finally, we observed that the the tetragonal *I4/mmm* phase for Sb$_2$Se$_3$ is only dynamically stable at 40 GPa (Figure 8), since there is a localized imaginary mode at the high-symmetry *M*-point, at any other pressure value. These results differ from the remaining two S-based compounds, and evidence that the disordered *Im-3m* phase will be the only (thermodynamically and dynamically) stable phase for the Sb$_2$Se$_3$ compound at very HP (since the *Pnma* is also unstable at 50 GPa, Figure 5c), and as discussed in the previous subsection.

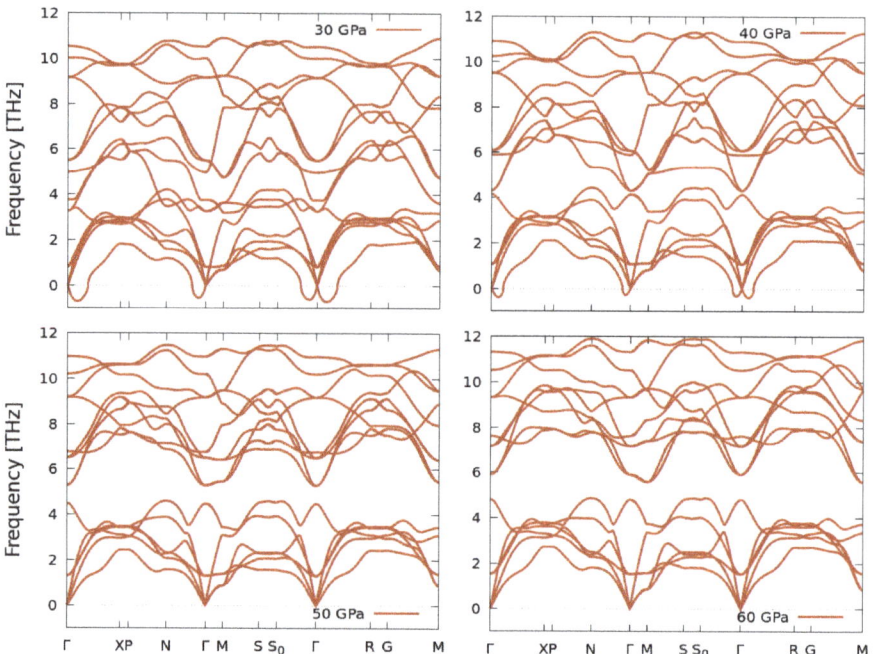

Figure 7. Harmonic phonon dispersion curves of the *I4/mmm* phase of the Bi$_2$S$_3$ compound between 30 and 60 GPa.

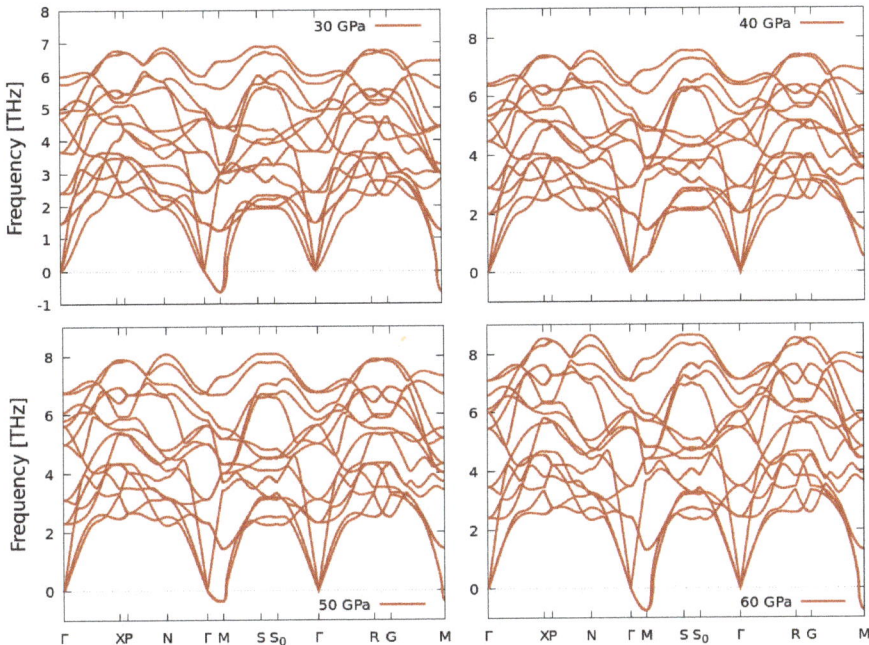

Figure 8. Harmonic phonon dispersion curves of the *I4/mmm* phase of the Sb_2Se_3 compound between 30 and 60 GPa.

4. Conclusions

We have carried out a comprehensive set of total energy and lattice dynamics calculations in order to investigate the stabilities of six possible structural phases, viz. *Pnma*, *R-3m*, *C2/m*, and *C2/c*; and disordered *Im-3m* and *I4/mmm* for the Sb_2S_3, Bi_2S_3, and Sb_2Se_3 sesquichalcogenides under hydrostatic pressures up to 60 GPa. Our theoretical results have been commented in the light of the available experimental data.

We find that the *Pnma* phase is energetically more stable at room pressure for the Sb_2S_3 and Bi_2S_3 compounds. Curiously, the trigonal *R-3m* phase is the most energetically favorable phase for Sb_2Se_3 at 0 GPa, although from an experimental perspective, this compound is synthesized in the *Pnma* phase. In fact, we find that the *Pnma* phase is dynamically stable for Sb_2S_3 and Bi_2S_3 up to 50 GPa, but not for Sb_2Se_3 at this pressure value.

From our calculations, we observe that the monoclinic *C2/m* and *C2/c* systems for the three compounds are energetically less favorable throughout the studied pressure range, and they are not expected to be observed at HP under hydrostatic conditions, except for Sb_2Se_3 between 10 and 20 GPa. Moreover, the *C2/m* phase of Sb_2Se_3 is dynamically stable at 20 GPa, so a pressure-induced phase transition from the *Pnma* to the *C2/m* system could be observed between 10 and 20 GPa under hydrostatic conditions.

The disordered bcc-like *Im-3m* phase is predicted to be the most energetically stable phase above 35, 30, and 22 GPa for Sb_2S_3, Bi_2S_3, and Sb_2Se_3, respectively; however, this structure is dynamically unstable for Sb_2S_3 and Bi_2S_3 up to 60 GPa, and for Sb_2Se_3 up to 50 GPa. Therefore, this HP phase is expected to occur for Sb_2Se_3 above 50 GPa, in good agreement with experiments, and perhaps above 60 GPa for the other two compounds.

With respect to the *I4/mmm* phase, we have learned that the structure is not energetically competitive with the *Im-3m* phase, throughout the studied pressure range, and for the three compounds; however, this phase is close in energy to the *Im-3m* phase for Bi_2S_3 at 30 GPa, and for Sb_2Se_3, close to 38 GPa. On the other hand, the *I4/mmm* system is dynamically stable for Bi_2S_3 above 50 GPa, and for Sb_2Se_3, quite close to 40 GPa. Therefore, this phase could potentially be observed for the former compound at HP; however, not for Sb_2Se_3.

Curiously enough, we must mention that through symmetry analysis, the *I4/mmm* → *Fmmm* → *C2/m* structural transition pathway may occur as a second-order phase transition. We have, however, not considered the intermediate *Fmmm* system, since, and to the best of our knowledge, such a phase has not yet been experimentally observed for sesquichalcogenides at HP, and therefore, respective analysis would be out the scope of the present work.

We hope that this work will stimulate further investigation of the sesquichalcogenides at HP in order to clarify which pressure-induced phase transitions are observed under hydrostatic and non-hydrostatic conditions.

Author Contributions: E.L.d.S., M.C.S., P.R.-H., and A.M. performed the ab initio calculations. F.J.M. supervised the work and comparison with experimental data. E.L.d.S. and F.J.M. wrote the manuscript. All authors have read and agreed to the published version of the manuscript.

Funding: This publication is part of the Project MALTA Consolider Team network (RED2018-102612-T), financed by MINECO/AEI/10.13039/501100003329; by I+D+i project PID2019-106383GB-42/43, financed by MCIN/AEI/10.13039/501100011033; by projects PROMETEO CIPROM/2021/075 (GREENMAT) and MFA/2022/025 (ARCANGEL), financed by Generalitat Valenciana; and by the European Union Horizon 2020 research and innovation programme under the Marie Sklodowska-Curie grant, agreement No. 785789-COMEX.

Data Availability Statement: The data that support the work presented within this paper are available from the corresponding author upon reasonable request.

Acknowledgments: E.L.d.S acknowledges the High Performance Computing Chair—an R&D infrastructure (based at the University of Évora; PI: M. Avillez), endorsed by Hewlett Packard Enterprise, and involving a consortium of higher education institutions, research centers, enterprises, and public/private organizations; and the Portuguese Foundation of Science and Technology with the CEEC individual fellowship, 5th edition, with Reference 2022.00082.CEECIND.

Conflicts of Interest: The authors declare no conflict of interest.

Appendix A. Comparison of Enthalpy vs. Pressure Behavior for Different Exchange-Correlation Functionals

In order to confirm the enthalpy variation as the pressure increases for the six studied phases presented in Figure 2, and obtained by employing the PBEsol functional, we observe that the energetic trends are functional-independent. In Figures A1 and A2, we present enthalpy results carried out by considering the PBE-D2 and the LDA functional, respectively. For PBE-D2 (Figure A1), the transition pressures are very close to those obtained with PBEsol. Moreover, for the Sb_2Se_3 system, the energetic trends are also comparable to those of PBEsol, namely, that the *R-3m* phase is energetically the most stable at room pressure, and that the *C2/m* phase shows a small interval range of stability before the disordered *Im-3m* phase takes over the energetic configurational space. The *I4/mmm* system is energetically competitive, with the disordered phase at a slightly lower pressure range when compared to the results obtained with PBEsol. For Sb_2S_3 and Bi_2S_3, the tetragonal *I4/mmm* phase becomes energetically the most stable at pressures close to 60 GPa. The results obtained with the LDA functional (Figure A2) also show similar trends; however, the transition pressures are slightly higher than for PBEsol. Moreover, a competition between the two monoclinic phases and *I4/mmm* seems to occur in Sb_2Se_3 slightly, before the latter phase becomes the most stable phase. At around 60 GPa, the disordered phase stabilizes, becoming the most stable structure for Sb_2Se_3. In addition, for the Bi_2S_3 compound, we observe a pressure range, at which the *I4/mmm* becomes energetically the most stable (~30–45 GPa), which then is energetically taken over by the disordered *Im-3m* phase at above 45 GPa.

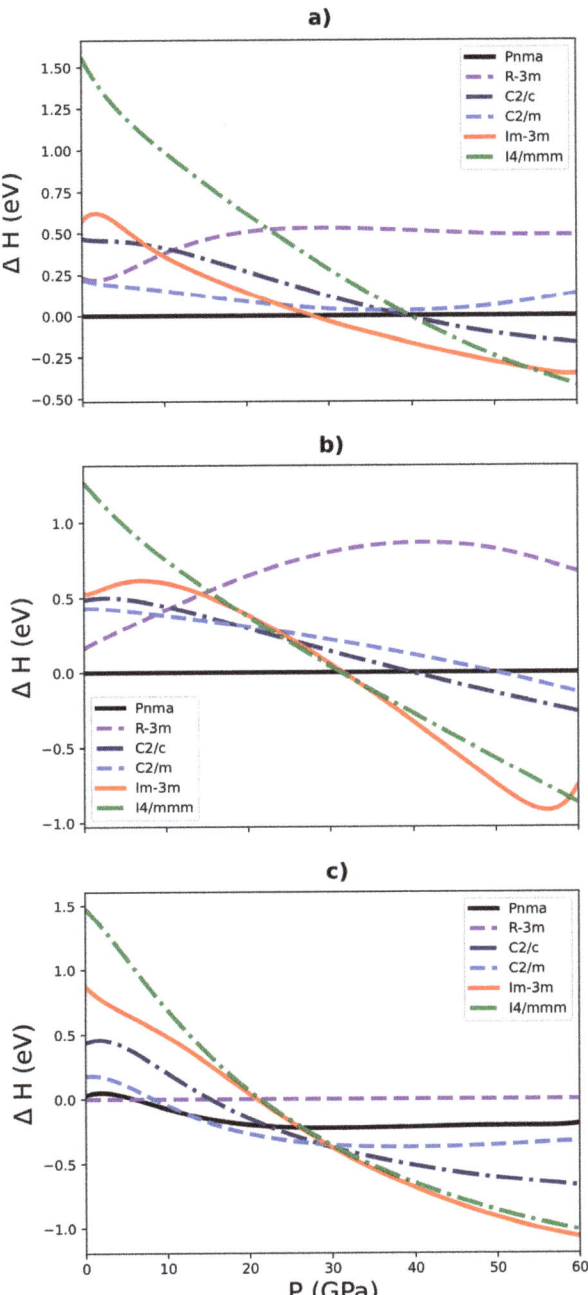

Figure A1. PBE+vdW enthalpy vs. pressure curves, for the different possible phases (shown in Figure 1) of Sb$_2$S$_3$ (**a**), Bi$_2$S$_3$ (**b**), and Sb$_2$Se$_3$ (**c**), relative to the lowest-energy phase at ambient pressure: the *Pnma* phase for Sb$_2$S$_3$ and Bi$_2$S$_3$, and the *R-3m* phase for Sb$_2$Se$_3$.

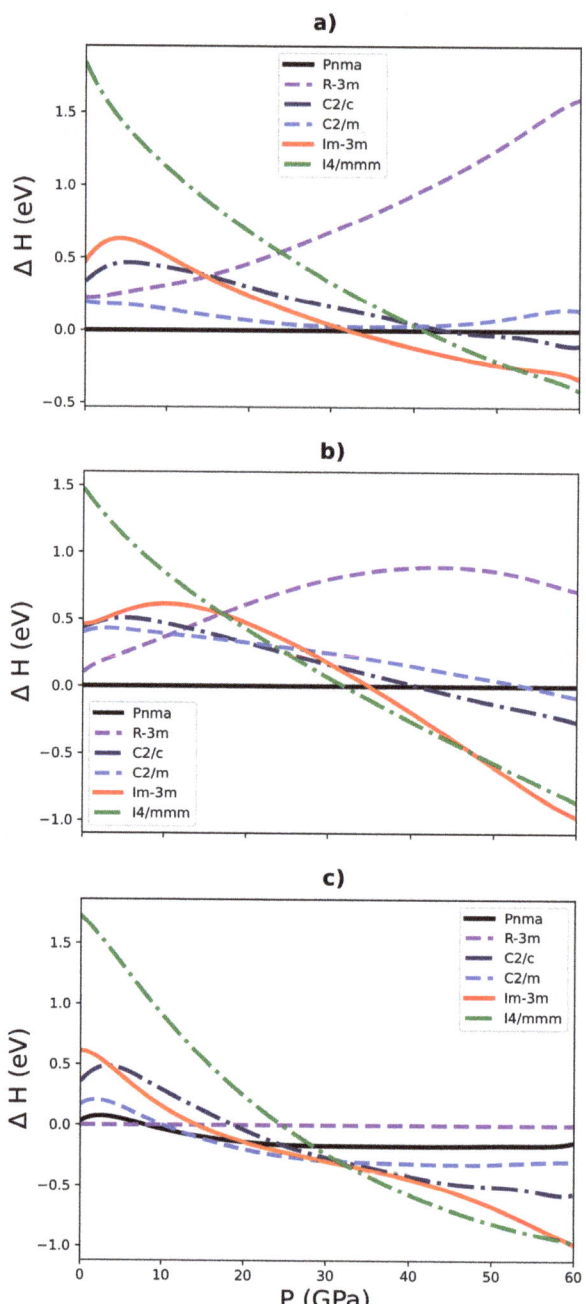

Figure A2. LDA enthalpy vs. pressure curves, for the different possible phases (shown in Figure 1) of Sb$_2$S$_3$ (**a**), Bi$_2$S$_3$ (**b**), and Sb$_2$Se$_3$ (**c**), relative to the lowest-energy phase at ambient pressure: the *Pnma* phase for Sb$_2$S$_3$ and Bi$_2$S$_3$, and the *R-3m* phase for Sb$_2$Se$_3$.

Appendix B. Evolution of the Lattice Parameters of Sb_2S_3, Bi_2S_3, and Sb_2Se_3 as a Function of Pressure

In Tables A1–A3, we present the lattice parameters of the different studied compounds at pressure values where stability (thermodynamic and dynamical) is evidenced. Since the $C2/c$ structure is never stable throughout the studied pressure range, we do not present the respective theoretical values for any of the compounds. As for Sb_2S_3 and Bi_2S_3, we only present the lattice parameters related to the $Pnma$ phase, at 0 and 30 GPa, since this is the pressure range for which the structure is energetically more favorable; and where none of the remaining phases satisfy both of the stability conditions that we mentioned previously.

Table A1. Theoretical estimation (PBEsol) of the lattice parameters of the Sb_2S_3 compound, for the $Pnma$ structural phase, for different pressure values, at which the system is both energetically and dynamically stable (presented in units of Å).

Sb_2S_3	0 GPa	10 GPa	20 GPa	30 GPa
a_0	11.24	10.30	10.02	9.82
b_0	3.83	3.72	3.63	3.58
c_0	10.91	10.24	9.83	9.53

Table A2. Theoretical estimation (PBEsol) of the lattice parameters of the Bi_2S_3 compound, for the $Pnma$ structural phase, for different pressure values, at which the system is both energetically and dynamically stable (presented in units of Å).

Bi_2S_3	0 GPa	10 GPa	20 GPa	30 GPa
a_0	11.19	10.59	10.38	10.22
b_0	3.96	3.82	3.74	3.67
c_0	10.94	10.32	9.94	9.64

Table A3. Theoretical estimation (PBEsol) of the lattice parameters of the Sb_2Se_3 compound, for the different structural phases that are both energetically and dynamically stable within the range of studied pressures (presented in units of Å).

Sb_2Se_3		
$Pnma$	10 GPa	$a_0 = 10.77$, $b_0 = 3.83$, $c_0 = 10.69$
	15 GPa	$a_0 = 10.63$, $b_0 = 3.77$, $c_0 = 10.48$
	20 GPa	$a_0 = 10.50$, $b_0 = 3.72$, $c_0 = 10.35$
$R-3m$	0 GPa	$a_0 = 4.01$, $c_0 = 28.16$
$C/2m$	20 GPa	$a_0 = 13.15$, $b_0 = 3.55$, $c_0 = 8.06$
disordered $Im-3m$	50 GPa	$a_0 = 13.55$, $b_0 = 4.49$, $c_0 = 5.50$
$I4/mmm$	40 GPa	$a_0 = 3.26$, $c_0 = 15.84$

Appendix C. Mode Mapping of the Disordered $Im-3m$ Phase of Sb_2Se_3 at 30 GPa

By observing the phonon dispersion curves of the disordered $Im-3m$ phase in Sb_2Se_3 at 30 GPa (Figure 4c), we may notice that the instability mode is localized at a high-symmetry point, the M-point, which is defined as being a zone-boundary instability (or anti-ferroelectric instability, resulting in an anti-phase periodic distortion). Since the negative mode is localized, it is therefore possible to map out the anharmonic potential energy surfaces by following the eigenvectors associated with this instability/displacement, and thus, to be able to obtain a lower energy structure corresponding to the minima of the potential energy surface. For these calculations, we use the open-source MODEMAP package [72,73]. A sequence of displaced structures in a commensurate supercell ($1 \times 2 \times 2$) expansion is generated by displacing along the phonon eigenvectors over a range of amplitudes of the normal-mode coordinate Q. The total energies of the "frozen phonon" structures are evaluated from single-point DFT calculations. The E(Q) curves are then fitted to a polynomial function, with the number of terms depending on the form of the potential well.

The displacements associated with the disordered bcc high-pressure structure result from the condensation of these soft M-modes, leading to a lower energy system and an enlarged primitive unit-cell. The displacement involves two Se atoms at the center of the cell, namely at (0.50000, 0.75000, 0.50000) and (0.50000, 0.25000, 0.50000). Both atoms evidence mild distortions along the x- and z-directions, minimizing the energy to 3.63 meV when compared to the original disordered $Im\text{-}3m$ phase at $Q = 0$ (Figure A3).

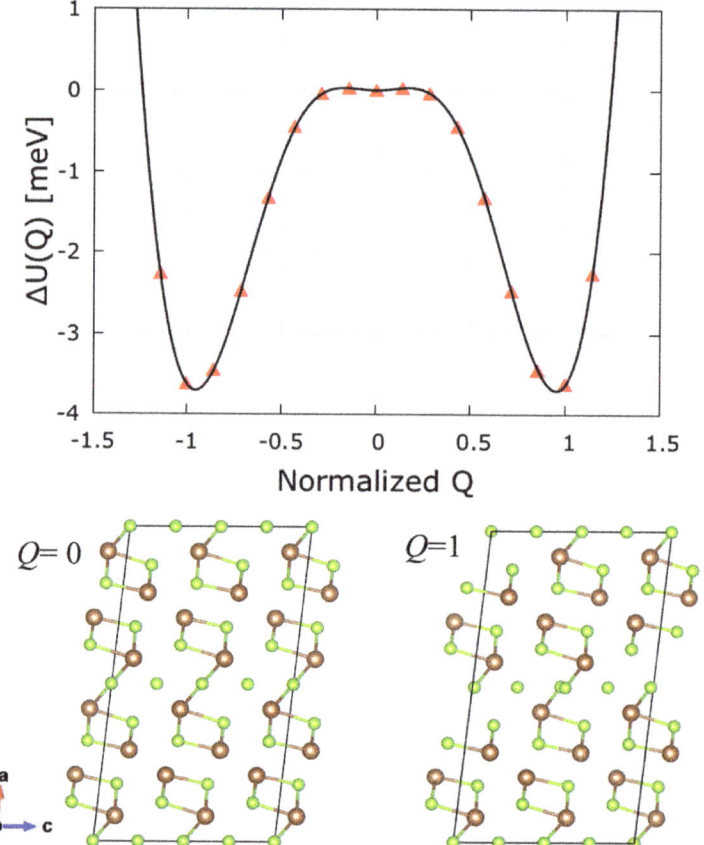

Figure A3. Double −well potential−energy surfaces for the phonon instabilities associated with the high-symmetry M −point negative modes (Figure 4). The normal-mode coordinates Q have been normalized so that the minima are located at $Q = \pm 1$, and the energy differences are those calculated in a $1 \times 2 \times 2$ supercell. The bottom figures represent the structure at $Q = 0$ (**left**) and at $Q = 1$ (**right**).

References

1. Chen, Y.L.; Analytis, J.G.; Chu, J.H.; Liu, Z.K.; Mo, S.K.; Qi, X.L.; Zhang, H.J.; Lu, D.H.; Dai, X.; Fang, Z.; et al. Experimental Realization of a Three-Dimensional Topological Insulator, Bi_2Te_3. *Science* **2009**, *325*, 178. [CrossRef]
2. Zhang, H.; Liu, C.X.; Qi, X.L.; Dai, X.; Fang, Z.; Zhang, S.C. Topological insulators in Bi_2Se_3, Bi_2Te_3 and Sb_2Te_3 with a single Dirac cone on the surface. *Nat. Phys.* **2009**, *5*, 438. [CrossRef]
3. Hasan, M.Z.; Kane, C.L. Colloquium: Topological insulators. *Rev. Mod. Phys.* **2010**, *82*, 3045. [CrossRef]
4. Venkatasubramanian, R.; Siivola, E.; Colpitts, T.; O'Quinn, B. Thin-film thermoelectric devices with high room-temperature figures of merit. *Nature* **2001**, *413*, 597. [CrossRef] [PubMed]

5. Martinez, J.C.; Lu, L.; Ning, J.; Dong, W.; Cao, T.; Simpson, R.E. The Origin of Optical Contrast in Sb_2Te_3-Based Phase-Change Materials. *Phys. Status Solidi B* **2020**, *257*, 1900289. [CrossRef]
6. Sa Moon, S.J.; Itzhaik, Y.; Yum, J.H.; Zakeeruddin, S.M.; Hodes, G.; Grätzel, M. Sb_2S_3-Based Mesoscopic Solar Cell using an Organic Hole Conductor. *J. Phys. Chem. Lett.* **2010**, *1*, 1524. [CrossRef]
7. Patrick, C.E.; Giustino, F. Structural and Electronic Properties of Semiconductor-Sensitized Solar-Cell Interfaces. *Adv. Funct. Mater.* **2011**, *21*, 4663. [CrossRef]
8. Zhou, Y.; Wang, L.; Chen, S.; Qin, S.; Liu, X.; Chen, J.; Xue, D.; Luo, M.; Cao, Y.; Cheng, Y.; et al. Thin-film Sb_2Se_3 photovoltaics with oriented one-dimensional ribbons and benign grain boundaries. *Nat. Photonics* **2015**, *9*, 409. [CrossRef]
9. Wang, L.; Li, D.; Li, K.; Chen, C.; Deng, H.X.; Gao, L.; Zhao, Y.; Jiang, F.; Li, L.; Huang, F.; et al. Stable 6%-efficient Sb_2Se_3 solar cells with a ZnO buffer layer. *Nat. Energy* **2017**, *2*, 17046. [CrossRef]
10. Rabin, O.; Perez, J.; Grimm, J.; Wojtkiewicz, G.; Weissleder, R. An X-ray computed tomography imaging agent based on long-circulating bismuth sulphide nanoparticles. *Nat. Mater.* **2006**, *5*, 118. [CrossRef]
11. Yao, K.; Zhang, Z.; Liang, X.; Chen, Q.; Peng, L.M.; Yu, Y. Effect of H_2 on the Electrical Transport Properties of Single Bi_2S_3 Nanowires. *J. Phys. Chem. B* **2006**, *110*, 21408. [CrossRef] [PubMed]
12. Cademartiri, L.; Scotognella, F.; O'Brien, P.; Lotsch, B.; Thomson, J.; Petrov, S.; Kherani, N.; Ozin, G. Cross-Linking Bi_2S_3 Ultrathin Nanowires: A Platform for Nanostructure Formation and Biomolecule Detection. *Nano Lett.* **2009**, *9*, 1482. [CrossRef] [PubMed]
13. A review on properties, applications, and deposition techniques of antimony selenide. *Sol. Energy Mater. Sol. Cells* **2021**, *230*, 111223. [CrossRef]
14. Chen, S.; Liu, T.; Zheng, Z.; Ishaq, M.; Liang, G.; Fan, P.; Chen, T.; Tang, J. Recent progress and perspectives on Sb_2Se_3-based photocathodes for solar hydrogen production via photoelectrochemical water splitting. *J. Energy Chem.* **2022**, *67*, 508. [CrossRef]
15. Li, Q.; Zhang, W.; Peng, J.; Yu, D.; Liang, Z.; Zhang, W.; Wu, J.; Wang, G.; Li, H.; Huang, S. Nanodot-in-Nanofiber Structured Carbon-Confined Sb_2Se_3 Crystallites for Fast and Durable Sodium Storage. *Adv. Funct. Mater.* **2022**, *32*, 2112776. [CrossRef]
16. Ma, X.; Chen, W.; Tong, L.; Liu, S.; Dai, W.; Ye, S.; Zheng, Z.; Wang, Y.; Zhou, Y.; Zhang, W.; et al. Experimental demonstration of harmonic mode-locking in Sb_2Se_3-based thulium-doped fiber laser. *Opt. Laser Technol.* **2021**, *143*, 107286. [CrossRef]
17. Zhu, L.; Wang, H.; Wang, Y.; Lv, J.; Yanmei, M.; Cui, Q.; Ma, Y.; Zou, G. Substitutional Alloy of Bi and Te at High Pressure. *Phys. Rev. Lett.* **2011**, *106*, 145501. [CrossRef]
18. Einaga, M.; Ohmura, A.; Nakayama, A.; Ishikawa, F.; Yamada, Y.; Nakano, S. Pressure-induced phase transition of Bi_2Te_3 to a bcc structure. *Phys. Rev. B* **2011**, *83*, 092102. [CrossRef]
19. Zhao, J.; Liu, H.; Ehm, L.; Chen, Z.; Sinogeikin, S.; Zhao, Y.; Gu, G. Pressure-Induced Disordered Substitution Alloy in Sb_2Te_3. *Inor. Chem.* **2011**, *50*, 11291. [CrossRef]
20. Vilaplana, R.; Santamaría-Pérez, D.; Gomis, O.; Manjón, F.J.; González, J.; Segura, A.; Muñoz, A.; Rodríguez-Hernández, P.; Pérez-González, E.; Marín-Borrás, V.; et al. Structural and vibrational study of Bi_2Se_3 under high pressure. *Phys. Rev. B* **2011**, *84*, 184110. [CrossRef]
21. Ma, Y.; Liu, G.; Zhu, P.; Wang, H.; Wang, X.; Cui, Q.; Liu, J.; Ma, Y. Determinations of the high-pressure crystal structures of Sb_2Te_3. *J. Phys. Condens. Matter* **2012**, *24*, 475403. [CrossRef]
22. Liu, G.; Zhu, L.; Yanmei, M.; Lin, C.; Liu, J.; Ma, Y. Stabilization of 9/10-Fold Structure in Bismuth Selenide at High Pressures. *J. Phys. Chem. C* **2013**, *117*, 10045. [CrossRef]
23. Zhao, J.; Liu, H.; Ehm, L.; Dong, D.; Chen, Z.; Gu, G. High-pressure phase transitions, amorphization, and crystallization behaviors in Bi_2Se_3. *J. Phys. Condens. Matter* **2013**, *25*, 125602. [CrossRef] [PubMed]
24. Yu, Z.; Wang, L.; Hu, Q.; Zhao, J.; Yan, S.; Yang, K.; Sinogeikin, S.; Gu, G.; Mao, H.K. Structural phase transitions in Bi_2Se_3 under high pressure. *Sci. Rep.* **2015**, *5*, 15939. [CrossRef]
25. Hao, X.; Zhu, H.; Guo, Z.; Li, H.; Gong, Y.; Chen, D. Local insight to the structural phase transition sequence of Bi_2Se_3 under quasi-hydrostatic and nonhydrostatic pressure. *J. Phys. Condens. Matter* **2021**, *33*, 215402. [CrossRef]
26. Tseng, Y.C.; Lin, C.M.; Jian, S.R.; Le, P.H.; Gospodinov, M.M.; Marinova, V.; Dimitrov, D.Z.; Luo, C.W.; Wu, K.H.; Zhang, D.Z.; et al. Structural and electronic phase transition in $Bi_2Se_{2.1}Te_{0.9}$ under pressure. *J. Phys. Chem. Solids* **2021**, *156*, 110123. [CrossRef]
60. Ovsyannikov, S.V.; Shchennikov, V.V. High-Pressure Routes in the Thermoelectricity or How One Can Improve a Performance of Thermoelectrics. *Chem. Mater.* **2010**, *22*, 635. [CrossRef]
60. Ovsyannikov, S.V.; Shchennikov, V.V.; Vorontsov, G.V.; Manakov, A.Y.; Likhacheva, A.Y.; Kulbachinskii, V.A. Giant improvement of thermoelectric power factor of Bi_2Te_3 under pressure. *J. App. Phys.* **2008**, *104*, 053713. [CrossRef]
29. Zhang, J.L.; Zhang, S.J.; Weng, H.M.; Zhang, W.; Yang, L.X.; Liu, Q.Q.; Feng, S.M.; Wang, X.C.; Yu, R.C.; Cao, L.Z.; et al. Pressure-induced superconductivity in topological parent compound Bi_2Te_3. *Proc. Natl. Acad. Sci. USA* **2011**, *108*, 24. [CrossRef]
30. Zhang, C.; Sun, L.; Chen, Z.; Zhou, X.; Wu, Q.; Yi, W.; Guo, J.; Dong, X.; Zhao, Z. Phase diagram of a pressure-induced superconducting state and its relation to the Hall coefficient of Bi_2Te_3 single crystals. *Phys. Rev. B* **2011**, *83*, 140504. [CrossRef]
31. Zhu, J.; Kong, P.P.; Zhang, S.J.; Yu, X.H.; Zhu, J.L.; Liu, Q.Q.; Li, X.; Yu, R.C.; Ahuja, R.; Yang, W.G.; et al. Superconductivity in Topological Insulator Sb_2Te_3 Induced by Pressure. *Sci. Rep.* **2013**, *3*, 2016. [CrossRef] [PubMed]
32. Kong, P.P.; Zhang, J.L.; Zhang, S.J.; Zhu, J.; Liu, Q.Q.; Yu, R.C.; Fang, Z.; Jin, C.Q.; Yang, W.G.; Yu, X.H.; et al. Superconductivity of the topological insulator Bi_2Se_3 at high pressure. *J. Phys. Condens. Matter* **2013**, *25*, 362204. [CrossRef]
33. Taguchi, T.; Ikeda, M.; Li, H.; Suzuki, A.; Yang, X.; Ishii, H.; Liao, Y.F.; Ota, H.; Goto, H.; Eguchi, R.; et al. Superconductivity of topological insulator $Sb_2Te_{3-y}Se_y$ under pressure. *J. Phys. Condens. Matter* **2021**, *33*, 485704. [CrossRef] [PubMed]

34. Crichton, W.A.; Bernal, F.L.M.; Guignard, J.; Hanfland, M.; Margadonna, S. Observation of the Sb_2S_3-type post-post-$GdFeO_3$-perovskite: A model structure for high density ABX_3 and A_2X_3 phases. *arXiv* **2014**, arXiv:1410.2783.
35. Crichton, W.A.; Bernal, F.L.; Guignard, J.; Hanfland, M.; Margadonna, S. Observation of Sb_2S_3-type post-post-perovskite in $NaFeF_3$. Implications for ABX_3 and A_2X_3 systems at ultrahigh pressure. *Mineral. Mag.* **2016**, *80*, 659. [CrossRef]
36. Efthimiopoulos, I.; Zhang, J.; Kucway, M.; Park, C.; Ewing, R.; Wang, Y. Sb_2Se_3 under pressure. *Sci. Rep.* **2013**, *3*, 2665. [CrossRef] [PubMed]
37. Efthimiopoulos, I.; Kemichick, J.; Zhou, X.; Khare, S.; Ikuta, D.; Wang, Y. High-pressure Studies of Bi_2S_3. *J. Phys. Chem. A* **2014**, *118*, 1713. [CrossRef]
38. Sorb, Y.A.; Rajaji, V.; Malavi, P.S.; Subbarao, U.; Halappa, P.; Peter, S.C.; Karmakar, S.; Narayana, C. Pressure-induced electronic topological transition in Sb_2S_3. *J. Phys. Condens. Matter* **2015**, *28*, 015602. [CrossRef] [PubMed]
39. Ibañez, J.; Sans, J.A.; Popescu, C.; López-Vidrier, J.; Elvira-Betanzos, J.J.; Cuenca-Gotor, V.P.; Gomis, O.; Manjón, F.J.; Rodríguez-Hernández, P.; Muñoz, A. Structural, Vibrational, and Electronic Study of Sb_2S_3 at High Pressure. *J. Phys. Chem. C* **2016**, *120*, 10547. [CrossRef]
40. Li, C.; Zhao, J.; Hu, Q.; Liu, Z.; Yu, Z.; Yan, H. Crystal structure and transporting properties of Bi_2S_3 under high pressure: Experimental and theoretical studies. *J. Alloys Compd.* **2016**, *688*, 329. [CrossRef]
41. Cheng, H.; Zhang, J.; Yu, P.; Gu, C.; Ren, X.; Lin, C.; Li, X.; Zhao, Y.; Wang, S.; Li, Y. Enhanced Structural Stability of Sb_2Se_3 via Pressure-Induced Alloying and Amorphization. *J. Phys. Chem. C* **2020**, *124*, 3421. [CrossRef]
60. Anversa, J.; Chakraborty, S.; Piquini, P.; Ahuja, R. High pressure driven superconducting critical temperature tuning in Sb_2Se_3 topological insulator. *Appl. Phys. Lett.* **2016**, *108*, 212601. [CrossRef]
43. Kong, P.; Sun, F.; Xing, L.; Zhu, J.; Zhang, S.; Li, W.; Wang, X.; Feng, S.; Yu, X.; Zhu, J.; et al. Superconductivity in Strong Spin Orbital Coupling Compound Sb_2Se_3. *Sci. Rep.* **2014**, *4*, 6679. [CrossRef] [PubMed]
44. Das, S.; Sirohi, A.; Kumar Gupta, G.; Kamboj, S.; Vasdev, A.; Gayen, S.; Guptasarma, P.; Das, T.; Sheet, G. Discovery of highly spin-polarized conducting surface states in the strong spin-orbit coupling semiconductor Sb_2Se_3. *Phys. Rev. B* **2018**, *97*, 235306. [CrossRef]
45. Zhang, K.; Xu, M.; Li, N.; Xu, M.; Zhang, Q.; Greenberg, E.; Prakapenka, V.B.; Chen, Y.S.; Wuttig, M.; Mao, H.K.; et al. Superconducting Phase Induced by a Local Structure Transition in Amorphous Sb_2Se_3 under High Pressure. *Phys. Rev. Lett.* **2021**, *127*, 127002. [CrossRef]
46. Cheng, Y.; Cojocaru-Mirédin, O.; Keutgen, J.; Yu, Y.; Küpers, M.; Schumacher, M.; Golub, P.; Raty, J.Y.; Dronskowski, R.; Wuttig, M. Understanding the Structure and Properties of Sesqui-Chalcogenides (i.e., V_2VI_3 or Pn_2Ch_3 (Pn = Pnictogen, Ch = Chalcogen) Compounds) from a Bonding Perspective. *Adv. Mater.* **2019**, *31*, 1904316. [CrossRef]
47. Efthimiopoulos, I.; Buchan, C.; Wang, Y. Structural properties of Sb_2S_3 under pressure: Evidence of an electronic topological transition. *Sci. Rep.* **2016**, *6*, 24246. [CrossRef]
48. Dai, L.; Liu, K.; Li, H.; Wu, L.; Hu, H.; Zhuang, Y.; Linfei, Y.; Pu, C.; Liu, P. Pressure-induced irreversible metallization accompanying the phase transitions in Sb_2S_3. *Phys. Rev. B* **2018**, *97*, 024103. [CrossRef]
49. Wang, Y.; Yanmei, M.; Liu, G.; Wang, J.; Li, Y.; Li, Q.; Zhang, J.; Ma, Y.; Zou, G. Experimental Observation of the High Pressure Induced Substitutional Solid Solution and Phase Transformation in Sb_2S_3. *Sci. Rep.* **2018**, *8*, 14795. [CrossRef]
50. Liu, G.; Yu, Z.; Liu, H.; Redfern, S.A.T.; Feng, X.; Li, X.; Yuan, Y.; Yang, K.; Hirao, N.; Kawaguchi, S.I.; et al. Unexpected Semimetallic BiS_2 at High Pressure and High Temperature. *J. Phys. Chem. Lett.* **2018**, *9*, 5785. [CrossRef]
51. da Silva, E.L.; Skelton, J.M.; Rodríguez-Hernández, P.; Muñoz, A.; Santos, M.C.; Martínez-García, D.; Vilaplana, R.; Manjón, F.J. A theoretical study of the Pnma and R-3m phases of Sb_2S_3, Bi_2S_3 and Sb_2Se_3. *J. Mater. Chem. C* **2022**, *10*, 15061. [CrossRef]
52. Hohenberg, P.; Kohn, W. Inhomogeneous electron gas. *Phys. Rev.* **1964**, *136*, B864. [CrossRef]
53. Kresse, G.; Furthmüller, J. Efficiency of ab-initio total energy calculations for metals and semiconductors using a plane-wave basis set. *Comput. Mater. Sci.* **1996**, *6*, 15. [CrossRef]
54. Perdew, J.P.; Ruzsinszky, A.; Csonka, G.I.; Vydrov, O.A.; Scuseria, G.E.; Constantin, L.A.; Zhou, X.; Burke, K. Restoring the density-gradient expansion for exchange in solids and surfaces. *Phys. Rev. Lett.* **2008**, *100*, 136406; Erratum in *Phys. Rev. Lett.* **2009**, *102*, 039902. [CrossRef] [PubMed]
55. Monkhorst, H.J.; Pack, J.D. Special Points for Brillouin Zone Integrations. *Phys. Rev. B* **1976**, *13*, 5188. . PhysRevB.13.5188. [CrossRef]
56. Murnaghan, F.D. The Compressibility of Media under Extreme Pressures. *Proc. Natl. Acad. Sci. USA* **1944**, *30*, 244. [CrossRef]
57. Birch, F. Finite Elastic Strain of Cubic Crystals. *Phys. Rev.* **1947**, *71*, 809. [CrossRef]
58. Togo, A.; Oba, F.; Tanaka, I. First-principles calculations of the ferroelastic transition between rutile-type and $CaCl_2$-type SiO_2 at high pressures. *Phys. Rev. B* **2008**, *78*, 134106. [CrossRef]
60. Gonzalez, J.M.; Nguyen-Cong, K.; Steele, B.A.; Oleynik, I.I. Novel phases and superconductivity of tin sulfide compounds. *J. Chem. Phys* **2018**, *148*, 194701. [CrossRef]
60. Nguyen, L.T.; Makov, G. GeS Phases from First-Principles: Structure Prediction, Optical Properties, and Phase Transitions upon Compression. *Cryst. Growth Des.* **2022**, *22*, 4956. [CrossRef]
61. Xu, M.; Jakobs, S.; Mazzarello, R.; Cho, J.Y.; Yang, Z.; Hollermann, H.; Shang, D.; Miao, X.; Yu, Z.; Wang, L.; et al. Impact of Pressure on the Resonant Bonding in Chalcogenides. *J. Phys. Chem. C* **2017**, *121*, 25447. [CrossRef]

62. Cuenca-Gotor, V.P.; Sans, J.A.; Gomis, O.; Mujica, A.; Radescu, S.; Muñoz, A.; Rodríguez-Hernández, P.; da Silva, E.L.; Popescu, C.; Ibañez, J.; et al. Orpiment under compression: Metavalent bonding at high pressure. *Phys. Chem. Chem. Phys.* **2020**, *22*, 3352. [CrossRef] [PubMed]
63. Grimme, S. Semiempirical GGA-Type Density Functional Constructed with a Long-Range Dispersion Correction. *J. Comp. Chem.* **2006**, *27*, 1787. [CrossRef] [PubMed]
64. Perdew, J.P.; Zunger, A. Self-interaction correction to density-functional approximations for many-electron systems. *Phys. Rev. B* **1981**, *23*, 5048. [CrossRef]
65. Born, M. On the stability of crystal lattices. I. In *Mathematical Proceedings of the Cambridge Philosophical Society*; Cambridge University Press: Cambridge, UK, 1940; Volume 36, p. 160. [CrossRef]
66. Dove, M.T. *Introduction to Lattice Dynamics*; Cambridge University Press: Cambridge, UK, 1993. [CrossRef]
67. Dove, M.T. *Structure and Dynamics: An Atomic View of Materials*; Oxford Master Series in Physics; Oxford University Press: Oxford, UK, 2003. [CrossRef]
68. Dove, M.T. Review: Theory of displacive phase transitions in minerals. *Am. Min.* **2015**, *82*, 213. [CrossRef]
69. Venkataraman, G. Soft modes and structural phase transitions. *Bull. Mater. Sci.* **1979**, *1*, 129. [CrossRef]
70. Di Gennaro, M.; Saha, S.; Verstraete, M. Role of Dynamical Instability in the Ab Initio Phase Diagram of Calcium. *Phys. Rev. Lett.* **2013**, *111*, 025503. [CrossRef]
71. Quadbeck-Seeger, H.J. *World of the Elements: Elements of the World*; Wiley-VCH: Weinheim, Germany, 2007. [CrossRef]
72. Skelton, J.M.; Burton, L.A.; Parker, S.C.; Walsh, A.; Kim, C.E.; Soon, A.; Buckeridge, J.; Sokol, A.A.; Catlow, C.R.A.; Togo, A.; et al. Anharmonicity in the High-Temperature *Cmcm* Phase of SnSe: Soft Modes and Three-Phonon Interactions. *Phys. Rev. Lett.* **2016**, *117*, 075502. [CrossRef]
73. Skelton, J.M. ModeMap. Available online: https://github.com/JMSkelton/ModeMap (accessed on 10 June 2019).

Disclaimer/Publisher's Note: The statements, opinions and data contained in all publications are solely those of the individual author(s) and contributor(s) and not of MDPI and/or the editor(s). MDPI and/or the editor(s) disclaim responsibility for any injury to people or property resulting from any ideas, methods, instructions or products referred to in the content.

Article

Pressure-Induced Structural Phase Transition of Co-Doped SnO$_2$ Nanocrystals

Vinod Panchal [1], Laura Pampillo [2], Sergio Ferrari [2], Vitaliy Bilovol [2,3], Catalin Popescu [4] and Daniel Errandonea [5,*]

1. Department of Physics, Royal College, Mumbai 401107, India; panchalvinod@yahoo.com
2. Consejo Nacional de Investigaciones Científicas y Técnicas, Instituto de Tecnología y Ciencias de la Ingeniería "Hilario Fernández Long" (INTECIN), Universidad de Buenos Aires, Av. Paseo Colón 850, Ciudad Autónoma de Buenos Aires C1063ACV, Argentina; lpampillo@fi.uba.ar (L.P.); lic.sergio.ferrari@gmail.com (S.F.); vbilovol@agh.edu.pl (V.B.)
3. Academic Centre for Materials and Nanotechnology, AGH University of Science and Technology, Al. Mickiewicza 30, 30-059 Krakow, Poland
4. CELLS-ALBA Synchrotron Light Facility, Cerdanyola del Vallés, 08290 Barcelona, Spain; cpopescu@cells.es
5. Departamento de Física Aplicada, Instituto de Ciencias de Materiales, MALTA Consolider Team, Universitat de Valencia, 46100 Valencia, Spain
* Correspondence: daniel.errandonea@uv.es

Abstract: Co-doped SnO$_2$ nanocrystals (with a particle size of 10 nm) with a tetragonal rutile-type (space group $P4_2/mnm$) structure have been investigated for their use in in situ high-pressure synchrotron angle dispersive powder X-ray diffraction up to 20.9 GPa and at an ambient temperature. An analysis of experimental results based on Rietveld refinements suggests that rutile-type Co-doped SnO$_2$ undergoes a structural phase transition at 14.2 GPa to an orthorhombic CaCl$_2$-type phase (space group $Pnnm$), with no phase coexistence during the phase transition. No further phase transition is observed until 20.9 GPa, which is the highest pressure covered by the experiments. The low-pressure and high-pressure phases are related via a group/subgroup relationship. However, a discontinuous change in the unit-cell volume is detected at the phase transition; thus, the phase transition can be classified as a first-order type. Upon decompression, the transition has been found to be reversible. The results are compared with previous high-pressure studies on doped and un-doped SnO$_2$. The compressibility of different phases will be discussed.

Keywords: high pressure; phase transition; synchrotron radiation; X-ray diffraction

1. Introduction

At present, nano-scale materials are having a great impact on human life. They are changing dental medicine, healthcare, and human life more profoundly than several scientific developments of the past decades. Research on nanomaterials has also become a striking area for applied and fundamental research due to the intriguing chemical and physical properties of nanomaterials. Nanomaterials are preferred for many technological applications over their bulk counterparts due to enhancements in their catalytic, optical, magnetic, and electrical properties [1–3]. In addition, the nature of pressure-induced structural phase transitions, pressure effects on elastic properties, and transition pressures in nanomaterials could be quite different when compared to bulk materials [4–9]. Tin dioxide (SnO$_2$) is a very technologically important material, and it is widely used as a wide bandgap semiconductor. It finds applications in the field of solar cells [10], in ultraviolet photodetectors [11], in short wavelength light-emitting diodes [12], in spintronic applications [13], as a gas sensor [14], and in lithium-ion batteries [15], among other applications. Recently, SnO$_2$ has also attracted substantial attention due to its wide bandgap energy, large exciton binding energy, and outstanding electrical properties and optoelectronic features. Due to

these characteristics, SnO$_2$ has been employed for building ultraviolet light-emitting and light-detecting devices based on a SnO$_2$/GaN heterojunction [16].

Under ambient conditions, SnO$_2$, which is also known as stannic oxide or by the mineralogical name Cassiterite, crystallizes into a crystal structure isomorphic to that of tetragonal rutile (space group $P4_2/mnm$, Z = 2). In this structure, the Sn atoms are in the corners and center of the tetragonal unit cell (see Figure 1a). They are six-fold coordinated by oxygen atoms, which are shared with the adjacent SnO$_6$ octahedral units, as shown in Figure 1a. The structure is also commonly described as a distorted hcp oxide array, with half of the octahedral sites occupied by Sn atoms. The oxygen atoms have a coordination number of three, resulting in a trigonal planar coordination.

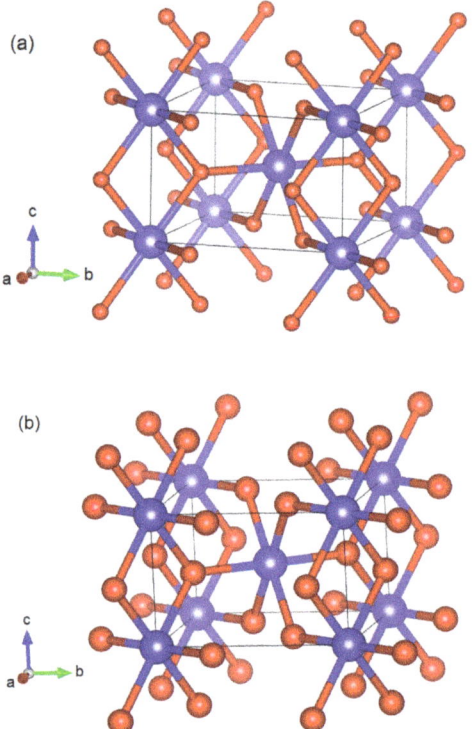

Figure 1. Schematic view of (**a**) the tetragonal rutile-type structure of SnO$_2$, space group $P4_2/mnm$, and (**b**) the high-pressure orthorhombic structure of SnO$_2$, space group $Pnnm$ (Sn^{+4} cations are in blue and O^{-2} anions are in red). In the figure, we include red, green, and blue arrows showing the directions of the unit-cell axes a, b, and c, respectively.

In the past, it was reported that SnO$_2$ undergoes the following sequence of pressure-induced structural phase transitions: rutile-type phase → CaCl$_2$-type phase → PbO$_2$-type phase → fluorite-type phase [17]. Earlier high-pressure (HP) powder X-ray diffraction (XRD) measurements [18,19], theoretical ab initio calculations [19–23], and HP Raman scattering measurements [24] on bulk as well as in nanocrystalline samples provided evidence of the occurrence of a second order-structural phase transition at pressures that vary from 7 GPa to 14 GPa. The structural phase transition is from the tetragonal rutile-type to the orthorhombic CaCl$_2$-type structure (space group $Pnnm$), which is schematically represented in Figure 1b. In the CaCl$_2$-type structure, the Sn atoms form a body-centered orthorhombic structure, and they are also octahedrally coordinated by oxygen atoms, as in the low-pressure phase. The SnO$_6$ octahedral units form chains of edge-sharing octahedra running along the c-axis of the crystal structure, while perpendicular to the c-axis, the

SnO$_6$ octahedra are linked by sharing one corner with each neighboring SnO$_6$ octahedron. However, there are a few exceptions for the rutile to CaCl$_2$-type phase transition. For example, high-pressure powder XRD measurements carried out on nanocrystalline 5 nm sized samples [25] and 8 nm sized samples [26,27] have indicated the occurrence of a direct tetragonal rutile to cubic fluorite first-order structural phase transition beyond 18 GPa. The transition pressure has been also found to increase when decreasing the particle size [27]. On the other hand, the doping of SnO$_2$ has been shown to cause alterations in the high-pressure behavior, elastic properties, and optical properties of the material. For instance, in Fe-doped nanoparticles (size 18 nm and 10 at % doping) of SnO$_2$, increases in the transition pressure and bulk modulus have been reported [28]. On the contrary, in V-doped nanoparticles (size 10–30 nm and 5–12.5 at % doping) of SnO$_2$, a decrease in the bulk modulus with an increasing concentration of the dopant has been found [29]. Hence, it is very interesting to know how the particle size, doping concentration, and the type of doping ion alter the high-pressure behavior and elastic properties of SnO$_2$, and further systematic investigations are required in this direction. In the current contribution, we report HP synchrotron powder XRD studies up to a pressure of 20.9 GPa on Co-doped nanocrystalline SnO$_2$ (crystallite size 15 nm and 10 at % doping) to contribute to the understanding of the effect of doping on the transition pressure and elastic properties. Such a study has not been reported yet.

2. Experiments

Cobalt-doped SnO$_2$ nanoparticles were prepared at Universidad de Buenos Aires via the wet chemical co-precipitation method, following the procedure described previously by Ferrari et al. [30]. To implement the cobalt doping, we used anhydrous cobalt chloride (CoCl$_2$) with a purity of 99.99%, which was obtained from Sigma-Aldrich (St. Louis, MI, USA). The synthesized nanoparticles were characterized at ambient conditions via powder XRD using Cu K$_\alpha$ radiation. The XRD measurements confirmed the formation of single-phase rutile-type nanoparticles with unit-cell parameters of a = 4.732(1) Å and c = 3.185(1) Å. These parameters are consistent with the parameters determined from single-crystal X-ray diffraction studies on bulk SnO$_2$ [31,32]. An average particle size of 15 nm was determined from the full-width-at-half-maximum of the XRD peaks, using the well-known Scherrer equation [33]. The cobalt concentration (10 at %) in our samples was determined via energy-dispersive X-ray spectroscopy (EDXS).

Angle-dispersive HP powder XRD measurements were carried out at the MSPD-BL04 beamline of the ALBA synchrotron, using a monochromatic beam with a wavelength of 0.4642 Å. Wavelength selection was achieved using a silicon (111) double-crystal monochromator with a resolution of 2×10^{-4} [34]. The X-ray beam was focused down to a 20 µm full-width-at-half-maximum spot using Kirkpatrick–Baez mirrors. A Rayonix charge-coupled device (CCD) detector was used to collect the XRD patterns. The CCD detector was calibrated using LaB$_6$ as a standard. Two-dimensional diffraction rings obtained from the detector were transformed into one-dimensional diffractograms using Dioptas. The measurements were taken at room temperature under compression with a membrane diamond anvil cell (DAC) equipped with 500 µm culet diameter diamond anvils. We used a 200 mm thick stainless-steel gasket pre-indented to a thickness of 40 µm, with a centered hole 200 µm in diameter, as a pressure chamber. As a pressure-transmitting medium we used a 16:3:1 methanol–ethanol–water mixture, which is known to remain quasi-hydrostatic up to approximately 10 GPa [35]. The pressure was determined using the equation of state (EOS) of copper (Cu), as reported by Dewaelle et al. [36], with a precision of ±0.05 GPa. During the XRD measurements, a rocking of ±3° of the DAC was used to reduce the influence of preferred orientations and to improve the homogeneity of the Debye rings. The structural analysis was carried out by employing the Rietveld technique via the Fullprof suite [37]. In the structural refinements, the background was fitted with a Chebyshev polynomial function of first kind with six coefficients, and the peak profiles were modeled using a pseudo-Voigt function. In addition to the unit-cell parameters, we also refined the atomic positions. The overall displacement factor was fixed to 0.5 Å.

3. Results and Discussion

Figure 2 shows the powder XRD pattern at the lowest pressure measured for the nanocrystalline Co-doped SnO_2 in the DAC (pressure = 0.6 GPa). The figure includes the results of the Rietveld refinement. As shown in the figure, the refinement is good, and the diffraction pattern can be undoubtedly assigned to the tetragonal rutile-type structure. In the XRD pattern, there is an extra weak peak, denoted by the asterisk symbol (*), which is assigned to the copper grain used as a pressure marker. It corresponds to the (111) reflection of copper. The unit-cell parameters obtained for the nanocrystalline Co-doped SnO_2 at 0.6 GPa are a = 4.728(1) Å and c = 3.184(1) Å. The goodness-of-fit parameters obtained from the structural refinement are Rwp = 6.16% and Rp = 3.4%.

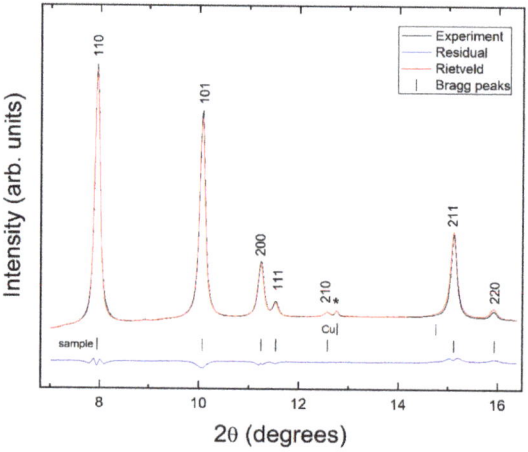

Figure 2. The Rietveld profile refinement for the XRD pattern of the tetragonal rutile-type phase of nanocrystalline Co-doped SnO_2 at 0.6 GPa and room temperature. The asterisk (*) is the (111) diffraction peak of copper used to determine pressure. The vertical bars indicate the calculated positions of diffraction peaks of the sample and copper (Cu). Miller indexes are shown in the figure.

Figure 3 shows a selection of XRD patterns of nanocrystalline Co-doped SnO_2 at representative pressures. There are no noticeable changes in the diffraction patterns up to 13 GPa aside from the shift of peaks to higher angles due to the typical contraction of unit-cell parameters under compression. All the XRD diffraction peaks up to 13 GPa could be successfully indexed considering only the low-pressure (LP) rutile-type phase and copper. In addition to the peaks assigned to SnO_2, a peak due to copper (marked by *), which was used to determine the pressure, can be identified in all the X-ray diffraction profiles. We observed a systematic shift in all the diffraction peaks to a higher 2θ due to lattice compression. At 14.2 GPa, we observed few discernible changes in the XRD diffraction profile, which we consider to be indicative of a phase transformation. In addition, we found that at 14.2 GPa, the cell metric is no longer consistent with $P4_2/mnm$ symmetry, and another crystallographic phase with $Pnmm$ symmetry occurs, corresponding to an orthorhombic $CaCl_2$-type structure. In particular, at the pressure of 14.2 GPa, the broadening of the (101), (200), and (111) diffraction peaks of rutile, in addition to the splitting of the (211) diffraction peak of rutile, were observed. The XRD diffraction profiles at 14.2 GPa and higher pressures could not be well indexed to tetragonal rutile-type phase. To make the changes in the XRD patterns indicating the structural phase transition more evident for the readers, we provide in Figure 4 an enhanced view of the XRD patterns measured at 13.0, 14.2, and 20.2 GPa. In Figure 4, it is clear that the (211) peak of rutile gradually splits into two different peaks. The same happens with peak (111). Another fact highlighted by Figure 4 that could not be explained by assuming the tetragonal rutile-type structure is the evolution of the (200) peak of rutile towards low angles as the pressure increased beyond 13.0 GPa. This can only be explained by assuming a decrease in the crystal symmetry from tetragonal to orthorhombic. Indeed, we found that the orthorhombic $CaCl_2$-type structure provides the correct positions of all diffraction peaks, as indicated by unbiased Rietveld refinements. An example of this can be seen in Figure 5, where we show the Rietveld refinement we performed at 14.2 GPa. Furthermore, the proposed rutile-$CaCl_2$ structural phase transition agrees with earlier investigations on rutile SnO_2 [17–20].

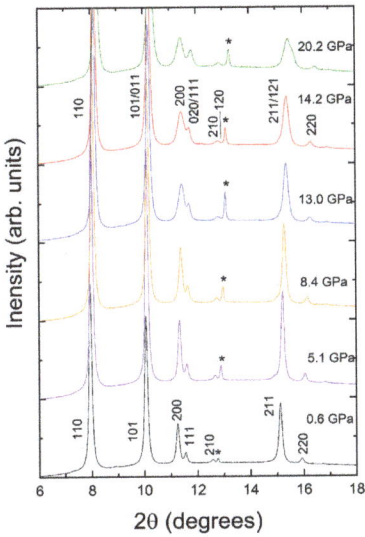

Figure 3. Room-temperature X-ray powder diffraction patterns of nanocrystalline Co-doped SnO_2 at representative pressures. Pressures are indicated in the figure. The asterisks (*) denote the (111) diffraction peak of copper at each pressure. The diffractogram of the low-pressure phase at 0.6 GPa and that of the high-pressure phase at 14.2 GPa have been indexed with the tetragonal rutile-type and the orthorhombic $CaCl_2$-type structure, respectively. Miller indexes are shown.

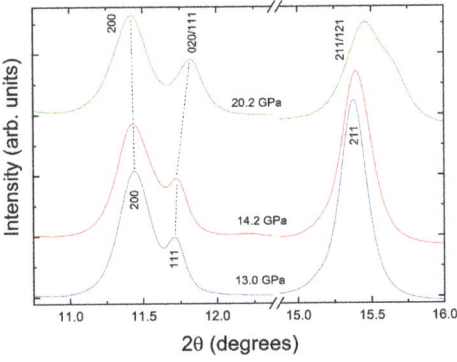

Figure 4. Enhanced view of the regions of XRD patterns measured in nanocrystalline Co-doped SnO_2 at 13.0, 14.2, and 20.2 GPa, highlighting the changes that evidence the phase transition discussed in the text.

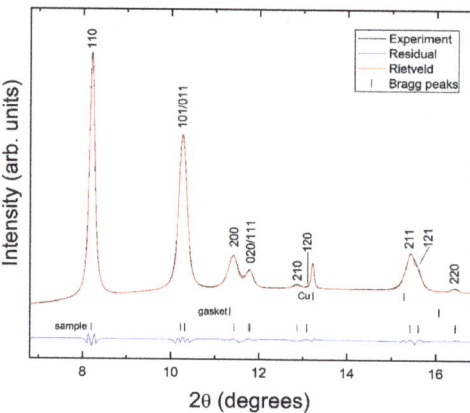

Figure 5. The Rietveld profile refinement of the XRD pattern measured at 14.2 GPa and room temperature for the orthorhombic $CaCl_2$-type phase of nanocrystalline Co-doped SnO_2. The vertical bars indicate the calculated positions of diffraction peaks of sample. The contributions from the gasket and copper (Cu) are also indicated.

Upon a further increase in pressure, we found that the orthorhombic $CaCl_2$-type structure continued to be stable in the nanocrystalline Co-doped SnO_2 up to 20.9 GPa, which was the highest pressure measured in this investigation. Upon the release of the pressure, we found that the phase transition was fully reversible. In the experiments, we did not observe any coexistence of phases during the compression and decompression cycles. This fact, and the group–subgroup relationship existing between space groups $P4_2/mnm$ and $Pnnm$, could be an indication that the observed phase transition is second-order in nature, which agrees with the conclusions extracted from previous powder XRD measurements on doped and un-doped nanocrystalline SnO_2 [28]. However, as we will explain below, in nanocrystalline Co-doped SnO_2 at the phase transition, we detected a volume discontinuity, which undoubtedly supports a first-order nature for the phase transition in Co-doped SnO_2. Notice that up to now, this is the only SnO_2 nanomaterial in which such a volume discontinuity, larger than experimental uncertainties, has been detected at the phase transition. We speculate that the observed volume discontinuity could be related to changes in the strong anti-ferromagnetic super-exchange interaction between the Co ions that exist in the Co-doped SnO_2 [38]. Another hypothesis to explain the observed abrupt volume decrease is the influence of structural defects that are induced by Co doping [39]. Additional studies are needed to fully understand the distinctive behavior of Co-doped SnO_2. The Rietveld refinement of the XRD patterns' profile measured for the HP phase at 14.2 GPa is shown in Figure 5. In addition to the peaks from the sample and Cu, we observed a weak contribution of the gasket material, which causes the peak (200) of the HP phase of Co-doped SnO_2 to become asymmetric. The unit-cell parameters determined for the HP phase of the Co-doped SnO_2 at 14.2 GPa are a = 4.675(2) Å, b = 4.572(3) Å, and c = 3.151(1) Å. The goodness-of-fit parameters of the Rietveld refinement shown in Figure 5 are Rwp = 9.84% and Rp = 7.79%.

The pressure evolution of the lattice parameters of the Co-doped nanocrystalline SnO_2 in the tetragonal rutile-type and orthorhombic $CaCl_2$-type phases is shown in Figure 6. The pressure dependence of the unit-cell volume is reported in Figure 7. We observed a discontinuity in the volume ($-\Delta V/V \sim 1.0\%$) at the transition pressure. The change in the volume was larger than the experimental error (which is smaller than the symbols), indicating the first-order nature of the structural phase transition. Using the EOSFIT7 software [40], the linear compressibility values of the axes of each phase were calculated. We observed that in the rutile-type phase of SnO_2, the axial compressibility values were highly anisotropic. In particular, the a-axis was more compressible than the c-axis, as is evident from the increase in the c/a ratio from 0.673 at ambient pressure to 0.681 at 13 GPa. The linear compressibility of the c-axis was found to be $K_c = 1.02(4) \times 10^{-3}$ GPa^{-1}, which is almost half of the compressibility of the a-axis, $K_a = 1.72(6) \times 10^{-3}$ GPa^{-1}. This anisotropic behavior in compressibility is in quite good agreement with the earlier reported values of bulk SnO$_2$ samples [31,41,42] and even for un-doped, Fe-doped, and V-doped nanocrystalline SnO$_2$ [28,29]. However, the of axial compressibility vales obtained in this investigation are slightly higher compared to earlier investigations [28,29,31,41,42]. In the case of the orthorhombic CaCl$_2$-type phase of the Co-doped SnO$_2$, the axial compressibility values were found to be highly anisotropic as well. In particular, the b-axis was more compressible when compared to other two axes. The linear compressibility values of all three axes were found to be $K_a = 1.65(4) \times 10^{-3}$ GPa^{-1}, $K_b = 1.90(1) \times 10^{-3}$ GPa^{-1}, and $K_c = 1.05(2) \times 10^{-3}$ GPa^{-1}.

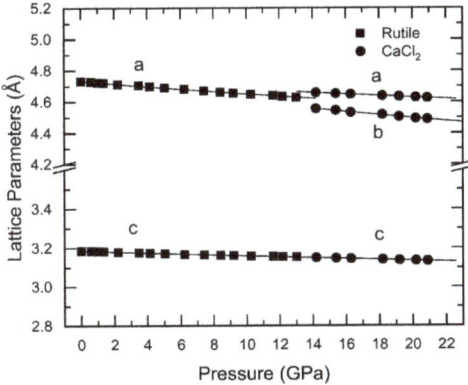

Figure 6. Pressure dependence of lattice parameters of nanocrystalline Co-doped SnO$_2$. Solid squares represent the tetragonal rutile-type phase and solid circles represent the orthorhombic CaCl$_2$-type structure. Solid lines are linear fits to lattice parameters. Error bars are smaller than the size of the symbols. The unit-cell axes, a, b, and c are identified in the figure.

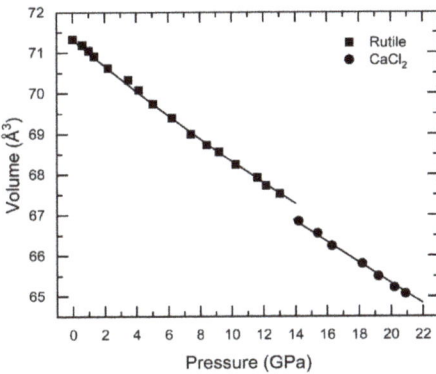

Figure 7. Volume versus pressure data for nanocrystalline Co-doped SnO$_2$. Solid squares represent the tetragonal rutile-type phase and solid circles represent the orthorhombic CaCl$_2$-type structure. Solid lines represent the second-order Birch–Murnaghan fit to the data for both phases (see text for details). Error bars are smaller than the size of the symbols.

The P–V data of the rutile-type phase of nanocrystalline Co-doped SnO$_2$ in the tetragonal phase, fitted to the second-order Birch–Murnaghan equation of state, provides a bulk modulus at zero pressure of $B_0 = 213(9)$ GPa, with its pressure derivative provided as

$B_0' = 4$. During this fit, the bulk modulus B_0 and ambient pressure volume V_0 were kept as free variables. This value of the bulk modulus is in good agreement with previously reported values for the same parameter in bulk SnO_2 [17] and even in un-doped and in Fe-doped nanocrystalline SnO_2 [28]. The values of the bulk moduli are summarized in Table 1. However, in V-doped nanocrystalline samples, the bulk modulus was found to decrease with increases in doping concentrations in the range from 142 GPa to 185 GPa [29]. We can then conclude that in contrast with V-doping, a 10% doping with Co and Fe does not affect the mechanical properties of SnO_2, which is good for technological applications. A possible reason for this it could be the fact that the V atom has larger ionic radii than the Fe and Co atoms, which would necessarily drive changes in the crystal structure more easily. For the high-pressure phase of the Co-doped SnO_2, the bulk modulus we obtained was B_0 = 228(9) GPa. The slight increase in the bulk modulus at the phase transition is consistent with the fact that the HP phase is denser than the low-pressure phase due to the volume contraction observed at the transition. However, both bulk moduli agree within error bars: 213(9) and 228(9) GPa. This fact is consistent with the displacive mechanism proposed for the rutile-CaCl2 phase transition [43]. In this regard, the behavior of the Co-doped SnO_2 is more similar to the behavior of bulk SnO_2 [17], for which no changes have been detected in the bulk modulus at the transition (see Table 1), than to the behavior of the bulk modulus for Fe-doped a nanocrystalline SnO_2 [29], for which a 15% increase in the bulk modulus has been reported after the phase transition. At present, it is not clear that the difference in behavior is inherent to the difference in doping or to the difference in non-hydrostatic stresses in the experiments [44]. Future studies are needed to clarify this issue. On the other hand, it should be noted that in the equation of state, the volume at zero pressure (V_0) and B_0 are correlated parameters. Thus, large uncertainties on V_0 could affect the value of B_0 [45]. This could be the case for nanocrystalline Co-doped SnO_2 for which the phase transition is reversible and therefore no data are available for the HP phase at pressures close to the ambient pressure. This means that the determination of V_0 comes from the extrapolation of data measured above 14.2 GPa, which could lead to large errors in the determination of V_0. This fact will necessarily propagate to the determination of B_0, whose value should be taken with caution.

Table 1. Bulk modulus determined for different phases of nanocrystalline of SnO_2. We include results from this work and for the literature [17,18,25,27–29].

Sample	Bulk Modulus (GPa) Rutile-Type SnO_2	Bulk Modulus (GPa) $CaCl_2$-Type SnO_2	Ref.
Bulk	205	204	[18]
Bulk	252	----	[27]
Bulk	205	204	[17]
Nanocrystalline (5 nm)	217	---	[25]
Nanocrystalline (3 nm)	233	---	[27]
Nanocrystalline (30 nm)	210	252	[28]
Nanocrystalline Fe-doped (18 nm)	213	256	[28]
Nanocrystalline V-doped (13 nm)	185	--	[29]
Nanocrystalline Co-doped (15 nm)	213(9)	228(9)	This work

To conclude the discussion, we would like to mention that the Co-doping of nanocrystalline SnO_2 does not affect the structural stability and mechanical properties, making Co-doped SnO_2 more attractive than SnO_2 doped with other metals for practical applications for which doped SnO_2 is deposited as a thin film on substrates such as amorphous solar cells and cadmium telluride solar cells [46]. Thin films are usually subjected to epitaxial strain, which gives rise to misfit stress. It is therefore better to use SnO_2 films in which the structural stability and mechanical properties are not affected by doping, as is the case in Co-doped SnO_2.

4. Conclusions

In summary, we performed an in situ synchrotron powder X-ray diffraction study on nanocrystalline Co-doped SnO_2 at room temperature up to a pressure of 20.9 GPa. This study shows that the studied material undergoes a first-order structural phase transition from the tetragonal rutile-type to the orthorhombic $CaCl_2$-type phase at ~14.2 GPa. The crystal structure of the high-pressure phase is orthorhombic, and it is isomorphic to the high-pressure structure observed previously in bulk, doped, and un-doped nanocrystalline samples of SnO_2. We did not observe any coexistence of phases during the phase transition under conditions of both compression and decompression. The compressibility behavior was found to be highly anisotropic in both phases, and the value of bulk modulus for the tetragonal rutile-type phase was in good agreement with earlier measurements. However, for the orthorhombic high-pressure $CaCl_2$-type phase, the bulk modulus of the Co-doped SnO_2 was 10% smaller than the same parameter in Fe-doped SnO_2. The high-pressure phase was observed to remain stable up to 20.9 GPa. Upon the release of the pressure, the phase transition was found to be reversible. The reported results could be relevant for the implementation of practical applications using Co-doped thin films prepared via deposition methods.

Author Contributions: Conceptualization, D.E.; methodology and validation, V.P. and D.E. formal analysis, V.P., S.F. and D.E.; investigation, V.P., L.P., S.F., V.B., C.P. and D.E.; writing—original draft preparation, V.P.; writing—review and editing, V.P., L.P., S.F., V.B., C.P. and D.E.; supervision, D.E. All authors have read and agreed to the published version of the manuscript.

Funding: D.E. thanks the financial support from Generalitat Valenciana under the grant PROMETEO CIPROM/2021/075-GREENMAT, the Advanced Materials Programme, supported by MCIN and GVA, with funding from the European Union, NextGenerationEU (PRTR-C17.I1), under grant MFA/2022/007, and from the Spanish Research Agency (AEI) and the Spanish Ministry of Science and Investigation (MCIN) under projects PID2019-106383GB-C41 (DOI: 10.13039/501100011033) and RED2018-102612-T (MALTA Consolider-Team Network).

Data Availability Statement: Data are available upon reasonable request to the corresponding author at daniel.errandonea@uv.es (D.E.).

Acknowledgments: The authors are grateful for financial support from the Spanish Research Agency (AEI) and the Spanish Ministry of Science and Investigation (MCIN) under grants PID2019-106383GB-41/43 (http://dx.doi.org/10.13039/501100011033) and RED2018-102612-T (MALTA Consolider Team Network) and from Generalitat Valenciana under Grant PROMETEO CIPROM/2021/075-GREENMAT. This study forms part of the Advanced Materials program and was supported by MCIN with funding from the European Union Next Generation EU (PRTR-C17.I1) and by Generalitat Valenciana, Grant MFA/2022/007. The authors thank ALBA synchrotron for providing beam time for the XRD experiments (Proposal 2021085226).

Conflicts of Interest: The authors declare no conflict of interest.

References

1. Gleiter, H. Nanocrystalline materials. *Prog. Mater. Sci.* **1989**, *33*, 223–315. [CrossRef]
2. Siegel, R. Cluster-assembled nanophase materials. *Annu. Rev. Mater. Sci.* **1991**, *21*, 559–578. [CrossRef]
3. Kodama, R. Magnetic Nanoparticles. *J. Magn. Magn. Mater.* **1999**, *200*, 359–372. [CrossRef]
4. Lv, H.; Yao, M.; Li, Q.; Li, Z.; Liu, B.; Liu, R.; Lu, S.; Li, D.; Mao, J.; Ji, X.; et al. Effect on grain size on pressure-induced structural transition in Mn_3O_4. *J. Phys. Chem. C* **2012**, *116*, 2165–2171. [CrossRef]
5. Srihari, V.; Verma, A.; Pandey, K.; Vishwanadh, B.; Panchal, V.; Garg, N.; Errandonea, D. Making $Yb_2Hf_2O_7$ Defect Fluorite Uncompressible by Particle Size Reduction. *J. Phys. Chem. C* **2021**, *125*, 27354–27362. [CrossRef]
6. Jiang, J.; Gerward, L.; Frost, D.; Secco, R.; Peyronneau, J.; Olsen, J. Grain-size effect on pressure-induced semiconductor-to-metal transition in ZnS. *J. Appl. Phys.* **1999**, *86*, 6608–6610. [CrossRef]
7. Wang, Z.; Tait, K.; Zhao, Y.; Schiferl, D.; Zha, C.; Uchida, H.; Downs, R. Size-induced reduction of transition pressure and enhancement of bulk modulus of AlN Nanocrystals. *J. Phys. Chem. B* **2004**, *108*, 11506–11508. [CrossRef]

8. Wang, Z.; Saxena, S.; Pischedda, V.; Liermann, H.; Zha, C. In situ X-ray diffraction study of the pressure-induced phase transformation in nanocrystalline CeO_2. *Phys. Rev. B* **2001**, *64*, 012102. [CrossRef]
9. Zvoriste-Walters, C.; Heathman, S.; Jovani-Abril, R.; Spino, J.; Janssen, A.; Caciuffo, R. Crystal size effect on the compressibility of nano-crystalline uranium dioxide. *J. Nucl. Mater.* **2013**, *435*, 123–127. [CrossRef]
10. Bouras, K.; Schmerber, G.; Rinnert, H.; Aureau, D.; Park, H.; Ferblantier, G.; Colis, S.; Fix, T.; Park, C.; Kim, W.; et al. Structural, optical and electrical properties of Nd-doped SnO_2 thin films fabricated by reactive magnetron sputtering for solar cell devices. *Sol. Energy Mater. Sol. Cells* **2016**, *145*, 134–141. [CrossRef]
11. Wu, J.; Kuo, C. Ultraviolet photodetectors made from SnO_2 nanowires. *Thin Solid Film.* **2009**, *517*, 3870–3873. [CrossRef]
12. Tsai, M.; Bierwagen, O.; Speck, J. Epitaxial Sb-doped SnO_2 and Sn-doped In_2O_3 transparent conducting oxide contacts on GaN-based light emitting diodes. *Thin Solid Film.* **2016**, *605*, 186–192. [CrossRef]
13. Ogale, S.; Choudhary, R.; Buban, J.; Lofland, S.; Shinde, S.; Kale, S.; Kulkarni, V.; Higgins, J.; Lanci, C.; Simpson, J.; et al. High Temperature Ferromagnetism with a Giant Magnetic Moment in Transparent Co-doped $SnO_{2-\delta}$. *Phys. Rev. Lett.* **2003**, *91*, 077205. [CrossRef] [PubMed]
14. Tadeev, A.; Delabouglise, G.; Labeau, M. Influence of Pd and Pt additives on the microstructural and electrical properties of SnO_2-based sensors. *Mater. Sci. Eng. B* **1998**, *57*, 76–83. [CrossRef]
15. Chandra, A.; Kalpana, D.; Thangadurai, P.; Ramasamy, S. Synthesis and characterization of nanocrystalline SnO_2 and fabrication of lithium cell using nano-SnO_2. *J. Power Sources* **2002**, *107*, 138–141. [CrossRef]
16. Xu, T.; Jiang, M.; Wan, P.; Tang, K.; Shi, D.; Kan, C. Bifunctional ultraviolet light-emitting/detecting device based on a SnO_2 microwire/p-GaN heterojunction. *Photonics Res.* **2021**, *9*, 2475–2485. [CrossRef]
17. Haines, J.; Leger, J. X-ray diffraction study of the phase transitions and structural evolution of tin dioxide at high pressure: Relationships between structure types and implications for other rutile-type dioxides. *Phys. Rev. B* **1997**, *55*, 11144. [CrossRef]
18. Shieh, S.; Kubo, A.; Duffy, T.; Prakapenka, V.; Guoyin, G. High pressure phases in SnO_2 to 117 GPa. *Phys. Rev. B* **2006**, *73*, 14105. [CrossRef]
19. Parlinski, K.; Kawazoe, Y. Ab Initio study of phonons in the rutile structure of SnO_2 under pressure. *Eur. Phys. J. B* **2000**, *13*, 679–683. [CrossRef]
20. Hassan, F.; Alaeddine, A.; Zoaeter, M.; Rachidi, I. First-principles investigation of SnO_2 at high pressure. *Int. J. Mod. Phys. B* **2005**, *19*, 4081–4092. [CrossRef]
21. Gracia, L.; Beltran, A.; Andres, J. Characterization of high-pressure structures and phase transformations in SnO_2. A density functional theory study. *J. Phys. Chem. B* **2007**, *111*, 6479–6485. [CrossRef] [PubMed]
22. Casali, R.; Lasave, J.; Caravaca, M.; Koval, S.; Ponce, C.; Migoni, R. Ab initio and shell model studies of structural, thermoelastic and vibrational properties of SnO_2 under pressure. *J. Phys. Condens. Matter* **2013**, *25*, 135404. [CrossRef]
23. Yang, L.; Weiliu, F.; Yanlu, L.; Wei, L.; Zhao, X. Pressure induced ferroelastic phase transition in SnO_2 from density functional theory. *J. Chem. Phys.* **2014**, *140*, 164706. [CrossRef] [PubMed]
24. Hellwig, H.; Goncharov, A.; Gregoryanz, E.; Mao, H.; Hemley, R. Brillouin and Raman spectroscopy of the ferroelastic rutile-to-$CaCl_2$ transition in SnO_2 at high pressure. *Phys. Rev. B* **2003**, *67*, 174110. [CrossRef]
25. Garg, A. Pressure induced volume anomaly and structural phase transition in nanocrystalline SnO_2. *Phys. Status Solidi B* **2014**, *251*, 1380–1385. [CrossRef]
26. Jiang, J.; Gerward, L.; Olsen, J. Pressure induced phase transformation in nanocrystal SnO_2. *Scr. Mater.* **2001**, *44*, 1983–1986. [CrossRef]
27. He, Y.; Liu, J.; Chen, W.; Wang, Y.; Wang, H.; Zeng, Y.; Zhang, G.; Wang, L.; Liu, J.; Hu, T.; et al. High pressure behavior of SnO_2 nanocrystals. *Phys. Rev. B* **2005**, *72*, 212102. [CrossRef]
28. Grinblat, F.; Ferrari, S.; Pampillo, L.; Saccone, F.; Errandonea, D.; Santamaria-Perez, D.; Segura, A.; Vilaplana, R.; Popescu, C. Compressibility and structural behavior of pure and Fe-doped SnO_2 nanocrystals. *Sol. State Sci.* **2017**, *64*, 91–98. [CrossRef]
29. Ferrari, S.; Bilovol, V.; Pampillo, L.; Grinblat, F.; Saccone, F.; Errandonea, D. Characterization of V-doped SnO_2 nanoparticles at ambient and high pressures. *Mater. Res. Express* **2018**, *5*, 125005. [CrossRef]
30. Ferrari, S.; Pampillo, L.; Saccone, F. Magnetic properties and environment sites in Fe doped SnO_2 nanoparticles. *Mater. Chem. Phys.* **2016**, *177*, 206–212. [CrossRef]
31. Hazen, R.; Finger, L. Bulk moduli and high-pressure crystal structures of rutile-type compounds. *J. Phys. Chem. Solids* **1981**, *42*, 143–151. [CrossRef]
32. Bauer, W. Rutile type compounds. V. Refinements of MnO_2 and MgF_2. *Acta Crystallogr. B* **1976**, *32*, 2200–2204. [CrossRef]
33. Zhang, Z.; Zhou, F.; Lavernia, E. On the analysis of grain size in bulk nanocrystalline materials via X-ray diffraction. *Metall. Mater. Trans. A* **2003**, *34*, 1349–1355. [CrossRef]
34. Chang, Y.Y.; Tsai, Y.W.; Weng, S.C.; Chen, S.L.; Chang, S.L. Integrated optical chip for a high-resolution, single-resonance-mode X-ray monochromator system. *Opt. Lett.* **2021**, *46*, 416–419. [CrossRef] [PubMed]
35. Klotz, S.; Chervin, J.; Munsch, P.; Marchand, G. Hydrostatic limits of 11 pressure transmitting media. *J. Phys. D Appl. Phys.* **2009**, *42*, 075413. [CrossRef]
36. Dewaele, A.; Loubeyre, P.; Mezouar, M. Equation of state of six metals above 94 GPa. *Phys. Rev. B* **2004**, *70*, 094112. [CrossRef]
37. Rodriguez-Carvajal, J. Recent advances in in magnetic structure determination by neutron powder diffraction. *Phys. B* **1993**, *192*, 55–69. [CrossRef]

38. Liu, X.F.; Gong, W.M.; Iqbal, J.; He, B.; Yu, R.H. Structural defects-mediated room-temperature ferromagnetism in Co-doped SnO$_2$ insulating films. *Thin Solid Film.* **2009**, *517*, 6091–6095. [CrossRef]
39. Gao, Y.; He, J.; Guo, J. Effect of co-doping and defects on electronic, magnetic, and optical properties in SnO$_2$: A first-principles study. *Phys. B* **2022**, *639*, 413924. [CrossRef]
40. Gonzalez-Platas, J.; Alvaro, M.; Nestola, F.; Angel, R. *EosFit7-GUI*: A new graphical user interface for equation of state calculations, analyses and teaching. *J. Appl. Cryst.* **2016**, *49*, 1377–1382. [CrossRef]
41. Haines, J.; Leger, J.; Schulte, O. The high-pressure phase transition sequence from the rutile-type through to the cotunnite-type structure in PbO$_2$. *J. Phys. Condens. Matter* **1996**, *8*, 1631. [CrossRef]
42. Ross, N.; Shu, J.; Hazen, R.; Gasparik, T. High-pressure crystal chemistry of stishovite. *Am. Mineral.* **1990**, *75*, 739–747.
43. Hyde, B.G. The effect of non-bonded, anion-anion interactions on the CaCl$_2$/rutile transformation and on the bond lengths in the rutile type. *Z. Krist.* **1987**, *179*, 205–213. [CrossRef]
44. Errandonea, D.; Muñoz, A.; Gonzalez-Platas, J. Comment on high pressure X-ray diffraction study of YBO$_3$/ Eu^{3+}, GdBO$_3$, and EuBO$_3$: Pressure induced amorphization in GdBO$_3$. *J. Appl. Phys.* **2014**, *115*, 216101. [CrossRef]
45. Anzellini, S.; Errandonea, D.; MacLeod, S.; Botella, P.; Daisenberger, D.; De'Ath, M.; Gonzalez-Platas, J.; Ibáñez, J.; McMahon, M.; Munro, K.; et al. Phase diagram of calcium at high pressure and high temperature. *Phys. Rev. Mater.* **2018**, *2*, 083608. [CrossRef]
46. Bouabdalli, E.M.; El Jouad, M.; Garmim, T.; Louardi, H.; Hartiti, M.; Monkade, M.; Touhtouh, S.; Hajjaji, A. Elaboration and characterization of Ni and Al co-doped SnO$_2$ thin films prepared by spray pyrolysis technique for photovoltaic applications. *Mater. Sci. Eng. B* **2022**, *286*, 116044. [CrossRef]

Disclaimer/Publisher's Note: The statements, opinions and data contained in all publications are solely those of the individual author(s) and contributor(s) and not of MDPI and/or the editor(s). MDPI and/or the editor(s) disclaim responsibility for any injury to people or property resulting from any ideas, methods, instructions or products referred to in the content.

Article

High-Pressure Vibrational and Structural Studies of the Chemically Engineered Ferroelectric Phase of Sodium Niobate

Sanjay Kumar Mishra [1,2,*], Nandini Garg [2,3,*], Smita Gohil [4], Ranjan Mittal [1,2,*] and Samrath Lal Chaplot [1,2]

1. Solid State Physics Division, Bhabha Atomic Research Centre, Mumbai 400085, India; chaplot@barc.gov.in
2. Homi Bhabha National Institute, Anushaktinagar, Mumbai 400094, India
3. High Pressure Physics Division, Bhabha Atomic Research Centre, Mumbai 400085, India
4. Department of Condensed Matter Physics and Materials Science, Tata Institute of Fundamental Research, Mumbai 400005, India; smitagohil@tifr.res.in
* Correspondence: skmsspd@barc.gov.in (S.K.M.); nandini@barc.gov.in (N.G.); rmittal@barc.gov.in (R.M.)

Abstract: Pure NaNbO$_3$ has an antiferroelectric phase at ambient pressure. The structural behaviour of the chemically engineered ferroelectric phase of sodium niobate, NNBT05: [(0.95) NaNbO$_3$-(0.05) BaTiO$_3$], under high-pressure has been studied using Raman scattering and angle-dispersive synchrotron X-ray diffraction techniques. At pressure > 1 GPa, noticeable changes in the Raman spectra can be seen in the low wavenumber modes (150–300 cm^{-1}). Large changes in the positions and intensities of the Raman bands as a function of pressure provide evidence for structural phase transition. The results indicate significant changes in the bond-lengths and the orientation of the NbO$_6$ octahedra at ~1 GPa, and a transition to the paraelectric phase at ~5 GPa, which are at lower pressures than previously found in pure NaNbO$_3$. The powder X-ray diffraction pattern shows an appreciable change in the peak profile in terms of position and width on increasing pressure. The pressure dependences of the structural parameters show that the response of the lattice parameters to pressure is strongly anisotropic. By fitting the pressure–volume data using the Birch–Murnaghan equation of state, the isothermal bulk modulus was estimated. The experimental results suggest that on doping BaTiO$_3$ in NaNbO$_3$, the bulk modulus increases. The bulk modulus of NNBT05 has been estimated to be 164.5 GPa, which is fairly close to 157.5 GPa, as previously observed in NaNbO$_3$.

Keywords: ferroelectric; antiferroelectric; niobate; Raman scattering; high pressure diffraction

1. Introduction

Materials with perovskite structures exhibit diverse crystal structures and physical properties such as (anti)ferroelectricity, relaxor, magnetic, multiferroic properties, etc., and remain pertinent in next generation applications [1–14]. The perovskite structure consists of two types of polyhedra, the octahedron surrounding the B-cation, and the A-cation, coordinated by 12 oxygen atoms forming a cuboctahedron. In the perovskite family, alkaline niobates and their derivatives have attracted a great deal of attention, and are promising candidates for eco-friendly lead-free piezoceramics [10,11,15–17].

The application of hydrostatic pressure strongly modifies the short-range interatomic and long-range Coulomb interactions, which are responsible for the structural stability of ferroelectric and antiferrodistortive phases. These phases are associated with the freezing of zone-centre and zone-boundary phonon instabilities. Sodium niobate (NaNbO$_3$) has both these competing instabilities simultaneously. At ambient conditions, sodium niobate crystallizes in the antiferroelectric orthorhombic phase. Sodium niobate perovskite is a textbook example for understanding a complicated series of structural phase changes that occur with temperature and pressure [17–28]. Using high pressure Raman scattering techniques, Shiratori et al. [28] have reported transitions at around 2, 6, and 9 GPa, respectively. The pressure-dependent neutron diffraction studied showed that the antiferroelectric (*Pbcm*) phase transformed to a paraelectric (*Pbnm*) phase at 8 GPa [19]. Neutron diffraction also

revealed changes in the baric behaviour of Nb-O-Nb bond angles as a result of complex reorientations of NbO$_6$ octahedra at high pressures of ~2 GPa [19]. Recent, high pressure X-ray diffraction data suggest that the paraelectric (*Pbnm*) phase is stable up to 30 GPa [29].

In addition to temperature and pressure, the structure and physical properties of materials can also be tuned by chemical doping. Specific doping elements are used to produce subtle distortions in crystal structures and to regulate changes in their physical properties. For example, when doping the BaTiO$_3$ in the NaNbO$_3$ matrix, there is a significant enhancement in the dielectric and piezoelectric response of the material [30–34]. The structural, dielectric, ferroelectric and piezoelectric properties of NNBTx [(1 − x) NaNbO$_3$ − xBaTiO$_3$] ceramics have been investigated by various researchers [10,16,30,31,33–36]. This solid solution is similar to K$_x$ Na$_{1-x}$NbO$_3$ (KNN), except that the KNbO$_3$ has been replaced by BaTiO$_3$. The main problem with KNN is the poor densification of the ceramic due to the high volatility of alkaline oxides, which significantly affects the functional performance of this ceramic [37]. As suggested by Zeng et al. [37], the problem of the poor densification of KNN can be resolved by doping BaTiO$_3$ instead of KNbO$_3$ in the NaNbO$_3$ matrix, because both show similar structural phase transitions. It can also be seen that BaTiO$_3$ based piezo-ceramics show very good electromechanical properties. Raveskii et al. [32] and other researchers have studied the various compositions in the full composition range. They have reported extensive electro-physical properties of this solid solution, with expected structures. It is reported that on increasing the BaTiO$_3$ composition, the dielectric maximum shifts to the lower temperatures and reaches near room temperature for x = 0.25. They also found P-E hysteresis loops with good saturation for x = 0.10. For the same sample, they observed the coercive field E$_C$ (11 kV/cm), the spontaneous polarization P$_S$ (15 µC/cm^2) and residual polarization P$_R$ (13 µC/cm^2) at 50 Hz. However, the electromechanical properties of the NNBTx in the NaNbO$_3$ are exceptionally remarkable. The literature contains very little information about the structural phase transition with composition and temperature. Most of the reports contain only knowledge of expected room temperature structures probed using X-ray diffraction studies. Raveskii et al. [32] have proposed the monoclinic structure of the compositions x~0.05–0.07 at room temperature.

Earlier, we reported the crystal structure and phase stability of the NNBTx system for the small doping of BaTiO$_3$ (x = 0.0 to 0.15). We reported an orthorhombic antiferroelectric phase (space group: *Pbcm*) for x < 0.02, an orthorhombic ferroelectric phase (space group: *Pmc2$_1$*) for 0.02 < x < 0.10, and an orthorhombic ferroelectric phase (space group: *Amm2*) for x $\geq$ 0.10. We also reported the phase transitions as a function of temperature for NNBT03 and 05 using powder X-ray and neutron diffraction studies in conjunction with dielectric and Raman scattering measurements [30,31,35,36]. Apart from this, we provided evidence of the existence of a functional monoclinic phase at low temperatures. This monoclinic (*Cc*) phase is expected to provide easy polarization rotation at low temperatures for the compositions x = 0.03 (NNBT03) and x = 0.05 (NNBT05).

The application of pressure either distorts or symmetrises the structure and causes the alteration of the crystal symmetry with huge consequences on the physical properties. The present study gives insight into the unique pressure-dependent phase stability of chemically engineered ferroelectric NaNbO$_3$, i.e., NNBT05, and aims to deepen the understanding of structural distortion in this material. It also helps in the understanding of the role of doping on pressure responses and the phase stability of the ferroelectric phase. The pressure dependence of the powder X-diffraction and Raman scattering experiments will be used to understand the microscopic origin of the structural and vibrational properties of the sample. At pressure > 1 GPa, noticeable changes in the Raman spectra are seen in the low wavenumber modes (150–300 cm^{-1}). These observations are most likely associated with octahedral reorientations in the parent ferroelectric phase. A detailed analysis of the high-pressure X-ray diffraction data suggests that the response of the lattice parameters to pressure is strongly anisotropic. By fitting the pressure–volume data using the Birch–Murnaghan equation of state, the isothermal bulk modulus can be estimated. We found that on doping BaTiO$_3$ in NaNbO$_3$, the bulk modulus increases.

2. Materials and Methods

The solid solution NNBT05 was prepared with the solid-state reaction method. Dried powders of high purity Na_2CO_3, $BaCO_3$, Nb_2O_5 and TiO_2 were taken in the stoichiometric ratios and were mixed in a planetary ball mill for 12 h in acetone medium. The mixtures were calcined in the air at 1173 K. The phase purity of the solid solutions was confirmed with powder diffraction data. The calcined powders were then sintered at 1273 K for 6 h. The sintered pellets were crushed into fine powders and were used for the high-pressure X-ray diffraction and Raman scattering measurements.

Angle-dispersive X-ray powder diffraction patterns at high pressures up to 30 GPa and at ambient temperature were measured at the BL-11 beamline at INDUS-2 India. The diffraction images were collected with the wavelength λ = 0.79997 Å on the MAR345 detector located at a distance of 190 mm from the sample. The two-dimensional X-ray diffraction (XRD) images were converted to one-dimensional diffraction patterns using the FIT2D program. The data at each pressure level were analysed with the Rietveld refinement method using the FULLPROF program [38]. A Thompson–Cox–Hastings pseudo-Voigt with Axial divergence asymmetry function was used to model the peak profiles. The background was fitted using a Chebychev polynomial. Except for the occupancy parameters of the atoms, which were fixed corresponding to the nominal composition, all other parameters, i.e., scale factor, zero displacement, and isotropic profile parameters, lattice parameters, isotropic thermal parameters, and positional coordinates, were refined.

High pressure Raman experiments were performed using an in-house-made Mao-Bell type of diamond anvil cell. Raman spectra were recorded at increasing and decreasing pressures using a 514.5 nm excitation wavelength from a mixed gas laser (model: Stabilite 2018 from Spectra Physics) and a triple grating Raman spectrometer (T64000 from Horiba Jobin Yvon) equipped with a liquid nitrogen cooled charge coupled device. We used the holographic grating with 1800 g/mm and a 20-micron exit slit with a spectrometer focal length of 1 m, which resulted in 0.7 cm^{-1} optical resolution of the spectra.

High pressure X-ray diffraction and Raman experiments were carried out using a Mao Bell kind of diamond anvil cell, which was equipped with diamonds with a culet size of ~400 microns. The sample was loaded into a hole of 100 μm and drilled in a pre-indented 50 μm thick tungsten gasket. Methanol/ethanol in the ratio of 4:1, which is known to be quasi hydrostatic upto 10 GPa [39], was used as the pressure transmitting medium, and florescence peaks from ruby sphere were used to estimate the pressure inside the sample chamber of the diamond anvil cell. Care was taken to load a very tiny sample so that the environment around the sample remained quasi hydrostatic [40]. The limitations of the present Raman spectroscopic setup made it difficult to investigate Raman spectra below 150 cm^{-1}.

3. Results and Discussion

3.1. High Pressure Raman Scattering Study

The high symmetry phase of the ABO_3 perovskite compound had a cubic structure. In these compounds, the condensation of different phonon instabilities led to phase transitions. The A atoms (Na/Ba) were located at the cubic structure's eight corners, while the B atoms (Nb/Ti) were situated in the body's centre. The oxygen (six atoms) was arranged at the face-centred site and forms an octahedron. The Raman modes, also known as translational (Tr), librational (L), bending (B), and stretching (S) modes, were produced as a result of these atoms' vibrations. The roto-vibrational motion of the oxygen octahedra, the translational motion of the A ions against the BO_6 octahedra, and the bending of the BO_6 octahedra linkages gave rise to these modes. For systems of the ABO_3 type, these are collectively referred to as external modes. The bending and stretching motion of the oxygen octahedra's O-B-O bonds caused the internal modes of the octahedra, known as the bending (B) and stretching (S) modes.

As said earlier, the NNBT05 ambient phase structure had orthorhombic symmetry (space group: $Pmc2_1$; tilt system: $a^-a^-c^+$) with unit cell dimensions $2a_P \times \sqrt{2}a_P \times \sqrt{2}a_P$

and four numbers of formula units, i.e., a total of 20 atoms (see Figure 1a), and it was ferroelectric in nature. Here, a_P is the lattice constant of the parent cubic phase. The group theoretical analysis gave 57 Raman active modes, represented by $\Gamma_{Raman} = 16A_1 + 13A_2 + 12B_1 + 16B_2$. Due to the non-centrosymmetric structure, except for the A_2 mode, these Raman active modes were simultaneously IR active. The number of observed modes was lower compared to the expected number of modes due to the peaks in the experimental Raman spectrum not being properly resolved, and due to accidental degeneracy. Generally, at elevated temperatures and pressures, the cation displacements from the centre of the octahedra (which results in nonzero polarization) may become vanishingly small and the structure may transform to the centrosymmetric orthorhombic paraelectric (*Pnma*) phase. This phase also had similar cell dimensions to ferroelectric (*Pmc2$_1$*) and in total 20 atoms in the primitive cell (Figure 1b). The number of Raman active modes in this phase (in another setting of *Pbnm*) was 24, $7A_g + 5B_{2g} + 7B_{2g} + 5B_{3g}$.

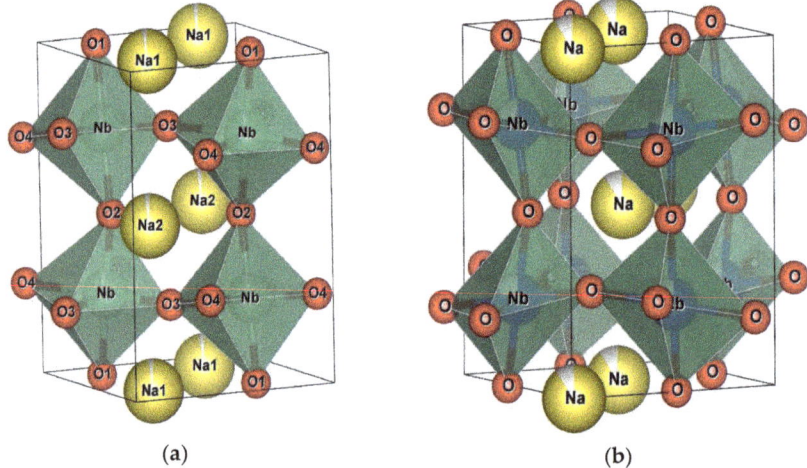

Figure 1. Crystal structure of (**a**) ferroelectric phase (space group: *Pmc2$_1$*) and (**b**) paraelectric phase (space group: *Pbnm*).

Figure 2 depicts the evolution of the high-pressure Raman spectra of NNBT05 at selected pressures and room temperature. For pressure > 1 GPa, distinct changes in the low wavenumber modes (150–300 cm^{-1}) region of the Raman spectra were observed. The peaks containing the most prominent changes were denoted ν_1, ν_2, ν_3, and ν_4 (bending mode: 150–300 cm^{-1}) and showed drastic changes in their relative intensity in the pressure range 1 to 5.4 GPa. These observations at ~1 Gpa are most likely associated with octahedral reorientation in the parent ferroelectric phase, as has also been found earlier in pure NaNbO$_3$. The change in Raman spectra at ~4.6 GPa may be a signature of structural phase transition from the ferroelectric phase (*Pmc2$_1$*) to the paraelectric phase (*Pbnm*). We noted that the paraelectric transition was observed in pure NaNbO$_3$ at ~8 GPa. On further increasing the pressure above 5.4 GPa, the intensities of these peaks reduced and the stretching modes (500–750 cm^{-1}) centred around 584 cm^{-1} became merged. Figure 2b shows the fitted positions of the Raman bands. In literature [41], it was shown that even the 4:1 methanol/ethanol transmitting medium had Raman active modes assigned to the C-C stretch at ~450 and 880 cm^{-1}. However, we would like to mention here that these modes were very weak and were not discernible in the presence of relatively stronger Raman scatterers. In the Raman spectra depicted in Figure 2a, we can clearly see that there was no interference from these modes.

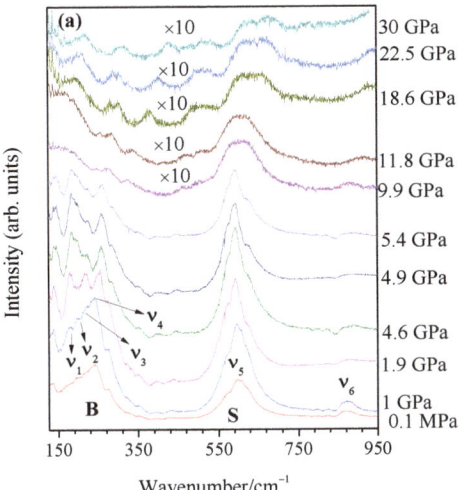
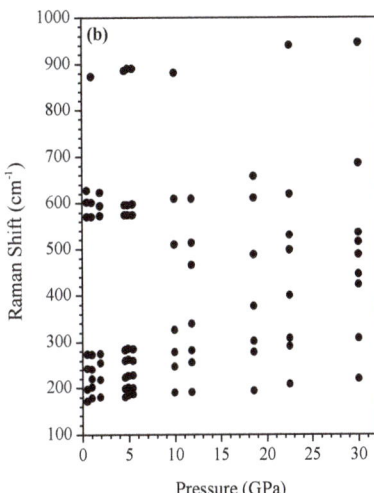

Figure 2. (**a**) Evolution of the Raman spectra collected at selected pressures for NNBT05. The pressure dependencies of the vibration modes (Raman shift) are shown in (**b**). The error bar on the pressure is the size of symbol. The intensities of Raman spectra at higher pressure >5.4 GPa are expanded 10 times.

The frequency evolution showed a blue-shift trend with increasing pressure. It is evident from Figure 2 that the frequencies of Nb-O stretching modes at ~500–700 cm^{-1} increased with the pressure, implying a shortening of the Nb–O bond length. For pressure above 5.4 GPa, the Raman modes became very broad and weak, which could be due to an onset of disorder or due to a phase with a weak Raman response. In nanoporous silicon, it has been observed that though the Raman spectra showed very weak broad Raman modes and were termed a glassy phase, it actually was a crystalline phase which had very weak Raman activity, also because it was metallic [42]. At this point, we would also like to mention that even though the sample environment was quasi-hydrostatic, even up to ~10 Gpa, we cannot rule out the presence of some small pressure inhomogeneities, which could lead to the broadening of the Raman modes. However, on the release of pressure, the Raman modes of the parent phase were observed and so the broadening and reduction in the intensity of the Raman modes could be attributed to a reversible phase transition and not to the pressure inhomogeneities. To ascertain if indeed the high-pressure phase was disordered or it was a new crystalline phase, we carried out high pressure X-ray diffraction studies.

3.2. High Pressure X-ray Diffraction Study

High pressure X-ray diffraction patterns of NNBT05 were collected at 11 pressures up to 40 GPa in a pressure-increasing cycle. Figure 3 depicts the quality of the powder X-ray diffraction patterns at selected pressure levels. It is evident from this figure that on increasing the pressure, the diffraction peaks systematically shift towards higher angles, indicating the compression of the lattice parameters. To investigate the structural parameters with pressure, we carried out a detailed Rietveld analysis of the high-pressure X-ray diffraction data.

The detailed Rietveld refinement of the powder–diffraction data showed that the synchrotron powder diffraction pattern at 300 K could be indexed using the orthorhombic structure (space group $Pmc2_1$) in a low pressure range. Figure 3b shows the quality of Rietveld refinement of high-pressure powder diffraction data at 4.1 GPa. For a pressure range P > 4.1, the ferroelectric ($Pmc2_1$) structural model in the refinements gave highly correlated atomic positions.

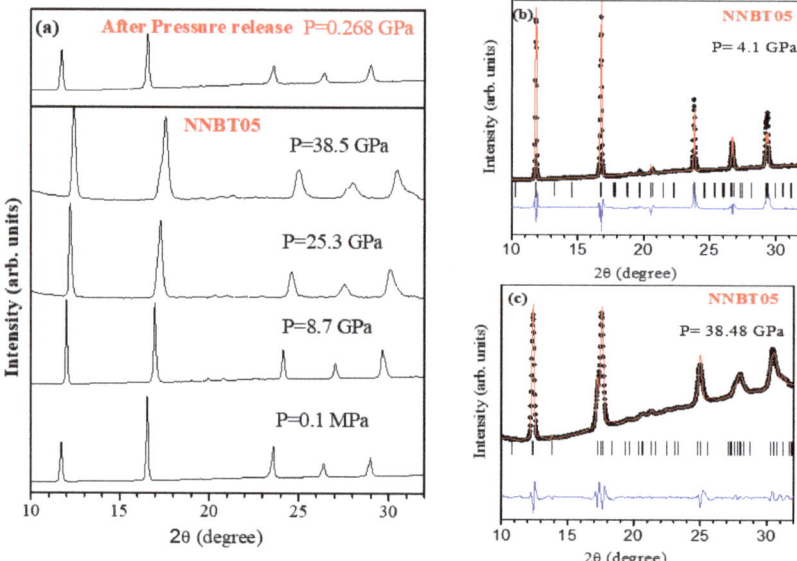

Figure 3. (Colour online) (**a**) Evolution of the powder—X-ray diffraction patterns of NNBT05 at selected pressures. calculated (continuous red line) and difference (bottom blue line) profiles obtained from Rietveld refinement using an orthorhombic ferroelectric phase (space group: $Pmc2_1$) at 4.1 GPa (**b**), and paraelectric (space group: $Pbnm$) at 38.5 GPa (**c**).

The high pressure neutron diffraction study on pure sodium niobate clearly revealed a structural phase transition from the orthorhombic antiferroelectric to paraelectric phase ($Pbnm$) above 8 GPa. In view of this, we refined the high-pressure synchrotron X-ray diffraction data using the paraelectric orthorhombic phase $Pbnm$, which accounted for all the reflections and gave nearly similar goodness-of-fit parameters compared with the space group $Pmc2_1$. However, it should be noted that the total number of variable structural parameters for $Pmc2_1$ was doubled (20) compared to the $Pbnm$ space group (10 parameters). Even doubling the variable parameters did not help to improve the fitting. Hence, the latter phase has been preferred over the ferroelectric ($Pmc2_1$) phase. The Rietveld refinements using the orthorhombic ($Pbnm$) phase proceeded smoothly, revealing a monotonic decrease in lattice constants and cell volume with the pressure increasing up to 38.5 GPa. Structural parameters obtained after the Rietveld refinement for the compound are given in Table 1. The X-ray powder diffraction results indicate that the broadening and weakening of Raman modes beyond 5.4 GPa cannot be attributed to a disordered phase. It is possible that this high-pressure phase is weakly Raman active.

The pressure dependences of the lattice parameters and unit cell volume obtained after the Rietveld refinement of high-pressure synchrotron diffraction data are shown in Figure 4. For easy comparison with Figure 2, we used pseudocubic lattice parameters for the orthorhombic phase using the relations $a_p = A_o/\sqrt{2}$, $b_p = B_o/\sqrt{2}$, and $c_p = C_o/2$, where a_p, b_p, c_p, A_o, B_o, and C_o are lattice parameters corresponding to the equivalent pseudocubic and orthorhombic phases, respectively. It is clear from Figure 5 that on increasing the pressure, the lattice parameters monotonically decreased in the entire range of our measurements for NNBT05. However, unlike the 'b' and 'c' lattice constants, 'a' showed a sudden decrease around 23 GPa. This could not be attributed to the non-hydrostatic stresses caused due to the freezing of the pressure transmitting medium. If we look at the structure of sodium niobate along the b and c axes, we can see that there was only one octahedra, whereas along the 'a' axes there were two corner-shared octahedra. With pressure, there is a possibility of bending of the octahedral hinge, thus drastically reducing the lattice constant along the 'a' axes.

Table 1. Structural parameters obtained through Rietveld refinement of high-pressure X-ray diffraction of NNBT05 using orthorhombic ferroelectric (space group: $Pmc2_1$) and paraelectric (space group: $Pbnm$).

Atom	Pressure = 4.1 GPa Space Group $Pmc2_1$				Pressure = 38.48 GPa Space Group $Pbnm$			
	Positional Coordinates				Positional Coordinates			
	x	y	z	B (Å^2)	x	y	z	B (Å^2)
Na1/Ba1	0.0000	0.2525(4)	0.7637(7)	1.98(5)	−0.059(5)	0.505 (5)	0.2500	2.81(5)
Na2/Ba2	0.5000	0.2514(9)	0.7691(5)	0.35(7)				
Nb/Ti	0.7507(9)	0.7480(8)	0.7925(3)	0.72(2)	0.0000	0.0000	0.0000	1.30 (2)
O1	0.0000	0.2185(7)	0.3217(8)	1.48(1)	0.010 (3)	0.019 (6)	0.2500	1.23 (5)
O2	0.5000	0.3109(9)	0.2986(9)	1.08(8)	0.281(7)	0.307 (4)	0.039 (3)	1.557(2)
O3	0.2265(8)	0.4482(6)	0.2140(6)	2.18(8)				
O4	0.2667(6)	−0.086(5)	0.6128(2)	1.55(3)				
	A_o = 7.7425(4) Å; B_o = 5.4571(5) Å; C_o = 5.5018(5) Å, Unit cell volume (V) = 232.46(9) Å^3 Rp = 3.68 Rwp = 6.61; Rexp = 5.45; χ2 = 1.22				A_o = 5.2147(7) Å; B_o = 5.3295(6) Å; C_o = 7.3933(5) Å, Unit cell volume (V) = 205.41(9) Å^3 Rp = 5.68 Rwp = 8.03; Rexp = 7.65; χ2 = 1.10			

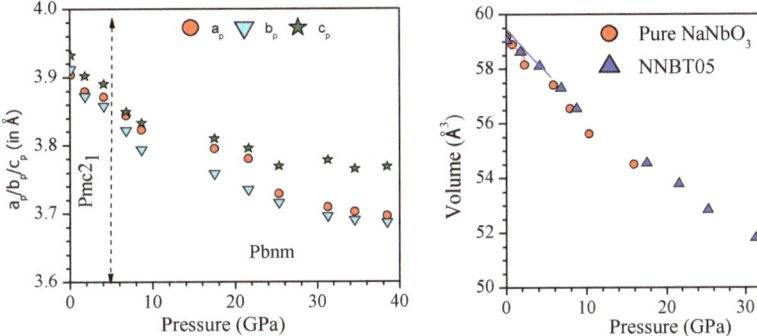

Figure 4. Evolution of lattice parameters and unit cell volume of NNBT05 as functions of pressure. For comparison, the pressure dependence of the volume of pure sodium niobate is also plotted. The solid lines are fitted to the experimental data using the Birch–Murnaghan equation of state. The error bar on the pressure is the size of the symbol.

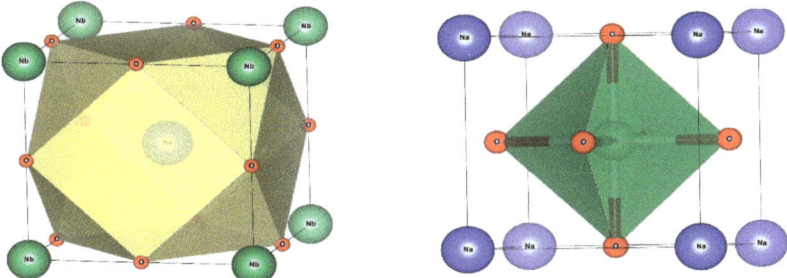

Figure 5. Schematic representation of A-containing cuboctahedra (**left side**) and BO_6 octahedra (**right side**).

From Figure 4, it can be seen that the lattice compression of the orthorhombic phase of NNBT05 was anisotropic. The compressibility of the 'c' parameter was smaller in comparison with those of the 'a' and 'b' parameters. Compressibility (at ambient conditions (1/l (dl/dP)) along a, b, and c was 0.0028, 0.0025 and 0.0020 GPa^{-1}, respectively. This may be compared with the values for pure $NaNbO_3$, which showed the compressibility along a, b, and c as 0.0027, 0.0022, and 0.0018 GPa^{-1}, respectively. The compressibility in the high-pressure paraelectric phase in NaNbO$_3$ was very different from that in the low-pressure phase. However, in NNBT05, we did not find any significant discontinuous change in the compressibility as a function of pressure. Fitting the pressure versus volume data with a third-order Birch–Murnaghan equation of state gave a bulk modulus B ($\approx$164.5 $\pm$ 1.0) GPa (with B' = 4 fixed) for the orthorhombic phase at room temperature. This experimental bulk modulus in NNBT05 was fairly close to that in NaNbO$_3$ (B $\approx$ 157.5 GPa) at room temperature.

It is evident from the foregoing discussion that the doping of BaTiO$_3$ in the NaNbO$_3$ matrix results in reduced compressibility, because on doping the free space in the A-containing cuboctahedron reduces. It was found that in perovskites in which the A cation has a lower formal charge than the B cation, the AO$_{12}$ sites are more compressible than the BO$_6$ octahedra (see Figure 5). To reduce the unit cell volume, the tilts of the BO$_6$ octahedra increased with increasing pressure [43]. Thus, the application of pressure drives this type of perovskite structure away from the phase transition boundary to higher-symmetry structures.

On the basis of the Raman experimental results, we can conclude that NNBT05 showed changes in the Raman spectra from 1 to 1.9 GPa, 1.9 to 4.6 GPa, and above 5.4 GPa. Below 5.4 GPa, the spectra evolved continuous changes. It can be seen that the intensity of phonons located at 150–350 cm^{-1} gradually changed with increasing pressure, up to 5.4 GPa. Then, these Raman peaks merged and became broad peaks as the pressure reached 9.9 GPa. Meanwhile, BO$_6$ octahedra showed a growing distortion under higher pressure, as a result of the volume shrinkage. Hence, additional changes to the Raman spectra occurred when the pressure held at more than 10 GPa. As shown for pure NaNbO$_3$, the observed anomalies in Raman spectra at a low pressure of ~2 GPa were mainly attributed to significant changes in Nb-O-Nb bond angles as a result of complex reorientations of NbO$_6$ octahedra [19,29]. On the other hand, the lattice beyond 8 GPa belonged to a paraelectric orthorhombic phase with space group *Pbnm*. Increasing the pressure resulted in a shrinking of the polyhedral volume, and the cation was displaced towards the symmetric position to avoid over bonding. In addition to this, the ab-initio calculation of the enthalpy in the various phases of NaNbO$_3$ [19] also supported the stabilization of the paraelectric (*Pbnm*) phase at high pressure. In view of this, we may conclude that on application of pressure, the ferroelectric phase of NNBT05 may transform to the paraelectric phase at a high pressure of ~5 GPa, which is lower than the transition pressure of ~8 GPa in pure NaNbO$_3$. Since structural changes in the structure of the dodecahedra are very small due to changes in the average ionic radii in A and B positions, at 5% substitution, the lowering of pressure could be attributed to the chemical pressure introduced due to the substitution of BaTiO$_3$.

4. Conclusions

In summary, we have investigated the effect of pressure on the crystal structure and structural phase transition behaviour in the engineered ferroelectric phase of sodium niobate using Raman scattering and powder X-ray diffraction techniques. At pressure > 1 GPa, noticeable changes in the Raman spectra were seen in the low wavenumber modes (150–300 cm^{-1}). These observations are most likely associated with significant changes in Nb-O-Nb bond angles as result of complex reorientations of NbO$_6$ octahedra in the parent ferroelectric phase, as in pure NaNbO$_3$. Further changes were observed at ~4.6 GPa, which may be due to a transition to the paraelectric phase, which has a lower pressure than that of ~8 GPa in pure NaNbO$_3$. For pressure levels greater than 5.4 GPa, the Raman spectrum became diffusive and may be associated with an increase in inhomogeneities in the crystal. We have determined the bulk modulus of NNBT05 to be 164.5 GPa, which is fairly close to that of 157.5 GPa observed in NaNbO$_3$.

Author Contributions: S.K.M., R.M., N.G. and S.L.C. formulated the problem. N.G. carried out the high-pressure Raman and X-ray diffraction experiments. S.G. performed high-pressure Raman experiments. S.K.M. and N.G. analyzed experimental data. All the authors discussed the results and wrote the manuscript. All authors have read and agreed to the published version of the manuscript.

Funding: This research received no external funding.

Data Availability Statement: The data that support the findings of this study are available from the corresponding author upon reasonable request.

Acknowledgments: S.L.C. thanks the Indian National Science Academy for the financial support of the INSA Senior Scientist position.

Conflicts of Interest: The authors declare that they have no known competing financial interest or personal relationship that could have appeared to influence the work reported in this paper.

References

1. Lines, M.E.; Glass, A.M. *Principles and Applications of Ferroelectrics and Related Materials*; Oxford University Press: Oxford, UK, 2001.
2. Fujimoto, M. *The Physics of Structural Phase Transitions*; Springer: New York, NY, USA, 2005. [CrossRef]
3. Cross, E. Lead-free at last. *Nature* **2004**, *432*, 24–25. [CrossRef] [PubMed]
4. Xu, Y. *Ferroelectric Materials and Their Applications*; Sole distributors for the USA and Canada, Elsevier Science Pub. Co.: Amsterdam, The Netherlands; New York, NY, USA, 1991.
5. Saito, Y.; Takao, H.; Tani, T.; Nonoyama, T.; Takatori, K.; Homma, T.; Nagaya, T.; Nakamura, M. Lead-free piezoceramics. *Nature* **2004**, *432*, 84–87. [CrossRef] [PubMed]
6. Rabe, K.M.; Ahn, C.; Triscone, J.M. *Physics of Ferroelectrics: A Modern Perspective*; Springer: Berlin, Germany, 2007; Volume 105.
7. Rödel, J.; Webber, K.G.; Dittmer, R.; Jo, W.; Kimura, M.; Damjanovic, D. Transferring lead-free piezoelectric ceramics into application. *J. Eur. Ceram. Soc.* **2015**, *35*, 1659–1681. [CrossRef]
8. Scott, J.F. Applications of Modern Ferroelectrics. *Science* **2007**, *315*, 954. [CrossRef] [PubMed]
9. Ye, Z.-G. *Handbook of Advanced Dielectric, Piezoelectric and Ferroelectric Materials: Synthesis, Properties and Applications*; Elsevier: Amsterdam, The Netherlands, 2008.
10. Zheng, T.; Wu, J.; Xiao, D.; Zhu, J. Recent development in lead-free perovskite piezoelectric bulk materials. *Prog. Mater. Sci.* **2018**, *98*, 552–624. [CrossRef]
11. Htet, C.S.; Nayak, S.; Manjón-Sanz, A.; Liu, J.; Kong, J.; Sørensen, D.R.; Marlton, F.; Jørgensen, M.R.V.; Pramanick, A. Atomic structural mechanism for ferroelectric-antiferroelectric transformation in perovskite $NaNbO_3$. *Phys. Rev. B* **2022**, *105*, 174113. [CrossRef]
12. Vilarinho, R.; Bouvier, P.; Guennou, M.; Peral, I.; Weber, M.C.; Tavares, P.; Mihalik, M.; Mihalik, M.; Garbarino, G.; Mezouar, M.; et al. Crossover in the pressure evolution of elementary distortions in RFe_3 perovskites and its impact on their phase transition. *Phys. Rev. B* **2019**, *99*, 064109. [CrossRef]
13. Xiang, H.J.; Guennou, M.; Íñiguez, J.; Kreisel, J.; Bellaiche, L. Rules and mechanisms governing octahedral tilts in perovskites under pressure. *Phys. Rev. B* **2017**, *96*, 054102. [CrossRef]
14. Zhou, J.S. Structural distortions in rare-earth transition-metal oxide perovskites under high pressure. *Phys. Rev. B* **2020**, *101*, 224104. [CrossRef]
15. Mishra, S.K.; Choudhury, N.; Chaplot, S.L.; Krishna, P.S.R.; Mittal, R. Competing antiferroelectric and ferroelectric interactions in $NaNbO_3$: Neutron diffraction and theoretical studies. *Phys. Rev. B* **2007**, *76*, 024110. [CrossRef]
16. Xie, A.; Qi, H.; Zuo, R.; Tian, A.; Chen, J.; Zhang, S. An environmentally-benign NaNbO3 based perovskite antiferroelectric alternative to traditional lead-based counterparts. *J. Mater. Chem. C* **2019**, *7*, 15153–15161. [CrossRef]
17. Xu, Q.; Li, T.; Hao, H.; Zhang, S.; Wang, Z.; Cao, M.; Yao, Z.; Liu, H. Enhanced energy storage properties of $NaNbO_3$ modified $Bi_{0.5}Na_{0.5}TiO_3$ based ceramics. *J. Eur. Ceram. Soc.* **2015**, *35*, 545–553. [CrossRef]
18. Mishra, S.K.; Mittal, R.; Pomjakushin, V.Y.; Chaplot, S.L. Phase stability and structural temperature dependence in sodium niobate: A high-resolution powder neutron diffraction study. *Phys. Rev. B* **2011**, *83*, 134105. [CrossRef]
19. Mishra, S.K.; Gupta, M.K.; Mittal, R.; Chaplot, S.L.; Hansen, T. Suppression of antiferroelectric state in $NaNbO_3$ at high pressure from in situ neutron diffraction. *Appl. Phys. Lett.* **2012**, *101*, 242907. [CrossRef]
20. Mishra, S.K.; Gupta, M.K.; Mittal, R.; Zbiri, M.; Rols, S.; Schober, H.; Chaplot, S.L. Phonon dynamics and inelastic neutron scattering of sodium niobate. *Phys. Rev. B* **2014**, *89*, 184303. [CrossRef]
21. Lanfredi, S.; Lente, M.H.; Eiras, J.A. Phase transition at low temperature in $NaNbO_3$ ceramic. *Appl. Phys. Lett.* **2002**, *80*, 2731–2733. [CrossRef]
22. Lima, R.J.C.; Freire, P.T.C.; Sasaki, J.M.; Ayala, A.P.; Melo, F.E.A.; Mendes Filho, J.; Serra, K.C.; Lanfredi, S.; Lente, M.H.; Eiras, J.A. Temperature-dependent Raman scattering studies in $NaNbO_3$ ceramics. *J. Raman Spectrosc.* **2002**, *33*, 669–674. [CrossRef]
23. Lin, S.J.; Chiang, D.P.; Chen, Y.F.; Peng, C.H.; Liu, H.T.; Mei, J.K.; Tse, W.S.; Tsai, T.R.; Chiang, H.P. Raman scattering investigations of the low-temperature phase transition of $NaNbO_3$. *J. Raman Spectrosc.* **2006**, *37*, 1442–1446. [CrossRef]
24. Shen, Z.X.; Wang, X.B.; Kuok, M.H.; Tang, S.H. Raman scattering investigations of the antiferroelectric–ferroelectric phase transition of $NaNbO_3$. *J. Raman Spectrosc.* **1998**, *29*, 379–384. [CrossRef]

25. Shen, Z.X.; Wang, X.B.; Tang, S.H.; Kuok, M.H.; Malekfar, R. High-pressure Raman study and pressure-induced phase transitions of sodium niobate NaNbO$_3$. *J. Raman Spectrosc.* **2000**, *31*, 439–443. [CrossRef]
26. Shiratori, Y.; Magrez, A.; Dornseiffer, J.; Haegel, F.-H.; Pithan, C.; Waser, R. Polymorphism in Micro-, Submicro-, and Nanocrystalline NaNbO$_3$. *J. Phys. Chem. B* **2005**, *109*, 20122–20130. [CrossRef]
27. Shiratori, Y.; Magrez, A.; Fischer, W.; Pithan, C.; Waser, R. Temperature-induced Phase Transitions in Micro-, Submicro-, and Nanocrystalline NaNbO$_3$. *J. Phys. Chem. C* **2007**, *111*, 18493–18502. [CrossRef]
28. Shiratori, Y.; Magrez, A.; Kato, M.; Kasezawa, K.; Pithan, C.; Waser, R. Pressure-Induced Phase Transitions in Micro-, Submicro-, and Nanocrystalline NaNbO$_3$. *J. Phys. Chem. C* **2008**, *112*, 9610–9616. [CrossRef]
29. Kichanov, S.E.; Kozlenko, D.P.; Belozerova, N.M.; Jabarov, S.H.; Mehdiyeva, R.Z.; Lukin, E.V.; Mammadov, A.I.; Liermann, H.P.; Morgenroth, W.; Dubrovinsky, L.S.; et al. An intermediate antipolar phase in NaNbO$_3$ under compression. *Ferroelectrics* **2017**, *520*, 22–33. [CrossRef]
30. Jauhari, M.; Mishra, S.K.; Mittal, R.; Sastry, P.U.; Chaplot, S.L. Effect of chemical pressure on competition and cooperation between polar and antiferrodistortive distortions in sodium niobate. *Phys. Rev. Mater.* **2017**, *1*, 074411. [CrossRef]
31. Jauhari, M.; Mishra, S.K.; Mittal, R.; Chaplot, S.L. Probing of structural phase transitions in barium titanate modified sodium niobate using Raman scattering. *J. Raman Spectrosc.* **2019**, *50*, 1177–1185. [CrossRef]
32. Raevskii, I.P.; Proskuryakova, L.M.; Reznichenko, L.A.; Zvorykina, E.K.; Shilkina, L.A. Obtaining solid solutions in the NaNbO$_3$-BaTiO$_3$ system and investigation of its properties. *Sov. Phys. J.* **1978**, *21*, 259–261. [CrossRef]
33. Zuo, R.; Qi, H.; Fu, J.; Li, J.; Shi, M.; Xu, Y. Giant electrostrictive effects of NaNbO$_3$-BaTiO$_3$ lead-free relaxor ferroelectrics. *Appl. Phys. Lett.* **2016**, *108*, 232904. [CrossRef]
34. Zuo, R.; Qi, H.; Fu, J.; Li, J.-F.; Li, L. Multiscale identification of local tetragonal distortion in NaNbO$_3$-BaTiO$_3$ weak relaxor ferroelectrics by Raman, synchrotron x-ray diffraction, and absorption spectra. *Appl. Phys. Lett.* **2017**, *111*, 132901. [CrossRef]
35. Jauhari, M.; Mishra, S.K.; Poswal, H.K.; Mittal, R.; Chaplot, S.L. Evidence of low-temperature phase transition in BaTiO$_3$-modified NaNbO$_3$: Raman spectroscopy study. *J. Raman Spectrosc.* **2019**, *50*, 1949–1955. [CrossRef]
36. Mishra, S.K.; Mrinal, J.; Mittal, R.; Krishna, P.S.R.; Reddy, V.R.; Chaplot, S.L. Evidence for existence of functional monoclinic phase in sodium niobate based solid solution by powder neutron diffraction. *Appl. Phys. Lett.* **2018**, *112*, 182905. [CrossRef]
37. Zeng, J.T.; Kwok, K.W.; Chan, H.L.W. Ferroelectric and Piezoelectric Properties of Na$_{1-x}$Ba$_x$Nb$_{1-x}$Ti$_x$O$_3$ Ceramics. *J. Am. Ceram. Soc.* **2006**, *89*, 2828–2832. [CrossRef]
38. Rodríguez-Carvajal, J. Recent advances in magnetic structure determination by neutron powder diffraction. *Phys. B Condens. Matter* **1993**, *192*, 55–69. [CrossRef]
39. Errandonea, D.; Meng, Y.; Somayazulu, M.; Häusermann, D. Pressure-induced $\alpha \rightarrow \omega$ transition in titanium metal: A systematic study of the effects of uniaxial stress. *Phys. B Condens. Matter* **2005**, *355*, 116–125. [CrossRef]
40. Errandonea, D.; Muñoz, A.; Gonzalez-Platas, J. Comment on "High-pressure x-ray diffraction study of YBO$_3$/Eu^{3+}, GdBO$_3$, and EuBO$_3$: Pressure-induced amorphization in GdBO$_3$". *J. Appl. Phys.* **2014**, *115*, 216101. [CrossRef]
41. Wang, X.B.; Shen, Z.X.; Tang, S.H.; Kuok, M.H. Near infrared excited micro-Raman spectra of 4:1 methanol–ethanol mixture and ruby fluorescence at high pressure. *J. Appl. Phys.* **1999**, *85*, 8011–8017. [CrossRef]
42. Garg, N.; Pandey, K.K.; Shanavas, K.V.; Betty, C.A.; Sharma, S.M. Memory effect in low-density amorphous silicon under pressure. *Phys. Rev. B* **2011**, *83*, 115202. [CrossRef]
43. Angel, R.J.; Zhao, J.; Ross, N.L. General Rules for Predicting Phase Transitions in Perovskites due to Octahedral Tilting. *Phys. Rev. Lett.* **2005**, *95*, 025503. [CrossRef]

Disclaimer/Publisher's Note: The statements, opinions and data contained in all publications are solely those of the individual author(s) and contributor(s) and not of MDPI and/or the editor(s). MDPI and/or the editor(s) disclaim responsibility for any injury to people or property resulting from any ideas, methods, instructions or products referred to in the content.

MDPI
St. Alban-Anlage 66
4052 Basel
Switzerland
www.mdpi.com

Crystals Editorial Office
E-mail: crystals@mdpi.com
www.mdpi.com/journal/crystals

Disclaimer/Publisher's Note: The statements, opinions and data contained in all publications are solely those of the individual author(s) and contributor(s) and not of MDPI and/or the editor(s). MDPI and/or the editor(s) disclaim responsibility for any injury to people or property resulting from any ideas, methods, instructions or products referred to in the content.

www.ingramcontent.com/pod-product-compliance
Lightning Source LLC
LaVergne TN
LVHW070159100526
838202LV00015B/1965